MATERIALS SCIENCE AND TECHNOLOGIES

COMPREHENSIVE GUIDE FOR MESOPOROUS MATERIALS

VOLUME 4

APPLICATION AND COMMERCIALIZATION

MATERIALS SCIENCE AND TECHNOLOGIES

Additional books in this series can be found on Nova's website
under the Series tab.

Additional e-books in this series can be found on Nova's website
under the e-book tab.

MATERIALS SCIENCE AND TECHNOLOGIES

COMPREHENSIVE GUIDE FOR MESOPOROUS MATERIALS

VOLUME 4

APPLICATION AND COMMERCIALIZATION

MAHMOOD ALIOFKHAZRAEI
EDITOR

nova publishers
New York

Library of Congress Cataloging-in-Publication Data

ISBN: 978-1-63482-092-9

Published by Nova Science Publishers, Inc. † New York

CONTENTS

PREFACE

This is the final volume of the four volume set of Comprehensive Guide for Mesoporous Materials. Mesoporous materials are found abundant in biological systems and minerals and some of these materials have long been used in various industries. Recent advances in scientific understanding and manipulating in nanoscale caused an increase in the usage of these materials. One of the most important applications of mesoporous membranes are due to the special ability of pores. For example, controls of specific DNA structures and molecules filtration are possible. This volume mainly discusses the application and commercialization of mesoporous materials.

Chapter 1 - In this Chapter some attention is focused on the various traditional adsorbents, a few of the local repositories, which prove enhanced performance for environmental pollutants removal. Due to their low cost and local availability, natural materials such as zeolite, montmorillonite, low rank coal lignite, dolomite, clay and iron oxyhydroxides were chosen to evaluate their adsorption performance in inorganic nutrients and some other environmental pollutants removal. Except those inexpensive natural adsorbents, some opportunities to synthesize some hybrid organic-inorganic materials and multifunctional design adsorbents is emerging in water decontamination. Many of these strategies are evident in biological materials. Thus, they are inspiring exciting new directions in materials design and synthesis, based on the molecular level, using the soft and biomimetic materials, simultaneously. The process of natural biomaterials implementation for the intensive improvement of the adsorption properties of the traditional materials is advantageous, in hierarchy over the ordinary adsorption, as published in the literature. Natural zeolites have been considered for the most popular, inexpensive and alternative adsorbents in water cleanup since the several decades. They are approx. 15-times cheaper than, e.g. the other worldwide broaden natural product chitosan. On the other hand, if the chitosan starts to be used more often for water treatment as coated or as host substance in embedded hybrid adsorbents, this product may significantly improve its removal performance, profitability and economic feasibility including its considerable price decrease. By using such a strategy, the environment may also benefit. Nevertheless, the low-cost adsorbents, in general, may in the near future, contribute to the more efficient protection and sustainability of our surrounded environment. Some practical examples for clinoptilolite-rich tuff application, deposited at the Eastern Slovakian repository Nižný Hrabovec, which has been used for inland water treatment and purification processes in the last 25 years are mentioned, too. Since this field is wide, the overview is limited and highlighting only those water purification and treatment processes, which have been realized in pilot applications.

Chapter 2 - The reduced dimensions and increased integration in ultralarge scale integrated (ULSI) circuits cause an increased resistance on back end of the line metallization and increased interlevel and intralevel capacitances, which cause an increased signal delay. In order to improve the switching performance of future ULSI circuits, it is necessary to reduce these capacitances by introducing new dielectrics with dielectric constants significantly lower than that of silica. The urgent requirement of low dielectric constant less than 2.5 inevitably introduces mesopores in the dielectric film, which may cause mechanical failure, thermal/chemical instabilities, etc. Ordered mesoporous silica films fabricated by the surfactant-templating method are possible candidates for low-k materials with excellent mechanical/electrical properties. In this chapter, physiochemical properties of ordered mesoporous silica film and its applicability to low-k dielectric were introduced.

Chapter 3 - Fundamental developments in energy conversion and storage (ECS) are essential to meeting the challenge of global warming and the limited supply of nonrenewable resources, such as fossil fuels. In line with this need, mesoporous materials have attracted much attention in the research field of ECS systems given their special structures, large surface-to-volume ratios, favorable transport properties, short diffusion paths, and reduced diffusion effect. Thus, an updated review of the applications of mesoporous materials in ECS is presented in this chapter. The factors involved in energy conversion, such as dye-sensitized solar cells, are also discussed. Insights into the various underlying mechanisms of each case are listed to highlight the advantages of introducing mesoporous materials into energy conversion devices. Moreover, energy storage technologies such as lithium–ion batteries based on mesoporous materials are investigated. The relationship between mesoporous materials and their effectiveness in the presented ECS devices ultimately paves the way for the design of novel mesoporous structures with advanced properties.

Chapter 4 - Due to the depletion of fossil fuels and the serious environmental problems accompanying their combustion, a new form of energy that is clean, renewable, low-cost and a viable alternative to fossil fuels is urgently needed. The direct conversion of sunlight into fuels (i.e., solar fuels) with the aid of an artificial photocatalyst is an attractive prospect, considering that in a single hour the sun delivers energy enough for human activities in an entire year. Solar fuels can take the form of hydrogen and hydrocarbons such as methane and methanol. These fuels may be formed by photocatalytic water splitting and by photoreduction of CO_2 with water, respectively. In this chapter, the authors discuss recent developments in high-surface-area hybrid photocatalysts for the production of hydrocarbon from CO_2 reduction with water using sunlight energy. In order to improve the efficiency of sun-light driven photocatalysts, the factors that govern the photoactivity such as the structure, morphology at nanometric scale, charge transfer mechanism between semiconductor surfaces and cocatalysts will be also briefly discussed.

Chapter 5 - Functionalized mesoporous silica nanoparticles (FMSNs) have been proposed as efficient drug delivery systems, able to target a large quantity of a biologically active molecule to specific tissues and cells with no effect on other cell types, avoiding side effects that many times limit their therapeutic application. In this work, the authors review the main features of a new generation of FMSN-based nanodrugs, with specific examples in preclinical research. Here, the control of particle size, morphology and surface properties is crucial to ensure the stability in biological fluids, as well as a progressive and sustained release of the delivered compound, allowing monitoring of the therapeutic agent diffusion. Furthermore, the incorporation of stimuli-responsive triggers in nanoparticles promotes specific intracellular release in target cells, which improves therapeutic activity and minimizes possible side effects.

Chapter 6 - The application of mesoporous molecular sieves, such as MCM-41, SBA-15, MSU-X and others in the synthesis of fragrances is discussed. Mesoporous materials by definition have pores of diameters from the range 2-50 nm. The fragrance chemistry is an interesting blend of the chemistry of natural products, synthetic, physical and analytical chemistry to a certain degree open to original invention in composing new scents.

Especially terpenes that are a large class of organic compounds, are natural and sustainable feedstocks for the fine fragrance industry. These compounds can be used as intermediates and components of flavors, fragrances and pharmaceuticals. Transformation of several terpenes and terpenoids (geraniol, limonene, α-terpineol) by different chemical reactions over molecular sieves as catalysts is described, for example the epoxidation of terpenes is an attractive chemical transformation as such epoxides are versatile building units. Products of hydrogenation, cyclization, oxidation or isomerization are important compounds in the flavor/fragrance industry, pharmaceutical industry and in general synthesis.

Chapter 7 - Mesoporous materials are attractive adsorbents in analytical chemistry thanks to their specific properties such as large surface area, pore structure, possibility of surface functionalization to get hydrophobic/hydrophilic or positively or negatively charged surface. A combination of the size-exclusion effect of mesopores against large size molecules and the adsorption of small size analytes inside the mesopores actually mimics the effect of sample preparation before MS analysis. The use of mesoporous materials, such as ordered mesoporous silica and carbon, improves the efficiency and selectivity for profiling complicated samples, from human serum to cosmetic formulations. In this chapter a new applicability of mesoporous materials in sample analysis by using matrix assisted laser desorption/ionization mass spectrometry (MALDI-MS) is presented.

Chapter 8 - This chapter addresses recent trends, advances, and challenges of the design and synthesis of mesoporous materials as drug delivery vehicles. Due to their ability to adsorb a drug or cocktail of drugs, be covalently linked to a drug delivering functionality, penetrate the cell membrane, and release the drug inside the cell upon changing pH, mesoporous materials are attracting increasing attention from both biomedicine and nanotechnology research communities. While structural, functional, and morphological versatility of silica nanoparticles have placed them into mainstream of research endeavors in the field and to the focus of the chapter, other mesoporous materials are also discussed as promising materials for drug delivery. This chapter overviews current synthetic strategies for assembling functional elements into an efficient drug delivery nanodevice and suggests potential directions for the field to grow in the future, which includes commercial applications.

Chapter 9 - Research on mesoporous silica nanoparticles (MSNs) has explored an immense opportunity to develop an innovative type of nano-drug vehicle. Both systemic delivery and implantable local delivery devices were reported to be achievable with mesoporous silica material.

Mesoporous silica shows itself advantageous for drug loading and controlled release, due to a tunable morphological parameter, surface modification/functionalization capability, and biocompatibility properties. Functionalization of mesoporous silica can alter the drug loading capacity and release profile by changing the pore size and surface property.

In this chapter, research works on MSNs' properties influencing drug loading and drug release profile, and functionalization of mesoporous materials for applications in drug delivery systems are reviewed to illustrate the significance of MSNs as nano-drug carriers and how surface modification can improve the drug loading and controlled drug delivery.

Chapter 10 - Periodic Mesoporous Organosilicas or commonly named "PMOs" constitute of a new family of functional hybrid materials where organic groups are covalently integrated within the pore walls of a silica framework. With well-defined pore structures and unique properties, they combine the structural characteristics of ordered mesoporous silicas with the chemical functionalization possibilities of organic groups. These advantages convert these materials into promising candidates in areas such as catalysis, adsorption, nanoelectronics, and chromatography, among others. In this chapter, the authors will give an overview of all these applications where PMOs with specific functional groups have been successfully reported.

Chapter 11 - A promising field of application for ordered mesoporous (organo-)silica materials is the separation sciences, *i.e.,* HPLC (High Performance/Pressure Liquid Chromatography). From an economic point of view, chromatographic columns posses small amounts of (organo-)silica materials. This infers that the material's cost, which is higher for ordered mesoporous (organo-)silica materials than for their non-ordered equivalents, is not an issue.

The higher surface area of the materials strongly increases the retentive power of the stationary phase, up to four times when compared to one of the best retaining commercial columns.

Their uniform pore size on the other hand can mean a huge advantage in size exclusion chromatography, where well-defined pore systems can be combined to generate a high performance size exclusion column.

Chapter 12 - Periodic Mesoporous Organosilicas (PMOs) were first described in 1999 and gained immediately a great interest due to their high potential of application in catalysis, adsorption, and separation fields. These materials are formed by condensation of a bissilylated organic bridge precursor in the presence of a structure-directing agent to create porosity. They present unique structural features such as narrow pore distribution, high specific surface areas, alternating hydrophobic and hydrophilic character and adjustable hydrophobicity. PMOs offer the possibility of tuning the material properties by modifying the organic bridge of the precursors or by post-synthesis method. Furthermore, the publication of the phenylene PMO in 2002 presenting, not only meso-scale order but also, molecular-scale periodicity gave to this family an additional stability feature.

This Chapter presents a systematic revision of what has been made in the last fifteen years on the development of acid and basic PMOs for catalytic and adsorption purposes. Considering the complexity of the PMOs and the need of using computation tools to understand and predict experimental findings, the authors revise for the first time all the work performed concerning the modeling of the PMOs structure and to improve the understanding of their adsorption and catalytic properties.

Looking ahead, the Chapter ends with a roadmap of PMO development towards functionality for specific applications.

Chapter 13 - Resorcinol–formaldehyde gels can be prepared under a great variety of synthesis conditions leading to the production of organic and carbon gels such as aerogels, cryogels or xerogels – differing in the gel drying process: super-critically, freeze drying or sub-critically. If the synthesis conditions are appropriately chosen and the gel is mechanically strong enough to withstand the capillary pressures, subcritical-conventional drying can be carried out without involving major changes in the structure of the organic material. Moreover, it is possible to fine-tune the textural properties of these materials during its sol–gel process based synthesis.

Therefore, synthesis-tailored carbon xerogels of high porosity (>80%), high surface area (400-800 m^2/g) and controlled pore size distribution (1 to 50 nm), can be easily prepared.

These properties are fundamental requirements in the application of carbon materials as supports in the preparation of catalytic systems, especially within the field of fuel cell technologies. Pore structure in fuel cell electrodes affects mass transportation of fuel and exhausts, highly affecting the fuel cell performance. In fact, several studies point out the key importance of a mesopore-enriched porous structure in carbon materials for fuel cells mass transportation processes.

The present chapter aims to summarize the influence of the different synthesis parameters, such as reactants ratio (precursor/formaldehyde), the catalyst concentration (precursor/catalyst ratio) and type (basic and acid), and the type of solvent, on the final carbon xerogel properties, in view of their utilization as supports in the preparation of Pt catalysts for their application as electrodes in direct methanol fuel cells (DMFCs).

Chapter 14 - Transforming from inorganic-organic hybrid precursors into inorganic compounds is a newly emerged but versatile and efficient route to achieve inorganic porous materials, as the extraction of organic components in hybrid precursors is helpful for generating porous structure. This novel route has triggered increasing attention in recent years, as one can easily prepare porous materials with good control of morphology, structure and composition, as well as enhanced properties. This chapter reviews progress in this field, first introducing several methods to synthesize inorganic-organic hybrids, especially those that can be used as the starting materials for this transformation reaction, then dividing all transformation strategies into several categories: solid state decomposition, organic component stripping, couple of chemical transformation and organic component stripping, sonication, acidolysis and light irradiation, followed by highlight of superior properties according to different unique (mainly porous) structures of as-prepared products. Also, some prospects will be put forward briefly at the end of this chapter.

Chapter 15 - SBA-16 is an interesting mesoporous silica material having nanopores and a large surface area, which is widely employed as catalysts supports, absorbents, drug delivery materials, etc. This material has attracted attention of catalytic researchers because of its 3D structure consisting of ordered and interconnected spherical mesopores which facilitates the transport of reactants and products without pore blockage. Since it has lack of functionality, heteroatoms have been incorporated by direct or post-synthesis methods in order to modify their functionality.

The aim of this chapter is to show the state-of-the-art related to the SBA-16-based mesoporous systems as catalytic supports for hydrodesulfurization (HDS), to take stock of knowledge about current formulations and applications in this field.

Chapter 16 - This overview covers recent achievements in the application of mesoporous molecular sieves as catalysts for oxidation reactions. Main structural and textural features are outlined in relation to the novel synthetic approaches to mesoporous molecular sieves and understanding of their properties. Different ways of introduction of active sites of mesoporous molecular sieve surfaces are discussed in view of their application as catalysts.

A state-of-the-art review of transition metal containing mesoporous materials as promising redox, shape-selective catalysts is presented, focusing on the relationship of the preparation methods to the type of heteroatom species present and their special role in catalysis. The main part describes selected catalytic reactions catalyzed by mesoporous molecular sieves and it is based on very recent papers.

In: Comprehensive Guide for Mesoporous Materials, Volume 4 ISBN: 978-1-63482-092-9
Editor: Mahmood Aliofkhazraei

Chapter 1

ADVANCES IN WATER CLEANUP PROCESSES USING THE ZEOLITIC AND OTHER RELATIVE ADSORPTION MATERIALS

Eva Chmielewská*

Faculty of Natural Sciences, Comenius University, Bratislava, Slovakia

ABSTRACT

In this Chapter some attention is focused on the various traditional adsorbents, a few of the local repositories, which prove enhanced performance for environmental pollutants removal. Due to their low cost and local availability, natural materials such as zeolite, montmorillonite, low rank coal lignite, dolomite, clay and iron oxyhydroxides were chosen to evaluate their adsorption performance in inorganic nutrients and some other environmental pollutants removal. Except those inexpensive natural adsorbents, some opportunities to synthesize some hybrid organic-inorganic materials and multifunctional design adsorbents is emerging in water decontamination. Many of these strategies are evident in biological materials. Thus, they are inspiring exciting new directions in materials design and synthesis, based on the molecular level, using the soft and biomimetic materials, simultaneously. The process of natural biomaterials implementation for the intensive improvement of the adsorption properties of the traditional materials is advantageous, in hierarchy over the ordinary adsorption, as published in the literature. Natural zeolites have been considered for the most popular, inexpensive and alternative adsorbents in water cleanup since the several decades. They are approx. 15-times cheaper than, e.g. the other worldwide broaden natural product chitosan. On the other hand, if the chitosan starts to be used more often for water treatment as coated or as host substance in embedded hybrid adsorbents, this product may significantly improve its removal performance, profitability and economic feasibility including its considerable price decrease. By using such a strategy, the environment may also benefit. Nevertheless, the low-cost adsorbents, in general, may in the near future, contribute to the more efficient protection and sustainability of our surrounded environment. Some practical examples for clinoptilolite-rich tuff application, deposited at

* Corresponding author: Water Research Institute, L. Svobodu 5, 842 15 Bratislava, Slovakia - address of the author when the pilot experiments were accomplished, Email: chmielewska@fns.uniba.sk.

the Eastern Slovakian repository Nižný Hrabovec, which has been used for inland water treatment and purification processes in the last 25 years are mentioned, too. Since this field is wide, the overview is limited and highlighting only those water purification and treatment processes, which have been realized in pilot applications.

Keywords: Water treatment, adsorption, ion exchange, air-stripping, zeolite regeneration, clinoptilolite-rich tuff, carbonized clinoptilolite-rich tuff, Fe-alginate/zeolite pellets, hydrophobized ODA - zeolite, slovakite, Fe – oxyhydroxide GEH, montmorillonite, active coke, ammonium, phosphate, halogenide, sulphate, acid red removal from water

Clinoptilolite - One of the Most Useful Natural Zeolite

Zeolites are crystalline solids, structure of which is made from silicon, aluminum and oxygen atoms, that form a characteristic framework with cavities and channels inside, where cations, water and/or small molecules may reside. According to a new definition, zeolites are clathrates or inclusion compounds, able to host various guest substances in their versatile structure. They are often referred to as molecular sieves. Many of them occur naturally as minerals, and are extensively mined in many parts of the world, finding applications in industry and medicine. However, most zeolites have been prepared synthetically, some of them for commercial use, while others have been created by scientists to study their chemistry. At present, there are 213 unique zeolite topologies identified, and 67 naturally occurring zeolite species are described [1-3].

Certainly, the most abundant and frequently studied zeolite is clinoptilolite, a mineral of the heulandite group. The crystal structure of the HEU-type zeolites is characterized by a 3-dimensional aluminosilicate framework consisting fundamentally of secondary building units (SBU) of the 4-4-1 type, formed by tetrahedral primary building units (PBU). Its characteristic tabular morphology shows an open reticular structure of easy acces, formed by open channels of 8–10 membered rings. Clinoptilolite framework contains narrow 4- and 5-membered rings in parallel sheets arrangement, as well as broad 8- and 10-membered rings constituting intraframework micropores (channels with the size of 3.3×4.6 Å; 3.0×7.6 Å and 2.6×4.7 Å) perpendicularly to these parallel sheets, capable of hosting extra-framework (exchangeable) cations in association with mobile water molecules (Figure 1) [2].

The origin and use of the name *clinoptilolite* (*"oblique feather stone"*) has a convoluted history. The crystals were first identified by Pirsson in 1890 as "*crystallized mordenite*", based solely on chemical analysis. The platy habit and optical properties led Schaller in 1932 to conclude that it was a *ptilolite (*ptilo- from Greek alludes downy, finaly fibrous nature of mineral), what was named by Cross and Eakins already in 1886 for material occurring at the Table Mountain in Colorado (USA) and later found to be mordenite. Schaller named these crystals *clinoptilolite*, despite that its morphology was very similar to heulandite. Based on X-ray diffraction data later on, the heulandite and clinoptilolite were determined as isostructural. Finally, Mason and Sand in 1960 proposed a new definition for clinoptilolite, based on alkali-dominant and high Si/Al compositions regarding to the isostructural heulandite. According to them, thereafter the minerals with the same framework topologies, but with the Si/Al less than 4.0 were classified as heulandite and with the Si/Al ratio higher than 4 as clinoptilolite [1-5].

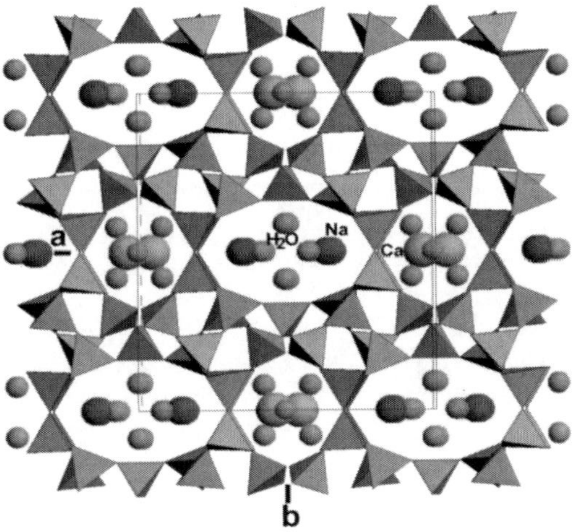

Figure 1. Schematic topology and framework structure of clinoptilolite with the exchangable cations (with permission of Marantos et al., 2012 [6]).

Clinoptilolite was formed mostly as a devitrification product of volcanic glass in tuffs, which mean basically the consolidated pyroclastic rocks. This devitrification usually occurred when the glass was in contact with saline waters. This mineral is mostly found in the form of small crystals 0.1 – 100 μm associated with clays and other silicate and aluminosilicate phases of similar density. However, even centimeter scale simple crystals can grow in cavities of igneous rocks from natural hydrothermal solutions [6, 7].

The relatively "open" structure of HEU-type zeolites, with a total pore volume of 35%, enhances the sorption properties of these microporous solids, especially towards gaseous molecules such as dinitrogen, water, carbon monoxide, ammonia, methanol, methane and pyridine, which are small enough to enter the channels. However, the chemical composition of framework (the Si/Al ratio, in the range 3-5) modulates the ion-exchange properties of the zeolites expressed as cation-exchange capacity (CEC, which can theoretically reach 330 meq/100 g). The rather low Si/Al ratio and the consequently increased CEC render HEU-type zeolites advantageous for binding dissolved cations from aqueous solutions [8-12].

Clinoptilolite is not the most well known zeolite species so far, however it is one of the most useful natural zeolite. The occurrence of mineral as fine-grained crystals in massive beds allows it to be mined and used in applications where large amounts of material are required. Zeolites are becoming widely used as alternative materials in areas where adsorptive applications as well as thermal and radiation resistance are required (clinoptilolite proves high ability to withstand thermal regeneration and resistance to attrition losses during the handling). They have been intensively studied because of their high efficiency in removing trace quantities of pollutants such as heavy metal ions, ammonium, radionuclides, some colour entities and many others noxious substances. Clinoptilolite is used in many applications, such as chemical sieve, gas absorber, dust and flue gas cleaning agent, feed and food additives, odour control agent, filtration cartrige for aquarium and as water filter or adsorbent in municipal or surface waters treatment, filler by rubber, paper and wood processing and as a component for cement and concrete manufacturing. Clinoptilolite has been successfully used for several decades more or less as an additive to feed for cows, pigs, horses and chickens, because is able to absorb toxins out of feed originating from molds and microscopic parasites, during the animal digestion. Many applications exploit the ion exchange properties, which are attractive for agronomy, horticulture and soil remediation, where the zeolite may be added to chemical fertilizers to improve the soil chemical and physical properties for plant growth, to increase fertilizer efficiency as well as to reduce the leaching of nutrients – Table 1 [13, 14].

Slovakian Natural Zeolites

Estimated zeolite reserves in the country reach the order of 150 mil. tons for the whole Nižný Hrabovec-Majerovce-Kučín formations. This clinoptilolite-rich tuff has begun to be mined industrially even in former Czechoslovakia, approximately 50 years ago by Zeocem Bystré. The company is currently producing 90 000 metric tons per year and consequently is approaching the top European Union position. Based upon the encouraging results of Slovakian natural clinoptilolite producer, recently achieved, it is convenient to characterize its basic ion exchange and selectivity properties in more detail.

The Eastern Slovakian basin with its basement of graben-synclinal structure has been filled by Neogene clastic sediments to about 7 km thickness, volcanogenic rocks and evaporites. Diverse lithofacies have been affected by volcanic activities of acidic or intermediate nature, thoroughly during the whole Miocene time-periode. Zeolitic and clay minerals occur hereby as the major and/or minor constituents of tuffs in all the Miocene stratigraphic phases. Zeolitization characterized as stratiformed with vertical mineralogical zoning, enriched beside analcime, especially with clinoptilolite, has been associated exclusively with the bedded marine Lower Badenian sequence of rhyodacitic volcanoclastics, i.e., so-called Hrabovec tuffs [15-17].

The common mineralogical and chemical compositions of clinoptilolite-rich tuff from Nižný Hrabovec repository including some other important data and of the mordenite-rich tuff from the vicinity of Bartošova Lehôtka – Jastrabá has been presented Table 2 [18].

Zeocem Company has produced plenty of clinoptilolite-based agricultural, environmental and industrial products since the late 1970s for the inland as well as for the foreign markets. This company has prepared various modified (micronized into 20 – 200 μm particle-sized powder) and into 0.2-5 mm grain-sized zeolite for water and wastewater treatment, respectively (Table 1).

Current Progress in Water Cleanup by Advanced Filtration or Adsorption Materials vs. Zeolites

Today, some update of zeolitic applications related to water treatment and purification processes are available by hundreds. However, recent literature reports the state of the art mainly in zeolite surface modification using the hydrophobization (sol-gel technique for coating the zeolitic surface by different surfactants, moreover with zero valent iron ZVI-addition), metal doping or pelletization of zeolite matrix with some biopolymeric eco-friendly carbohydrates. Since the last decade by our investigations, several new surface modified or hybridized zeolite adsorbents have been succesfully synthesized whose performance are favourable, including much extended adsorption properties towards aqueous pollutants [19-22].

Table 1. Some review of practical applications of clinoptilolite-rich tuff from Slovakian deposit in environmental protection

Practical illustration	Functional description
	Powderized zeolite use to be applied in biological wastewater treatment (BWWT) to improve performance and capacity of nitrification-denitrification (N-D) within activated sludge proces by existing plants and minimal capital expenditure.
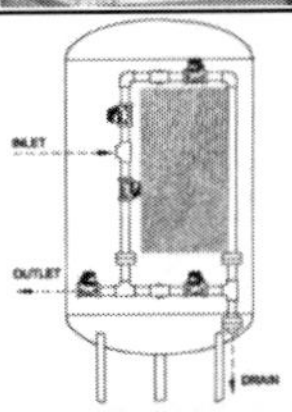	Clinoptilolite-rich tuff posses high efficiency in removing trace quantities of pollutants, such as heavy metal ions, ammonium, radionuclides, based upon its selectivity according to following sequence: Ag^+, Ba^{2+}, $Cs^+ > Pb^{2+} > NH_4^+ > Cu^{2+} > Co^{2+}$ and in surfactant modified form some colour entities, phenol, antibiotics and inorganic anions.
	Zeolite powder is usually dosed into activation (biological stage of WWTP) , mixed with biomass of activated sludge and regarding to aluminum or iron salts addition, able to precipitate phosphorus, what reduces the consumption of above salts or other polymers finally.
	Zeolite powder attracts dispersed particulate matter, provides proper conditions for microorganisms grow and such mixture posses better or improved settling and dewatering characteristics.
	Clinoptilolite is used in many applications, such as chemical sieve, gas absorber, dust and flue gas cleaning agent based upon its high melting point, which reaches the temperature over 1000°C.
	Low cost surfactant modified or biopolymeric pelletized zeolite, but alternativelly also native form of zeolite can be applied in multi-ton quantities for the use as subsurface permeable barriers against ground water contaminant migration.
	Other beneficial effects of natural zeolite in environmental protection are odour control based upon zeolite high selectivity towards ammonia, but also towards hydrogen sulphide and excess of plenty other odouring gases.
	Clinoptilolite-rich tuff may be applied as filtration cartrige for aquaria (incl. some marine ones) to keep nutrient concentration sufficient low for corals and as water filter or adsorbent for various type of contaminated waters, i.e., sewage, municipal, surface or wastewaters runoff, also to reduce the leaching of nutrients (ammonia and phosphate) and pesticides washed out in agriculture fields and plants.

Table 2. The mineralogical and chemical composition of the clinoptilolite-rich tuff from the repository Nižný Hrabovec (left) and of the mordenite-rich tuff from Bartošova Lehôtka – Jastrabá (right)

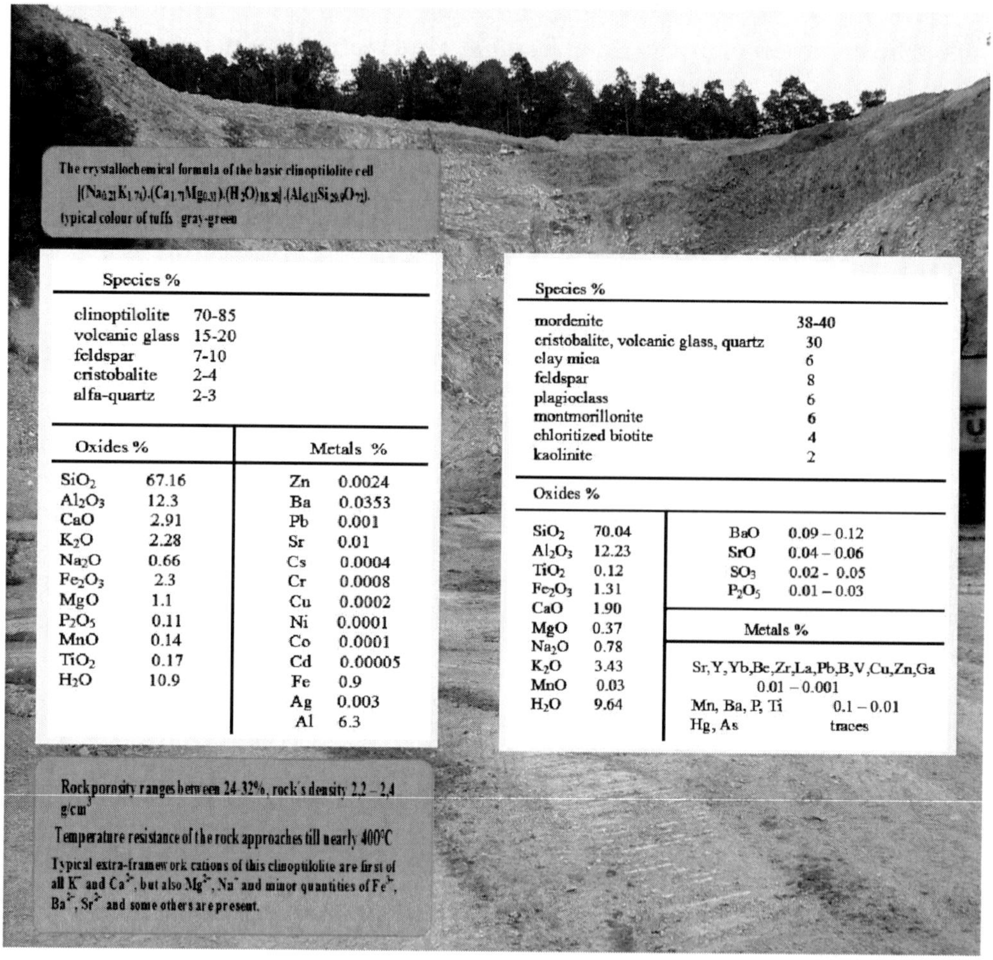

The crystallochemical formula of the basic clinoptilolite cell
$[(Na_{0.23}K_{1.76}).(Ca_{1.77}Mg_{0.31}).(H_2O)_{18.28}].(Al_{6.11}Si_{29.9}O_{72})$.
typical colour of tuffs gray-green

Species %	
clinoptilolite	70-85
volcanic glass	15-20
feldspar	7-10
cristobalite	2-4
alfa-quartz	2-3

Oxides %		Metals %	
SiO_2	67.16	Zn	0.0024
Al_2O_3	12.3	Ba	0.0353
CaO	2.91	Pb	0.001
K_2O	2.28	Sr	0.01
Na_2O	0.66	Cs	0.0004
Fe_2O_3	2.3	Cr	0.0008
MgO	1.1	Cu	0.0002
P_2O_5	0.11	Ni	0.0001
MnO	0.14	Co	0.0001
TiO_2	0.17	Cd	0.00005
H_2O	10.9	Fe	0.9
		Ag	0.003
		Al	6.3

Rock porosity ranges between 24-32%, rock's density 2.2 – 2.4 g/cm^3
Temperature resistance of the rock approaches till nearly 400°C
Typical extra-framework cations of this clinoptilolite are first of all K^+ and Ca^{2+}, but also Mg^{2+}, Na^+ and minor quantities of Fe^{2+}, Ba^{2+}, Sr^{2+} and some others are present.

Species %	
mordenite	38-40
cristobalite, volcanic glass, quartz	30
clay mica	6
feldspar	8
plagioclass	6
montmorillonite	6
chloritized biotite	4
kaolinite	2

Oxides %			
SiO_2	70.04	BaO	0.09 – 0.12
Al_2O_3	12.23	SrO	0.04 – 0.06
TiO_2	0.12	SO_3	0.02 - 0.05
Fe_2O_3	1.31	P_2O_5	0.01 – 0.03
CaO	1.90	**Metals %**	
MgO	0.37	Sr,Y,Yb,Be,Zr,La,Pb,B,V,Cu,Zn,Ga	0.01 – 0.001
Na_2O	0.78	Mn, Ba, P, Ti	0.1 – 0.01
K_2O	3.43	Hg, As	traces
MnO	0.03		
H_2O	9.64		

Research interests for the designing of alternative adsorbents to replace costly commercial grade active coke and synthetic ion exchange resins, which are derived from petroleum-based raw materials and thus using the non-environmentally friendly fabrication process, have intensified in recent years. Attention has been focused on the various adsorbents rich on functional groups able to bind a broad range of pollutants from contaminated streams at low cost. Based on their economic feasibility and local availability, natural materials such as chitosan, zeolites, clay, certain industrial or agricultural waste products, i.e., fly ash, low rank coal, lignite, natural metal oxides, peat, sawdust, waste slurry, lignine, shungite, coconut shell charcoal, waste biomass, cork (*Quercus suber L.*) seem to be perspective and sustainable adsorption materials, respectively. Among the current commercial prices, natural zeolites are classified far as the most inexpensive alternative adsorbents in regard to other commercial-trade materials (e.g. the price of chitosan, a biodegradable linear polysaccharide obtained upon partial alkaline deacetylation of chitin, a structural element in the exoskeleton of crustaceans and insects, is on the market approximately 15-times higher) [23-27].

Simultaneously in recent years, there has been pronounced tendency to utilize mechanically stable synthetic or natural solid matrices in many applications, such as chemically bonded phase in chromatography, extraction of cations from aqueous media via the membrane processes, ion preconcentration, electronics, ceramics, environmental control and also bioengineering. The solvent extractants or chelating groups used to be either impregnated into the pores of the porous solids or covalently bonded to the surfaces to improve their specific properties. Surface modification via functional group immobilization provides thus unique opportunity to engineer the interfacial properties of various solid matrices, while retaining their basic geometry and mechanical strength. By other words, such surface engineering used to be achieved either by physically adsorbing or chemically grafting functional groups onto a suitable matrix. In order to promote the removal performance of natural zeolites, i.e., native cation exchangers, towards the broadest range and versatility of environmental pollutants, their surface has been chemically modified, too [19-27].

Usually, the inorganic component of such hybrids provides mechanical, thermal or structural stability of the new product (adsorbent) and the organic component functionalizes the adsorbent´s surface. Such composed (hybridized) natural zeolites also potentially improve or broaden their adsorption performance towards the selected, even hydrophobic or anionic pollutants (while upgrading to amphiphilic properties) and remain still cost-effective. As proven so far, surface charging or templating with carbonaceous substances including application of sol-gel methods had the major influence for broadening of zeolite adsorption properties [23-26].

In general for water and wastewater treatment, a significant majority of recent developments relate to biological processes and treatment technologies using advanced adsorption and filtration media. Recently, dendrimers as highly branched organic molecules with specific functional groups to be variously combined with other traditional adsorbents incl. zeolites onto their surface, are applied as bactericide in nanotechnology processes to inhibit growth of microbes, but also in many other uses. Synthetic nanofiltration membranes are often fouling due to deposition of materials such as calcium and magnesium ions and salts, whereby reducing the membrane flux and purification efficiency, therefore waters use to be pretreated, currently by various advanced filtration cartridges composed of several specific filtration layers, e.g. layer of granulated zinc and copper alloy for chlorine and residual heavy metals removal, a fine mesh carbon or carbon nanotubes immobilized carrier block for chlorine, odour, colour, trihalomethanes and other organic contaminants and pharmaceuticals removal, a layer of granulated ion exchange resin or alternativelly zeolite upgraded with surface functionalization for all ionic pollutants removal, and layer of granulated activated calcite to adjust the pH [28].

Membranes are thin films of synthetic organic or inorganic materials which selectively separate a fluid from its components through the application of external power. Polymeric membranes are currently state of the art, with wide application in food and drinking water production, however zeolite nanofiltration membranes are expected to be a technological breakthrough with respect to polymeric ones in term of improved selectivity (through tunable pore size and hydrophobicity), higher fluxes (due to open porous texture) and resistance [29].

Evidently, the photocatalytic degradation of organic pollutants by TiO_2 supported solid matrices like, e.g. zeolites belong to the latest progress in field. Titanium dioxide has been widely used as photocatalytic material in removal of toxic chemicals from waters, but due to its fine sized grains (4-30 nm), it aggregates rapidly by loosing efficiency. TiO_2 is not porous and its surface exhibits polar property, however as immobilized on porous solids (semiconductor graded solids) using mostly the sol-gel route provides higher specific surface area and facilitates more effective adsorption sites than bare TiO_2. Some authors combine clay and zeolite with iron oxide to achieve magnetic particle modified adsorbents, while enhancing the metal removal efficiency. The same concerns also biopolymers such as chitin (second most abundant biopolymer after cellulose) and alginate (extracted from various species of brown seaweed), which by integrating with magnetic nanoparticles of Fe_3O_4 improve hydrodynamic properties as well as promote the adsorption ability towards broader spectrum of pollutants [30, 31].

Constructed wetlands (CWs) are a relative new, low-cost, alternative technology in the field of wastewater treatment. They are designed to exploit the various processes occurring in natural wetlands within a more controlled environment. CWs are appropriate systems for small communities, rural or mountainous areas, where the application of conventional wastewater systems is not economically and often technically feasible. In order to investigate possible ways to improve CW performance, natural zeolites have been proposed to be utilized in CWs. Their primary applications have to do with their use as substrates inside the wetlands, as pretreatment systems before the wastewater input and as an appropriate filter media for further treatment of CW effluent, i.e., as final polishing step [32, 33].

At last but not least has to be mentioned the development of wastewater treatment systems using oyster shells as the biological growth media that prove enhanced affinity to microorganisms or for even trace concentrations of POPs in water, development of biomimetic adsorbents imitating lipids from white whales, arctic wolves, South African fur seals, marketable fish and mussels, based upon the high accumulation affinity of organism´s lipids towards POPs. Thus, a new type of biocomposite embedded-cellulose acetate have been developed for membrane processes, respectively, in uptake of very toxic organochlorine pesticides and dioxins, while as supporting materials for biomolecules have been frequently used zeolites [28, 29, 34, 35].

Consequently to above latest literature excurse and on the base of encouraging results of Slovakian clinoptilolite-rich tuff producer (Zeocem Company) next part of this Chapter presents the author´s most recognized efforts contributing to water cleanup innovation by using Slovakian natural clinoptilolite.

Ammonia Removal from Drinking Water Using the Ion Exchange Technology with Clinoptilolite-Rich Tuff Including Regenerants Recovery

The zeolite ion exchange pilot installation (ZIEPI) with a hydraulic loading rate of 900 liter per hour (~ 13 Bed Volumes BV per hour) was situated at the field experimental facility of Water Research Institute in Vajnory, the closed vicinity of Bratislava, during the autumn 1986. This pilot facility treated the tap water which was enriched by ammonium chloride to the initial concentration of 1 mg ammonium per liter. Two pressure steel columns operating in series, each one filled with 70 liter or 56 kg of clinoptilolite-rich tuff of size-granulation 0.4 – 1 mm, as the part of above desribed pilot facility, provided the drinking water purification with the high efficiency for several weeks. Bottom of each column was filled with 14 liter of about 3 mm grain-sized sand as the drainage support layer.

The zeolite filters were operated in downflow mode during the common working service and upflow mode during the chemical regeneration. For the chemical regeneration of the loaded zeolite filter 2% NaCl solution of pH ~ 9 was used. Two hours lasted chemical regeneration required approximately 25 BV of 2% NaCl solution. To regenerate a zeolite filter by 85% or to eluate the ammonia out of the loaded zeolite bed by this effectivity was satisfactorily for maintaining the solid time schedule and provide the required water quality.

To provide a permanent running of the whole facility (ZIEPI) and the uninterrupted water purification, ammonia exhausted regenerants thereafter needed to get recovered by means of air stripping, to be able again to use for filter regeneration (Figure3). The HDPP plastic tower of the total high 6.5 meter, assembled from 6 modules of the ground-plane measures 980 x 650 mm, was applied for ammonia stripping. To strip ammonia out of the regenerant brine a large quantity of air through the tower was necessary, therefore the design of stripping tower was improved by installation of two lateral air blowers at its bottom module (each one with the blowing capacity of 450 liters per second). Stripped ammonia out of the exhausted regenerant solutions was then processed in countercurrent tower configuration, by which the entire air flow entered the bottom, while the treated regenerant was pumped to the top (with the same hydraulic loading rate as into the ZB, i.e., 900 liter per hour) and falled across the vertical waves-like shaped laminated slats to the bottom of the tower. The whole regenerant volume was air stripped without the mass closed loop operation, because the ZIEPI with the rather low operation capacity did not affect the surrounded air quality significantly. The entire volume of regenerants (1800 liters) was recycled through the tower during 150 minutes to decrease the initial ammonia concentration of 50 mg per liter to less than 10 mg per liter. In the first stripping cycle 47% of ammonia was removed, therefore also the second stripping cycle was necessary. The major factors which affected design and process performance of air stripping tower were the tower configuration, the air flow and the pH. For recovery of 1 liter of regenerant solution the consumption of 3.8 cubic meters air was necessary. To maintain pH ~ 11 of regenerant solutions in stripping process and to compensate some loss of sodium ions in regenerant effluents, regularly addition of NaOH into effluents was provided during the operation of ZIEPI.

Under above described operation, the one zeolite column was able to treat 85 cubic meters of drinking water with enhanced ammonium concentration up to the limited value of 0.5 mg per liter, what means that the 4 days operated zeolite filter removed from the whole treated volume of water 81 g of ammonia. The operation principle of two pilot zeolite columns (zeolite beds ZB) was as follows: Single or one ZB was under the operation service till the ammonia concentration in effluent reached the limited value of 0.5 mg per liter and then the second new ZB was connected with the first one in series to load the capacity of the first ZB thoroughly. When the first ZB was loaded to its maximum capacity, the column was disconnected from the operation in series and started to get regenerate, while the second ZB operated as single column. There was sufficient time to get regenerated the by ammonia loaded ZB and to prepare it for the sequenced operation in the next process of drinking water purification. The longer time operated ZB was used as joined in series always at the inffluent end, backed up by another ZB with lower loading level. Using this strategy of operation, the number of bed volumes throughput during the operation service increased effectivelly by about 50%. A half hour lasted ZB backwashing, using the tap water, with about 30% bed expansion followed the chemical regeneration with alkalic NaCl solution additively, due to necessity of pH decrease of the treated water filtered through ZB, left after its contact with the alkalic brines [36, 37].

Ammonia Removal from Tannery Wastewater Using the Clinoptilolite-Rich Tuff with Chemical Regeneration and Regenerant Recovery by Air/Steam Stripping

Similar operation with the same hydraulic loading rates through the zeolite beds as by the drinking water purification units was used for ammonia removal of mixed tannery and sewage sludge wastewaters at Shoe Manufactury Wastewater Reclamation Plant in Otrokovice (former Czechoslovakia) in the 1987 (Tables 3, 4) [37-39]. A higher ammonia concentration in wastewater (in average 50 mg per liter) increased the ion exchange capacity of zeolite considerably (approximately 10-times higher in regard to the drinking water purification).

Three columns operation (two ZB under stream while the third ZB being regenerated) as well as much more complicated management of the entire ZIEPI valving and piping was required at this wastewater treatment facility. Comparing to the sufficient long, i.e., 4 days lasting drinking water service of ZB, ammonia removal from the wastewater by ZB reached the limited value of 3 mg per liter in effluent already after 15 hours. Consequently, the time of regeneration including ammonia stripping was necessary to shorten, to keep the entire operation system permanently running. The 3-fractional regeneration and adequate ammonia stripping process used for regenerant recovery and recycling in this zeolite ion exchange pilot installation was therefore proposed. Only the highest, on ammonia mostly concentrated elutriant (spent regenerant) solution was stripped in air stripping tower. Ammonia stripping process was intensified and considerably shorten by the injection of the mixed hot waste steam and air (1: 3) at the bottom of the tower against the spent regenerant solution pumped to the top of this tower. The stripped ammonia was proposed to get absorbed into water to produce the technological solution used in tannery process. For the recovery of 1 liter of highly concentrated elutriants (spent regenerant solutions) the consumption of 20 cubic meters air was necessary (Table 4).

After the several months of ZIEPI operation, the Centroproject Zlin provided, based on the submitted pilot experimental data, some economic calculations for model capacity of treated wastewater 6000 cubic meters per day [40]. At that time the investment for the construction of zeolite ion exchange and regeneration & air stripping recovery units for the proposed capacity was calculated to 550 000 Euro, whereas the investment for ammonia removal by means of biological nitrification-denitrification method was estimated to 700 000 Euro. Nevertheless, the operation cost of ion exchange technology using the natural zeolite per one cubic meter of treated water was a bit higher than the operation cost of the comparable biological nitrogen removal using the nitrification-denitrification method. Concluding, the examined technology of zeolitic ion exchange appeared technically and financially competitive to the traditional biological methods.

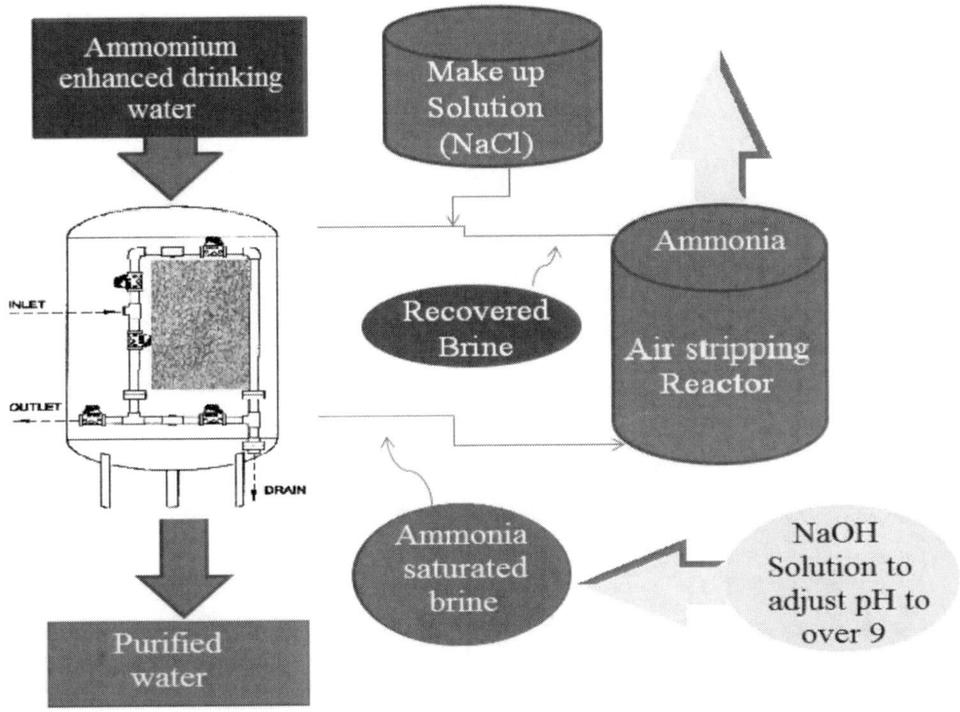

Figure 2. Principle sketch of ion exchange technology for ammonia removal from water using the zeolitic filtration incl. its chemical regeneration and the used regenerant recovery by air stripping.

Table 3. Chemical composition of communal wastewater in ZIEPI Otrokovice in various treatment stages

Index, mg/L	Influent	Effluent of mechanical treatment	Effluent of mechanical-biological treatment	Influent to clinoptilolite column after experimental unit pretreatment
dissolved substances	1997	1906	1317	1200
inorganic solids	1500	1482	1117	1026
BOD_5	488	310	55	17
COD	2044	827	199	108
NH_4^+	61	55	43	40
Cl^-	482	508	465	301
SO_4^{2-}	278	259	206	156
Cr^{3+}	12	7.2	2.8	2.4
Ca^{2+}	138	141	135	139
Mg^{2+}	46	21	16	17
petroleum compounds	-	2.5	-	-
pH	9	8.9	7.2	7.4
total hardness, mmol/L	5.3	4.4	4.0	4.2

Table 4. Technological data for pilot plant in Vajnory and Wastewater treatment facility in Otrokovice

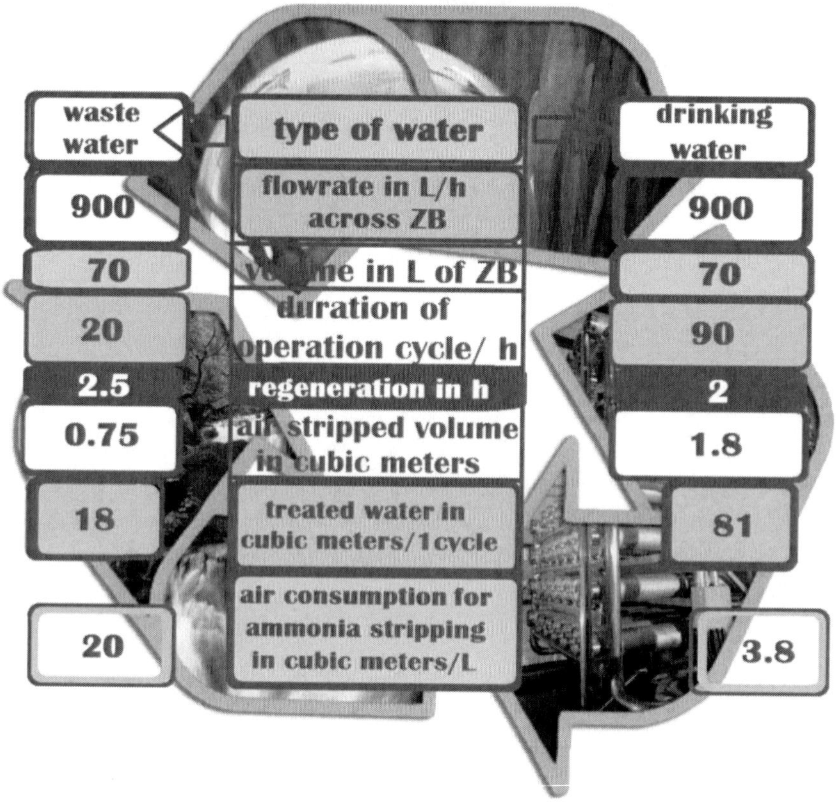

waste water	type of water	drinking water
900	flowrate in L/h across ZB	900
70	volume in L of ZB	70
20	duration of operation cycle/ h	90
2.5	regeneration in h	2
0.75	air stripped volume in cubic meters	1.8
18	treated water in cubic meters/1cycle	81
20	air consumption for ammonia stripping in cubic meters/L	3.8

TOMA industrial facility and the whole municipality in Otrokovice delivered up to 55 000 cubic meters of wastewaters per day, while conventionally treated a volume of 18 000 cubic meters per day (200 liters per second), therefore the process of zeolite ion exchange including regeneration and air stripping by mass closed loop (merry-go-round) operation was designed for the entire capacity [39]. The proposed wastewater treatment plant consisted of three series of zeolite beds (ZB) or pressure columns, each of sets composed from five ZB (1 pentade), with the zeolite weight per one bed 8 tons. The set of 5 ZB was designed to operate simultaneously. A large air stripping tower (total volume of 140 m^3) with the same elutriant loading rate as the ZB filtration rates, i.e., 40 liters per second, was proposed to provide elutriants recovery, while consuming 80 m^3 air per second, by regeneration & air stripping of 1 ZB lasting not longer than 3 hours.

Inorganic Phosphate Removal Potentially for Water Bodies with Nutrients Overburden

Currently, many physicochemical techniques like chemical precipitation, membrane processes, filtration or coagulation for phosphate removal are available, however they suffer by many disadvantages like high cost, salt content in the effluent, necessities of chemical sludge disposal and the other. Based on that disadvantages, several commercial (Fe-oxyhydroxide GEH and slovakite) as well as domestic natural (clinoptilolite-rich tuff and montmorillonite) adsorbents for phosphate removal in laboratory-scale examination have been selected. The domestic natural clinoptilolite (repository Nižný Hrabovec) was chosen on the base of its low-price availability in the local market (15-35 Euro per ton for size-granulation of 0.3 – 1 mm), cost effectivness and due to its sufficiently large surface area (~ 60 m^2/g), the highest one among the other natural products, rigidity and surface functionality [41, 42] – Table 5.

Continuous or well mixed batch reactor was firstly used to study and compare the adsorption performance of above four adsorption materials. Such experimental configuration offers a number of potential advantages like long contact time, upon which a high particle load is derived and controlled the mixing to reduce an external mass transfer effect. Based upon batch mode results, the highest effectivity in phosphate removal was reached by the montmorillonite and GEH products, while the uptake capacity of the slovakite and clinoptilolite-rich tuff reached about 30% lower values (Figure 3).

The differences in uptake capacities between above four selected adsorbents may be explained as follows: While the phosphate adsorption onto clinoptilolite-rich tuff proceeds by the mechanism of surface accessible extra-framework cations Ca^{2+} and Mg^{2+} undergoing then the calcium and magnesium phosphates precipitation, the adsorption onto GEH product proceeds by means of linkage of anionic phosphate with cationic Fe^{3+} of akaganeite structure, probably by much stronger covalent bonding. It must be pointed simultaneously, that volcanic rock of clinoptilolite type, which framework contains narrow 4- and 5-membered SiO_4^{4-} and AlO_4^{5-} tetrahedral rings as well as broad 8- and 10-membered tetrahedral rings constituting intraframework micropores (channels) of 0.33 x 0.46 nm, 0.3 x 0.76 nm, 0.26 x 0.47 nm dimensions, enriched by mobile H_2O molecules moreover, contains due to the presence of another minerals overgrowing the active matrix thoroughly (volcanic glass, feldspar, cristobalite, clay, quartz), much broader external pore openings and various interparticle voids, which probably worked as phosphate blocking centres, too. Based on the S(BET) analysis performed, the BJH average pore diameters of the examined natural clinoptilolite sample was determined in the range of 9.2 – 20 nm [23]. On the base of the ionic diameter of non-hydrated PO_4^{3-}, i.e., 1.196 nm, only interlamellar space of the montmorillonite structure (interlayer distance of about 1.2 nm) approaches to that dimensions and thus potentially supports phosphate intercalation mechanism by this process studied [43]. Finally, phosphate adsorption onto slovakite is assumed to proceed again by linkage of phosphate to plenty of alkalic earth cations present in its natural dolomite and clays constituents. Despite to predicted covalent bounds in this case respectivelly, much smaller surface area in regard to GEH product does not allow to consider the slovakite for such an effective adsorption material as the GEH product is.

Moreover, column, dynamic or packed bed mode examination of above adsorbents preffered mostly the GEH product, consequently to covalent bonding of pollutant onto this large surfaced oxyhydroxide.

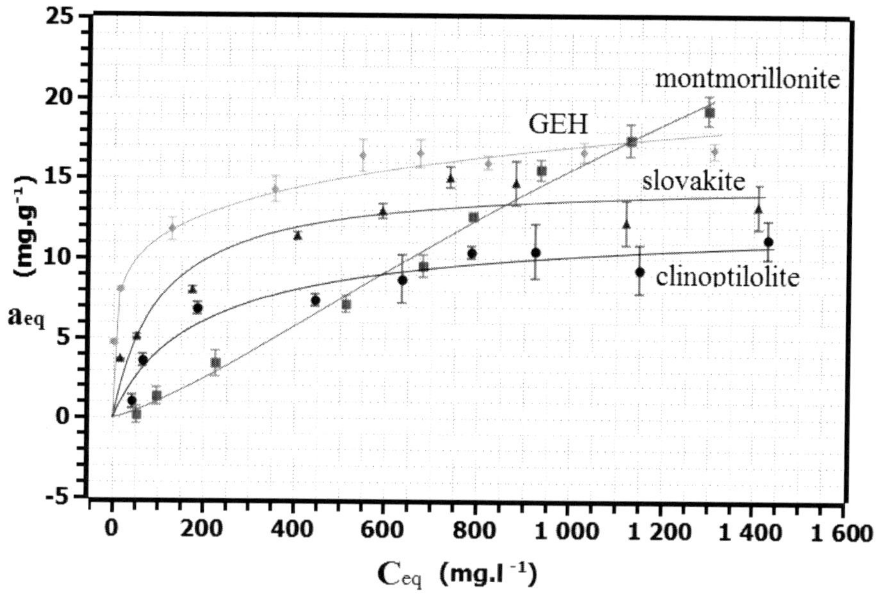

Figure 3. Adsorption isoterms measured for the system ortophosphate solution vs. GEH, clinoptilolite, montmorillonite or slovakite at T = 23 ± 0.2°C.

EFFORTS FOR BROADENING WATER POLLUTANTS REMOVAL WITH HYDROPHOBIZED AND PELLETIZED CLINOPTILOLITE-RICH TUFF

A laboratory setup was used to examine the uptake of mono- and polyatomic single or mostly double charged anions like chromate, arsenate, nitrate, sulphate, phosphate, halogenides and some organic substances like azodyes (acid red AR1, indigo carmine AB74) and phenol from aqueous model solutions by the octadecylammonium [hereafter ODA] modified, carbonized or Fe(III) alginate pelletized clinoptilolite-rich tuff. Removal efficiencies of such surface functionalized clinoptilolite-rich tuff towards above water pollutants were compared and evaluated with the other low-cost natural or commercial adsorbents, like active coke, pyrolysis char, lignite, waste product of petrochemical company, i.e., industrial ashes chezacarb and expanded perlite (under commercial name Vapex), respectively (Figure 4).

Surface functionalization of the native zeolite was done due to its cation exchange behaviour, which is usually not, or only negligible adsorbing aqueous anions or organic substances. Ammonium and some heavy metal's selectivity of clinoptilolite species is generally known, however for the current purpose, it has been lowered. The domestic natural clinoptilolite was chosen as the interface carrier as usually on the base of its low-price in the local market, sufficiently large surface area, rigidity and surface functionality [19-23, 36, 37].

Among the prepared postmodified zeolitic adsorbents (Figure 4), the highest capacity data proved Fe alginate gel – zeolitic adsorbent towards nitrate ion with maximum capacity of 89.3 mg/g, then this composite adsorbent, respectively, towards sulphate ion with maximum capacity of 79.4 mg/g. While Fe(III) cationic sites in Fe-alginate/zeolite composed adsorbent were responsible for electrostatic bounding forces between pollutant oxyanions and pelletized biosorbent in the proces studied, Ca ions were responsible for cations exchange in the Ca – alginate zeolitic pellets. The best adsorption performance towards the examined oxyanionic pollutants and adequate rigidity of the adsorbent proved the product consisted of alginate to zeolite weights ratio 1: 2 [22].

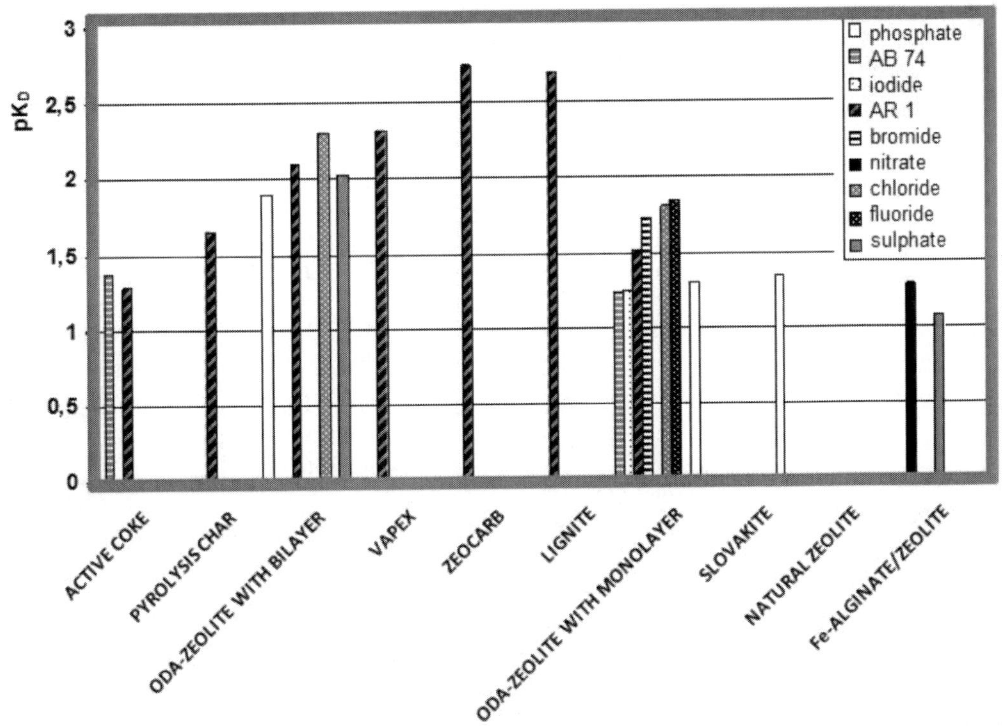

Figure 4. Comparable pK_D coefficient for various adsorbents and environmental pollutants.

Table 5. Illustration of some adsorbents used for water pollutants removal

Adsorbent illustration	Adsorbent characteristics
	The Al-Mg – montmorillonite-rich bentonite originates from the deposit Stará Kremnička – Jelšový Potok in SR, which is the most popular and long mined ore of the country. Montmorillonite from the bentonite ore was obtained after sedimentation and purification procedures. The price of the montmorillonite ranges in 30-80 Euro per ton. Its surface area S(BET) reaches the value 240 m^2/g, the bulk density of montmorillonite is about 2.0 g/cm^3.
	Vapex is expanded and hydrophobized volcanic perlite from Eastern Slovakian quarries with S(BET) equalled to 2.11 m^2/g. Perlite is actually a hydrated volcanic glass, i.e., aluminosilicate of characteristic light weight, porosity, gray-white coloured, which was thermally (by about 950 °C) and chemically (with silicone oil) treated at a local Company (Kerko Košice). Its price is 19 Euro per 100 liter.

Table 5. (Continued)

Adsorbent illustration	Adsorbent characteristics
	Carbonaceous fossil like lignite is of domestic origin. It possess some specific properties due to rich humic components, low toxic metal content, suitable physical structure, sorption properties and sensorial features, which support its usage as a fertiliser, soil conditioner, bio- and humidity regulator (to be considered as first product of coalification processes on the globe and potential intermediates between peat and bitumenous coal).
	Chitosan posses apropriate mechanical strength, rigidity and settling characteristics, which are a perfect alternative to other adsorbent – biomass, despite fragile structure and poor mechanical strength. Chitin is found in fungi, arthropods and marine invertebrates. It is one of the most abundant, easily obtainable and renewable natural biopolymers (polysaccharides), manufactured e.g. by Sigma-Aldrich (USA).
	Slovakite decodes a commercial adsorbent manufactured by IPRES inžiniering, Ltd. Bratislava from domestic dolomite, bentonite, diatomic clays, alginite and zeolite, justified only with clinker and final pressurizing. Slovakite is purchased for about 700 Euro per ton. The bulk density of this product equals 2.1 g/cm^3 and S(BET) surface to 23 m^2/g.
	Active coke of German provenience silcarbon is in size-granulation of 0.4 – 1.7 mm. It is manufactured from black coal as main component and coconut shells or some lignocellulose substances, using the activation with water steam. The product is accepted for drinking water purification fulfilling the European Norm EN 12 915. Its surface area S(BET) equals 1300 m^2/g. Silcarbon is purchased for about 13 000 Euro per ton.
	Granulated ferric hydroxide, (GEH[104]) developed at the Department of Water Quality Control in Technical University Berlin, is an approved commercial adsorbent manufactured by GEH Wasserchemie GmbH & Co. KG Osnabrück (Germany). The main components of GEH are akaganeite (ß–FeOOH) and goethite [α-FeO(OH)]. The product has the specific surface area about 220 m^2/g, water content of 45%, bulk density 1.2 g/cm^3 and the price of 3750 Euro per ton.

Thus, the recent advances in the development of zeolite/ biopolymeric pellets synthesis, so far reached, may provide perspectively an unprecedented opportunity to develop more effective and environmental sustainable zeolite composed adsorbents.

ACKNOWLEDGMENTS

The project of natural zeolites utilization for water treatment processes is currently funded by the Slovak Scientific Council VEGA (Project # 1/0185/12) and under the bilateral Slovak-Chinese Cooperation SK-CN-0033-12 "Liquid waste minimization and valorization by multi-dimensional hierarchical composites of natural resources".

REFERENCES

[1] Breck, D. W. (1974). Zeolite Molecular Sieves. Structure, Chemistry and Use, 771 p. New York, London, Sydney, Toronto: John Wiley & Sons.

[2] Baerlocher, C., Meier, W. H. & Olson, D. H. (2007). Atlas of zeolite framework types, 6th Edition. Elsevier (http://www.iza-structure.org/databa).

[3] Steed, J. W., Turner, D. R. & Wallace, K. J. (2007). Core Concepts in Supramolecular Chemistry and Nanochemistry, John Wiley & Sons, Ltd. ISBN: 978-0-470-85866-0.

[4] Pirsson, L. V. (1890). On mordenite. *Am. J. Sci.*, *140*, 232 – 237.

[5] Schaller, W. T. (1932). The mordenite – ptilolite group: clinoptilolite, a new species. *Am. Mineral.*, *17*, 128 – 134.

[6] Marantos, I., Christidis, G. E., Ulmanu, M. (2012). Zeolite formativ and deposits, Handbook of natural zeolites, Inglezakis, V. J. and Zorpas, A. (Eds.) Bentham Sci. Publ., 28-51.

[7] Gottardi, G. & Galli, E. (1987). Natural Zeolite. Minerals and Rocks. Springer Verlag Berlin Heidelberg New York Tokyo.

[8] Armbruster, Th. (2001) Clinoptilolite-heulandite: applications and basic research. Studies in Surface Science and Catalysis 135. Galarneau, A., Di Renco, F., Fajula F. and Vedrine J. (Eds.) Elsevier Sci. 14-26.

[9] Townsend, R. P. & Coker, E. N. (2001). Ion exchange in zeolites. Stud. Surf. Sci. Catal. 137, 467. The Database of Periodic Porous Structures (http://mmkvk1. fos. su. se/).

[10] Godelitsas, Ath. & Armbruster, Th. (2003). HEU-type zeolites modified by transition elements and lead. *Microporous and Mesoporous Materials*, *61*, 3-24.

[11] Hernandes, H. R., Rojas, F. & Lara, V. H. (2000). Nitrogen-Sorption Characterization of the microporous Structure of Clinoptilolite-Type Zeolites, *J. Porous Materials*, *7*, 443-454, Kluwer Academic Publ., Netherlands.

[12] Yamanaka, S., Malla, P. B. & Komarneni, S. (1989). *Water sorption and desorption isotherms of some naturely occurring zeolites*, *9*, 18-22.

[13] U. S. Geological Survay, Mineral Commodity Summaries, January 2011, prepared by Robert L. Virta, rvirta@usgs. gov (703)648-7726, 198 p. Reston (Virginia), ISBN 978 - 1-4113-3083-2.

[14] Zeolite ´93, Occurrence, Properties and Utilization of Natural Zeolites, (A Conference Program and Abstracts based on papers presented at Zeolite ´93 in Boise, Idaho, June 20-28, *1993*), ed. D. W. Ming and F. A. Mumpton, ICNZ Brockport, NewYork, 622.

[15] Kozáč, J., Očenáš, D., Derco, J., Rusnák, D. (1981). Nižný Hrabovec – zeolitic tuffs – applied technological research. Manuscript – Geofond, Bratislava 224 p. (in Slovak).

[16] Varga, I. (1984). The first zeolite deposit in Czechoslovakia: Results of Geological exploration of the clinoptilolite tuff in Nižný Hrabovec. Mineralia Slovaca 16 (4) 371-376 (in Slovak).

[17] Šamajová, E., Kraus, I. (1985). Distribution of smectites and zeolites in the West Carpathians. In Proc. 5th Meeting of European Clay Groups, Charles University, Praque, pp. 321-328.

[18] www.zeocem.sk (Production catalogue, Products for protecting the environment, Bystré, Slovakia) [accessed on July 2011].

[19] Chmielewská, E. (2012). An update of zeolitic and other traditional adsorption and ion exchange materials in water cleanup processes, Natural Zeolite Handbook, Chapter 3. 9, 274-285, Bentham Science Publ. NewYork.

[20] Chmielewská, E., Jesenák, K. & Gáplovská, K. (2003). Arsenate and chromate removal on cationic surfactant-loaded and cation-exchanged clinoptilolite-rich tuff vs. montmorillonite, *Collection of Czechoslovak Chemical Communications*, *68*, 4, 823-836.

[21] Chmielewská, E., Sabová, L., Sitek, J., Gáplovská, K. & Morvová, M. (2010). Removal of nitrates, sulfate and Zn(II) ions from aqueous solutions by using biopolymeric alginate/ clinoptilolite-rich tuff pellets, *Fresenius Environmental Bulletin*, *19*, 5, 884-891.

[22] Chmielewská, E., Sabová, L., Peterlik, H. & Wu, A. (2011). Batch-wise adsorption, SAXS and microscopic studies of zeolite pelletized with biopolymeric alginate, *Brazilian J. Chem. Eng.*, *28*, 01, 63-71.

[23] Chmielewská, E., Sabová, L., Gáplovská, K. & Čaplovičová, M. (2008). Functionally graded mesoporous zeolitic adsorbents in order to promote the pollutants removal, *Kuwait J. Sci. Eng.*, *35* (2B), 1-15.

[24] Mojumdar, S. C., Varshney, K. G. & Agrawal, A. (2006). Hybrid Fibrous Ion Exchange Materials: Past, Present and Future. *Res. J. Chem. Environ.*, *10*, 89-97.

[25] Maryuk, O., Pikus, S. & Olszewska, E. (2005). Benzyldimethyloctadecylammonium bentonite in chromate adsorption, *Materials Letters*, *59*, 16, 2015-2017.

[26] Faghihian, H. & Mousazadeh, M. H. (2007). Removal of PAHs from n-paraffin by modified clinoptilolite, *Iranian Journal of Chemistry and Chemical Eng.*, *26*, 3, 121- 127.

[27] Babel, S. & Kurniawan, T. A. (2003). Low-cost adsorbents for heavy metals uptake from contaminated water: A Review, *J. Haz. Mat.*, *B97*, 219- 243.

[28] Tansel, B. (2008). New technologies for water and wastewater treatment: A survay of recent patents. Recent patents on chemical engineering, *Bentham Sci. Publ. Ltd.*, *1*, 17- 26.

[29] Perego, C., Bagatin, R., Tagliabue, M. & Vignola, R. (2013). Zeolites and related mesoporous materials for multi-talented environmental solutions, *Microporous and Mesoporous Mater.*, *166*, 37-49.

[30] Šiljeg, M., Stefanovič, Š. C., Mazaj, M., Tušar, N. N., Arčon, I., Kovač, J., Margeta, K., Kaučič, V. & Logar, N. Z. (2009). Structure investigation of As(III)- and As(V)-species bound to Fe-modified clinoptilolite tuffs, Micropor. *Mesoporous Materials*, *118*, 408- 415.

[31] Qu, J. (2008). Research progress of novel adsorption processes in water purification: A review, *J. Environ. Sci.*, *20*, 1-13.

[32] Stefanakis, A. I., Akratos, Ch. S., Gikas, G. D. & Tsihrintzis, V. A. (2009). Effluent quality improvement of two pilot-scale, horizontal subsurface flow constructed wetlands using natural zeolite (clinoptilolite), *Microporous and Mesoporous Mater.*, *124*, 131-143.

[33] Woinarski, A. Z., Snape, I., Stevens, G. W. & Stark, S. C. (2003). The effects of cold temperature on copper ion exchange by natural zeolite for use in permeable reactive barrier in Antarctica, *Cold regions science and technology*, *37*, 159-168.

[34] Hårleman, C., Jacks, G. & Rybeck, B. (2009). The use of a clinoptilolite-based filter in emergency situations, *Desalination 248*, 629-635.

[35] Bhatnagar, A. (2013). Application of adsorbents for water pollution control. A note on the advances in adsorption technology for water treatment: Progress and challenges. eISBN:978-1-60805-269-1, 523-528, Bentham Sci. Publ. Ltd.

[36] Chmielewská, E. (2012). Natural Zeolite – A versatile commodity - Some retrospectives in water cleanup processes, *Desalination and Water Treatment, 41*, 335-341.

[37] Chmielewská - Horváthová, E., Konečný, J. & Bošan, Z. (1992). Ammonia Removal from Tannery Waste-waters by Selective Ion Exchange on Slovak Clinoptilolite, *Acta hydrochimica et hydrobiologica*, *20*, 5m 269 - 272.

[38] Chmielewská-Horváthová, E., Konečný, J. & Bošan, Z. (1989). Technology for wastewater treatment in tannery processing plants, CS-Patent 02 604-89, No. 274 068.

[39] Amicarelli, V. & Liberti, L. (1991). Zeolite ammonia removal at Manfredonia municipal plant, Atti 1st Convegno Nazionale di Scienza e Tecnologia delle Zeoliti, 139-146, C. Colella (Ed.), 147-155, L´Aquila 26-27 Settembre.

[40] Hegmon S. (1989). Technical-economic evaluation of technology for selective ion exchange of ammonium out of tannery waste waters using the clinoptilolite-rich tuff, Svit Zlín—Waste Water Treatment Facility in Otrokovice, Centroprojekt, 1989, 36 p.

[41] www.geh-wasserchemie.de (Physicochemical properties of Fe-oxyhydroxide GEH product incl. price list) [accessed 2011-07-30].

[42] Chmielewská, E. & Hodossyová, R. (2012). Removal of phosphate from aqueous solutions. Kinetics, Thermodynamics and Isotherm calculations, *Fresenius Environmental Bulletin*, *22*, 2a, 598-603.

[43] Barbalace, K. Periodic Table of Elements–Sorted by Ionic Radius. Environmental Chemistry.com, 1995 – 2010 http://klbprouctions.com (accessed 2010-7-03).

In: Comprehensive Guide for Mesoporous Materials, Volume 4 ISBN: 978-1-63482-092-9
Editor: Mahmood Aliofkhazraei

Chapter 2

ORDERED MESOPOROUS SILICA FILMS FOR INTERMETAL DIELECTRICS IN ELECTRONIC DEVICES

Hyung-Ho Park, Sang-Bae Jung, Chang-Sun Park and Min-Hee Hong
Department of Materials Science and Engineering,
Yonsei University, Seoul, Korea

ABSTRACT

The reduced dimensions and increased integration in ultralarge scale integrated (ULSI) circuits cause an increased resistance on back end of the line metallization and increased interlevel and intralevel capacitances, which cause an increased signal delay. In order to improve the switching performance of future ULSI circuits, it is necessary to reduce these capacitances by introducing new dielectrics with dielectric constants significantly lower than that of silica. The urgent requirement of low dielectric constant less than 2.5 inevitably introduces mesopores in the dielectric film, which may cause mechanical failure, thermal/chemical instabilities, etc. Ordered mesoporous silica films fabricated by the surfactant-templating method are possible candidates for low-k materials with excellent mechanical/electrical properties. In this chapter, physiochemical properties of ordered mesoporous silica film and its applicability to low-k dielectric were introduced.

1. INTRODUCTION

1.1. Ordered Mesoporous Silica Film

Ordered mesoporous materials have been widely investigated since 1992 [1] as electrochemical sensors [2], optical sensors [3], low-k dielectrics [4-8], thermal insulating layer [9-12], photonic devices [13], permselective membranes [14], gas sensor [15], photocatalysts [16], photovoltaic devices [17], and various other applications [18] which have been outlined in many review papers [19-22]. In particular, silica has broad applicability due to low cost, prevalence, compatibility with common processes, chemical and structural stability [23].

Preferential evaporation of the solvent during dip- or spin-coating drives cooperative self-assembly from a homogeneous solution of silica and a surfactant dissolved in alcohol below the critical micelle concentration [24]. This process, evaporation-induced self-assembly (EISA), enables the formation of highly-ordered mesoporous silica films with various pore structures. A mesostructure can be formed during the thin film formation process because self-assembly between inorganic and organic species occurs very quickly during solvent evaporation [25].

1.2. The Study of Porous Silica Film for Low-K Dielectric Film

The development of materials of a low dielectric constant is necessary to enhance the device performance for ultralarge scale integrated devices [26]. Low dielectric constant films are required to reduce power dissipation, propagation delay, and cross-talk noise of the interconnect structure.

It is generally accepted that extendibility to the ultralow dielectric region is not attainable without introducing the concept of porosity because air possesses the lowest dielectric constant of 1 [27].

Among the organic or inorganic porous materials, porous silica films are most promising due to their superior mechanical strength, thermal stability, and most importantly, compatibility with silicon wafers and related materials that are used in existing integrated circuit technology. A very attractive synthetic route for producing porous silica films is the sol–gel process, which easily enables control of the dielectric constant. Among the sol–gel-derived porous silica films, ordered mesoporous materials have been widely investigated for low dielectric film applications [28, 29].

1.3. Preparation of Ordered Mesoporous Silica Film

Ordered mesoporous silica films were synthesized as follows. The Brij-76 block copolymer ($C_{18}H_{37}(OCH_2CH_2)_{10}OH$, Aldrich) was dissolved in EtOH (Duksan, 99.9%) or mixed solution (EtOH and Acetone) and acidified H_2O was added. After stirring for 1 h, tetraethoxysilane, (TEOS, $Si(OC_2H_5)_4$, Fluka, 98%) were sequentially added to the solution as silica precursors.

The final molar ratio of TEOS : solvent : H_2O : HCl : Brij-76 was 1 : 20-70 : 5-20 : 0.01-0.15 : 0-0.2. After aging for 30 min, the sol was spin-coated onto a Si wafer at 3000 rpm for 30 sec. Ordered mesoporous silica films were calcined at 400°C for 2 h at a heating rate of 1°C/min in an oxygen atmosphere to remove the block copolymer. Experimental flow is shown in Figure 1.

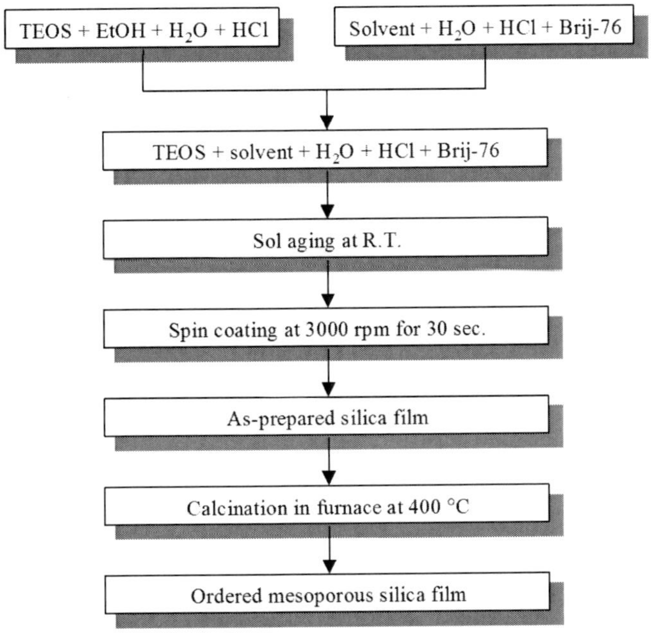

Figure 1. Experimental flow chart for the synthesis of ordered mesoporous silica film.

1.4. Analysis Procedure of Mesostructured Silica Film for Low-K Application

For the investigation of pore structure, X-ray powder diffraction (XRD) pattern was collected using Fe$K\alpha$ radiation with a wavelength of 1.9373 Å. The film porosity was calculated by measuring the critical angle of the film using specular X-ray reflectivity at the 3D beamline of Pohang Light Source (PLS) in Korea and ellipsometry with a He–Ne laser source of 632 nm wavelength [30]. The mesostructure of the film was investigated using grazing incidence small-angle X-ray scattering (GISAXS) measurements at a 3C beamline (λ = 1.54 Å and $\Delta\lambda/\lambda = 5 \times 10^{-4}$) of PLS [31]. The sample-to-detector distance was 75 cm.

Fourier transform infrared (FT-IR; Jasco 300Z) spectroscopic analysis was performed to reveal the chemical species of the silica wall. The spectra are shown only for wave numbers between 800 and 1300 cm^{-1} for clarity. X-ray photoelectron spectroscopy (XPS, VG Scientific, ESCALAB 220i-XL) measurement was carried out with excitation source of monochromatic Al $K\alpha$ radiation.

The mechanical properties of ordered or disordered mesoporous silica films were investigated by continuous stiffness measurements using a MTS nanoindentor XP [32]. In this technique, a small oscillating force is applied to the Berkovich indentor and the amplitude and phase shift of the oscillations are measured. This allows for a continuous measurement of hardness and modulus even for extremely thin layered materials.

For the electrical properties measurement, circular Al dots were deposited on calcined ordered mesoporous films using an e-beam evaporator. The thickness is 150 nm and the diameter is 0.5 mm, much larger than the spacing of the striations. Aluminum was also deposited on the backside of the Si for ohmic contact. After the deposition of upper and backside Al electrodes, post-metal-annealing was performed in a rapid thermal annealing at

300°C for 1 min. Capacitance-voltage characteristics of the metal-insulator-semiconductor structure were obtained using a HP 4284A, impedance/gain-phase analyzer at 1 MHz, and current-voltage characteristics were measured using a HP 4145B semiconductor parameter analyzer.

CONCLUSION

Study of Characteristic Changes by Various Aging Time

In acidic silica sol, hydrolysis reaction is completed right after the addition of the catalyst while polycondensation reaction proceeds slowly [33]. In this experiment, the polycondensation reaction of silicic acid proceeds simultaneously with sol aging. In order to investigate the effect of sol-aging time on the formation of the mesostructure, the d spacing of major diffraction peak was measured for silica films prepared by varying sol aging time. The XRD result of the as-prepared mesoporous silica films with Brij-76/TEOS molar ratio of 0.05 is given in Figure 2(a) [34]. The interplanar spacing of the as-prepared film gradually increases with sol-aging time. Mesoporous silica film shows the same behavior with anisotropic shrinkage of d-spacing after calcination. It is well known that the micelle structure templated by the block copolymer is made up of a hydrophobic core surrounded by hydrophilic polyethylene oxide (PEO) chains and silica wall. Due to the solubility of PEO chains into silica framework, a part of PEO chains may form an intermediate region between the hydrophobic block and silica, corona region [35]. The polycondensation reaction reduces the charge density of the silicic acid and the interaction between silica and block copolymer. As a result, PEO chains are driven into the hydrophobic core region and the hydrocarbon chains elongate in order to reduce the interaction with PEO chains. Therefore, increased d-spacing in Figure 2(a) was attributed to the elongated hydrocarbon chains [36]. Figure 2(b) represents the changes in the XRD peak intensities of the as-prepared and mesoporous silica films as a function of sol aging. It was observed that the diffraction peak intensity was maximized with the film prepared using 1 day aged sol. And after 1 day aging, the charge density of the silicic acid was reduced. Mesostructured silica films in acidic conditions with the Brij-76 block copolymer can be synthesized by the following mechanism [37].

$$Si(OEt)_4 + H_2O + HCl \rightarrow \approx Si(OH^{2+}) + EtOH + Cl \quad (1)$$

$$\sim EOC_{18} + H_2O + HCl \rightarrow \sim [(EO)\cdot H_3O^+]C_{18} + Cl \quad (2)$$

$$\approx Si(OH^{2+}) \text{ ---- } Cl^- \text{ ---- } \sim [(EO)\cdot H_3O^+]C_{18} \quad (3)$$

First, the acid catalysts accelerate the hydrolysis reaction of silicon alkoxide, and then the hydrolyzed species are positively charged [reaction (1), denoted as I^+]. The ethylene oxide headgroup is hydrated by hydronium ions through hydrogen bonds [reaction (2), denoted as S^0H^+]. And finally, Cl^- counterions (denoted as X^-) have a mediation role of electrostatic repulsion forces for cooperative self-assembly [reaction (3), $I^+X^-(S^0H^+)$ mechanism]. In Eq. (3), it can be estimated that hydrated EO chains can interact with the silica framework.

Mesostructure was determined by Eq. (3). This structure affects the charge density balance between hydrated copolymer and silicic acid. When TEOS was used as silica precursor, optimum Mesostructure could be formed before Q_3 bonding was formed [38]. Therefore, from big intensity of XRD, thin film aged for 1 day has ordered pore structure because of charge density balance. With increased aging time, condensation reaction was progressed and ordered pore structure was collapsed because charge density of silica was decreased.

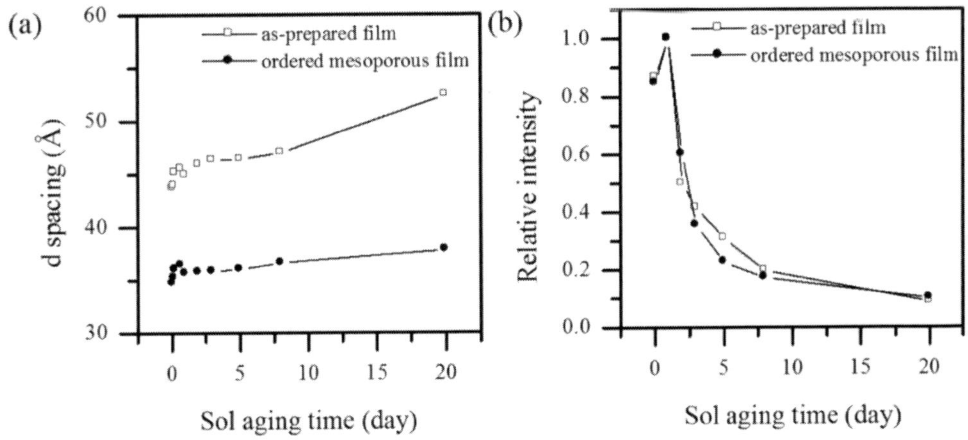

Figure 2. (a) d-spacing and (b) relative intensity of major diffraction peak in the XRD measurements of as-prepared and mesoporous silica films with Brij-76/TEOS molar ratio of 0.05 by varying sol-aging time.

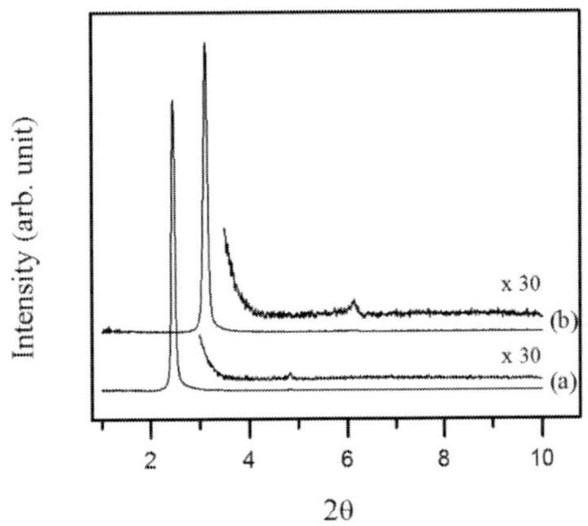

Figure 3. XRD patterns of (a) as-prepared and (b) calcined mesoporous silica film at 400°C.

XRD pattern of thin film which synthesized optimum sol aging time was shown in Figure 3 [5]. As prepared film exhibited a sharp diffraction peak with plane spacing of 4.49 nm. Any other diffraction peak was not observed except second harmonic in the magnified XRD pattern of Figure 3(a). So, as-prepared film exhibited a highly textured pore structure. It was reported that mesoporous silica using Brij-76 surfactant forms the pore structure of polytypic intergrowth of hexagonal close-packed (hcp) and cubic close-packed (ccp) structure [39]. So, diffraction peak in Figure 3(a) could be corresponded to (002) plane of hcp or (111) plane of ccp structure. Even after calcination, the texture characteristics were maintained except anisotropy shrinkage along normal direction to film surface as shown in Figure 3(b).

Although well-ordered structure could be found from XRD patterns, investigation of pore structure was very difficult because of limited peaks. GISAXS analyze was progressed to determine the pore structure of mesoporous silica thin film, and results in Figure 4 [21]. Figure 4(a) shows the GISAXS pattern of ordered mesoporous silica film using a Brij-76/TEOS molar ratio of 0.05 and EtOH/TEOS molar ratio of 20. In Figure 4(a), two diffraction spots with vertical alignment can be seen which correspond to diffraction spots due to the transmitted and reflected beams. When analyzing the GISAXS pattern, anisotropic shrinkage during calcination and random orientation of domains in the in-plane direction should be considered. In Figure 4(a), the existence of in-plane diffraction excludes the possible presence of a 2D hexagonal mesophase. Further, characteristic GISAXS patterns for ordered silica films with Pm3n, Fm3m, and $P6_3$/mmc symmetries exclude assignment of these symmetries [40]. From the existence of in-plane diffraction and the angle relationship between the first five spots close to the beam center, we can assume that the pores are arranged in a distorted body-centered cubic structure (bcc, Im3m) with anisotropic shrinkage for the normal direction to the film surface, where the (011) plane is parallel to the substrate. Figure 4(b) shows the 2D reciprocal space of ordered mesoporous silica film based on the bcc structure. This Figure 4 was constructed with the assumption that the d spacing of the (011) plane was the same as the d spacing of the (01-1) plane, indicating that no anisotropic shrinkage occurred. Each diffraction plane in Figure 4(a) was indexed on the spot in Figure 4(b) assuming a bcc structure. In order to confirm the above assumption related to the mesophase of the film, TEM analysis was performed.

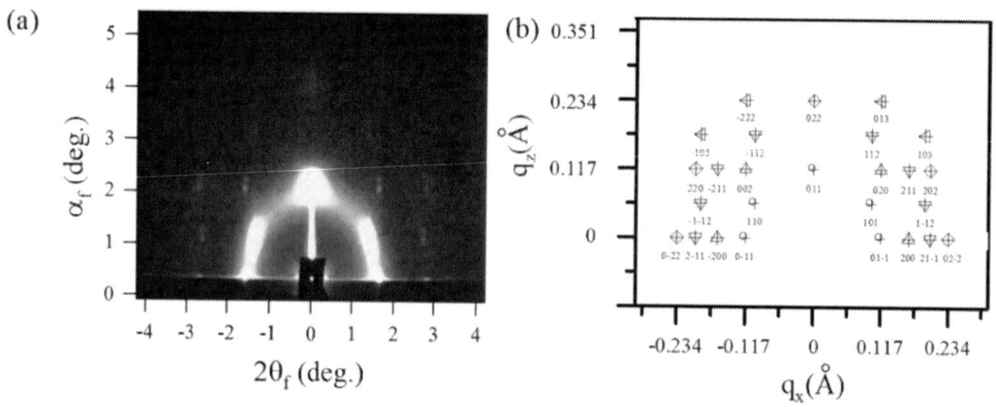

Figure 4. (a) GISAXS pattern of ordered mesoporous silica film from a 0.05 Brij-76/TEOS and 20 EtOH/TEOS molar ratio, (b) simulated 2D reciprocal space assuming no out-of-plane d spacing shrinkage in (a). The incident angle of the X-ray was set at 0.19°.

Figure 5(a, b) shows TEM images of an ordered mesoporous silica film created using a Brij-76/TEOS molar ratio of 0.05 and EtOH/TEOS molar ratio of 20 with [1-11] and [100] zone axes, respectively [21]. The plane relationship is denoted on the TEM image. However, d spacing obtained from TEM images is smaller than that from GISAXS patterns. Meanwhile, in-plane d spacing from TEM images is comparable with that from GISAXS patterns. This difference might be related to sample deformation during Ar ion milling due to low mechanical property or focused high-energy electrons during TEM analysis. Figure 5 (c, d) shows the simulated pore arrangement of ideal bcc lattice with [1-11] and [100] zone axes, respectively. Although there were artifacts during TEM analysis, we can clearly observe the bcc pore arrangement of ordered mesoporous silica film by comparing Figure 5 (a, c) or (b, d). Based on these findings, ordered mesoporous silica film with a distorted bcc arrangement was synthesized with a 0.05 molar ratio of Brij-76/TEOS and 20 molar ratio of EtOH/TEOS.

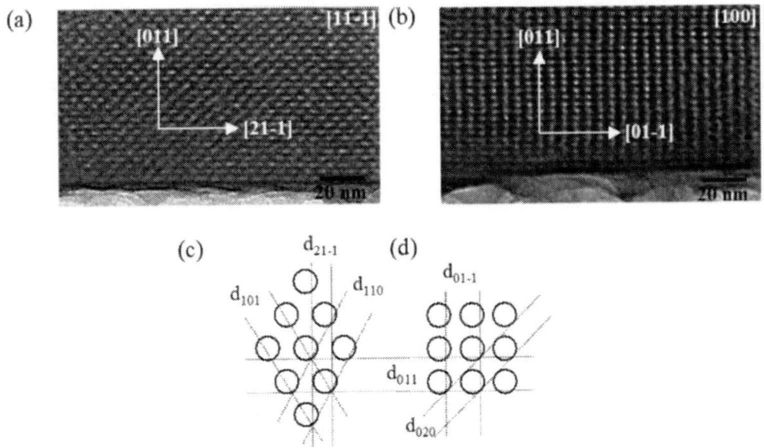

Figure 5. TEM image of ordered mesoporous silica film from a 0.05 Brij-76/ TEOS and 20 EtOH/TEOS molar ratio with a zone axis of (a) [1-11] and (b) [100], and ideal pore arrangement in bcc structure when zone axis is (c) [1-11], and (d) [100].

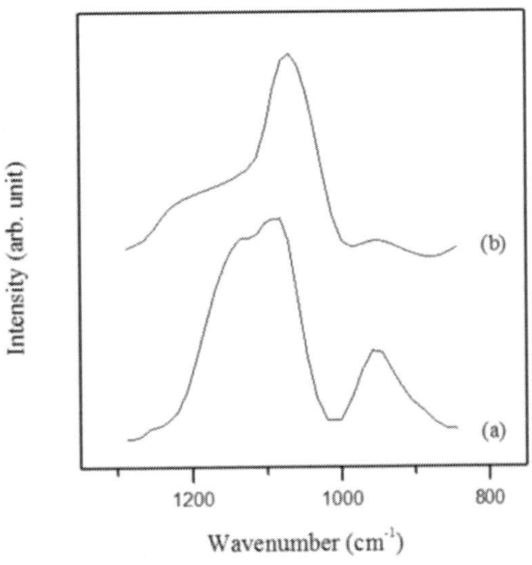

Figure 6. FT-IR spectra of (a) as-prepared film and (b) calcined mesoporous silica film at 400°C.

Figure 6 shows FT-IR spectra of as-prepared and calcined mesoporous silica films at 400°C [5]. As shown in Figure 6(a), FT-IR spectrum of as-prepared film shows characteristic absorption behavior near 1080 cm^{-1} by Si–O–Si of TO_3, 1130 cm^{-1} by cyclic Si–O–Si and 960 cm^{-1} by Si–OH vibration [41]. In acid-catalyzed silica sol, a lot of hydroxyl group and cyclic species exist due to low polycondensation reaction rate [41]. Organic species was completely removed from the observation of FT-IR spectrum in region of 2800-3000 cm^{-1} after calcination at 400°C. In Figure 6(b), cyclic species of siloxane bonds reduced due to thermal activation and activation of longitudinal optical vibration mode around 1230 cm^{-1} occurred from the porous nature after calcination [42]. Unfortunately, silanol group remained even after calcination at 400°C.

Figure 7 represents leakage current density of mesoporous silica film as a function of calcination temperature [5]. As the calcination temperature increased up to 500°C, leakage current density of the film reduced. The reduction of leakage current was attributed to the effective removal of hydroxyl group as confirmed already with the measurement of dielectric constant. Leakage current density of mesoporous silica calcined at 500°C was measured less than 2×10^{-6} A/cm^2 up to 1.6 MV/cm. So, it would be concluded that surface functional group has to be controlled for improvement of the electrical properties of mesoporous silica film.

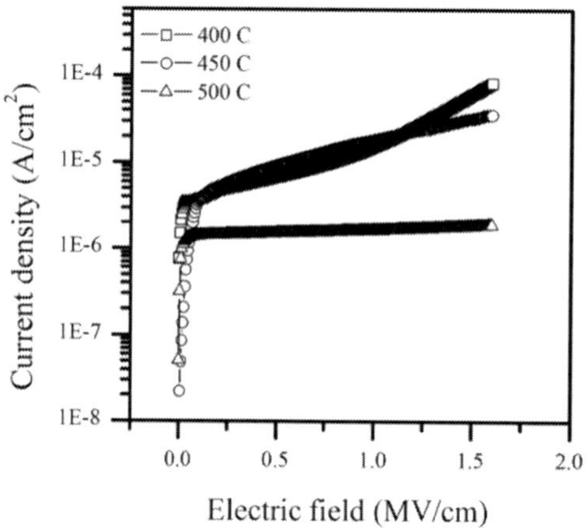

Figure 7. Leakage current behaviors of calcined mesoporous silica films at 400°C, 450°C, and 500°C.

Study of Characteristic Changes by Various Surfactant Concentration

Figure 8 shows the GISAXS patterns of the as-prepared films and mesoporous silica film using a sol aged for 1 day with varying the block copolymer concentration [34]. In GISAXS pattern, elliptic spots on faint ring could be clearly seen. The faint ring corresponds to domains with the same structure but random orientation and elliptic spots show the mesostructure of the film. The extra spot located in the upper part of main spot was caused by the reflection effect at substrate surface [43]. In Figure 8(a, b), GISAXS pattern was indexed with BCC structure [44]. This BCC mesostructure was not maintained with block copolymer concentration of 0.1. Worm-like micelle and lamellar phase coexist as given in Figure 8(c, d) corresponds to the mesoporous silica film with molar ratio of 0.05 and the pattern could be indexed with BCC structure as in the case of the as-prepared film. Up to now, hexagonal-closed packed structure and face-centered cubic structure have been proposed as a possible three-dimensional micellar structure for mesoporous silica prepared using the Brij-76 block copolymer [45]. In the case of the mesoporous film with one-dimensional shrinkage of the BCC structure along <011>, a structural transformation into face-centered orthorhombic

(FCO) or base-centered monoclinic could be induced [46]. For example, (011), (101), and (01-1) in Figure 8(a) can be assigned as (020), (111), and (020) of FCO with lattice parameters of a =77.5Å, b =89.4Å, and c =114.2Å. In Figure 8(d), the lattice parameter was obtained as a =74.1Å, b =73.4Å, c =108.5Å with FCO structure. *a* and *c* axis values parallel to the film surface were slightly different while *b* axis values normal to the film surface were greatly different, as much as 22%, which proves the anisotropic shrinkage of the film during the calcination.

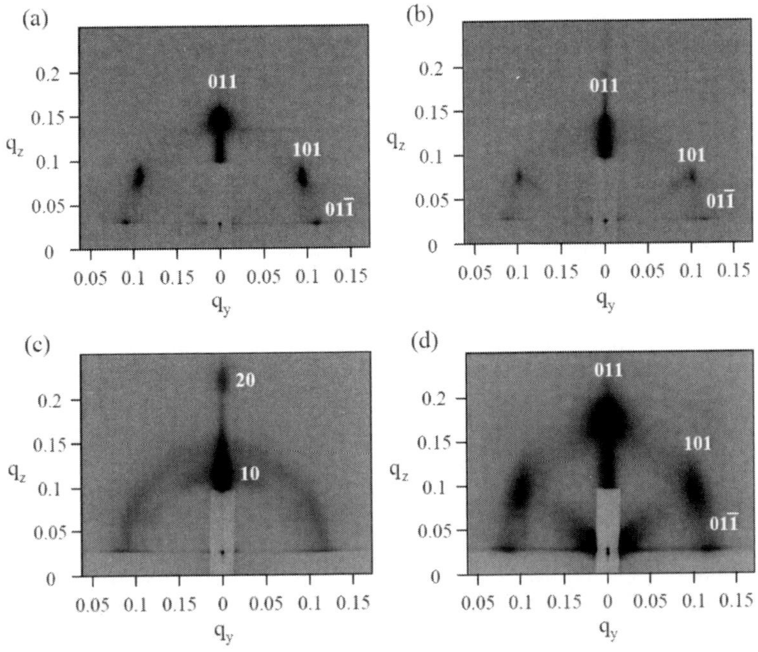

Figure 8. GISAXS patterns of as-prepared silica films with Brij-76/TEOS molar ratio of (a) 0.05, (b) 0.07 and (c) 0.1 by using the sol aged for 1 day, and (d) after calcination of film (a), i.e., mesoporous silica film.

Figure 9 shows the XRD spectra of the as-prepared and mesoporous silica films with various Brij-76/TEOS molar ratio [34]. In Figure 9(a), d-spacing of (011) peak increased with a molar ratio from 0.05 to 0.07 and maintained constant over 0.07. The increase in d-spacing was mainly attributed to the length of the hydrocarbon chain when the molar ratio increased from 0.05 to 0.07. In the case of the molar ratio of 0.09, it was considered that thinning of the silica wall and lengthening of the hydrocarbon chain simultaneously affected the d-spacing. In Figure 9(b), interplanar spacing for mesoporous silica film increases with the increase in molar ratio from 0.05 to 0.07 due to the increase of pore diameter. Due to the thermal contraction of the thin silica wall during the calcination, a decrease of d-spacing with the molar ratio of 0.09 was found.

Figure 10 shows the porosity variation depending on the Brij-76/TEOS molar ratio [34]. Porosity values of mesoporous silica film fabricated with molar ratios of 0.05 and 0.07 were 37.6% and 51.2%, respectively. This increased porosity could be related to the increased mesopore size as given in Figure 9(b). Although the d-spacing of the mesoporous film with molar ratio of 0.09 was smaller than that of 0.07, the film shows a slightly increased porosity of 52.8% due to the thinning of the silica wall as described previously.

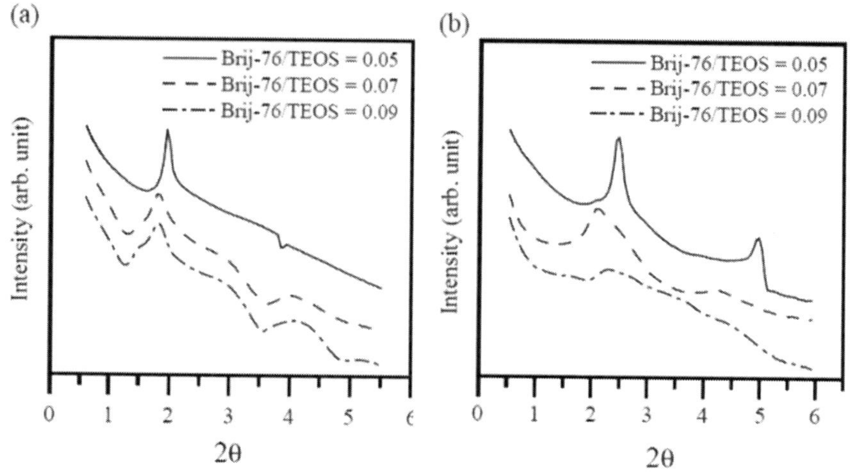

Figure 9. XRD spectra of (a) as-prepared and (b) mesoporous silica films for various Brij-76/TEOS molar ratio.

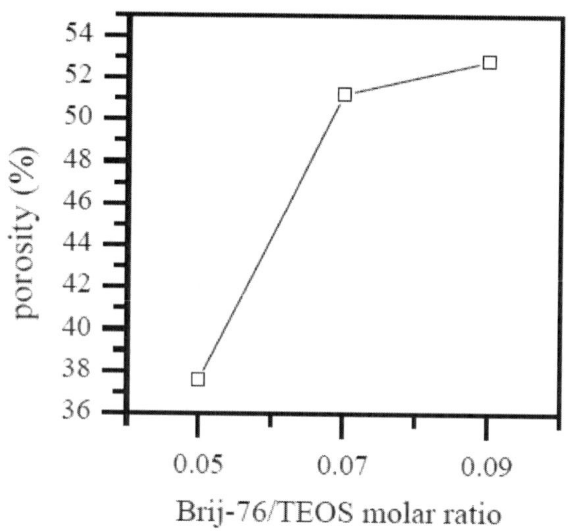

Figure 10. Porosity of mesoporous silica films as a function of Brij-76/TEOS molar ratio.

Figure 11 shows the elastic modulus and hardness of the mesoporous silica film according to Brij-76/TEOS molar ratio [34]. Due to the hard substrate effect, elastic modulus and hardness steadily increased as indentation depth increased. So, mechanical properties were extracted from the flat region in Figure 11. The elastic modulus and hardness values of mesoporous silica films were reduced with the increase in the molar ratio. Mechanical properties of the porous material are strongly dependent on porosity, pore wall rigidity, and the degree of pore ordering. When the molar ratio increases from 0.05 to 0.07, a degradation of mechanical properties is mainly caused by the increase in porosity. However, with the case of molar ratio of 0.09, a decrease in silica wall thickness and an increase in pore disordering as shown in Figure 9(b) might be the main causes of degradation. Mesoporous silica films with the same mesostructure but different pore size and ordering were prepared by adjusting the Brij-76/TEOS molar ratio. A long aging period before spin coating increases the hydrocarbon chain length in the micelle and destroys pore ordering. However, structural transition does not occur even after 20 days of sol aging before spin coating. GISAXS patterns of mesoporous silica film with Brij-76/TEOS molar ratio of 0.05–0.09 were indexed with a BCC structure while a structural transition into mixed lamellar structure was observed with the molar ratio over 0.1. When the molar ratio increases from 0.05 to 0.07, the porosity of the film was increased with pore size.

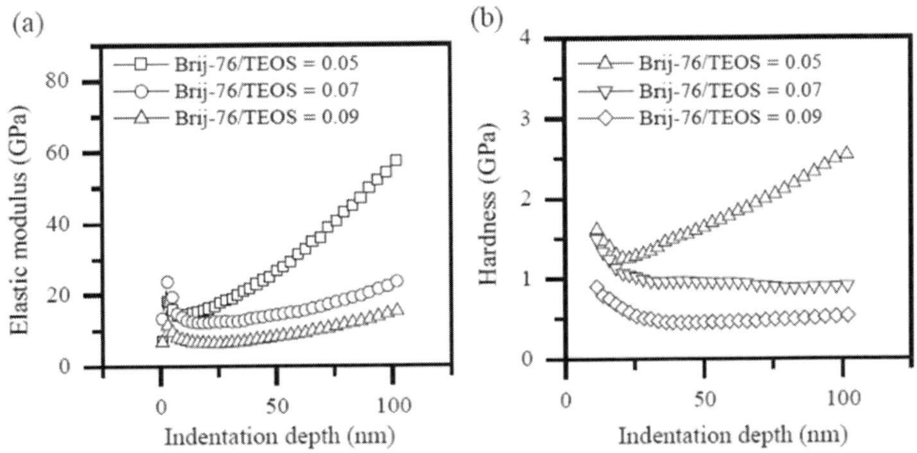

Figure 11. (a) Elastic modulus and (b) hardness of mesoporous silica films with various Brij-76/TEOS molar ratio.

Control of Roughness and Pore Structure by Using Co-Solvent of Acetone and Ethanol

There are four distinct stages in the spin coating process. The first stage is the deposition of the coating fluid onto the wafer, and the second stage corresponds to the period approaching the desired rotation speed. The third stage is the fluid thinning period at a constant spinning rate, and finally at the fourth stage thinning behavior is observed due to solvent evaporation. The key point is that the surface tension of the remaining solution should be lower during the solvent evaporation stage when the highly volatile alcohol leaves the fluid surface [47]. However, this does not usually happen. So, it is necessary to introduce another solvent as acetone with a higher vapor pressure and surface tension than the common solvent, EtOH. Acetone was added as a co-solvent to the EtOH-based sol and the enhancement of film surface roughness was verified. Acetone was selected because of its solubility, vapor pressure, and surface tension compared with EtOH, and because it does not form an azeotropic mixture with EtOH [48]. Figure 12 represents the relationship between vapor pressure and surface tension of water, EtOH, and acetone. A possible evaporation sequence is given by the arrow in Figure 12 [6]. Table 1 shows the physical properties of EtOH and acetone.

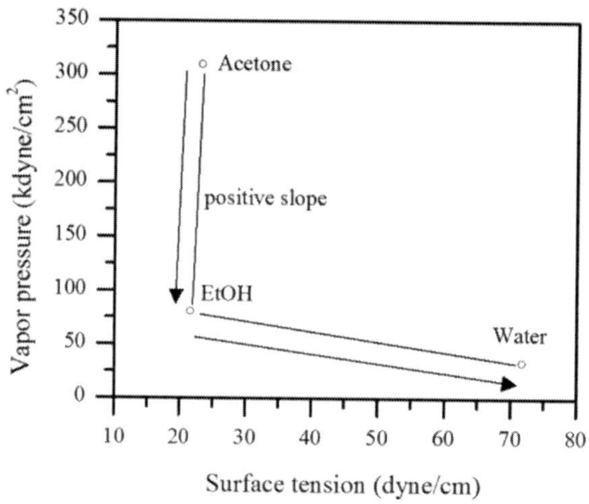

Figure 12. Possible evaporation sequences shown by the arrow from the consideration of vapor pressure and surface tension of liquids during spin coating.

Table 1. Physical properties of EtOH and acetone

	EtOH	Acetone
Labile proton	O	×
Density[a] (g/cm^3)	0.7893	0.7899
Viscosity[b] (cP)	1.074	0.306
Vapor pressure[b] ($kdyne/cm^2$)	78.7	308
Surface tension[b] (dyne/cm)	21.97	23.46
Evaporation rate (BuAc=1)	2.80	11.6
Thermal conductivity[b] (W/cm/K)	0.00169	0.00161
Dipole moment (D)	1.69	2.88
Specific heat capacity[b] (C_p)	2.44	2.18
Dielectric constant[c]	25.3	21.01

[a]measured at 20 °C.
[b]measured at 25 °C.
[c]measured at 20.2 °C.

Figure 13 shows surface profilometer scans of as-prepared EtOH-based silica films according to the probing direction [6]. The spin coating speed was 3000 rpm, and the sol aging time was 1 day, where the structure of the film was highly organized in the case of the EtOH-based silica sol [34]. For comparison, the surface profilometer data of thermally grown silica films are given in Figure 13(a) and the RMS roughness was 3 nm. The surface roughness of the as-prepared EtOH-based silica film was found to be dependent on probing direction as shown in Figs. 13(b, c). When the tip probes the sample in the transverse direction as shown in Figure 13(b), the film surface is very rough. But, the wavy characteristic caused by the Marangoni effect was not seen. When the tip probes the sample in the longitudinal direction, the wavy characteristic of the film surface was clearly observed as shown in Figure 13(c). The RMS roughnesses in Figs. 13(b, c) were measured as 6.1 and 10.5

nm, respectively, and the average was 8.5 nm. From the above results, it can be concluded that the as-prepared EtOH-based silica film shows a very rough surface and the roughness is dependent on the fluid flow direction. Even after calcination, rough and wavy characteristics were still observed. The Marangoni instability is summarized by the dimensionless Marangoni number as follows: [46]

$$\mathrm{Ma} = \frac{(\frac{\delta\sigma}{\delta T})\Delta T h}{\eta\alpha} \quad (4)$$

where Ma is the Marangoni number, $\frac{\delta\sigma}{\delta T}$ is the variation of the surface tension with temperature, h is the fluid thickness, ΔT is the temperature difference between the Si bottom end and the top fluid surface, η is the viscosity, and α is the diffusivity. For stable fluid flow, the Marangoni number must be as low as possible. From the equation, it is clear that the film thickness is an important factor in striation formation for the same sol composition. So, the spin coating speed was changed and physical changes were investigated in order to check the possibility of controlling the surface roughness by the reduction of film thickness.

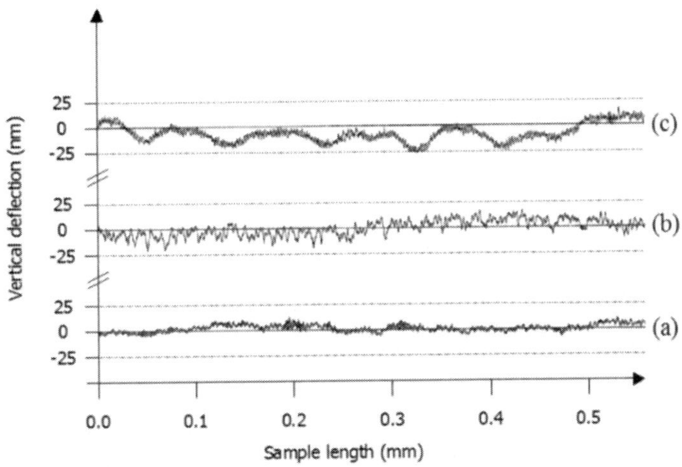

Figure 13. Surface profilometer scans of (a) thermally grown silica film and as-prepared EtOH-based silica film probed in the (b) transverse, and (c) longitudinal directions to the striations.

Figure 14(a, b) show the XRD patterns and FWHM of the main peak of an as-prepared EtOH-based silica film according to the rotation speed [6]. The peaks were (011) plane of BCC structure. Except for the film spun at 2000 rpm, the peak positions were not dependent on rotation speed. The maximum XRD peak intensity with minimum FWHM value was observed in the cases of 3000 and 3500 rpm. The structural ordering of the thin film by surfactant templating is determined by competing reactions between structuration by organic and inorganic species (thermodynamics) and gelation of the silica framework (kinetics) [50]. During structuration, surfactant micelles are surrounded by silica and as a result, the flexible nature of mesophase depends on the degree of silica condensation. Constant interplanar spacing means that the evaporation rate of the solvent does not affect micelle aggregation behavior. If the gelation rate is too slow, the evaporation rate of the solvent slows down and the tendency for organic-organic or inorganic-inorganic agglomeration increases. However, if the gelation rate is too fast at the topmost surface relatively, such as in the case of the 5000 rpm rotation speed, the network of silica is too rigid to organize organic-inorganic species with regular and periodic spaces in the whole film thickness and wormlike micelles may form in the interior of the film [25]. As a result, the structural ordering of the films is disrupted for high rotation speeds compared with that for 3000 rpm due to the loss of flexibility.

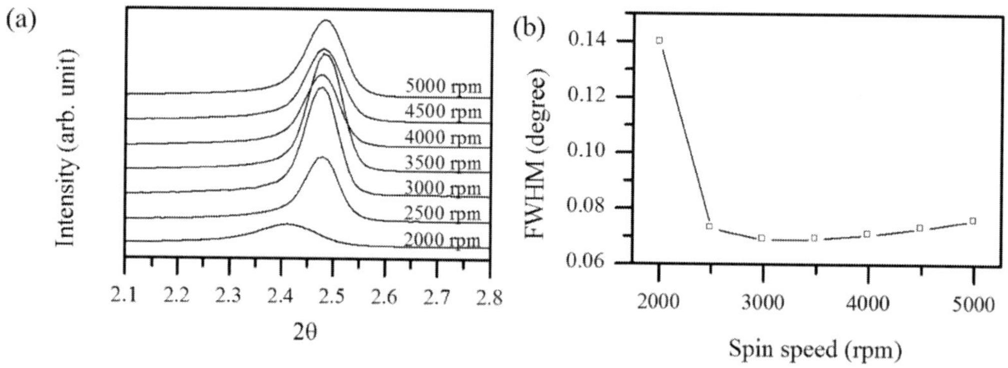

Figure 14. (a) XRD patterns and (b) FWHM of the corresponding main XRD peak of as-prepared EtOH-based silica films according to rotation speed.

Figure 15(a) shows the thickness variation with rotation speed in a log-log scale [6]. In Figure 15(a), surface roughness data were inserted as an error bar, shown in Figure 15(b). Final film thickness (h_f) for the evaporating fluid by spin coating is determined by the following equation: [46]

$$h_f = x\left[\frac{e}{2(1-x)K}\right]^{1/3} \quad (5)$$

$$e = C\sqrt{\omega}, K = (\rho\omega^2)/3\eta \quad (6)$$

Here x is the solid concentration in solution, e is the evaporation constant, K is the flow constant, ω is rotation speed, ρ is the density of the solution, and C is a proportionality constant related to the volatility of the solvent. From the above equation, film thickness will ideally be reduced according to rotation speed with a $-1/2$ power. But film thickness is generally proportional to ω^{-b}, for b other than 1/2. From the log-log relationship of film thickness and rotation speed in Figure 15(a), the thickness relationship of the as-prepared EtOH-based film has *a b* value of 0.43. Deviation from the ideal may originate from the increase in viscosity during spin coating. From the reduction of film thickness, it is anticipated that surface roughness can be reduced. The roughness of the films according to rotation speed measured using a surface profilometer is given in Figure 15(b). The roughness of the films was enhanced after increasing the rotation speed over 3000 rpm. However, striation formation cannot be prevented by this procedure.

Figure 16 represents the surface profilometer data of as-prepared EtOH-based silica films spin coated at 3500 and 5000 rpm, respectively [6]. The probing direction is longitudinal to the striations. Striation spacing decreases with increasing rotation speed. However, striations on the film surface still exist even though the rotation speed is 5000 rpm. From the above results, we concluded that increasing the rotation speed is not the right choice for the structural ordering and roughness of the film.

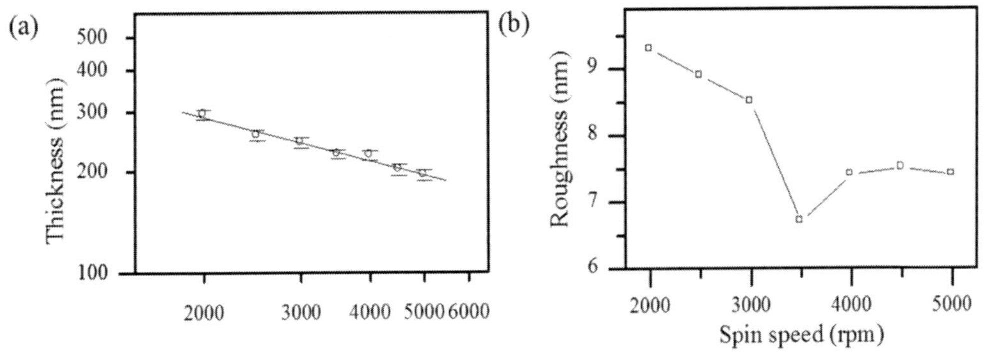

Figure 15. (a) Thickness and (b) surface roughness of as-prepared EtOH-based silica films according to rotation speed.

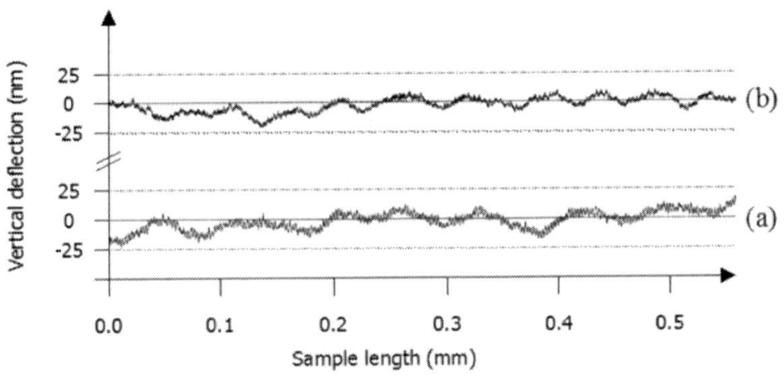

Figure 16. Surface profilometer scans of as-prepared EtOH-based silica films probed in the longitudinal direction with rotation speeds of (a) 3500 and (b) 5000 rpm.

Figure 17 shows the surface profilometer scans of as-prepared silica films according to acetone molar ratio [6]. The probing direction is longitudinal to the striations. When the acetone molar ratio is 0.15 in Figure 17(a), a slightly wavy characteristic appeared. However, the film surface is very flat and wavy characteristics are not observed when the acetone molar ratio is between 0.3 and 0.45. From consideration of the molar ratio, it was estimated that the optimized molar ratio is close to the EtOH-rich region for the reduction of surface tension of the remaining solution during the evaporation of the co-solvent. Although an acetone molar ratio over 0.45 makes the film surface rougher, wavy characteristics were not clearly observed. The minimum roughness of the as-prepared film is 5.1 nm when the acetone molar ratio is 0.3. However, there was no noticeable roughness difference for acetone molar ratios between 0.3 and 0.45. From this result, it can be concluded that adding the co-solvent with an optimized molar ratio enhances the film surface characteristics. During the hydrolysis and condensation reactions, EtOH can be released from the precursor silica source [51]. So, the exact acetone molar ratio may be lower than the given molar ratio. Also, if the total molar ratio of the solvent is not fixed at 20 or if the sol composition is changed, the optimized molar ratio range may change. For example, the molar ratio range may be narrowed when the total molar ratio is below 20 due to the film thickness. However, this fact does not underestimate the usefulness of the co-solvent for enhancing surface characteristics.

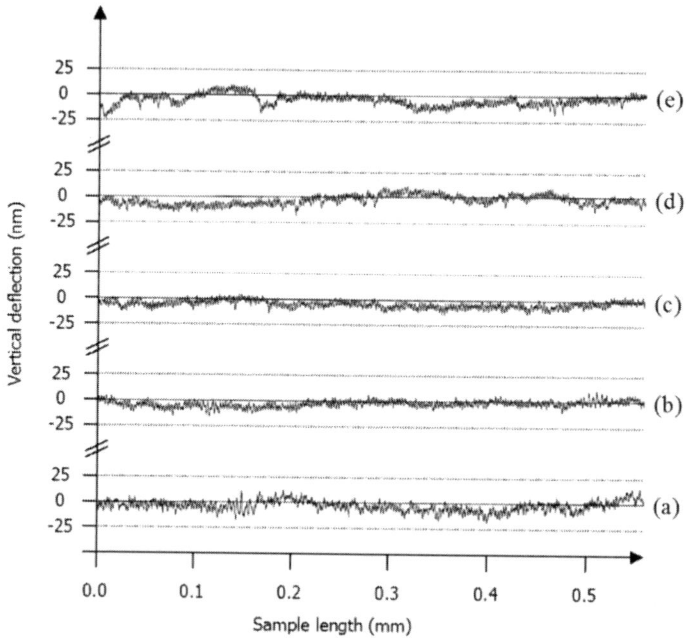

Figure 17. Surface profilometer scans of as-prepared silica films using the sol aged for 1 day as a function of acetone molar ratio: (a) 0.15, (b) 0.3, (c) 0.45, (d) 0.6, and (e) 0.75.

Figure 18(a) shows the interplanar spacing and FWHM of the (011) major peak from XRD patterns for as-prepared silica films with various acetone molar ratios, and Figure 18(b) shows the effect of sol aging time on interplanar spacing and FWHM for as-prepared EtOH-based films and as-prepared films using a 0.45 acetone molar ratio [6]. It can be observed that the increase of acetone molar ratio expands the interplanar spacing and increases the FWHM as shown in Figure 18(a). In acidic conditions (pH<2), it is well known that the rate of condensation reaction increases with pH reduction [33]. However, a high acetone molar ratio retards the condensation rate due to the inert characteristics of acetone for the sol-gel reaction. Previous work revealed that the unit cell of as-prepared EtOH-based silica films is enlarged with silica sol aging time as shown in Figure 18(b) [49]. This increase is due to the enlarged micelle diameter induced by elongated hydrophobic chains. If the increased interplanar spacing with acetone addition simply originates from the degree of condensation reaction, the interplanar spacing of the film using a 0.45 acetone molar ratio may vary with aging time following the fitted line for EtOH-based as-prepared silica films as shown in Figure 18(b). However, the interplanar spacing of as-prepared silica films using a 0.45 acetone molar ratio did not increase in this fashion. This means that the degree of condensation reaction is not the origin of the interplanar spacing variation in Figure 18(a). Meanwhile, acetone has a higher dipole moment than EtOH, which means that the headgroup repulsion between EO groups can be reduced by the electric shielding effect of acetone [51]. As a result, the effective headgroup area is decreased and the aggregation number is increased. Moreover, the aprotic solvent can penetrate into the hydrophobic chain during EISA, which increases the hydrophobic volume [51, 52]. These overall effects increase the g factor and therefore increases the micelle diameter. The increased FWHM in Figure 18(a) is attributed to the existence of acetone in the region of the hydrophobic chains and the retarded condensation rate [53]. Also, in Figure 18(b), the FWHM of as-prepared EtOH-based films was minimum for a 1 day-aged sol while the FWHM of as-prepared films using a 0.45 acetone molar ratio was minimum for a 3 days-aged sol. The increase in optimum sol aging time with acetone addition is a manifestation of the retarded condensation rate. However, the FWHM is slightly higher even for the optimized aging time, which means that a misaligned domain exists more when acetone is added to the sol.

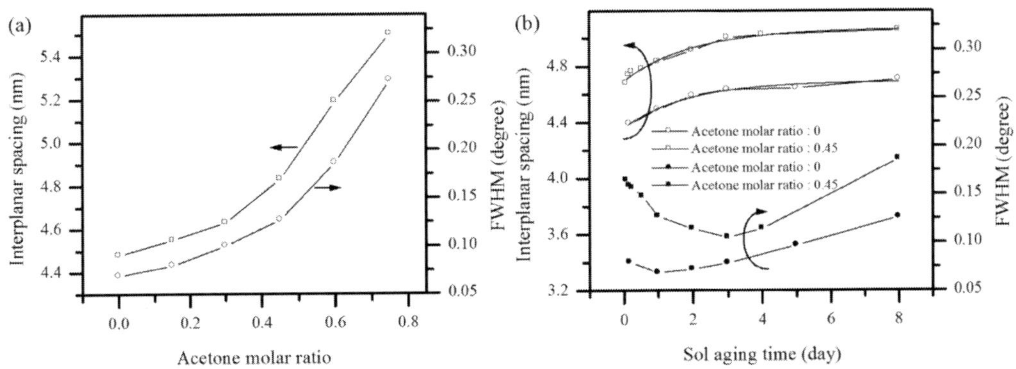

Figure 18. (a) Interplanar spacing and FWHM of (011) peak from XRD patterns for as-prepared silica films with various acetone molar ratios using the sol aged for 1 day and (b) interplanar spacing and FWHM of (011) peak for as-prepared EtOH-based silica films and as-prepared films using 0.45 acetone molar ratio as a function of sol aging time. Interplanar spacing in was fitted to a sinusoidal pattern for guidance.

Figure 19 shows the porosity, interplanar spacing and dielectric constants of calcined mesoporous films according to the acetone molar ratio and sol aging time [6]. The filled symbols denote the values of the mesoporous silica film using an acetone molar ratio of 0.45 and 3 days of sol aging time. From the refractive index of the film and the Lorentz-Lorentz equation, the porosity of the films can be estimated [26]. The porosity of the films increased slightly up to a 0.45 acetone molar ratio and then decreased due to the collapse of pores. The interplanar spacing of the films gradually increased with acetone molar ratio. In the case of 0.6 and 0.75 acetone molar ratios, the interplanar spacing and thickness show different behaviors. Except for mesoporous films using the sols of 0.6 and 0.75 acetone molar ratios, the dielectric constants of the films were measured as 2.48-2.55. The dielectric constant of the mesoporous film is dependent on porosity, water absorption, surface chemical species, etc. The slight reduction of the dielectric constant after the addition of acetone is due to the slightly increased pore radius/lattice parameter, which induces an increase in mesoporosity [54]. Using the Bruggeman effective medium approximation, the dielectric constant of the film can be calculated from the refractive index [50]. However, the dielectric constant calculated from the refractive index is higher than the measured one, which means that the microporosity of the film in the silica framework contributes to the total porosity to some extent.

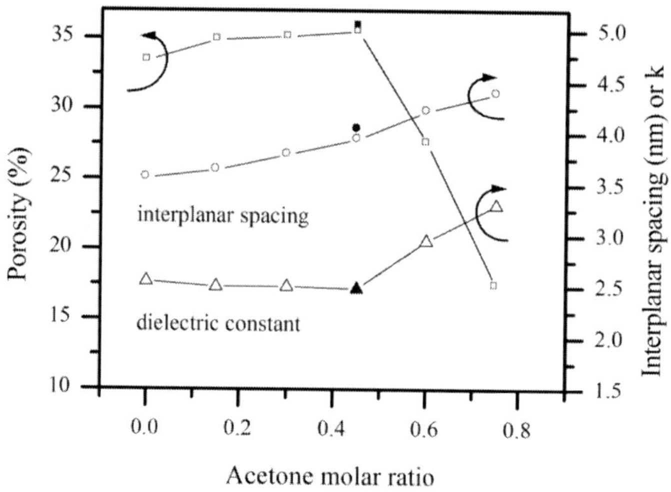

Figure 19. Porosity, interplanar spacing, and dielectric constant of mesoporous silica films according to the acetone molar ratio and aging time. The filled symbols denote the value corresponding to the mesoporous silica films using the sol aged for 3 days with an acetone molar ratio of 0.45.

Figure 20 shows the leakage current property of calcined mesoporous films according to the acetone molar ratio and sol aging time [6]. When the acetone molar ratio is 0 and 0.45, the leakage current behavior shows almost the same tendency and becomes 2.35×10^{-5} and 3.58×10^{-5} at 1.5 MV/cm as shown in Figure 20(a, b). This shows that there is no merit in the leakage current property, even though the surface property of the mesoporous film is enhanced using the sol aged for 1 day with a 0.45 acetone molar ratio, due to the degree of pore disordering. Furthermore, the leakage current property was greatly degraded when the acetone molar ratio is 0.75, as shown in Figure 20(c). It is estimated that some microcracks or conduction paths may form during the calcination process. Improved leakage current property was observed when the acetone molar ratio is 0.45 and the sol aging time is 3 days as shown in Figure 20(d). The leakage current behavior of porous films is bulklimited conduction and depends on the defect density such as silanol groups and strained or cyclic Si–O bonds. It is considered that the surface roughness of the film causes an electric field concentration, which induces the degradation of the electrically insulating properties. The enhanced surface roughness of the film therefore increases the leakage current behavior by preventing uneven electric field distribution between the electrode and the film [55]. The benefits of structural ordering and using a co-solvent for the synthesis of mesostructured silica films were elucidated. As-prepared silica films showed a wavy and rough surface when only EtOH was used as a solvent. Well-ordered as-prepared EtOH-based films show an average roughness of 8.5 nm from the Marangoni effect when the rotation speed is 3000 rpm. By increasing the rotation speed during spin coating, a reduction of film thickness modulation was anticipated. However, this was not the right choice from the point of view of structural ordering and roughness. When the acetone molar ratio is between 0.3 and 0.45, the film surface is extremely flat, and wavy characteristics were not observed irrespective of sol aging time. From consideration of the molar ratio, the optimized molar ratio should be close to the EtOH-rich region in order to reduce the surface tension of the remaining solution during the evaporation of the co-solvent. As a result, the optimized sol aging time increased with higher acetone molar ratio. When the sol aging time is not properly controlled for structural ordering, some of the collapsed pore structure is derived after calcination.

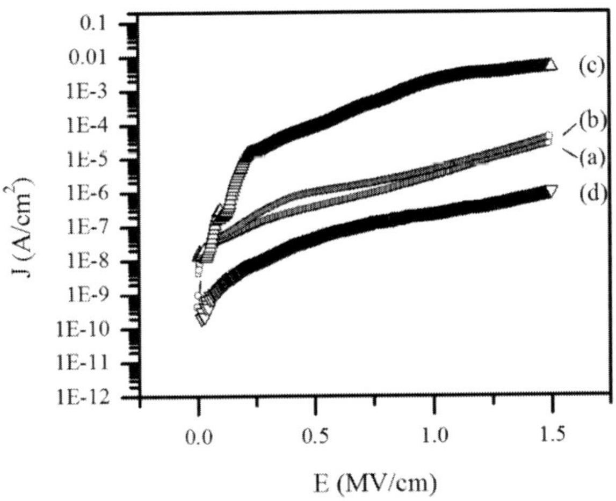

Figure 20. Leakage current behavior of ordered mesoporous silica films using the sol aged for one day with Acetone molar ratio of (a) 0, (b) 0.45, (c) 0.75, and (d) using the sol aged for 3 days with acetone molar ratio of 0.45.

Control Effect of MTES/TEOS Co-Precursor on the Mechanical Properties of Mesoporous SiO_2

Generally, ordered mesoporous silica films have been prepared by using TEOS as a hybrid framework source and surfactant as template, where the mesopores are distributed uniformly with some degree of order. In this work, methyltriethoxysilane (MTES) was introduced as a silica precursor and the physiochemical properties of the films were investigated according to the mixing ratio with TEOS. Figure 21 shows the FT-IR spectra of ordered mesoporous silica films for various MTES/TEOS molar ratios [7]. The peaks at 1067 and 1109 cm^{-1} correspond to the TO_3 and cyclic Si–O vibrations vibrations, respectively [56]. The peak corresponding to the cyclic Si–O bond originates from the Si–O bonds at the pore surface, which signify the porous nature of the film, and the peak at 1279 cm^{-1} corresponds to the Si–C vibrations. Si–C bonds were not dissociated during calcination at 400°C. It was observed that the intensity of the Si–C bond increased with respect to the MTES molar ratio. However, an absorption appeared at 960 cm^{-1} due to the residual Si–OH group, even though there were framework modifications with hydrophobic Si–C bonds. The peak intensity of the cyclic Si–O bonds increased with increasing MTES/TEOS molar ratios. This is related to the fact that the terminal Si–C bonds in the framework make a short chain Si–O–Si ring structure rather than a 6-member ring structure.

Synthesis of ordered mesoporous silica films by co-condensation of TEOS and MTES was reported by some authors [57]. However, investigations on the mechanical properties of the ordered mesoporous silica film by organic modification have been scarce. In this work, enhancement of the mechanical properties by MTES addition was focused on. Figure 22 shows the elastic modulus and hardness of the ordered mesoporous silica films with respect to

MTES/TEOS molar ratio [7]. Due to the hard substrate effect, the elastic modulus and hardness steadily increased as the indentation depth increased. Because of this, mechanical properties were extracted from the flat region in Figure 22 [58]. Surprisingly, it was observed that the elastic modulus and hardness of the film were increased with all MTES/TEOS molar ratios except 0.2. In order to verify the benefits of pore structure ordering, disordered mesoporous silica films were prepared at each MTES/TEOS molar ratio.

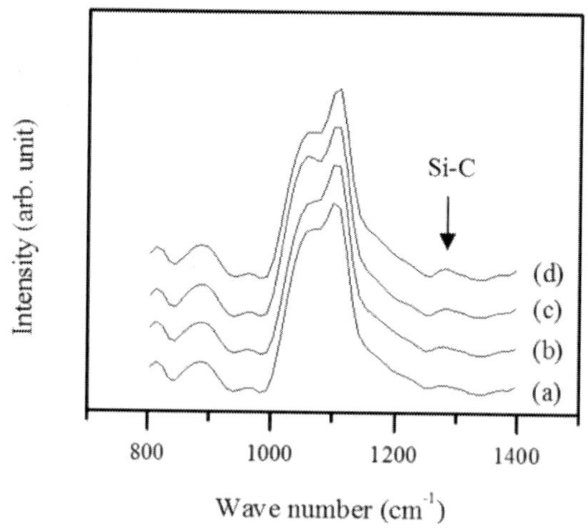

Figure 21. FT-IR spectra of ordered mesoporous silica films with MTES/TEOS molar ratios of (a) 0.1, (b) 0.15, (c) 0.2, and (d) 0.25.

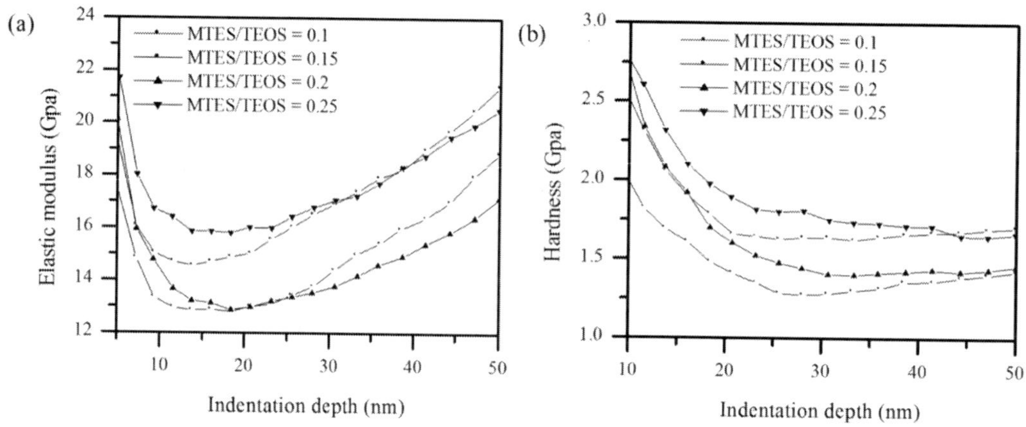

Figure 22. (a) Elastic modulus and (b) hardness of ordered mesoporous silica films according to the MTES/TEOS molar ratio.

Figure 23 shows the elastic modulus and hardness of the disordered mesoporous silica films with respect to the MTES/TEOS molar ratio [7]. The mechanical properties of the film deteriorated as the MTES/TEOS molar ratio increased. This mechanical behavior is opposite from that observed in ordered mesoporous films. In fact, pore collapse and densification during calcination occurred in the case of a 0.3 MTES/TEOS molar ratio.

The mechanical properties of a porous film are largely dependent on its density. Generally, the elastic modulus and density of a film have a linear relationship in log-scale with positive slope of 2 to 4 [59]. The value without an underline in Figure 24 represents the MTES/TEOS molar ratio of ordered mesoporous silica films, but the underlined value belongs to the disordered mesoporous silica film. The disordered mesoporous silica films showed normal density-dependent mechanical properties [7]. However, the mechanical properties were especially enhanced, even though there was a reduction of the density in the case of the ordered mesoporous silica films. The terminal Si–C bonds may make the silica framework weaker, the mechanical properties of the film are more strongly controlled by the micropore volume in the film.

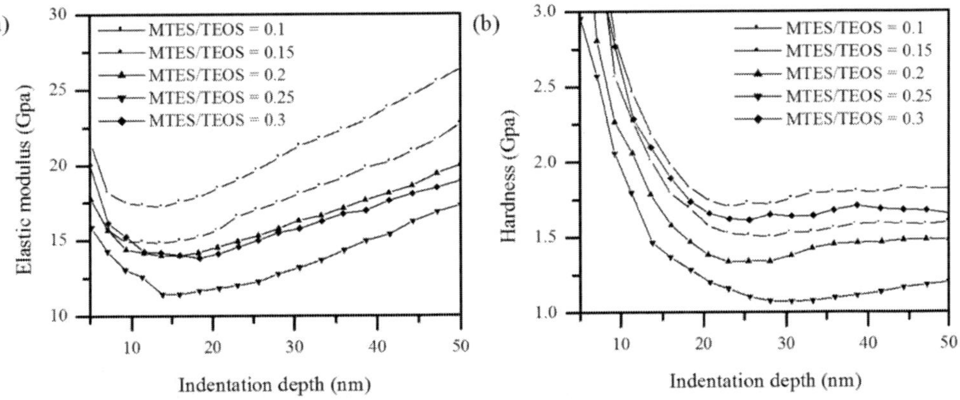

Figure 23. (a) Elastic modulus and (b) hardness of disordered mesoporous silica films according to the MTES/TEOS molar ratio.

The ordered mesoporous silica film with a 0.25 MTES/TEOS molar ratio has a body centered cell (BCC) structure. An abrupt increase in the elastic modulus for the ordered mesoporous silica film with a 0.25 MTES/TEOS molar ratio could be attributed to the fact that cubic structures with spherical pores are more mechanically robust than hexagonally packed structures [60]. This result highlights the benefits of BCC geometry and pore structure ordering for the application of ordered mesoporous silica films, such as low-k dielectrics.

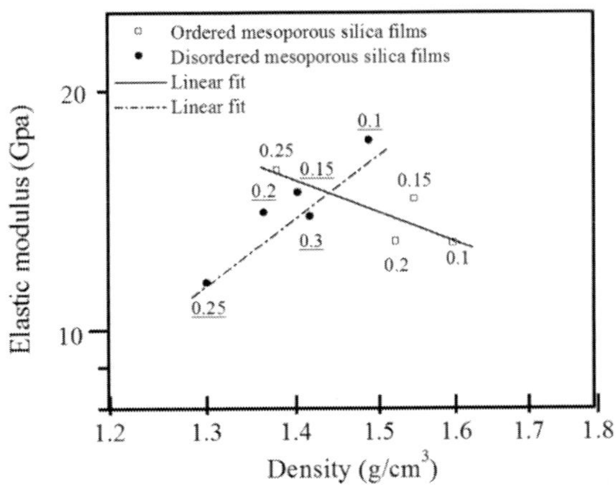

Figure 24. Relationship between density and elastic modulus in log–log scale for ordered and disordered mesoporous silica films. The numerical number around each data represents the MTES/TEOS molar ratio.

Figure 25 shows the variations in the dielectric constant of the ordered mesoporous silica films with respect to the MTES/TEOS molar ratio [7]. The dielectric constant gradually decreased due to a reduction of the density. The dielectric constant is a little smaller than the calculated one by using the Bruggeman effective medium approximation. This is attributed to the existence of Si–C bonds with low molar polarizability [61]. In particular, the ordered mesoporous silica film with a 0.25 MTES/TEOS molar ratio has the lowest dielectric constant. Namely, the film with BCC pore structure shows the lowest dielectric constant and highest mechanical strength, as shown in Figs. 24, 25. So it can be concluded that film with low micropore volume fraction and BCC pore ordering is more applicable to a low-k material due to low dielectric constant and high mechanical strength. A modification of a silica framework with Si–C bonds was successfully performed using TEOS and MTES as silica precursors. The organic modification of the framework by methyl ligands could provide mechanical stability and a low dielectric property to an ordered mesoporous silica film for low-k applications due to reduction of micropore volume and BCC pore ordering.

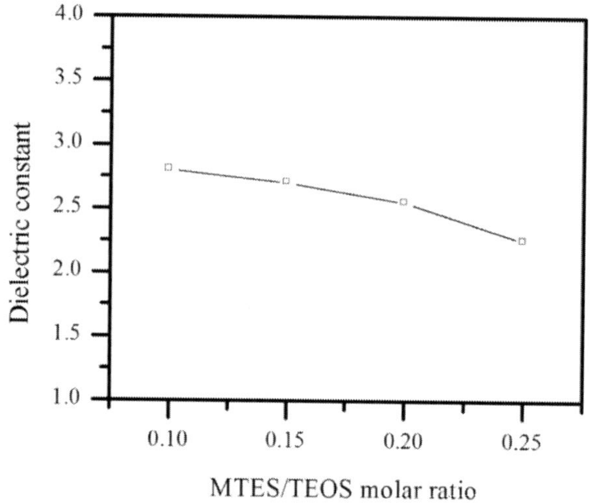

Figure 25. The dielectric constant of ordered mesoporous silica films according to the MTES/TEOS molar ratio.

References

[1] C. T. Kresge, M. E. Leonowicz, W. J. Roth, J. C. Vartuli, J. S. Beck, *Nature* 359 (1992) 710.

[2] K. Domansky, J. Liu, L. Q. Wang, M. H. Engelhard, S. Baskaran, *J. Mater. Res.* 16 (2001) 2810.

[3] H. Y. Fan, Y. F. Lu, A. Stump, S. T. Reed, T. Baer, R. Schunk, L. V. Perez, G. P. Lopez, C.J. Brinker, *Nature* 405 (2000) 56.

[4] R. D. Miller, *Science* 286 (1999) 421.

[5] S. B. Jung, C. K. Han, H. H. Park, *Applied Surface Science* 244 (2005) 47.

[6] S. B. Jung, T. J. Ha, H. H. Park, *Journal of Applied Physics* 101 (2007) 024109.

[7] S. B. Jung, T. J. Ha, H. H. Park, *Journal of Colloid and Interface Science* 320 (2008) 527.

[8] T. J. Ha, S. G. Choi, S. B. Jung, B. G. Yu, H. H. Park, *Ceramic International* 34 (2008) 947.

[9] T. J. Ha, H. K. Kim, D. J. Choi, S. Shin, H. H. Cho, H. W. Jang, S. J. Yoon, H. H. Park, *Microporous and Mesoporous Materials* 163 (2012) 321.

[10] T. J. Ha, S. Shin, H. K. Kim, M. H. Hong, C. S. Park, H. H. Cho, D. J. Choi, H. H. Park, *Applied Physics Letters* 102 (2013) 144102.

[11] T. J. Ha, S. Y. Jung, J. H. Bae, H. L. Lee, H. W. Jang, S. J. Yoon, S. Shin, H. H. Cho, H. H. Park, *Microporous and Mesoporous Materials* 144 (2011) 191.

[12] T. J. Ha, S. J. Yoon, J. S. Kim, H. H. Park, *Applied Surface Science* 256 (2009) 1073.

[13] P. D. Yang, G. Wirnsberger, H. C. Huang, S. R. Cordero, M. D. McGehee, B. Scott, T. Deng, G. M. Whitesides, B. F. Chmelka, S. K. Buratto, G. D. Stucky, *Science* 287 (2000) 465.

[14] N. Nishiyama, A. Krpmoide, Y. Egashira, K. Ueyama, *Chem. Commun.* (1998) 2147.

[15] T. J. Ha, M. H. Hong, C. S. Park, H. H. Park, *Sensors and Actuators B* 181 (2013) 874.

[16] J. W. Tang, Y. Y. McFarland, E. W. Stucky, G. D. *Chem. Commun.* (2004) 1670.

[17] E. Lancelle-Beltran, P. Prene, C. Boscher, P. Belleville, P. Buvat, S. Lambert, F. Guillet, C. Boissière, D. Grosso, C. Sanchez, C. *Chem. Mater.* (2006) 6152.

[18] M.E. Davis, *Nature* 417 (2002) 813.

[19] C. Sanchez, C. Boissière, D. Grosso, C. Laberty, L. Nicole, *Chem. Mater.* 20 (2008) 682.

[20] S. Mann, S. L. Burkett, S. A. Davis, C. E. Fowler, N. H. Mendelson, S. D. Sims, D. Walsh, N. T. Whilton, *Chem. Mater.* 9 (1997) 2300.

[21] S. B. Jung, T. J. Ha, J. B. Seon, H. H. Park, *Microporous and Mesoporous Materials* 111 (2008) 188.

[22] T. J. Ha, H. W. Jang, S. J. Yoon, S. Shin, H. H. Cho, H. H. Park, *Microporous and Mesoporous Materials* 158 (2012) 123.

[23] P. W. Lee, S. Mizuno, A. Verma, H. Tran, B. Nguyen, *J. Electrochem. Soc.* 143 (1996) 2015.

[24] D. Grosso, F. Cagnol, G. J. de A. A. Soler-Illia, E. L. Crepaldi, H. Amenitsch, A. Brunet-Bruneau, A. Bourgeois, C. Sanchez, *Adv. Funct. Mater.* 14 (2004) 309.

[25] D. A. Doshi, A. Gibaud, V. Goletto, M. Lu, H. Gerung, B. Ocko, S.M. Han, C.J. Brinker, *J. Am. Chem. Soc.* 125 (2003) 11646.

[26] K. Maex, M. R. Baklanov, D. Shamiryan, F. Lacopi, S. H. Brongersma, Z. S. Yanovitskaya, *J. Appl. Phys.* 93 (2003) 8793.

[27] Z. Chen, K. Prasad, *Appl. Phys. Lett.* 84 (2004) 2242.

[28] N. Hata, C. Negoro, K. Yamada, T. Kikkawa, *Jpn. J. Appl. Phys.* 43 (2004) 1323.

[29] A. R. Balkenende, F. K. de Theije, J. C. K. Kriege, *Adv. Mater.* 15 (2003) 139.

[30] B. J. Park, S. Y. Rah, Y. J. Park, K. B. Lee, *Rev. Sci. Instrum.* 66 (1995) 1722.

[31] B. D. Lee, I. S. Park, J. H. Yoon, S. J. Park, J. H. Kim, K. W. Kim, T. H. Chang, M. H. Ree, *Macromolecules* 38 (2005) 4311.

[32] A. Jain, S. Rogojevic, W. N. Gill, J. L. Plawsky, I. Mattew, M. Tomozawa, E. Simonyi, *J. Appl. Phys.* 90 (2001) 5832.

[33] C. J. Brinker, G. W. Scherer, Sol-Gel Science, 2nd ed. (Academic, San Diego, 1990), pp. 213, 795, 108, 127, and 104.

[34] S. B. Jung, H. H. Park, *Thin Solid Films* 494 (2006) 320.
[35] M. Imperor-Clerc, P. Davidson, A. Davidson, *J. Am. Chem. Soc.* 122 (2000) 11925.
[36] K. Flodstrom, V. Alfredsson, *Microporous Mesoporous Mater.* 59 (2003) 167.
[37] K. J. Edler, S. J. Roser, *Int. Rev. Phys. Chem.* 20, 387 (2001).
[38] P. C. A. Alberius, K. L. Frindell, R. C. Hayward, E. J. Kramer, G. D. Stucky, B. F. Chmelka, *Chem. Mater.*, 14 (2002) 3284.
[39] Y. Sakamoto, I. Dı´az, O. Terasaki, D. Zhao, J. P. Pariente, J. M. Kim, G.D. Stucky, *J. Phys. Chem. B* 106 (2002) 3118.
[40] N. Liu, R. A. Assink, B. Smarsly, C. J. Brinker, *Chem. Commun.* (2003) 1146.
[41] T. M. Parrill, *J. Mater. Res.* 7 (1992) 2230.
[42] R. M. Almeida, C. G. Pantano, *J. Appl. Phys.* 68 (1990) 4225.
[43] H. Miyata, T. Suzuki, A. Fukuoka, T. Sawada, M. Watanabe, T. Noma, K. Takada, T. Mukaide, K. Kuroda, *Nat. Mater.* 3 (2004) 651.
[44] B. Smarsly, D. Grosso, T. Brezesinski, N. Pinna, C. Boissiere, M. Antonietti, C. Sanchez, *Chem. Mater.* 16 (2004) 2948.
[45] Y. Sakamoto, I. Dı´az, O. Terasaki, D. Zhao, J. P. Pariente, J. M. Kim, G. D. Stucky, *J. Phys. Chem., B* 106 (2002) 3118.
[46] S. Besson, C. Ricolleau, T. Gacoin, C. Jacquiod, J. P. Boilot, *Microporous Mesoporous Mater.* 60 (2003) 43.
[47] D. P. Birnie III, J. *Mater. Res. 16* (2001) 1145.
[48] M. J. Lee, C. H. Hu, *Fluid Phase Equilib.* 109 (1995) 83.
[49] D. R. Lide, *Handbook of Chemistry and Physics*, 76th ed. (CRC, New York, 1995), pp. 6–224, I-3, and I-13.
[50] Q. Huo et al., *Chem. Mater.* 6 (1994) 1176.
[51] M. T. Anderson, J. E. Martin, J. G. Odinek, P. P. Newcomer, *Chem. Mater.* 10 (1998) 311.
[52] S. Liu, P. Cool, O. Collart, P. V. D. Voort, E. F. Vansant, O. I. Lebedev, G. V. Tendeloo, and M. Jiang, *J. Phys. Chem.* B 107 (2003) 10405.
[53] Q. Huo, D. I. Margolese, G. D. Stucky, *Chem. Mater.* 8 (1996) 1147.
[54] P. I. Ravikovitch, A. V. Neimark, *Langmuir* 18 (2002) 1550.
[55] Y. S. Kim, Y. H. Lee, K. M. Lim, and M. Y. Sung, *Appl. Phys. Lett.* 74 (1999) 2800.
[56] P. Innocenzi, P. Falcaro, D. Grosso, F. Babonneau, *J. Phys. Chem. B* 107 (2003) 4711.
[57] Y. E. Lee, B. S. Kang, S. H. Hyun, C. H. Lee, *Sep. Sci. Technol.* 39 (2004) 3541.
[58] Y. H. Wang, M. R. Moitreyee, R. Kumar, S. Y. Wu, J. L. Xie, P. Yew, B. Subramanian, L. Shen, K. Y. Zeng, *Thin Solid Films* 462 (2004) 227.
[59] N. Chemin, M. Klotz, V. Rouessac, A. Ayral, E. Barthel, *Thin Solid Films* 495 (2006) 210.
[60] R. A. Pai, R. Humayun, M. T. Schulberg, A. Sengupta, J. N. Sun, J. J. Watkins, *Science* 303 (2004) 507.
[61] C. M. Yang, A. T. Cho, F. M. Pan, T. G. Tsai, K. J. Chao, *Adv. Mater.* 13 (2001) 1099.

In: Comprehensive Guide for Mesoporous Materials, Volume 4 ISBN: 978-1-63482-092-9
Editor: Mahmood Aliofkhazraei

Chapter 3

MESOPOROUS MATERIALS IN ENERGY CONVERSION AND STORAGE

***Yumin Liu*[1,2,*] *and Xing-zhong Zhao*[1,3*]**
[1]Institute for Interdisciplinary Research (IIR), Jianghan University, Wuhan, China
[2]Key Laboratory of Optoelectronic Chemical Materials and Devices of Ministry of Education, Jianghan University, Wuhan, China
[3]Department of Physics and Technologies, Key Laboratory of Artificial Micro/Nano Structure of Ministry of Education, Wuhan University, Wuhan, China

ABSTRACT

Fundamental developments in energy conversion and storage (ECS) are essential to meeting the challenge of global warming and the limited supply of nonrenewable resources, such as fossil fuels. In line with this need, mesoporous materials have attracted much attention in the research field of ECS systems given their special structures, large surface-to-volume ratios, favorable transport properties, short diffusion paths, and reduced diffusion effect. Thus, an updated review of the applications of mesoporous materials in ECS is presented in this chapter. The factors involved in energy conversion, such as dye-sensitized solar cells, are also discussed. Insights into the various underlying mechanisms of each case are listed to highlight the advantages of introducing mesoporous materials into energy conversion devices. Moreover, energy storage technologies such as lithium–ion batteries based on mesoporous materials are investigated. The relationship between mesoporous materials and their effectiveness in the presented ECS devices ultimately paves the way for the design of novel mesoporous structures with advanced properties.

* Corresponding author: Tel.: +86-27-84223798. E-mail: ymliu_phy@whu.edu.cn; xzzhao@whu.edu.cn

1. Introduction

Energy conversion and storage (ECS) is among the greatest challenges in the 21st century. Hence, research in this field is progressing. In response to rapid economic globalization, the increase in the world population, and environmental deterioration, ECS devices that are clean, low-cost, and environmental-benign must be developed. The performance of these energy-related devices is strongly dependent on the properties of the materials employed. Various materials have been innovated and assessed for application in ECS devices over the past two decades, including 1D nanomaterials that generate an effective electron pathway and accommodate significant volume changes [1–5], 2D nanostructures with large contact surface areas and short paths for electrons/ion transport [6–10], and 3D constructed nanomaterials with good conductivity and connections [11–15]. These nanostructured materials possess unusual mechanical, electrochemical, and photovoltaic properties endowed with the accurate control of material dimensions and combinations. Thus, they enhance the performance of ECS devices considerably.

Among the nanostructured materials that are promising candidates for ECS application, mesoporous materials have attracted the attention of researchers in this field given their special structures, large surface-to-volume ratios, favorable transport properties, short diffusion paths, reduced diffusion effect, and excellent connection. These features are the result of meticulous and ingenious material structure design. Therefore, recent advancements in mesoporous materials for use in dye-sensitized solar cells (DSSCs) and lithium–ion batteries (LIBs) are described. The challenges, approaches, advantages, and disadvantages of employing mesoporous materials for ECS application in recent research are also highlighted instead of providing a general overview. Furthermore, we determine the relationship between mesoporous materials and their performance in ECS devices. The advantages and shortcomings of the mesoporous materials used in the examined ESC applications demonstrate that the careful design of mesoporous structures may either generate the desired functions in the materials or generate new mechanisms, thereby improving the overall performance of devices.

2. Mesoporous Materials for DSSCs

Since O'Regan and Grätzel published their seminal *Nature* paper entitled "A low-cost, high-efficiency solar-cell based on dye-sensitized colloidal TiO_2 films" in 1991 [16], DSSCs have attracted much attention as one of the most credible alternatives to traditional silicon-based photovoltaic devices because of their high power conversion efficiency (PCE) and low production cost. A typical DSSC consists of a transparent conductive oxide on glass, a nanocrystalline semiconductor film with a wide bandgap, a sensitizing dye molecule, an electrolyte containing a redox couple, and a counter electrode. Dye molecules in DSSCs are anchored on the surface of a semiconductor with a wide bandgap (such as TiO_2, ZnO, and SnO_2). Upon illumination, the photogenerated electrons transfer rapidly from photo-excited sensitizers into the conduction band of the semiconductor. This band carries the electrons to a collector, such as fluorine-doped tin oxide, on glass. The oxidized dye molecule can be reduced to its neutral state through electron donation by a redox couple. This couple is regenerated at the counter-electrode to complete the circuit. The observed cell voltage depends on the difference between the quasi-Fermi level of the semiconductor under illumination and the electrochemical potential of the redox couple in the electrolyte. The overview and the major electron transfer processes of DSSCs are shown in Figure 1.

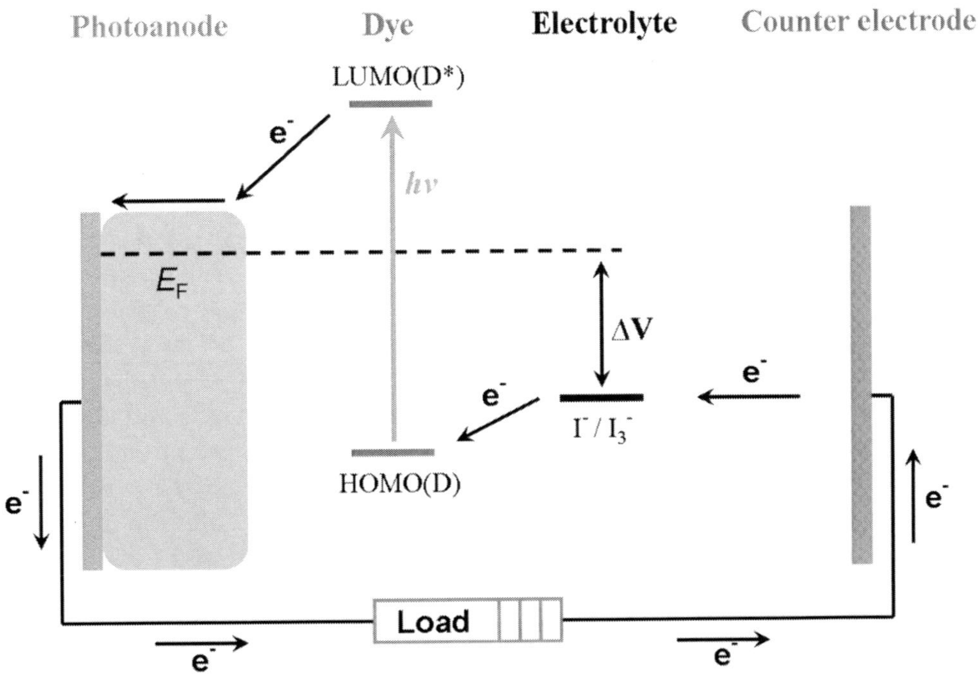

Figure 1. Schematic overview and major electron transfer processes of DSSCs.

2.1. Mesoporous Materials for Photoanodes

The DSSC is the sole photovoltaic device that separates the two functions of light harvesting and charge-carrier transport. Dye sensitizers that are adsorbed on the semiconductor act as antennas for light harvesting; thus, dye molecules have been studied intensively in the past two decades to increase the extinction coefficient and to extend the light-harvesting region into the near-infrared region [17, 18]. By contrast, charge-carrier transport mainly depends on the semiconductor film, and dye loading is influenced by the specific surface areas of photoanodes. Hence, a compatible photoanode nanostructure must be design and fabricated [19]. Mesoporous TiO_2 nanocrystalline films have been most widely studied for use in DSSCs to form photoanodes using state-of-the-art technology although various nanoarchitectured photoanodes have been investigated, including 1D nanostructures that generate an effective electron pathway [20–22], hybrid nanowire/nanotube structures that are combined with nanoparticles [23, 24], and 3D nanostructured photoanodes with multiple functions [25–27]. The mesoporous film that consists of TiO_2 nanocrystals (~20 nm) has a large surface area and produces an interconnected crystalline framework that can adequately facilitate dye loading and efficient electron diffusion. These features assist in improving power conversion efficiency (PCE). Considerable effort has also been exerted to optimize the performance of this mesoporous film further.

Several laboratories have investigated the doping effect of TiO_2 nanocrystals with various doping levels. Highly Nb-doped (from 2.5 mol% to 10 mol%) anatase TiO_2 has been examined by Lü et al., and their results indicate that the photocurrent of DSSC devices increases [28]. However, the enhanced doping of TiO_2 renders it unsuitable for electrochemical applications because of the increased electron recombination at the surface of the mesoporous TiO_2 film. To control the junction characteristics and enhance charge transport further [29], Nikolay et al. synthesized lightly Nb-doped (from 0.5 mol% to 3.0 mol%) TiO_2 through a sol–gel method that was followed by hydrothermal treatment. Their systematic study demonstrated that the electronic structures of TiO_2, such as the positions of Fermi energy level (E_F), shifted away after Nb-doping. They also noted that the E_F of the Nb-doped TiO_2 was lower than that of the undoped TiO_2 when doping level ranged from 0.5 mol% to 1.5 mol%. However, the E_F of the former was higher when the doping level was 2.5 mol% and more. As a result of the improved charge transport and the injection of a mesoporous TiO_2 film induced by Nb-doping, the PCE of the device was enhanced. Huang et al. prepared Zn-doped mesoporous TiO_2 electrodes with agarose gel as a template and examined the doping effects on DSSC performance [30]. Their results indicated that increased amounts of dye are adsorbed and recombination rate is reduced with low levels of Zn doping. Hence, photocurrent and overall PCE are improved. Other elements, such as N [31], B [32, 33], Cr [34], and Mg [35] have been introduced for TiO_2 doping. Furthermore, the performance of DSSC devices based on the mesoporous TiO_2 film can be enhanced through the accurate control of doping amounts.

The introduction of TiO_2 single crystals with tunable exposed reactive facets to form mesoporous films effectively increases device PCE. Anatase TiO_2 has been considered the ideal crystal structure for DSSC application given its outstanding performance in relation to dye loading and electron transport [36]. Most available anatase TiO_2 crystals are dominated by thermodynamically stable (101) facets rather than the more reactive (001) facets [37]. Nonetheless, dye loading is enhanced with the latter, which benefits DSSC application. Wu et al. synthesized nanosized TiO_2 single crystals with controllable percentages of exposed (001) facets using hydrofluoric acid as a morphology controlling agent [38]. Their results showed that the PCE of DSSC was enhanced by over 40% after anatase TiO_2 single crystals with 80% exposed (001) facets were introduced to form the mesoporous film, unlike in the device based on Degussa P25.

Through dark current potential and open-circuit voltage decay characterizations, the researchers indicated that the increased proportion of exposed (001) facets is not only favorable for dye loading but it also retards charge recombination in DSSCs. Yu et al. synthesized anatase TiO_2 nanosheets with exposed (001) facets through a one-pot hydrothermal route with hydrofluoric acid as a morphology controlling agent and $Ti(OC_4H_9)_4$ as a precursor. They assessed the performance of DSSC based on these TiO_2 nanosheets and reported that PCE can be enhanced because of increased light scattering and the high activity of (001) facets [39]. Hao et al. recently synthesized anatase TiO_2 single crystals with exposed (001) facets using a green hydrothermal method without the highly corrosive and poisonous hydrofluoric acid as a morphology controlling agent. This process is depicted in Figure 2. They also employed titanate nanotubes as titanium precursors for hydrothermal reaction and NH_4F as shape-capping reagents. The anatase TiO_2 single crystals with exposed (001) facets displayed an enhanced dye-loading capacity and charge collection efficiency. In the process, short-circuit photocurrent density and PCE are significantly improved [40].

Apart from aforementioned strategies, a self-assembled route has been developed to synthesize highly ordered mesoporous TiO_2. Crossland et al. employed a diblock copolymer as a sacrificial template to obtain an ordered bicontinuous TiO_2 mesoporous film. They integrated this film into solid-state DSSCs (SDSCs) and liquid DSSCs [41, 42]. The diblock copolymer used in their seminal work is poly(4-fluorostyrene)-*b*-poly(D, L-lactide) (PFS-b-PLA). Moreover, various self-assemble morphologies of mesoporous TiO_2 have been identified by controlling the molecular weight of PFS-b-PLA and PFS content, as illustrated in Figures 3a–3f. Crossland et al. reported that the devices based on 1D ordered photoanodes exhibited enhanced charge transport and significantly reduced charge recombination. However, the internal surface area of this 1D structure is small because of the lack of structural integrity. As a result, the photocurrent and PCE of the solar cells are significantly lower than those of the devices based on 3D gyroid mesoporous structures (Figure 3i).

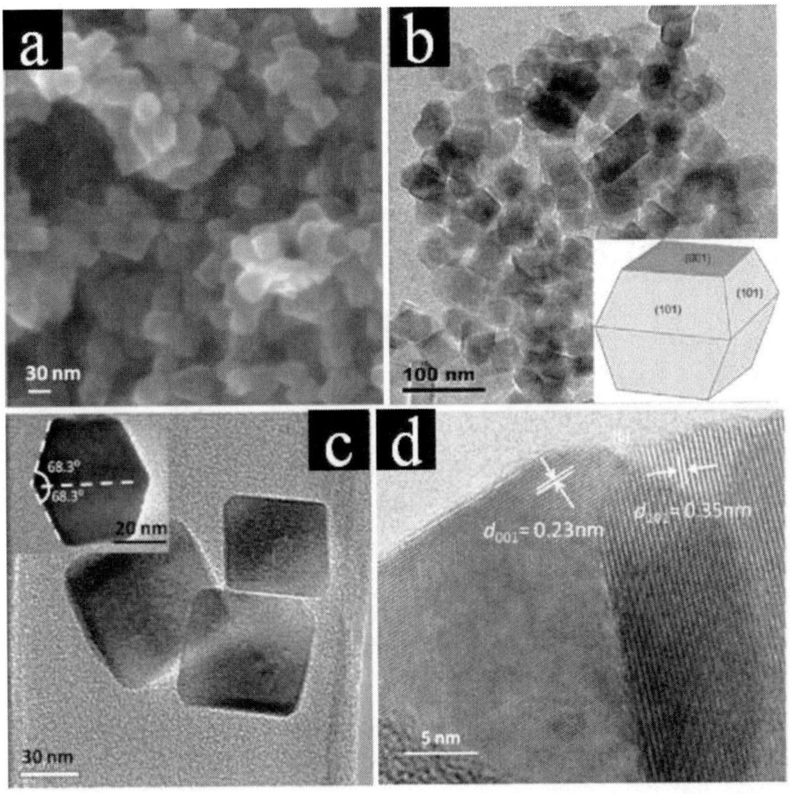

Figure 2. (a) SEM, (b, c) TEM and (d) HR-TEM images of as-synthesized anatase TiO_2 single crystals with exposed (001) facets. The inset of (b) shows the representative sketch of the faceted nanoparticle, and the inset of (c) shows the side view of a representative single crystal. Reprinted with permission from reference 40. Copyright (2012) American Chemical Society.

Guldin et al. coupled this high-surface mesoporous layer with a 3D photonic crystal (PC) through the self-assemble route at two length scales that facilitated electrical and pore connectivity at the mesoporous and microporous levels (Figures 4a–4b) [43]. Their results indicated that this construct induces effective dye sensitization, electrolyte infiltration, and charge collection for DSSC application. Moreover, the device based on the self-assembled double-layer photoanode displays higher photocurrent density and PCE than the reference single-layer cell does (Figure 4d).

In the subsequent research conducted by Guldin et al. [44], they improved on the high temperature stability of mesoporous TiO_2 further by employing an amphiphilic block copolymer polyisoprene-*block*-polyethylene oxide (PI-*b*-PEO) to prevent the collapse of

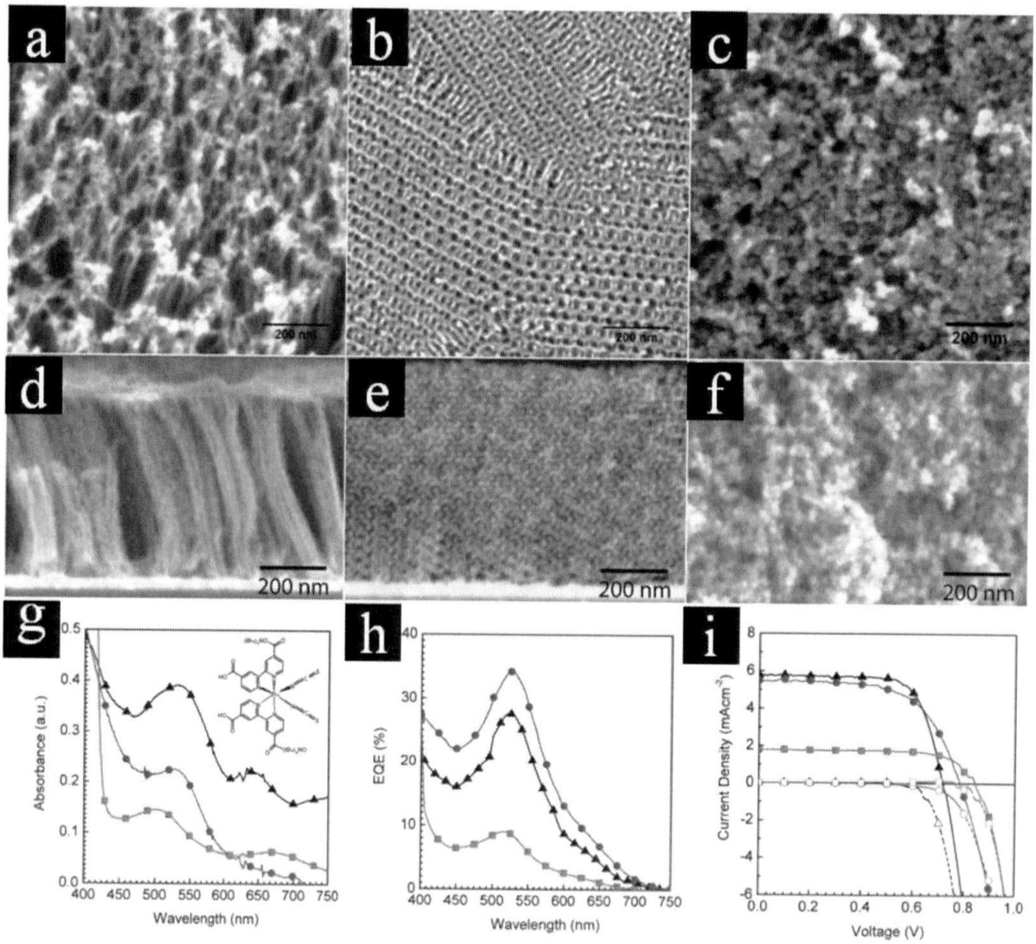

Figure 3. (a-f) Top-view and cross-sectional SEM images of nanostructured TiO_2 array with different morphologies: (a, d) TiO_2 nanowires; (b, e) Gyroid network; (c, f) sintered TiO_2 nanoparticles mesoporous layer. (g-i) Solar cells characteristics of 1.2 μm thick nanowires (solid box), 1.2 μm thick gyroid network (solid triangle) and 1.4 μm thick nanoparticles (solid circle) TiO_2 arrays: (g) Absorption spectra after sensitized with N719 (inset); (h) Spectral response of DSSCs with liquid electrolyte; (i) Current-voltage curves under a simulated AM 1.5 G illumination intensity of 100 mW cm^{-2} (full lines) and in the dark (dashed lines). Reprinted with permission from reference 42. Copyright (2009) American Chemical Society.

porosity at temperatures of up to 700 °C. The systematic study of temperature dependence on DSSC performance benefits from the enhanced high temperature stability. Furthermore, the results demonstrated that mesoporous TiO_2 conductivity improves remarkably through high temperature annealing, thus reducing the specific surface areas of mesoporous scaffolds. Mesoporous TiO_2 conductivity affects the PCE of SDSCs more significantly than that of liquid DSSCs; therefore, this block-copolymer-confined mesoporous TiO_2 is more promising for application in SDSCs. Sun et al. fabricated a highly ordered mesoporous TiO_2 photoanode using an evaporation-induced self-assembly method [45]. They enhanced cell PCE by ~200 % compared with the reference device; this enhancement is attributed to the improved dye loading, light scattering, and electron transport.

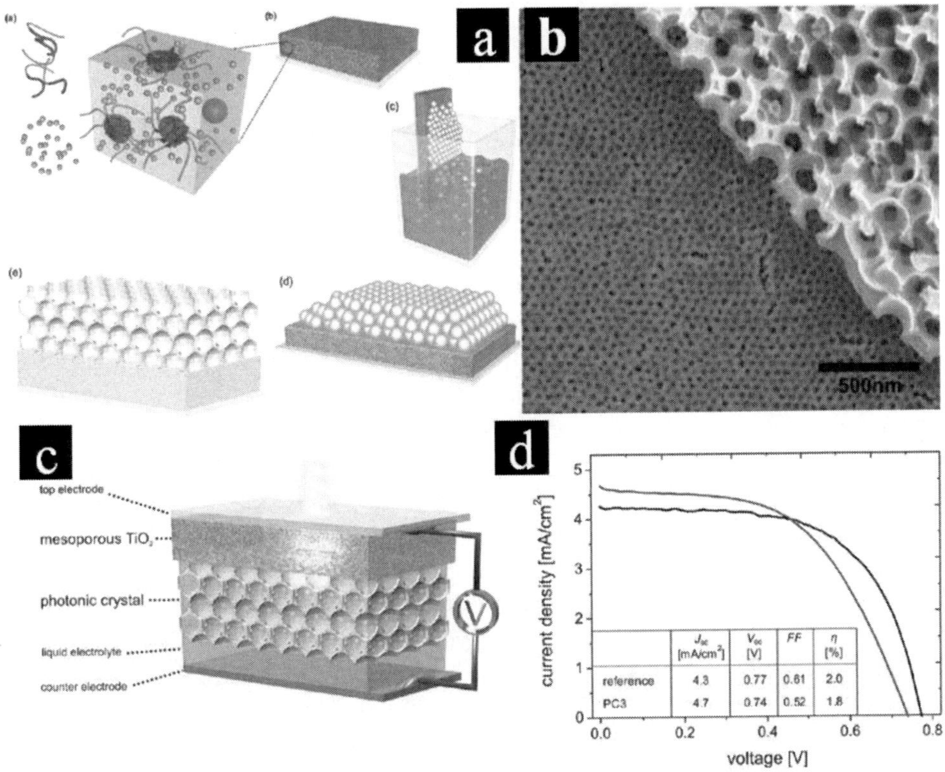

Figure 4. (a) Schematic of coupling TiO_2 mesoporous layer with 3D photonic crystal; (b) SEM micrograph of the inversed TiO_2 opal photo crystal on top of a TiO_2 mesoporous layer. (c) Schematic of a DSSC device based on the self-assembled double layer photoanode. (d) Current-voltage curves of the double layer device (black) and a reference single layer cell (gray). Reprinted with permission from reference 43. Copyright (2010) American Chemical Society.

Well-interconnected mesoporous anatase TiO_2 beads of controllable sizes with large surface areas and strong scattering capability have recently been developed and applied in DSSCs. They achieved a high PCE of over 10% [46]. Lee et al. synthesized highly monodispersed TiO_2 beads with a diameter of ~250 nm via a two-step method involving controlled hydrolysis and a hydrothermal process for TiO_2 crystallization [47]. Following the subsequent solvothermal treatment, the crystallized mesoporous TiO_2 beads exhibited ultralarge surface areas and well-developed nanopore structures. Electrolyte diffusion throughout the film has been enhanced remarkably because of the generation of large external pores from the interstitial voids among the 250 nm mesoporous TiO_2 beads. The diffusion of redox couples into the internal pores has also been improved and shortens the diffusion distance from the outside of the mesoporous beads to their interior [47]. Therefore, both dye loading and electrolyte diffusion have been enhanced effectively by incorporating these mesoporous TiO_2 beads into DSSC photoanodes. Furthermore, the photoanode based on these mesoporous TiO_2 beads exhibited a fairly strong scattering effect to enhance light harvesting capability without sacrificing the surface area for dye absorption. Chen et al. also synthesized monodispersed mesoporous anatase TiO_2 beads with large surface areas, as well as tunable pore sizes and diameters, through a process that is similar to that of Lee et al. as described above. However, Chen et al. introduced hexadecylamine as a structure-directing agent [48, 49], whereas Lee et al. used methylamine. The latter fabricated mesoporous TiO_2 beads with a diameter of ~1150 nm and a surface area of 108.0 $m^2\ g^{-1}$ [48], as displayed in Figures 5a–5c. DSSCs based on these mesoporous beads display lengthened electron diffusion and extended electron lifetime because of the well-interconnected TiO_2 nanocrystallines. Moreover, they achieved a maximum overall PCE of 10% after $TiCl_4$ post-treatment (Figure 5d). This hierarchical mesoporous nanostructure improves the diffusion of electrons among the densely packed TiO_2 nanocrystallines in the beads considerably; however, its electron diffusion is restricted because of the narrow connections among the mesoporous TiO_2 beads. Sauvage et al. systematically investigated the electron diffusion in photoanodes based on these mesoporous TiO_2 beads. Their results indicated that the devices based on such photoanodes report a lower electron diffusion coefficient than DSSCs based on a P25 film do at a relatively high charge density given the narrow connection among the beads. This connection constrains rapid electron transfer (Figure 5e) [46].

To address this issue, a deformed structure was established in the seminal work conducted by Lee et al. to improve contact among the individual mesoporous TiO_2 beads [47]. Liu et al. recently reported that constructing hierarchical TiO_2 spheres through an organic

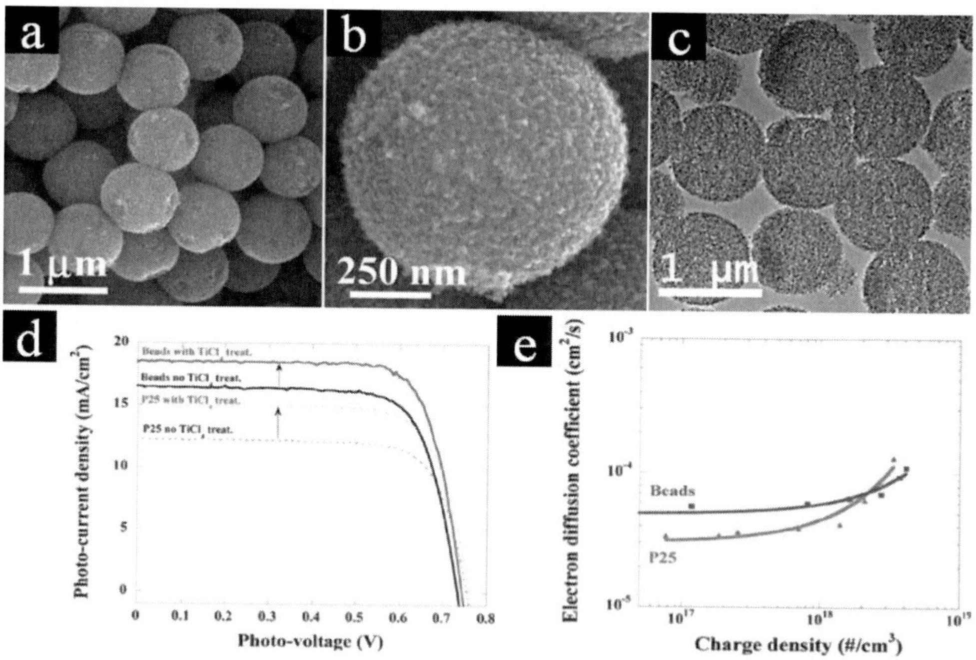

Figure 5. (a, b) SEM and (c) TEM images of mesoporous anatase TiO_2 beads obtained after solvothermal treatment and calcination. Reprinted with permission from reference 49. Copyright (2010) American Chemical Society. (d) *J-V* characteristics recorded at 100 mW/cm^2 of DSSCs based on P25 particles and TiO_2 beads; (e) Electron diffusion coefficient as a function of charge density for P25 and mesoporous TiO_2 beads. Reprinted with permission from reference 46. Copyright (2010) American Chemical Society.

colloid feasibly and effectively enhances the electron transfer between two neighboring TiO_2 spheres [50]. Furthermore, they presented a strategy to incorporate plasmon resonant metallic nanoparticles into the construction of mesoporous TiO_2 structures, thereby enhancing photocurrent and overall PCE. The diameter of mesoporous TiO_2 beads can be tuned accurately by altering the hydrolysis and condensation rates of titanium alkoxide [49]. Several groups have introduced size-controllable mesoporous TiO_2 beads as the scattering layer and have studied the effects of size dependence on DSSC performance [51–53]. Given the large surface areas of these mesoporous beads, the top layer exerted the dual effects of enhanced dye loading and light scattering and noticeably improved overall PCE. ZnO aggregates were also been synthesized by Zhang et al. to strengthen the dye absorption and light scattering effect [54]. They increased overall PCE by incorporating polydisperse ZnO aggregates into photoanodes; this PCE was higher than those of the devices based on other ZnO nanostructures, such as nanoparticles, nanowires, and monodispersed ZnO aggregates. These mesoporous TiO_2 or ZnO aggregates are suitable for application in high-performance DSSCs given their large surface areas, efficient electron transport, and strong light scattering effect. Nonetheless, their long-range electron conduction and structural coherence must be improved further, although mesoporous materials generate the desired highly accessible surfaces. Crossland et al. presented a general strategy to grow mesoporous single-crystal anatase TiO_2 through a seed nucleation process and within a mesoporous template that is immersed in a dilute reaction solution [55]. They employed this single-crystal TiO_2 with a large surface area in SDSCs and achieved a PCE of 7.3%. It also displayed a higher conductivity and electron mobility than conventional nanocrystalline TiO_2.

2.2. Mesoporous Materials for Counter Electrodes

Counter electrodes are important in DSSCs. They reduce I_3^- to I^- to complete the electron transfer cycle [56, 57]. Platinum-deposited and fluoride-doped tin oxide conductive glass is generally employed as the counter electrode in DSSC application because of the high catalytic activity and conductivity of platinum. However, it is a noble metal that is high cost and whose availability is limited. These disadvantages restrict the wide commercial application of platinum-based counter electrodes in DSSCs. In addition, the corrosion of platinum in electrolytes containing the I_3^-/I^- redox couple reduces the long-term stability of DSSCs [58, 59]. Therefore, much effort has been exerted to develop candidates for platinum as a material for counter electrodes in DSSCs.

Mesoporous carbon has recently been introduced as a counter electrode material for DSSCs. It infiltrates electrolytes rapidly and enlarges the effective catalysis area remarkably as a result of its high pore volume, accessible porosity, and large surface area [60, 61]. Mesoporous carbons can be synthesized through a hard-template method for DSSC application. However, this method has inherent drawbacks related to the wide preparation of mesoporous negative electrode materials, as detailed in section 3.2. Hence, several groups have employed the soft-template method instead. Wang et al. prepared mesoporous carbons with a specific surface area of 400 $m^2\ g^{-1}$, a pore diameter of 6.8 nm, and a pore volume of 0.63 $cm^3\ g^{-1}$ by carbonating the triblock copolymer F127/phloroglucinol-formaldehyde composite that was self-assembled in an acid medium [62]. The device based on this mesoporous carbon counter electrode exhibited high electrocatalytic activity to reduce I_3^- and achieved an overall PCE that is comparable to that of a conventional platinum-based electrode. Lee et al. employed a similar strategy to synthesize ordered mesoporous carbons (OMCs) as low temperature-processable counter electrodes for DSSCs [63, 64]. The device

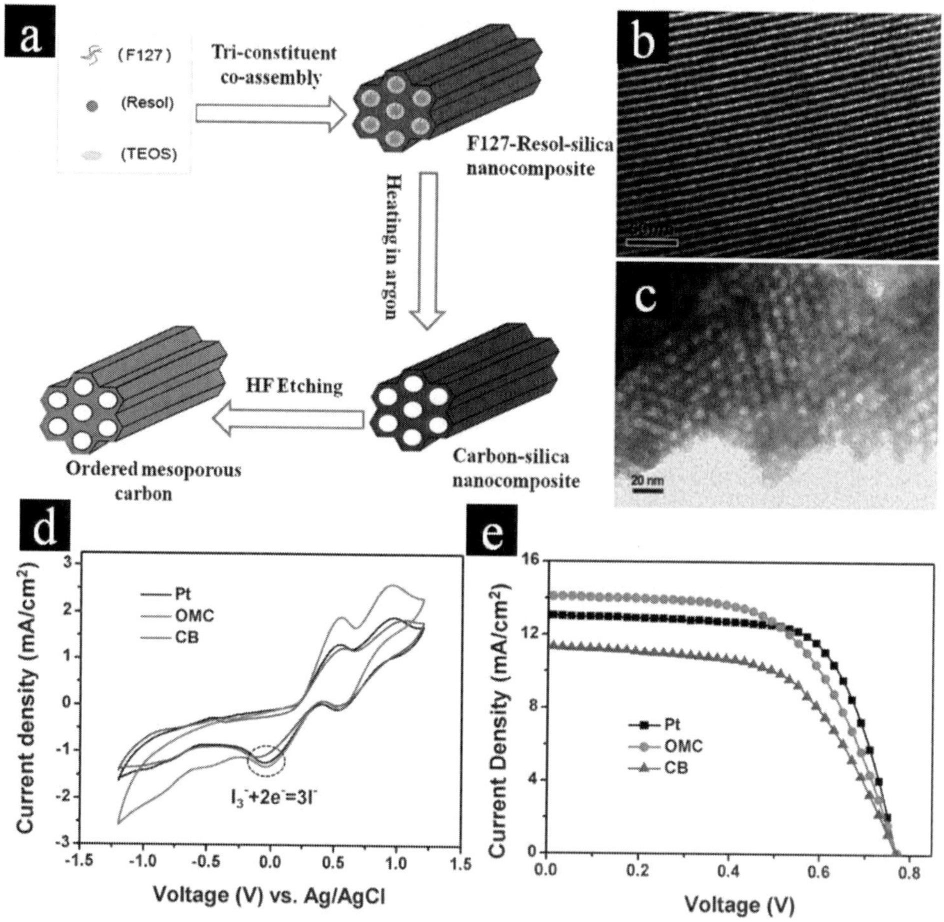

Figure 6. (a) Schematic representation of the preparation of ordered mesoporous carbon. (b, c) TEM image for the ordered mesoporous carbon viewed from (b) perpendicular to the pore channels, and (c) parallel to the pore channels. (d) Cyclic voltammetry of different counter electrodes. (e) Current-voltage curves of DSSCs employed Pt (box), OMC (circle) and CB (triangle) counter electrode. Reprinted with permission from reference 65. Copyright (2013) Royal Society of Chemistry.

attained an overall PCE of 7.46% with a low-temperature heat treatment at 125 °C given the interconnected mesoporous structure and large surface area of OMCs [63]. To optimize this soft-template method further for the synthesis of these carbons, Zhao et al. described a facile process to prepare high OMCs through the direct tri-constituent co-assembly method, which uses resol as the carbon precursor, tri-block copolymer F127 as the soft template, and tetraethoxysilane as the inorganic precursor (Figures 6a–6c) [65]. The obtained products with a highly ordered mesoporous structure possessed large surface areas measuring 1209 $m^2 g^{-1}$, a homogeneous pore size of 4.6 nm, and a high pore volume of 1.65 $cm^3 g^{-1}$. Cyclic voltammetry indicated that the OMC counter electrode exhibits superior catalytic activity (Figure 6d). The DSSC that employs this OMC achieves an overall PCE that is comparable to that of the device based on platinum and carbon black counter electrodes (Figure 6e). Mesoporous carbons have been also considered promising materials for LIBs, as described in the following section.

3. Mesoporous Materials for LIBs

LIBs are among the most successful power sources that have been regarded as prime candidates to power the next generation of plug-in hybrid electric and electric vehicles (P-HEVs and EVs) because of their high energy levels and power density [66, 67]. A typical LIB consists of a negative electrode for lithium–ion insertion, a separator, and a positive electrode for lithium–ion intercalation. The cell is filled with lithium–ion conducting electrolytes. LIBs have been successfully used commercially; however, their performance is difficult to improve further using current electrode and electrolyte materials [68]. Thus, the future challenge of energy storage must be met in addition to that of the design and fabrication of new materials, especially of novel nanomaterials.

Various nanostructured materials have recently been explored for application in LIB electrodes to improve cycling performance and to increase the rate/power of the batteries. These materials include 0D nanoparticles [69, 70], 1D nanowires/nanotubes [71–73], 2D nanosheets/nanoarrays [74, 75], core-shell nanostructures [76–78], and 3D ordered nanomaterials [79, 80]. Nanostructured positive and negative electrode materials can enhance intercalation/deintercalation rates given their reduced dimensions; thus, power output is high [68, 81]. However, these materials have inherent shortcomings, as does the employment of nanomaterials in LIBs. For example, nanostructured positive and negative electrodes are generally difficult to prepare, thereby increasing production cost. Furthermore, their dimensions/morphologies are hard to control [82]. Satisfactory electronic contact is also difficult to ensure among nanostructures as a result of expansion and contraction during intercalation/deintercalation [68]. Thus, nanostructured positive and negative electrodes with well-defined morphologies must be designed and fabricated for high electrochemical performance.

Mesoporous materials have been widely employed for application in LIBs because of their large surface areas, well-interconnected pores, and walls that are tens of nanometers thick. The rate performance of LIBs can be enhanced by applying mesoporous materials for electrodes because of the short distances in lithium ion transport. In addition, the well-interconnected wall structures of mesoporous materials can generate a continuous electron pathway and satisfactory electrical conductivity during charge and discharge because the lithium ions in electrolytes can transfer easily through the mesoporous structures [83]. Therefore, mesoporous materials should be among the most interesting materials in LIB application research.

3.1. Mesoporous Materials for Negative Electrodes

Current commercial LIBs are generally based on graphite negative electrodes and possess a theoretical capacity of 372 mAh g^{-1} to form LiC_6 [84]. However, the low specific capacity of graphite (LiC_6, 372 mAh g^{-1}) and its poor rate performance are results of its low lithium-ion diffusion coefficient. These disadvantages have hindered its further application in P-HEVs and EVs [85]. To increase the rate of lithium insertion/extraction and battery power in the process, much effort has been exerted since 1990 to develop high-capacity negative electrode materials that can replace graphite. The employment of mesoporous structures in negative electrode materials has improved them; it does not only shorten the solid-state diffusion of lithium ions but also generates fast transport channels for electrolytes. Hence, negative electrode materials with mesoporous structures display higher capacities and faster rate performances than their corresponding bulk counterparts [83].

Mesoporous carbons that have been synthesized using hard-template method, such as the inverse replicas of ordered mesoporous silica (OMS) templates, have been investigated in relation to negative electrode materials. Kim et al. prepared rod-shaped OMCs with the replication method using size-tunable SBA-15 silica with the same rod-like morphology as the templates [86]. This process is shown in Figures 7a–7f. By controlling the length of the rod-shaped OMCs precisely, the researchers obtained a reversible discharge capacity of 1012 mAh g^{-1} at 100 mA g^{-1}. They also achieved good cyclability by retaining 86.6% of initial capacity after 100 cycles (Figures 7g–7h). The LIBs based on these OMCs exhibited a significantly higher capacity, better cyclability, and higher rate performance than those based on commercial graphite. This result is attributed to high specific surface area and well-interconnected mesoporous channels of OMCs, which not only provided additional active sites for lithium adsorption and storage but also facilitated rapid lithium ion diffusion and electrolyte transport.

However, the use of the hard-template method to prepare numerous mesoporous negative electrode materials has drawbacks, such as the tedious steps required to synthesize the final materials and the use of toxic chemicals to etch the OMS scaffolds [87, 88]. The preparation process of phenolic resin-based, soft-templated mesoporous carbons has been investigated to address these issues given its facile preparation process, which contains fewer steps, and its employment of inexpensive and environmentally friendly templates [89, 90]. Guo et al. reported on the self-assembly synthesis of soft-template mesoporous carbon with multi-walled carbon nanotubes. These carbons displayed excellent electronic conductivity and rapid charge transfer kinetics as negative electrode materials for LIB application [91]. Mesoporous carbons have also been widely used as promising carbon matrices for combination with other negative electrode materials, such as TiO_2 [92], MoO_2 [93], ZnO [94], Si [95], and Sn [96] to form nanocomposites for LIB application, as depicted in Figures 7c–7d. LIBs based on these nanocomposites generally exhibited an excellent reversible capacity and rate performance because well-dispersed active materials are embedded into the OMC framework. This framework not only facilitates rapid lithium ion charge transport but also retains the special nanostructure with well-interconnected channels during charge and discharge [92–96].

Carbon can be used as negative electrode material because it is active in lithium insertion/extraction. The chemical reaction involved is expressed as follows:

$$6C + xLi^{+} + xe^{-} \leftrightarrow Li_xC_6.$$

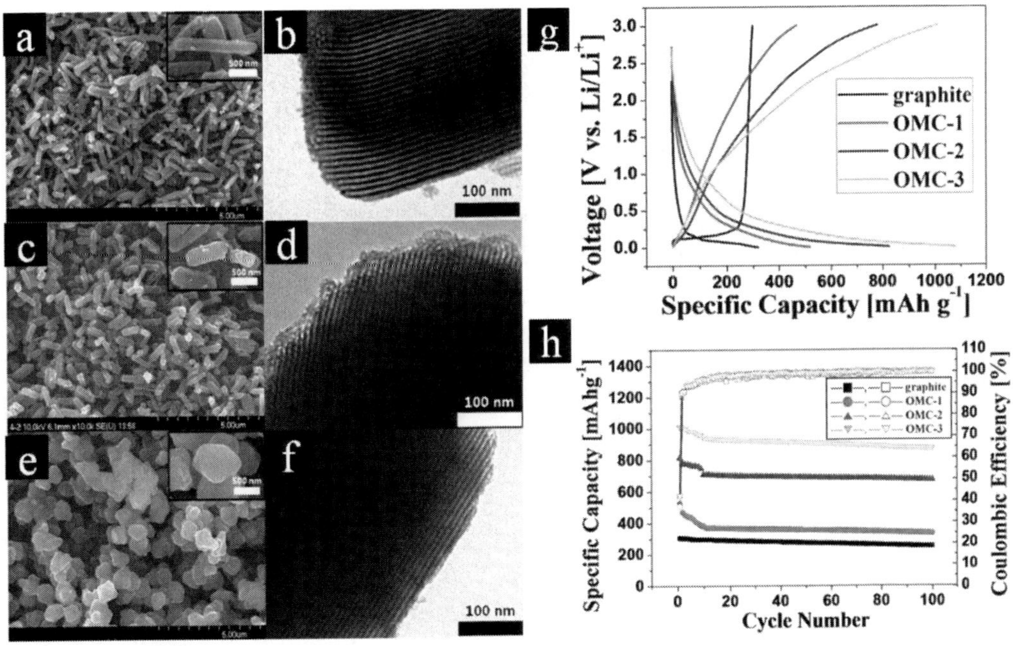

Figure 7. (a-f) SEM and TEM images of ordered mesoporous carbons with different morphology. (g) Charge-discharge galvanostatic curves of commercial graphite and various OMCs for second cycle at 100 mAh g^{-1}. (h) Cycling performance and Coulombic efficiency of commercial graphite and various OMCs at a specific current of 100 mA g^{-1}. Reprinted with permission from reference 86. Copyright (2013) American Chemical Society.

Given the limited theoretical capacity of carbon (372 mAh g^{-1}), other group IV elements that can form alloys with lithium rather than through a lithium insertion/extraction mechanism have been widely investigated as carbon replacements to increase reversible capacity and LIB power [97, 98]. Silicon has potential as a material for LIB negative electrodes because of its low discharge potential (< 0.5 V vs. Li/Li^{+}) and high theoretical lithium capacity of approximately 4200 mAh g^{-1} [99–102]. However, the poor cyclability caused by the low intrinsic electric conduction and the severe volume change (>300%) during lithium alloying and de-alloying hinders the large-scale commercial use of silicon as an ideal candidate material for negative electrodes. The heavy mechanical stress induced by these volume changes cracks and pulverizes the silicon-based negative electrode, thus generating electrically disconnected and tiny particles. Capacity eventually decreases in the process [103, 104]. Much effort has been exerted to limit the volume change of silicon and to improve its structural stability [105–107]. The construction of a silicon-based negative electrode material with a mesoporous structure is a feasible and effective approach. Jia et al. synthesized a mesoporous silicon-based material with a lotus root-like morphology using a magnesiothermic reduction method [108]. The well-dispersed mesopores accommodated volume expansion effectively and prevented the loss of capacity in the silicon-based negative electrode. The researchers also reported a stable capacity of 1500 mAh g^{-1} for 100 cycles at 1 C and a high rate performance that reaches 15 C by optimizing carbon coating.

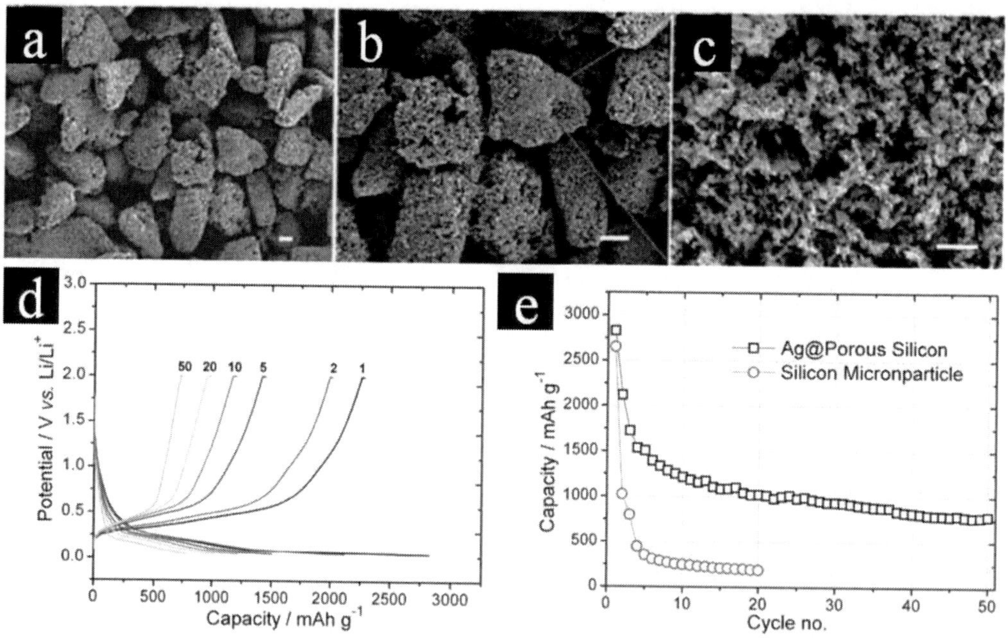

Figure 8. (a-c) SEM images of Ag embedded mesoporous silicon microparticles. (d) Charge-discharge galvanostatic curves for 3D Ag@porous silicon micronparticles. (e) Cycling performance of the batteries based on silicon and 3D Ag@porous silicon micronparticles. Reprinted with permission from reference 109. Copyright (2011) Royal Society of Chemistry.

This result is attributed to the uniformly distributed mesopores and the highly conductive carbon layer. Yuan et al. synthesized silicon microparticles with 3D silver-embedded mesoporous structures for LIB application through a metal-assisted chemical etching process, as displayed in Figures 8a–8c. The electrode based on this 3D mesoporous silicon material exhibited an improved electrochemical performance as a result of its unique interconnected 3D network structure and of the highly conductive silver nanoparticles embedded in the mesoporous silicon matrix [109]. The battery containing 3D silver-embedded porous silicon micronparticles retains a discharge capacity of 784 mAh g^{-1} after 50 cycles, which is more than twice that of graphite (~372 mAh g^{-1}). These capacities as illustrated in Figures 8d–8e. Silicon nanowires with mesoporous nanostructures have also been synthesized for LIB application (Figures 9a, 9b, and 9e) [110, 111]. A 1D nanostructure can accommodate great strain without being pulverized and can facilitate efficient electron transport and satisfactory contact; hence, mesoporous silicon nanowires exhibited a significantly higher reversible capacity than solid nanowires do. This scenario is shown in Figures 9c, 9d, and 9f.

Various metal oxides (MO_x) have been developed extensively as potential negative electrode materials for LIB application given their special chemical properties and high reversible capacities [112–114]. These oxides can be classified into three groups according to the different mechanisms of reaction with lithium, including lithium alloy, insertion, and conversion reaction mechanisms [112 114]. In this chapter, we focus only on the application of MO_x with mesoporous structures as negative electrode materials in LIBs. These oxides not only achieved high reversible capacity because of their large surface areas, but they also exhibited good cycle performance and rate capability because the uniform pores act as a structural buffer against volume changes and maintain favorable structural stability during cycling. SnO_2 has been widely investigated as an alternative to conventional graphite negative material given its high theoretical capacity of 782 mAh g^{-1} [115, 116]. Nevertheless, the severe volume change and the aggregation of Sn nanoparticles during cycles have restricted its practical use [117]. To overcome this drawback, mesoporous SnO_2 nanostructures have been synthesized and employed for LIB application [118, 119]. Mesoporous nanostructures can alleviate significant volume changes during cycling and protect the electrode materials from pulverization; therefore, the devices based on mesoporous SnO_2 displayed an enhanced reversible capacity. However, the mesoporous samples retain large irreversible capacities after several cycles because of the collapse of mesopore structures during the charge–discharge process [120, 121]. Yang et al. prepared a graphene-based mesoporous SnO_2 according to a solvothermal strategy by growing SnO_2 particles *in situ* on the graphene surface using cetyltrimethylammonium bromide as the structure-directing agent. The obtained graphene-based mesoporous SnO_2, which had a large surface area, good electrical conductivity, and a stable pore structure, exhibited the advantages of graphene nanosheets and mesoporous structural materials. Moreover, the devices based on this SnO_2 displayed higher reversible capacities, better cycling, and higher rate performance than LIBs based on pure mesoporous SnO_2 and graphene-based solid SnO_2 [122].

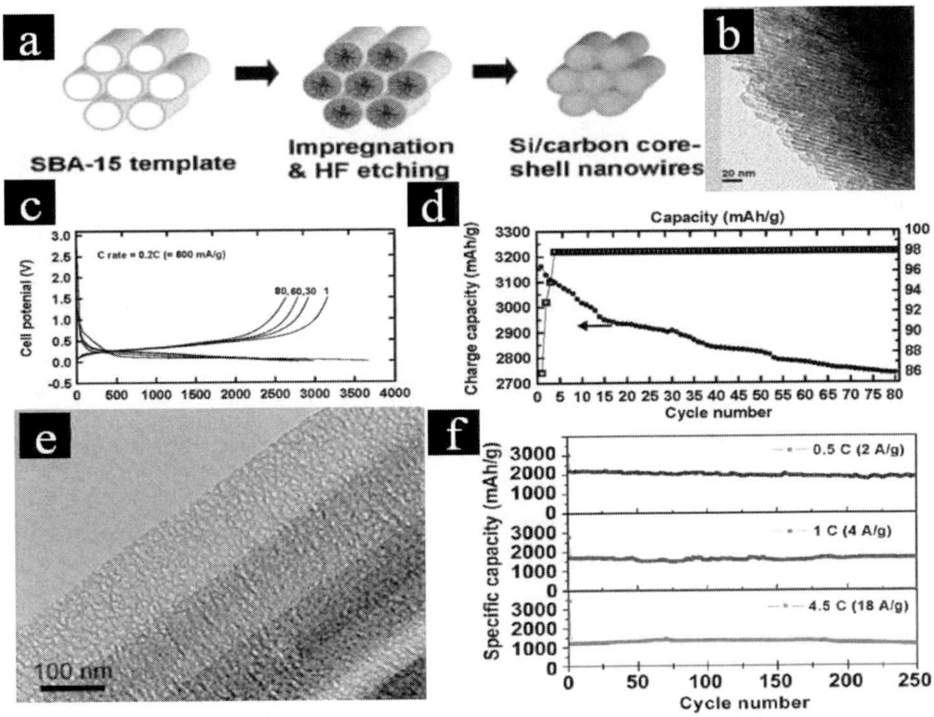

Figure 9. (a) Schematic view of the preparation of mesoporous Si@carbon core-shell nanowires; (b) TEM image of mesoporous Si@carbon core-shell nanowires. (c, d) Charge-discharge galvanostatic curves and cycling performance for mesoporous Si@carbon core-shell nanowires. Reprinted with permission from reference 110. Copyright (2008) American Chemical Society. (e) TEM of silicon nanowires with mesoporous structures. (f) Rate performance of batteries based on silicon nanowires with mesoporous structures. Reprinted with permission from reference 111. Copyright (2012) American Chemical Society.

TiO_2 has been widely investigated as materials for the negative electrodes of LIBs in their various polymorphs, such as anatase, rutile, and TiO_2-B, given their advantages of low cost, high safety, and excellent rate capability. The performance of negative electrodes based on TiO_2 depends strongly on the morphology, porosity, and crystalline phase of the employed TiO_2. Anatase TiO_2 has generally been considered the most active lithium-insertion host because of its low voltage for lithium insertion and its fast lithium insertion/extraction kinetics [123–124]. Zapien et al. synthesized highly porous wirelike TiO_2 with diameters ranging from 0.4 μm to 2 μm and lengths ranging between 2 and 20 μm (Figures 10a–10c) according to a two-step process [125]. The LIBs based on these TiO_2 nanostructures exhibited superior electrochemical performance with an effective combination of high specific capacity and remarkable rate capability as a result of its 1D mesoporous characteristics. Wang et al. have synthesized mesoporous TiO_2 microspheres with uniform nanocrystallines through a template-free hydrolytic precipitation route in normal solvent media [126], as shown in Figures 10d–10e. The battery based on these mesoporous TiO_2 microspheres exhibited an initial discharge capacity of 265 mAh g^{-1} at 10.2 mA g^{-1}, as well as a rate capacity of 151 mAh g^{-1} at 204 mA g^{-1} after 50 cycles. These findings indicated excellent cycling stability and reversibility. The improved electrochemical performance of this mesoporous TiO_2 in comparison with its bulk counterpart can be attributed to the short lithium diffusion and the large contact area between the electrolyte and the TiO_2 active materials. The nanosized mesoporous TiO_2 spheres shorten the diffusion of lithium ions from the surface into the bulk through the mesopores more than the mesoporous TiO_2 does on a scale of micrometers [127]. Li et al. reported on the template-free, self-assembly synthesis of well-dispersed mesoporous anatase TiO_2 nanospheres on graphene sheets. The integration of graphene sheets with uniform mesoporous TiO_2 nanospheres can improve the electron transport of the composites remarkably and prevent the aggregation of nanosized TiO_2 spheres given the large surface area, good flexibility, and high conductivity of graphene sheets [127]. The LIBs based on these unique structural nanocomposites achieved a discharge capacity of 332 mAh g^{-1} at 16.8 mA g^{-1} in the first cycle and improved significantly in terms of lithium-specific capacity from 1 C to 50 C (1 C is equal to 168 mA g^{-1}). They also obtained an excellent rate capacity of 97 mAh g^{-1} at 50 C, which substantially lowers charge-transfer resistance because of the unique structure of mesoporous TiO_2 nanospheres on graphene sheets. A TiO_2-B microsphere with mesoporous structure has recently been synthesized for LIB application by Liu et al. It displays superior rate performance with 165 mAh g^{-1} at 10 C and 116 mAh g^{-1} at 60 C (1 C is equal to 335 mAg^{-1}) given the rapid lithium transport kinetics of the TiO_2-B polymorph and the unique properties of mesoporous structures [128].

Lithium titanate $Li_4Ti_5O_{12}$ has also been widely investigated as an alternative for the negative electrode materials used in LIBs. Electrical insulation is the major drawback of $Li_4Ti_5O_{12}$; nonetheless, recent reports have demonstrated that the nanostructured element can improve electronic conductivity [129, 130]. The reduced dimension can significantly shorten the electron transport distance. Feckl et al. synthesized $Li_4Ti_5O_{12}$ with nanoscale porous frameworks by employing *tert*-butanol in a non-aqueous synthesis reaction [131]. The obtained spinel $Li_4Ti_5O_{12}$ mesoporous nanostructure exhibits an ultrafast lithium insertion rate because of the extremely small crystalline $Li_4Ti_5O_{12}$ nanoparticles and the large surface areas. Carbon-coating is an effective approach to improve the electrical conductivity of $Li_4Ti_5O_{12}$. Kang et al. synthesized a carbon-coated, mesostructured spinel $Li_4Ti_5O_{12}$ through block copolymer self-assembly by employing PI-*b*-PEO [132]. The pore walls of mesoporous

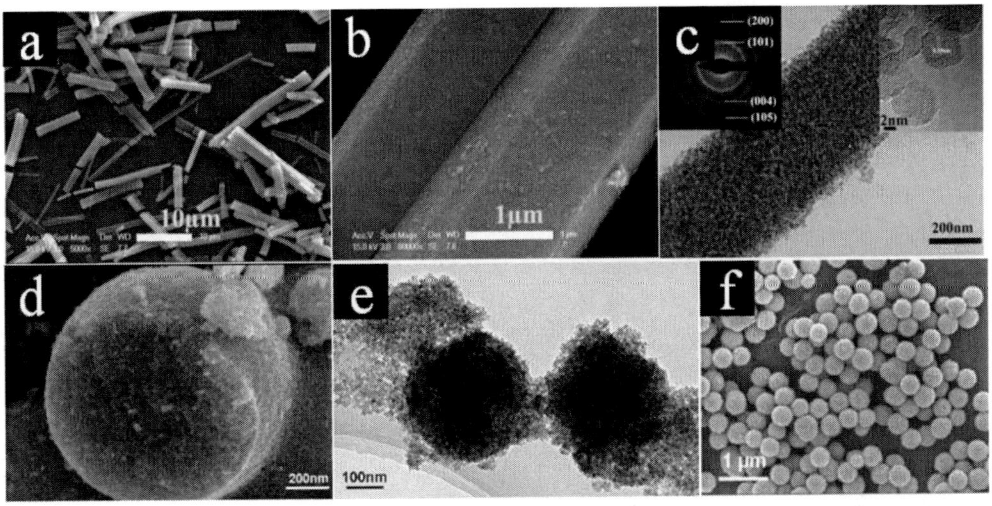

Figure 10. (a, b) SEM images of mesoporous wirelike TiO_2; (c) TEM micrograph, SAED pattern (left inset), and HR-TEM (right inset) of mesoporous wirelike TiO_2. Reprinted with permission from reference 125. Copyright (2012) American Chemical Society. (d) SEM and (e) TEM images of mesoporous anatase TiO_2 microspheres. Reprinted with permission from reference 126. Copyright (2011) American Chemical Society. (f) SEM image of mesoporous Au/$Li_4Ti_5O_{12}$ Spheres. Reprinted with permission from reference 135. Copyright (2012) American Chemical Society.

$Li_4Ti_5O_{12}$ are coated with conductive carbon. The battery based on this nanocomposite displays a high rate capacity of 115 mAh g^{-1} and retains 90% capacity after 500 cycles at 1600 mA g^{-1}. Lou et al. demonstrated a templating method to control the synthesis of hollow $Li_4Ti_5O_{12}$ spheres with mesoporous shells [133]. These structures provided a large surface area that facilitates the effective penetration of electrolytes into the active materials and prevents the aggregation of nanosized primary $Li_4Ti_5O_{12}$ particles. Other approaches have also been developed to prepare mesoporous $Li_4Ti_5O_{12}$ and overcome its electrical insulation to improve electrochemical performance [134]. An example of this product is the Au/$Li_4Ti_5O_{12}$ nanocomposite illustrated in Figure 10f [135].

Transitional MO_x (M = Fe, Co, Ni, Mn, Cr, Mo) have been widely investigated as negative electrode materials for LIB application. They generally exhibit high reversible capacities and energy densities because the oxidation state is maximized and because increased amounts of electrons are involved in the reaction [136–138]. Nonetheless, the initial Coulombic efficiency of the MO_x-based negative electrodes is usually low as a result of the formation of unstable solid electrolyte interface films, significant potential hysteresis, and poor capacity retention. To overcome this major limitation, much effort has been exerted to design nanostructured transitional MO_x for LIBs application [139–143]. In this chapter, we focus on the recent development of transitional MO_x with mesoporous structures. α-Fe_2O_3 has been considered a possible candidate material for negative electrodes because of its high theoretical capacity of 1007 mAh g^{-1}, its low cost, and non-toxicity. Cho et al. prepared spindle-like mesoporous α-Fe_2O_3 by employing an iron-based metal organic framework [MIL-88-Fe with a formula of $Fe_3O(H_2O)_2Cl(BDC)_3{\cdot}nH_2O$)] as a template (Figures 11a–11b) [144]. As shown in Figures 11c–11d, the resultant product exhibited a spindle-like mesoporous structure that is ~800 nm long and ~400 nm wide. The final mesoporous particles are composed of clustered α-Fe_2O_3 nanoparticles that are less than 20 nm in size. The LIBs based on this spindle-like mesoporous α-Fe_2O_3 exhibited a high capacity of 911 mAh g^{-1} after

50 cycles at 0.2 C (200 mA g^{-1}) and achieved an excellent rate capacity of 424 mAh g^{-1} at 10 C (10000 mA g^{-1}), as depicted in Figures 11e–11f. This superior performance is ascribed to the nanosized α-Fe_2O_3 cluster that facilitates fast and efficient lithium transport and to the internal pores that enable the α-Fe_2O_3 to effectively accommodate volume change during

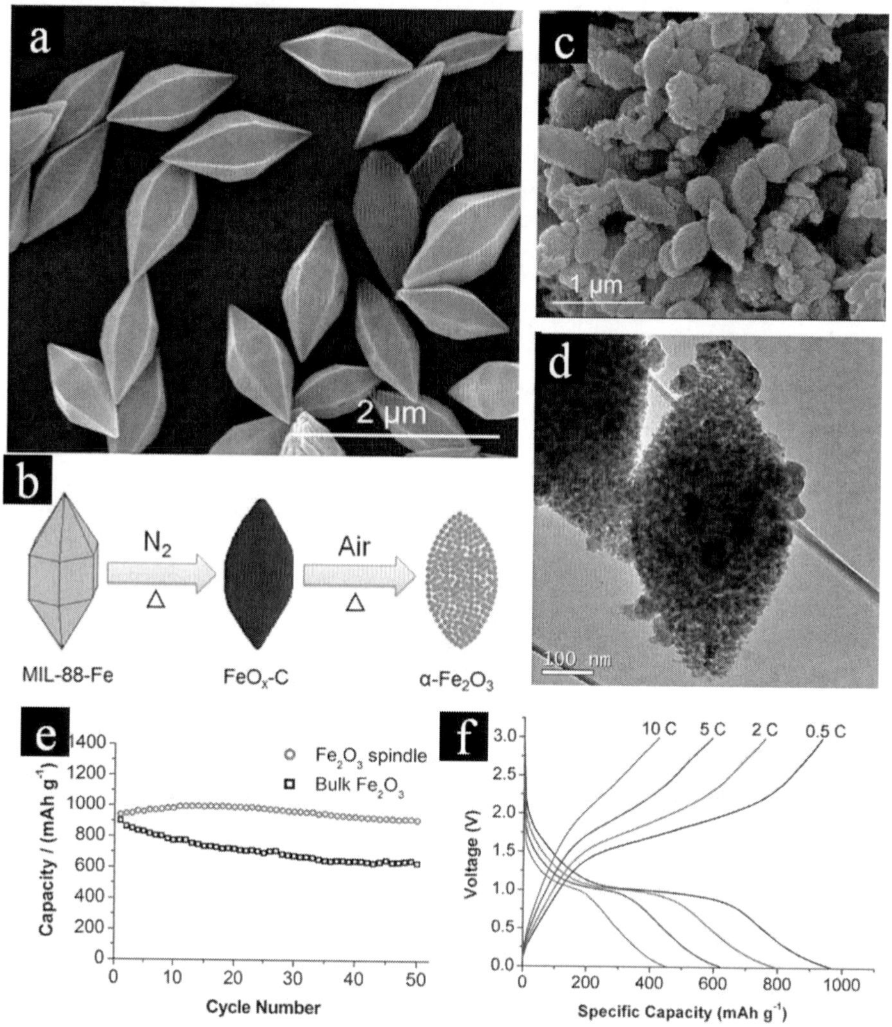

Figure 11. (a, b) SEM image of as-prepared MIL-88-Fe. (b) Schematic view of the fabrication of spindle-like porous α-Fe_2O_3. (c, d) SEM and TEM images of spindle-like mesoporous α-Fe_2O_3. (e) Comparative cycling performance of the final spindle-like α-Fe_2O_3 and bulk Fe_2O_3 at 0.2 C rate (200 mA g^{-1}). (f) Voltage profiles of spindle-like α-Fe_2O_3 at different rates from 0.5 C to 10 C (1 C = 1000 mA g^{-1}). Reprinted with permission from reference 144. Copyright (2012) American Chemical Society.

discharge and charge cycles. Other mesoporous α-Fe_2O_3 nanostructures have been prepared to improve the capacity retention and rate performance of LIBs, such as mesoporous nanorods [145] and 3D mesoporous structures [146].

Wang et al. synthesized a mesoporous Co_3O_4 nanobelt array on a conducting substrate for application as a negative electrode in LIBs (Figures 12a–12b). This substrate generated an efficient electron pathway. The battery based on this array displayed a specific capacity of 770 mAh g^{-1} after 25 cycles at 177 mA g^{-1}, as presented in Figures 12c–12d. Comparable capacities of 510 and 330 mAh g^{-1} can be achieved even during cycling at 1670 and 3350 mA g^{-1}, respectively (Figure 12e) [147]. Qiao et al. prepared highly ordered mesoporous Cr_2O_3 materials with large specific surface areas and a narrow pore size distribution through a vacuum-assisted route using 2D hexagonal SBA-15 and 3D cubic KIT-6 silica as templates [148]. The discharge capacities of the batteries based on 2D hexagonal Cr_2O_3 and 3D cubic Cr_2O_3 were 521 and 540 mAh g^{-1}, respectively, after 100 cycles. By contrast, the LIB based on bulk Cr_2O_3 exhibited a capacity of only 203 mAh g^{-1} under the same testing condition. Wang et al. synthesized polycrystalline mesoporous NiO nanowires according to a hydrothermal method for LIB application [149]. The diameters of the mesoporous NiO

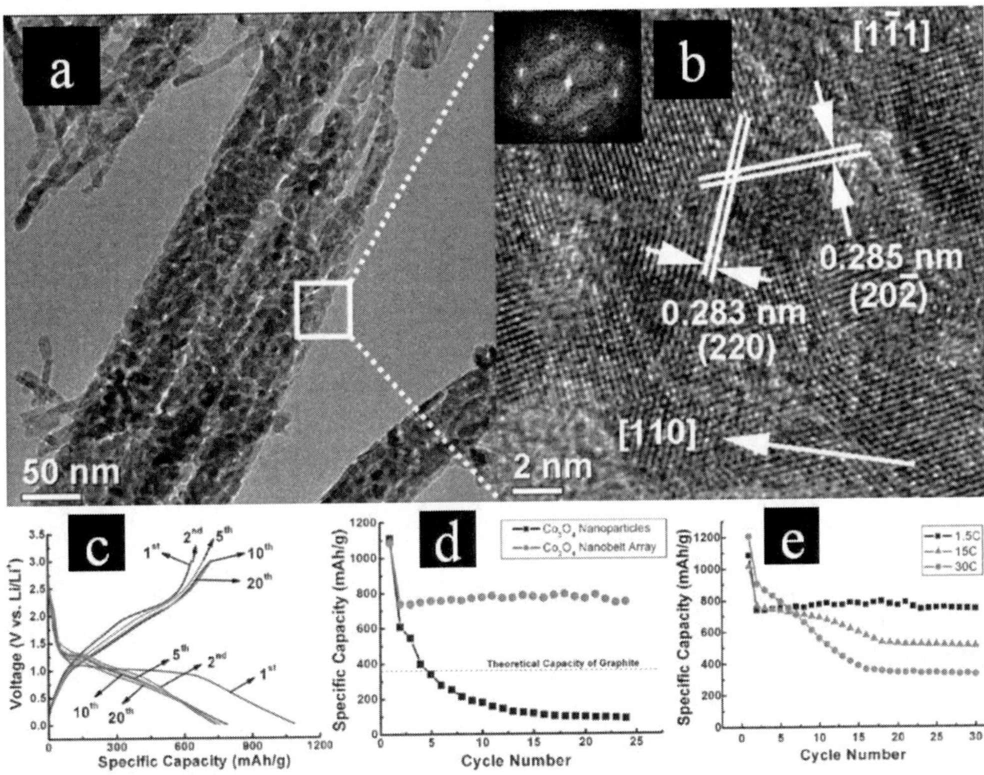

Figure 12. (a, b) TEM and HRTEM images of mesoporous Co_3O_4 nanobelts. (c) Charge-discharge galvanostatic curves of battery based on mesoporous Co_3O_4 nanobelts. (d) Comparative cycling performance of the mesoporous Co_3O_4 nanobelts and Co_3O_4 nanoparticles. (e) Rate performances of batteries based on mesoporous Co_3O_4 nanobelts. Reprinted with permission from reference 147. Copyright (2010) American Chemical Society.

nanowires measured 40–60 nm. Moreover, the nanowires were polycrystalline and consisted of aggregated nanoparticles. After 100 cycles, the battery based on these mesoporous NiO nanowires display a capacity of 1280 mAh g^{-1}. Other transitional MO_x with mesoporous nanostructures have also been synthesized and developed as potential negative electrode materials for LIB application, including MnO/MnO_2 [150, 151], MoO_3 [152], and CuO/Cu_2O [153, 154].

3.2. Mesoporous Materials for Positive Electrodes

$LiCoO_2$ has been investigated as a positive electrode material in rechargeable LIBs for nearly two decades because of its high Li^+ mobility and high electrochemical potential versus Li/Li^+ (~ 4 V), as well as its reversible, practical capacity of above 140 mAh g^{-1} [155]. However, cobalt compounds are toxic and are limited in terms of availability; thus, $LiCoO_2$ is a relatively high-cost material for LIBs. These inherent shortcomings have prevented the widespread use of the $LiCoO_2$ cathode material in rechargeable LIBs. Therefore, cheap, environmental-benign positive electrode materials must be identified as $LiCoO_2$ replacements [156]. Metallophosphates $LiMPO_4$ (M = Fe, Co, Mn, Ni) with an olivine structure have recently attracted attention [157–159]. Among these metallophosphate-based positive electrode materials, $LiFePO_4$ has emerged as a promising cathode material for LIBs given its low cost, long cycle life, environmental-friendliness, and high thermal stability [160]. However, the major disadvantage of the olivine $LiFePO_4$ is its poor electronic conductivity of approximately 3.7×10^{-9} S cm^{-1} in comparison with the conductivities of 10^{-3} and 10^{-5} S cm^{-1} for $LiCoO_2$ and $LiMn_2O_4$, respectively. This limitation results in low rate capacity [161, 162]. Furthermore, it restricts the commercialization of $LiFePO_4$, as well as its application in high-energy and high-power LIBs.

Much effort has been exerted to improve the electrochemical performance of $LiFePO_4$ over the past decade [163, 164]. The synthesis of $LiFePO_4$ with mesoporous structures is a facile and effective approach because the increased specific surface area can accelerate intercalation reaction, and because the interconnected mesoporous structures can strengthen interfacial contact and can enhance electron transfer. Qian et al. synthesized Mesoporous $LiFePO_4$ microspheres with a uniform size distribution of ~3 μm through a template-free hydrothermal process using low-cost Fe^{3+} salt as a starting iron source [165]. This process is shown in Figure 13. The obtained mesoporous $LiFePO_4$ microspheres are not only densely packed to enhance volumetric energy density, but they also facilitate the easy irrigation of electrolytes and generate a large surface area to improve the rate performance of LIBs. The device based on these microspheres exhibited a high rate capacity of 115 mAh g^{-1} at 1600 mA g^{-1} (Figures 13e–13f).

Bruce et al. employed carbon nanoparticles as a sacrificing template to prepare mesoporous $LiFePO_4$ that displayed a bimodal mesopore distribution (2.9 and 30 nm) [166]. The battery based on the prepared mesoporous $LiFePO_4$ also exhibited a high rate performance with specific capacities of 155, 137, and 110 mAh g^{-1} at discharge rates of 0.1, 1, and 5 C (1 C = 170 mAh g^{-1}), respectively. Apart from the microspheres, other $LiFePO_4$ morphologies with mesoporous structures have been developed to enhance performance, such as the mesoporous $LiFePO_4$ with a hierarchical spindle-like architecture established by Tao et al. [167]. To improve the electric conductivity of $LiFePO_4$ further, carbon integrated mesoporous $LiFePO_4$ materials have been synthesized for LIB application. Wang et al. prepared highly ordered mesoporous $LiFePO_4/C$ nanocomposites with a nanocasting technique by using mesoporous carbons (CMK-3) as a template [161]. $LiFePO_4$ nanoparticles were embedded into conductive and interconnected carbon frameworks; this process not only enables intimate contact between electrolytes and $LiFePO_4$ but also induces high electronic conductivity for both mass transfer and charge diffusion. The devices based on the mesoporous $LiFePO_4/C$ nanocomposite achieved a high capacity of 166 mAh g^{-1}, which approaches the theoretical capacity of 170 mAh g^{-1} at a low current rate. The cell also exhibited a satisfactory specific capacity of 118 mAh g^{-1} with an excellent cycle life at a high current rate.

Among these polyanion-type positive electrode materials, lithium orthosilicate Li_2MSiO_4 (M = Mn, Fe, Co and Ni) has also attracted considerable attention as a positive electrode material given its high potential capacity. This advantage is derived from the two lithium ions present per formula unit [168, 169]. Li_2MnSiO_4 has a high theoretical capacity of 333.1 mAh g^{-1}. However, its wide application has been restricted because it extracts/inserts only one lithium ion per formula unit and because of its low electroconductivity [170, 171]. To address this issue, Yoshitake et al. prepared mesoporous Li_2MnSiO_4/C nanocomposites by loading and dispersing Li_2MnSiO_4 nanoparticles into the frameworks of CMK-3 and CMK-8 mesoporous carbon [172]. The battery based on Li_2MnSiO_4-embedded CMK-3 exhibited a capacity of 330 mAh g^{-1}, which is close to the theoretical capacity for reaction with two lithium ions per formula unit.

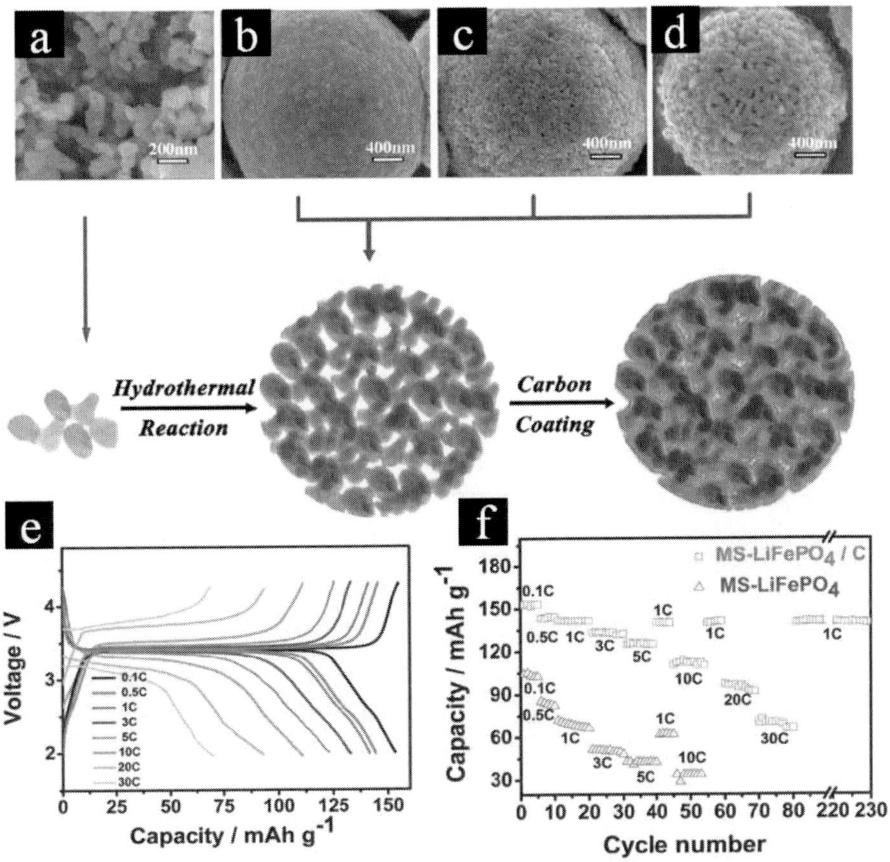

Figure 13. (a-d) Schematic illustration of the formation process of the mesoporous $LiFePO_4$ microspheres coated with carbon. Photographs denote the SEM images of the precursor produced hydrothermally at 180 °C for different times: (a) 1 h, (b) 2 h, (c) 6 h, and (d) 12 h. (e) Charge-discharge profiles of the MS-$LiFePO_4$/C at different current rates. (f) Rate and cycling performances of the MS-$LiFePO_4$/C. Reprinted with permission from reference 165. Copyright (2010) American Chemical Society.

Other inorganic compounds with mesoporous nanostructures have been investigated extensively as positive electrode materials, including V_2O_5 [173–175] and FeF_3 hydrate [176]. V_2O_5 has been considered a potential candidate for LIB application given its high capacity, high output voltage, and low cost [177]. However, the major disadvantages of bulk V_2O_5 materials include poor rate capability and cyclability as a result of its low lithium-ion diffusion coefficient, limited electron conductivity, and the irreversible phase transitions upon heavy discharge. These drawbacks have thus restricted the general application of V_2O_5 to LIBs [178]. Mesoporous architecture effectively solves this problem to enhance battery performance [179–180]. Chen et al. used electrostatic spray deposition technology to prepare V_2O_5 thin films with 3D mesoporous structures [181]. The LIBs based on these structures have significantly improved electrochemical properties. The devices exhibited stable reversible capacities, high Coulombic efficiency, and excellent rate performances. Cao et al. fabricated mesoporous V_2O_5 nanofibers by combining sol-gel processing with electrospinning [182]. The mesoporous nanofibers that consisted of orthorhombic V_2O_5 demonstrated a significantly enhanced capacity of ~370 mAh g^{-1} and an excellent cycling performance because of the improved mass and charge transport, with sufficient freedom to accommodate volume change.

CONCLUSION AND OUTLOOK

This chapter reviews recent developments in mesoporous materials for ECS application. Various materials with mesoporous structures have been investigated as potential candidates for application in DSSCs and LIBs given their special structures, large surface-to-volume ratios, favorable transport properties, short diffusion, and reduced diffusion effect. As the most widely used material, TiO_2 has been systematically developed to enhance photoelectrochemical performance. Representative samples of research conducted on mesoporous materials for photoanodes in DSSCs have been summarized (Section 2.1), such as studies on metal-doped TiO_2 nanocrystallines or those controlled by reactive facets to form mesoporous films with improved chemical and physical properties. Research on the synthesis of unique TiO_2 morphologies with mesoporous nanostructures is described as well, and the developments in mesoporous counter electrode materials are presented (Section 2.2). In particular, mesoporous carbons are introduced. These carbons infiltrate electrolytes quickly and enlarge the effective catalysis area considerably because of their high pore volume, accessible porosity and large surface area. Furthermore, mesoporous carbons have also attracted much research attention in terms of negative electrode application in LIBs (Section 3.1). Apart from carbon-based negative electrode materials, other group IV elements have been widely developed as replacements for carbon to increase reversible capacity and battery power. Furthermore, mesoporous architecture effectively solves the problem of poor cyclability caused by the low intrinsic electric conduction and the severe volume change during lithium alloying and de-alloying. The development of other MO_x with mesoporous structures for negative electrode application has also been discussed, and the relationship between these mesoporous materials and their capability in LIBs has been determined. This chapter also highlights the important advantages of mesoporous and structured positive electrode materials (Section 3.2), such as mesoporous $LiFePO_4$ microspheres or V_2O_5 thin films with mesoporous structures. Therefore, the development of mesoporous materials for ECS is currently a rapidly evolving field of interest. Table 1 summarizes these materials and their potential applications. As illustrated throughout this chapter, the performance of DSSCs and LIBs can be significantly enhanced by introducing electrode materials with mesoporous structures. The technology for material synthesis and chemical compositions is crucial to these materials. Thus, the progression of the design of novel morphologies with mesoporous structures will remain an important feature of future research.

Table 1. Summary of the mesoporous materials and their applications in energy conversion and storage devices

Structures	Preparation method	Materials	Applications and properties	References
Mesoporous particles	Hard-templates, soft-templates, such as silica, triblock copolymer	Carbon	Lower cost and enhanced electrocatalytic activity for DSSCs	[62-65]
	Hard template combining with hydrothermal process	TiO_2	Enhanced conductivity and electron mobility for photovoltaic device	[55]
	Iron-based metal organic framework template	α-Fe_2O_3	Negative materials for LIBs with enhanced capacity and rate performance	[144]
	Templates such as CMK-3, CMK-8	Li_2MnSiO_4	Positive materials for LIBs with enhanced capacity	[172]
	Synthesized by ionic liquid	$FeF_3 \cdot 0.33H_2O$	Positive materials for LIBs with good cycling and rate performance	[176]
Mesoporous spheres	Hydrolysis and hydrothermal process	TiO_2	Enhanced light harvesting and power conversion efficiency for DSSCs	[47-53]
	Hard templates, such as silica	TiO_2	Negative materials for LIBs with enhanced capacity and rate performance	[128]
	In situ conversion	$Li_4Ti_5O_{12}$	Negative materials for LIBs with enhanced capacity and rate performance	[135]
	Hard template combining with hydrothermal process	$Li_4Ti_5O_{12}$	Negative materials for LIBs with enhanced capacity and rate performance	[133]
	Hydrothermal method	Cu_2O/CuO	Negative materials for LIBs with enhanced capacity and cyclability	[154]
	Hydrothermal method	$LiFePO_4$	Positive materials for LIBs with enhanced capacity and rate performance	[165]

Table 1. (Continued)

Structures	Preparation method	Materials	Applications and properties	References
Mesoporous aggregates	Hydrolysis process	ZnO	Enhanced light harvesting and power conversion efficiency for DSSCs	[54]
Mesoporous 1D nanostructures	Microwave-assisted solvothermal process	TiO_2	Negative materials for LIBs with enhanced cyclability and rate performance	[125]
	Hard templates, such as SBA-15	Silicon	Negative materials for LIBs with enhanced capacity and rate performance	[110]
	Metal-assisted chemical etching	Silicon	Negative materials for LIBs with enhanced cyclablility and rate performance	[111]
	Hydrothermal reaction	NiO	Negative materials for LIBs with enhanced cyclablility and rate capacity	[149]
	Sol-gel with electrospining	V_2O_5	Positive materials for LIBs with enhanced capacity and rate performance	[182]
	Template-free solution growth	Co_3O_4	Negative materials for LIBs with enhanced cyclablility and rate performance	[147]
	Template-free hydrothermal method	α-Fe_2O_3/Fe_3O_4	Negative materials for LIBs with enhanced electrochemical reaction area and cyclablility	[145]
	SBA-15	Carbon	Negative materials for LIBs with enhanced cyclablility and rate capacity	[86]
Mesoporous film	Hydrothermal reaction	TiO_2 doping with Nb, N, B, Cr, Mg, etc.	Photoanodes for DSSCs with improved charge transport and power conversion efficiency	[28-35]
	Hydrothermal reaction	TiO_2 with exposed (001) facets	Photoanodes for DSSCs with enhanced power conversion efficiency	[36-40]

Structures	Preparation method	Materials	Applications and properties	References
	Electrodeposition	MoO_3	Negative materials for LIBs with enhanced cyclablility and rate capacity	[152]
	Electrodeposition	Ni/Fe_3O_4	Negative materials for LIBs with enhanced capacity and rate performance	[146]
Mesoporous 3D archtechtures	Soft-templates such as block-copolymer	TiO_2	Photoanodes for DSSCs with enhanced surface area and charge transport	[41-42], [44]
	Polystyrene template	TiO_2	Photoanodes for DSSCs with enhanced light harvesting and power conversion efficiency	[43]
	Metal-assisted chemical etching	Silicon	Negative materials for LIBs with enhanced cyclablility	[109]
	Hard templates combining with magnesiothermic reduction method	Silicon	Negative materials for LIBs with enhanced capacity and rate capacity	[108]
	Hard templates such as SBA-15, KIT-6	Cr_2O_3	Negative materials for LIBs with enhanced cyclablility	[148]
	KIT-6	β-MnO_2	Negative materials for LIBs with enhanced capacity and rate performance	[151]
	Solvothermal reaction	$Li_4Ti_5O_{12}$	Negative materials for LIBs with enhanced rate capacity	[131]
	Hydrothermal reaction	$LiFePO_4$	Positive materials for LIBs with enhanced cyclablility and rate performance	[167]
	Electrostatic spray depostion	V_2O_5	Positive materials for LIBs with enhanced cyclablility and rate performance	[181]
Mesoporous composites	Template-free self-assembly process	TiO_2/Graphene	Negative materials for LIBs with enhanced electron transport and Li storage performance	[127]

Table 1. (Continued)

Structures	Preparation method	Materials	Applications and properties	References
	In situ growth	SnO_2/Graphene	Negative materials for LIBs with enhanced electron transport and Li storage performance	[122]
	Soft templates	Silicon/Carbon	Negative materials for LIBs with enhanced cyclablility	[95]
	Hydrothermal reaction	Sn/Carbon	Negative materials for LIBs with enhanced cyclablility and rate capacity	[96]
	In situ growth and conversion	CuO/CNTs	Negative materials for LIBs with enhanced cyclablility and rate capacity	[153]
	CMK-3	$LiFePO_4$/Carbon	Negative materials for LIBs with enhanced capacity and rate performance	[161]
	Carbon thermal reduction route	MoO_2/Carbon	Negative materials for LIBs with enhanced cyclablility and rate capacity	[93]
	Solvothermal reaction	ZnO/Carbon	Negative materials for LIBs with enhanced cyclablility and rate capacity	[94]

References

[1] Xu, M.; Da, P. M.; Wu, H. Y.; Zhao, D. Y; Zheng, G. F. *Nano Lett.* 2012, 12, 1503.

[2] Xu, C. K.; Wu, J. M.; Desai, U. V.; Gao, D. *Nano Lett.* 2012, 12, 2420.

[3] Liao, J. Y.; Lei, B. X.; Chen, H. Y.; Kuang, D. B.; Su, C. Y. *Energy Environ. Sci.* 2012, 5, 5750.

[4] Lee, J. W.; Lim, S. Y.; Jeong, H. M.; Hwang, T. H.; Kang, J. K.; Choi, J. W. *Energy Environ. Sci.* 2012, 5, 9889.

[5] Yuan, F. W.; Yang, H. J.; Yuan, H. Y. *ACS Nano* 2012, 6, 9932.

[6] Rangappa, D.; Murukanahally, K. D.; Tomai, T.; Unemoto, A.; Honma, I. *Nano Lett.* 2012, 12, 1146.

[7] Sun, W. W.; Cao, F.; Liu, Y. M.; Zhao, X. Z.; Liu, X. G.; Yuan, J. K. *J. Mater. Chem.* 2012, 22, 20952.

[8] Shen, L. F.; Uchaker, E.; Zhang, X. G.; Cao, G. Z. *Adv. Mater.* 2012, 24, 6502.

[9] Cao, F. F.; Deng, J. W.; Xin, S.; Ji, H. X.; Schmidt, O. G.; Wan, L. J.; Guo, Y. G. *Adv. Mater.* 2011, 23, 4415.
[10] Yang, S. B.; Feng, X. L.; Müllen, K. *Adv. Mater.* 2011, 23, 3575.
[11] McCune, M.; Zhang, W.; Deng, Y. L. *Nano Lett.* 2012, 12, 3656.
[12] Liu, L. J.; Karuturi, S. K.; Su, L. T.; Tok, A. I. Y. *Energy Environ. Sci.* 2011, 4, 209.
[13] Wang, Z. Y.; Luan, D. Y.; Madhavi, S.; Hu, Y.; Lou, X. W. *Energy Environ. Sci.* 2012, 5, 5252.
[14] Magasinski, A.; Dixon, P.; Hertzberg, B.; Kvit, A.; Ayala, J.; Yushin, G. *Nat. Mater.* 2010, 9, 353.
[15] Chen, J. S.; Tan, Y. L.; Li, C. M.; Cheah, Y. L.; Luan, D. Y.; Madhavi, S.; Boey, F. Y. C.; Archer, L. A.; Lou, X. W. *J. Am. Chem. Soc.* 2010, 132, 6124.
[16] Regan, B. O'; Grätzal, M. *Nature* 1991, 353, 737.
[17] Yum, J. H.; Baranoff, E.; Wenger, S.; Nazeeruddin, M. K.; Grätzel, M. *Energy Environ. Sci.* 2011, 4, 842.
[18] Vougioukalakis, G. C.; Philippopoulos, A. I.; Stergiopoulos, T.; Falaras, P. *Coord. Chem. Rev.* 2011, 255, 2602.
[19] Tétreault, N.; Grätzel, M. *Energy Environ. Sci.* 2012, 5, 8506.
[20] Yang, L. J.; Leung, W. W. F. *Adv. Mater.* 2011, 23, 4559.
[21] Zhuge, F. W.; Qiu, J. J.; Li, X. M.; Gao, X. D.; Gan, X. Y.; Yu, W. D. *Adv. Mater.* 2011, 23, 1330.
[22] Lee, S.; Park, I. J.; Kim, D. H.; Seong, W. M.; Kim, D. W.; Han, G. S.; Kim, J. Y.; Jung, H. S.; Hong, K. S. *Energy Environ. Sci.* 2012, 5, 7989.
[23] Sheng, J.; Hu, L. H.; Xu, S. Y.; Liu, W. Q.; Mo, L. E.; Tian, H. J.; Dai, S. Y. *J. Mater. Chem.* 2011, 21; 5457.
[24] Joshi, P.; Zhang, L. F.; Davoux, D.; Zhu, Z. T.; Galipeau, D.; Fong, H.; Qiao, Q. Q. *Energy Environ. Sci.* 2010, 3, 1507.
[25] Sauvage, F.; Fonzo, F. Di; Bassi, A. Li; Casari, C. S.; Russo, V.; Divitini, G.; Ducati, C.; Bottani, C. E.; Comte, P.; Grätzel, M. *Nano Lett.* 2010, 10, 2562.
[26] Tétreault, N.; Arsenault, É.; Heiniger, L. P.; Soheilnia, N.; Brillet, J.; Moehl, T.; Zakeeruddin, S.; Ozin, G. A.; Grätzel, M. *Nano Lett.* 2011, 11, 4579.
[27] Guldin, S.; Hüttner, S.; Kolle, M.; Welland, M. E.; Müller-Buschbaum, P.; Friend, R. H.; Steiner, U.; Tétreault, N. *Nano Lett.* 2010, 10, 2303.
[28] Lu, X.; Mou, X.; Wu, J.; Zhang, D.; Zhang, L.; Huang, F.; Xu, F.; Huang, S. *Adv. Funct. Mater.* 2010, 20, 509.
[29] Nikolay, T.; Larina, L.; Shevaleevskiy, O.; Ahn, B. T. *Energy Environ. Sci.* 2011, 4, 1480.
[30] Huang, F. Z.; Li, Q.; Thorogood, G. J.; Cheng, Y. B.; Caruso, R. A. *J. Mater. Chem.* 2012, 22, 17128.
[31] Xie, Y. N.; Huang, N.; Liu, Y. M.; Sun, W. W.; Mehnane, H. F.; You, S. J.; Wang, L. Y.; Liu, W.; Guo, S. S.; Zhao, X. Z. *Electrochim. Acta* 2013, 93, 202.
[32] Tian, H. J.; Hu, L. H.; Li, W. X.; Sheng, J.; Xu, S. Y.; Dai, S. Y. *J. Mater. Chem.* 2011, 21, 7074.
[33] Tian, H. J.; Hu, L. H.; Zhang, C. N.; Mo, L.; Li, W. X.; Sheng, J.; Dai, S. Y. *J. Mater. Chem.* 2012, 22, 9123.
[34] Xie, Y. N.; Huang, N.; You, S. J.; Liu, Y. M.; Sebo, B.; Liang, L. L.; Fang, X. L.; Liu, W.; Guo, S. S.; Zhao, X. Z. *J. Power sources* 2013, 224, 168.

[35] Zhang, C. N.; Chen, S. H.; Mo, L.; Huang, Y.; Tian, H. J.; Hu, L. H.; Huo, Z. P.; Dai, S. Y.; Kong, F. T.; Pan, X. *J. Phys. Chem. C* 2011, 115, 16418.

[36] Hagfeldt, A.; Boschloo, G.; Sun, L. C.; Kloo, L.; Pettersson, H. *Chem. Rev.* 2010, 110, 6595.

[37] Yang, H. G.; Sun, C. H.; Qiao, S. Z.; Zou, J.; Liu, G.; Smith, S. C.; Cheng, H. M.; Lu, G. Q. *Nature* 2008, 453, 438.

[38] Wu, X.; Chen, Z. G.; Lu, G. Q.; Wang, L. Z. *Adv. Funct. Mater.* 2011, 21, 4167.

[39] Yu, J. G.; Fan, J. J.; Lv, K. L. *Nanoscale* 2010, 2, 2144.

[40] Hao, F.; Wang, X.; Zhou, C.; Jiao, X. J.; Li, X.; Li, J. B.; Lin, H. *J. Phys. Chem. C* 2012, 116, 19164.

[41] Crossland, E. J. W.; Kamperman, M.; Nedelcu, M.; Ducati, C.; Wiesner, U.; Smilgies, D. M.; Toombes, G. E. S.; Hillmyer, M. A.; Ludwigs, S.; Steiner U.; Snaith, H. J. A. *Nano Lett.* 2009, 9, 2807.

[42] Crossland, E. J. W.; Nedelcu, M.; Ducati, C.; Ludwigs, S.; Hillmyer, M. A.; Steiner, U.; Snaith, H. J. *Nano Lett.* 2009, 9, 2813.

[43] Guldin, S.; Hüttner, S.; Kolle, M.; Welland, M. E.; Müller-Buschbaum, P.; Friend, R. H.; Steiner, U.; Tétreault, N. *Nano Lett.* 2010, 10, 2303.

[44] Guldin, S.; Hüttner, S.; Tiwana, P.; Orilall, M. C.; Ülgüt, B.; Stefik, M.; Docampo, P.; Kolle, M.; Divitini, G.; Ducati, C.; Redfern, S. A. T.; Snaith, H. J.; Wiesner, U.; Edere, D.; Steiner, U. *Energy Environ. Sci.* 2011, 4, 225.

[45] Sun, Z. Q.; Kim, J. H.; Zhao, Y.; Bijarbooneh, F.; Malgras, V.; Dou, S. X. *J. Mater. Chem.* 2012, 22, 11711.

[46] Sauvage, F.; Chen, H.; Comte, P.; Huang, F. Z.; Heiniger, L. P.; Cheng, Y. B.; Caruso, R. A.; Grätzel, M. *ACS Nano* 2010, 4, 4420.

[47] Kim, Y. J.; Lee, M. H.; Kim, H. J.; Lim, G.; Choi, Y. S.; Park, N. G.; Kim, K.; Lee, W. I. *Adv. Mater.* 2009, 21, 3668.

[48] Chen, D. H.; Huang, F. Z.; Cheng, Y. B.; Caruso, R. A. *Adv. Mater.* 2009, 21, 2206.

[49] Chen, D.; Cao, L.; Huang, F.; Imperia, P.; Cheng, Y. B.; Caruso, R. A. *J. Am. Chem. Soc.* 2010, 132, 4438.

[50] Liu, Y.; Zhai, H.; Guo, F.; Huang, N.; Sun, W. W.; Bu, C.; Peng, T.; Yuan, J.; Zhao, X. Z. *Nanoscale* 2012, 4, 6863.

[51] Yu, I. G.; Kim, Y. J.; Kim, H. J.; Lee, C.; Lee, W. I. *J. Mater. Chem.* 2011, 21, 532.

[52] Roh, D. K.; Seo, J. A.; Chi, W. S.; Koh, J. K.; Kim, J. H. *J. Mater. Chem.* 2012, 22, 11079.

[53] Park, Y.-C.; Chang, Y.-J.; Kum, B.-G.; Kong, E.-H.; Son, J. Y.; Kwon, Y. S.; Parkc, T.; Jang, H. M. *J. Mater. Chem.* 2011, 21, 9582.

[54] Zhang, Q.; Chou, T. P.; Russo, B.; Jenekhe, S. A.; Cao, G. Z. *Angew. Chem. Int. Ed.* 2008, 47, 2402.

[55] Crossland, E. J. W.; Noel, N.; Sivaram, V.; Leijtens, T.; Alexander-Webber, J. A.; Snaith, H. J. *Nature* 2013, 495, 215.

[56] Tai, Q.; Chen, B.; Guo, F.; Xu, S.; Hu, H.; Sebo, B.; Zhao, X. Z. *ACS Nano* 2011, 5, 3795.

[57] Sun, W.; Sun, X.; Peng, T.; Liu, Y.; Zhu, H.; Guo, S.; Zhao, X. Z.; *J. Power Sources* 2012, 201, 402.

[58] Ham, D. J.; Ramasamy, E.; Lee, J.; Lee, J. S. *Chem. Commun.* 2010, 46; 8600.

[59] Syrrokostas, G.; Siokou, A.; Leftheriotis, G.; Yianoulis, P. *Sol. Energy Mater. Sol. Cells* 2012, 103, 119.
[60] Wu, M. X.; Ma, T. L. *ChemSusChem* 2012, 5, 1343.
[61] Brennan, L. J.; Byrne, M. T.; Bari, M.; Gun'ko, Y. K. *Adv. Energy Mater.* 2011, 1, 472.
[62] Wang, G.; Xing, W.; Zhuo, S. *J. Power Sources* 2009, 194, 568.
[63] Ramasamy, E.; Lee, J. *Carbon* 2010, 48 3715.
[64] Ramasamy, E.; Chun, J.; Lee, J. *Carbon* 2010, 48, 4556.
[65] Peng, T.; Sun, W.; Sun, X.; Huang, N.; Liu, Y.; Bu, C.; Guo, S.; Zhao, X.-Z. *Nanoscale* 2013, 5, 337.
[66] Armand, M.; Tarascon, J. M. *Nature* 2008, 451, 652.
[67] Whittingham, M. S. *Chem. Rev.* 2004, 104, 4271.
[68] Aricò, A. S.; Bruce, P. G.; Scrosati, B.; Tarascon, J.-M.; Schalkwijc, W. V. *Nat. Mater.* 2005, 4, 366.
[69] Zheng, X. F.; Shen, G. F.; Li, Y.; Duan, H. N.; Yang, X. Y.; Huang, S. Z.; Wang, H. E.; Wang, C.; Deng, Z.; Su, B. L. *J. Mater. Chem. A* 2013, 1, 1394.
[70] Hu, Y. S.; Kienle, L.; Guo, Y. G.; Maier, J. *Adv. Mater.* 2006, 18, 1421.
[71] Zhao, Y.; Li, J. X.; Wang, N.; Wu, C. X.; Ding, Y. H.; Guan, L. H. *J. Mater. Chem.* 2012, 22, 18797.
[72] Lee, H. W.; Muralidharan, P.; Ruffo, R.; Mari, C. M.; Cui, Y.; Kim, D. K. *Nano Lett.* 2010, 10, 3852.
[73] Tang, W.; Gao, X. W.; Zhu, Y. S.; Yue, Y. B.; Shi, Y.; Wu, Y. P.; Zhu, K. *J. Mater. Chem.* 2012, 22, 20143.
[74] Bak, S. M.; Nam, K. W.; Lee, C. W.; Kim, K. H.; Jung, H. C.; Yang, X. Q.; Kim, K. B. *J. Mater. Chem.* 2011, 21, 17309.
[75] Wang, B.; Li, X. L.; Zhang, X. F.; Luo, B.; Jin, M. H.; Liang, M. H.; Dayeh, S. A.; Picraux, S. T.; Zhi, L. J. *ACS Nano* 2013, 7, 1437.
[76] Shen, L. F.; Li, H. S.; Uchaker, E.; Zhang, X. G.; Cao, G. Z. *Nano Lett.* 2012, 12, 5673.
[77] Chen, Y.; Huang, Q. Z.; Wang, J.; Wang, Q.; Xue, J. M. *J. Mater. Chem.* 2011, 21, 17448.
[78] Zhou, W. W.; Zhu, J. X.; Cheng, C. W.; Liu, J. P.; Yang, H. P.; Cong, C. X.; Guan, C.; Jia, X. T.; Fan, H. J.; Yan, Q. Y.; Li, C. M.; Yu, T. *Energy Environ. Sci.* 2011, 4, 4954.
[79] Ryu, W. H.; Eom, J. Y.; Yin, R. Z.; Han, D. W.; Kim, W. K.; Kwon, H. S. *J. Mater. Chem.* 2011, 21, 15337.
[80] Gerasopoulos, K.; Pomerantseva, E.; McCarthy, M.; Brown, A.; Wang, C. S.; Culver, J.; Ghodssi, R. *ACS Nano* 2012, 6, 6422.
[81] Liu, D. W.; Cao, G. Z. *Energy Environ. Sci.* 2010, 3, 1218.
[82] Bruce, P. G.; Scrosati, B.; Tarascon, J. M. *Angew. Chem. Int. Ed.* 2008, 47, 2930.
[83] Li, Y.; Fu, Z.-Y.; Su, B.-L. *Adv. Funct. Mater.* 2012, 22, 4634.
[84] Dahn, J. R.; Zheng, T.; Liu, Y. H.; Xue, J. S. *Science* 1995, 270, 590.
[85] Persson, K.; Sethuraman, V. A.; Hardwick, L. J.; Hinuma, Y.; Meng, Y. S.; Ven, A. V.; Srinivasan, V.; Kostecki, R.; Ceder, G. *J. Phys. Chem. Lett.* 2010, 1, 1176.
[86] Kim, M.-S.; Bhattacharjya, D.; Fang, B.; Yang, D.-S.; Bae, T.-S.; Yu, J.-S. *Langmuir* 2013, 29, 6754.
[87] Ryoo, R.; Joo, S. H.; Kruk, M.; Jaroniec, M. *Adv. Mater.* 2001, 13, 677.
[88] Wan, Y.; Yang, H. F.; Zhao, D. Y. *Acc. Chem. Res.* 2006, 39, 423.

[89] Wang, X. Q.; Lee, J. S.; Tsouris, C.; DePaoli, D. W.; Dai, S. *J. Mater. Chem.* 2010, 20, 4602.
[90] Liang, C. D.; Dai, S. *J. Am. Chem. Soc.* 2006, 128, 5316.
[91] Guo, B.; Wang, X.; Fulvio, P. F.; Chi, M.; Mahurin, S. M.; Sun, X.-G.; Dai, S. *Adv. Mater.* 2011, 23, 4661.
[92] Ryu, M.-H.; Jung, K.-N.; Shin, K.-H.; Han, K.-S.; Yoon, S. *J. Phys. Chem. C* 2013, 117, 8092.
[93] Zeng, L.; Zheng, C.; Deng, C.; Ding, X.; Wei, M. *ACS Appl. Mater. Interfaces* 2013, 5, 2182.
[94] Shen, X.; Mu, D.; Chen, S.; Wu, B.; Wu, F. *ACS Appl. Mater. Interfaces* 2013, 5, 3118.
[95] Hu, Y.-S.; Adelhelm, P.; Smarsly, B. M.; Maier, J. *ChemSusChem* 2010, 3, 231.
[96] Qiu, Y.; Yan, K.; Yang, S. *Chem. Commun.* 2010, 46, 8359.
[97] Wang, G.; Wang, B.; Wang, X.; Park, J.; Dou, S.; Ahn, H.; Kim, K. *J. Mater. Chem.* 2009, 19, 8378.
[98] Park, M.-H.; Kim, K.; Kim, J.; Cho, J. *Adv. Mater.* 2010, 22, 415.
[99] Chan, C. K.; Peng, H. L.; Liu, G.; Mcilwrath, K.; Zhang, X. F.; Huggins, R. A.; Cui, Y. *Nat. Nanotechnol.* 2008, 3, 31.
[100] Zhang, S. C.; Du, Z. J.; Lin, R. X.; Jiang, T.; Liu, G. R.; Wu, X. M.; Weng, D. S. *Adv. Mater.* 2010, 22, 5378.
[101] Chan, C. K.; Patel, R. N.; O'Connell, M. J.; Korgel, B. A.; Cui, Y. *ACS Nano* 2010, 4, 1443.
[102] Park, M.-H.; Kim, M. G.; Joo, J.; Kim, K.; Kim, J.; Ahn, S.; Cui, Y.; Cho, J. *Nano Lett.* 2009, 9, 3844.
[103] Obrovac, M. N.; Christensen, L. Electrochem. *Solid-State Lett.* 2004, 7, A93.
[104] Weydanz, W. J.; Mehrens, M. W.; Huggins, R. A. *J. Power Sources* 1999, 81, 237.
[105] Gowda, S. R.; Pushparaj, V.; Herle, S.; Girishkumar, G.; Gordon, J. G.; Gullapalli, H.; Zhan, X.; Ajayan, P. M.; Reddy, A. L. M. *Nano Lett.* 2012, 12, 6060.
[106] Chen, W.; Fan, Z.; Dhanabalan, A.; Chen, C.; Wang, C. *J. Electrochem. Soc.* 2011, 158, A1055.
[107] Yu, Y.; Gu, L.; Zhu, C. B.; Tsukimoto, S.; Aken, P. A.; Maier, J. *Adv. Mater.* 2010, 22, 2247.
[108] Jia, H.; Gao, P.; Yang, J.; Wang, J.; Nuli, Y.; Yang, Z. *Adv. Energy Mater.* 2011, 1, 1036.
[109] Liu, Y.; Chen, B.; Cao, F.; Chan, H. L. W.; Zhao, X.; Yuan, J. *J. Mater. Chem.* 2011, 21, 17083.
[110] Kim, H.; Cho, J. *Nano Lett.*; 2008, 8, 3688.
[111] Ge, M.; Rong, J.; Fang, X.; Zhou, C. *Nano Lett.* 2012, 12, 2318.
[112] Park, C. M.; Kim, J. H.; Kim, H.; Sohn, H. J. *Chem. Soc. Rev.* 2010, 39, 3115.
[113] Cabana, J.; Monconduit, L.; Larcher, D. Palacin, M. R. *Adv. Mater.* 2010, 22, E170.
[114] Rumplecker, A.; Kleitz, F.; Salabas, E. L.; Schuth, F. *Chem. Mater.* 2007, 19, 485.
[115] Idota, Y.; Kubota, T.; Matsufuji, A.; Maekawa, Y.; Miyasaka, T. *Science* 1997, 276, 1395.
[116] Poizot, P.; Laruelle, S.; Grugeon, S.; Dupont, L.; Tarascon, J. *Nature* 2000, 407, 496.
[117] Kim, C.; Noh, M.; Choi, M.; Cho, J.; Park, B. *Chem. Mater.* 2005, 17, 3297.
[118] Guo, Z. P.; Du, G. D.; Nuli, Y.; Hassan, M. F.; Liu, H. K. *J. Mater. Chem.* 2009, 19, 3253.

[119] Kim, H.; Cho, J. *J. Mater. Chem.* 2008, 18, 771.
[120] Yue, W. B.; Lin, Z. Z.; Jiang, S. H.; Yang, X. J. *J. Mater. Chem.* 2012, 22, 16318.
[121] Yang, X. L.; Fan, K. C.; Zhu, Y. H.; Shen, J. H.; Jiang, X.; Zhao, P.; Li, C. Z. *J. Mater. Chem.* 2012, 22, 17278.
[122] Yang, S.; Yue, W.; Zhu, J.; Ren, Y.; Yang, X. *Adv. Funct. Mater.* 2013, 23; 3570.
[123] Guo, Y. G.; Hu, Y. S.; Maier, J. *Chem. Commun.* 2006, 2783.
[124] Guo, Y. G.; Hu, Y. S.; Sigle, W.; Maier, J. *Adv. Mater.* 2007, 19, 2087.
[125] Wang, H. E.; Lu, Z. G.; Xi, L. J.; Ma, R. G.; Wang, C. D.; Zapien, J. A.; Bello, I.; *ACS Appl. Mater. Interfaces* 2012, 4, 1608.
[126] Wang, J.; Zhou, Y.; Hu, Y.; O'Hayre, R.; Shao, Z. *J. Phys. Chem. C* 2011, 115, 2529.
[127] Li, N.; Liu, G.; Zhen, C.; Li, F.; Zhang, L.; Cheng, H.-M. *Adv. Funct. Mater.* 2011, 21, 1717.
[128] Liu, H.; Bi, Z.; Sun, X.-G.; Unocic, R. R.; Paranthaman, M. P.; Dai, S.; Brown, G. M. *Adv. Mater.* 2011, 23, 3450.
[129] Zhu, G.-N.; Liu, H.-J.; Zhuang, J.-H.; Wang, C.-X.; Wang, Y.-G.; Xia, Y.-Y. *Energy Environ. Sci.* 2011, 4, 4016.
[130] Yu, S.-H.; Pucci, A.; Herntrich, T.; Willinger, M.-G.; Baek, S.-H.; Sung, Y.-E.; Pinna, N. *J. Mater. Chem.* 2011, 21, 806.
[131] Feckl, J. M.; Fominykh, K.; Döblinger, M.; Fattakhova-Rohlfing, D.; Bein, T. *Angew. Chem. Int. Ed.* 2012, 51, 7459.
[132] Kang, E.; Jung, Y. S.; Kim, G.-H.; Chun, J.; Wiesner, U.; Dillon, A. C.; Kim, J. K.; Lee, J. *Adv. Funct. Mater.* 2011, 21, 4349.
[133] Yu, L.; Wu, H. B.; Lou, X. W. *Adv. Mater.* 2013, 25, 2296.
[134] Chen, J.; Yang, L.; Fang, S.; Hirano, S.; Tachibana, K. *J. Power Sources* 2012, 200, 59.
[135] Li, C. C.; Li, Q. H.; Chen, L. B.; Wang, T. H. *ACS Appl. Mater. Interfaces* 2012, 4, 1233.
[136] Villevieille, C.; Fraisse, B.; Womes, M.; Jumas, J. C.; Monconduit, L. *J. Power Sources* 2009, 189, 324.
[137] Liu, J. P.; Li, Y. Y.; Huang, X. T.; Ding, R. M.; Hu, Y. Y.; Jiang, J.; Liao, L. *J. Mater. Chem.* 2009, 19, 1859.
[138] Adelhelm, P.; Hu, Y. S.; Antonietti, M.; Maier, J.; Smarsly, B. M. *J. Mater. Chem.* 2009, 19, 1616.
[139] Goodenough, J. B.; Kim, Y. *Chem. Mater.* 2010, 22, 587.
[140] Fergus, J. W. *J. Power Sources* 2010, 195, 939.
[141] Wang, Y. G.; Li, H. Q.; He, P.; Hosono, E.; Zhou, H. S. *Nanoscale* 2010, 2, 1294.
[142] Liu, C.; Li, F.; Ma, L. P.; Cheng, H. M. *Adv. Mater.* 2010, 22, E28.
[143] Zhang, W. J. *J. Power Sources* 2011, 196, 13.
[144] Xu, X.; Cao, R.; Jeong, S.; Cho, J. *Nano Lett.* 2012, 12, 4988.
[145] Xiao, Z.; Xia, Y.; Ren, Z.; Liu, Z.; Xu, G.; Chao, C.; Li, X.; Shen, G.; Han, G. *J. Mater. Chem.* 2012, 22, 20566.
[146] Xiong, Q.; Tu, J.; Lu, Y.; Chen, J.; Yu, Y.; Wang, X.; Gu, C. *J. Mater. Chem.* 2012, 22, 18639.
[147] Wang, Y.; Xia, H.; Lu, L.; Lin, J. *ACS Nano* 2010, 4, 1425.
[148] Liu, H.; Du, X.; Xing, X.; Wang, G.; Qiao, S. Z. *Chem. Commun.* 2012, 48, 865.
[149] Su, D.; Kim, H.-S.; Kim, W.-S.; Wang, G. *Chem. Eur. J.* 2012, 18, 8224.
[150] Sun, Y.; Hu, X.; Luo, W.; Huang, Y. *J. Mater. Chem.* 2012, 22, 19190.

[151] Jiao, F.; Bruce, P. G. *Adv. Mater.* 2007 19, 657.
[152] Zhao, G.; Zhang, N.; Sun, K. *J. Mater. Chem. A* 2013, 1, 221.
[153] Ko, S.; Lee, J.-I.; Yang, H. S.; Park, S.; Jeong, U. *Adv. Mater.* 2012, 24, 4451.
[154] Zhang, Z.; Chen, H.; She, X.; Sun, J.; Teo, J.; Su, F. *J. Power Sources* 2012, 217, 336.
[155] Winter, M.; Besenhard, J. O.; Spahr, M. E.; Novák, P. *Adv. Mater.* 1998, 10, 725.
[156] Ammundsen, B.; Paulsen, J. *Adv. Mater.* 2001, 13, 943.
[157] Recham, N.; Chotard, J.-N.; Dupont, L.; Delacourt, C.; Walker, W.; Armand, M.; Tarascon, J.-M. *Nat. Mater.* 2010, 9, 68.
[158] Kang, B.; Ceder, G. *Nature*; 2009, 458, 190.
[159] Ellis, B. L.; Makahnouk, W. R. M.; Makimura, Y.; Toghill, K.; Nazar, L. F. *Nat. Mater.* 2007, 6, 749.
[160] Andersson, A. S.; Thomas, J. O.; Kalska, B.; Haggstrom, L. *Electrochem. Solid State lett.* 2000, 3, 66.
[161] Wang, G.; Liu, H.; Liu, J.; Qiao, S.; Lu, G. Q.; Munroe, P.; Ahn, H. *Adv. Mater.* 2010, 22, 4944.
[162] Zhou, N.; Wang, H.-Y.; Uchaker, E.; Zhang, M.; Liu, S.-Q.; Liu, Y.-N.; Cao, G. *J. Power Sources* 2013, 239, 103.
[163] Delacourt, C.; Poizot, P.; Levassear, S.; Masquelier, C. *Electrochem. Solid State lett.* 2009, 9, A352.
[164] Wang, Y. Q.; Wang, J. L.; Yang, J.; Nuli, Y. N. *Adv. Func. Mater.* 2006, 16, 2135.
[165] Qian, J.; Zhou, M.; Cao, Y.; Ai, X.; Yang, H. *J. Phys. Chem. C* 2010, 114, 3477.
[166] Ren, Y.; Bruce, P. G. *Electrochem. Commun.* 2012, 17, 60.
[167] Xia, Y.; Zhang, W.; Huang, H.; Gan, Y.; Tian, J.; Tao, X. *J. Power Sources* 2011, 196, 5651.
[168] Nyten, A.; Kamali, S.; Haeggstroem, L.; Gustafsson, T.; Thomas, J. O. *J. Mater. Chem.* 2006, 16, 2266.
[169] Lyness, C.; Delobel, B.; Armstrong, A. R.; Bruce, P. G. *Chem. Commun.* 2007, 43, 4890.
[170] Wang, F.; Chen, J.; Wang, C.; Yi, B. *J. Electroanal. Chem.* 2013, 688, 123.
[171] Muraliganth, T.; Stroukoff, K. R.; Manthiram, A. *Chem. Mater.* 2010, 22, 5754.
[172] Kawase, Takashi; Yoshitake, H. *Microporous Mesoporous Mater.* 2012, 155, 99.
[173] Cao, Z.; Wei, B. *Nano Energy* 2013, 2, 481.
[174] Lee, J.-K.; Kim, G.-P.; Song, I. K.; Baeck, S.-H. *Electrochem. Commun.* 2009, 11, 1571.
[175] Yu, L.; Zhao, C.; Long, X.; Chen, W. *Microporous Mesoporous Mater.* 2009, 126, 58.
[176] Li, C.; Gu, L.; Tong, J.; Tsukimoto, S.; Maier, J. *Adv. Funct. Mater.* 2011, 21, 1391.
[177] Zhou, S.; Yang, X.; Lin, Y.; Xie, J.; Wang, D. *ACS Nano* 2011, 6, 919.
[178] Tarascon, J. M.; Armand, M. *Nature* 2001, 414, 359.
[179] Lee, J.-K.; Kim, G.-P.; Song, I. K.; Baeck, S.-H. *Electrochem. Commun.* 2009, 11, 1571.
[180] Yu, L.; Zhao, C.; Long, X.; Chen, W. *Microporous Mesoporous Mater.* 2009, 126, 58.
[181] Wang, S.; Li, S.; Sun, Y.; Feng, X.; Chen, C. *Energy Environ. Sci.* 2011, 4, 2854.
[182] Yu, D.; Chen, C.; Xie, S.; Liu, Y.; Park, K.; Zhou, X.; Zhang, Q.; Li, J.; Cao, G. Z. *Energy Environ. Sci.* 2011, 4, 858.

In: Comprehensive Guide for Mesoporous Materials, Volume 4 ISBN: 978-1-63482-092-9
Editor: Mahmood Aliofkhazraei

Chapter 4

RECENT ADVANCES IN THE CONVERSION OF CARBON DIOXIDE TO FUELS USING HIGH-SURFACE-AREA PHOTOCATALYSTS

Cao-Thang Dinh, Minh-Hao Pham and Trong-On Do
Department of Chemical Engineering, Laval University, Quebec, Canada

ABSTRACT

Due to the depletion of fossil fuels and the serious environmental problems accompanying their combustion, a new form of energy that is clean, renewable, low-cost and a viable alternative to fossil fuels is urgently needed. The direct conversion of sunlight into fuels (i.e., solar fuels) with the aid of an artificial photocatalyst is an attractive prospect, considering that in a single hour the sun delivers energy enough for human activities in an entire year. Solar fuels can take the form of hydrogen and hydrocarbons such as methane and methanol. These fuels may be formed by photocatalytic water splitting and by photoreduction of CO2 with water, respectively. In this chapter, we discuss recent developments in high-surface-area hybrid photocatalysts for the production of hydrocarbon from CO2 reduction with water using sunlight energy. In order to improve the efficiency of sun-light driven photocatalysts, the factors that govern the photoactivity such as the structure, morphology at nanometric scale, charge transfer mechanism between semiconductor surfaces and cocatalysts will be also briefly discussed.

1. INTRODUCTION

Today the global energy supply is dominated by fossil energy that accounts for ca.80% of the used energy. However, with increasing energy use all over the world the fossil resources are dwindling. The rate of global energy use is projected to increase, from its current level of about 15 TW to 27 TW by 2050 and may reach 43 TW by 2100. Alternative energy supplies are urgently needed to limit our dependence on fossil fuel [1-3]. On the other aspect, the increase in CO_2 emission has resulted in an increase in atmospheric CO_2 during the last 200 years from approximately 270 ppm to 385 ppm. It is apparent that atmospheric CO_2 concentration will continue to increase monotonically in the future due to fossil fuel consumption. This increase is thought to cause atmospheric warming, due to the prevention of infrared re-emission. To avoid increasing atmospheric CO_2 concentrations, efforts must be increased to capture and sequester CO_2 and to recycle carbon dioxide as a fuel feedstock [4, 5].

The chemical reduction of CO_2 to fuels offers a way of recycling CO_2 and could therefore ameliorate associated climate change and increase the sustainability of oil reserves for other purposes [6-8]. There are several ways to achieve this goal including electrocatalytic, photocatalytic and thermocatalytic strategies. However, the reduction of CO_2 is an endergonic process requiring a hydrogen source and/or energy. The most abundant sources of energy and hydrogen are sunlight and water, respectively. Solar-driven photocatalytic reduction of CO_2 to produce fuels using water as the hydrogen source thus has the potential to be a means to store intermittent solar energy and to recycle CO_2 while decreasing the use of fossil fuels [6, 9]. Currently, the major obstacle preventing efficient conversion of CO_2 into energy-bearing products is the lack of photocatalysts that can readily couple an abundant energy source, solar, with inexpensive reducing agents, water, to achieve rapid and selective cleavage of C–O bonds in CO_2 and formation of C–H bonds in the product [6]. In general, photocatalysis involves three processes: the excitation, bulk diffusion and surface transfer of photoinduced charge carriers. Thus, an efficient photocatalyst must satisfy several critical requirements related to its semiconducting and chemical properties, its crystalline structure and surface characteristics. Significant challenges remain in developing a sunlight driven photocatalyst with high efficiency and photo-chemical stability [5, 10, 11].

Several strategies have been developed for the design of efficient photocatalytsts in general and photocatalyts for CO_2 reduction in particular. They include doping large bandgap semiconductors with metal or non-metal elements, coupling semiconductors and using cocatalysts. In addition, because the photocatalytic reactions take place on the surface of the semiconductors, the surface area of the photocatalysts, therefore, also plays a crucial role in their photocatalytic performance. In this chapter, we will first review some strategies towards efficient photocatalysts. Next, we will focus on discussing recent developments in high-surface-area photocatalysts with different structures for the production of hydrocarbon from CO_2 reduction with water using sunlight energy.

2. CO_2 PHOTOREDUCTION

Reduction of CO_2: The energy grade of CO_2 is low from a thermodynamic perspective, which is why any transformation of CO_2 to hydrocarbon requires energy infusion. For example, based on thermodynamics, converting one mole of CO_2 into methanol requires 228 kJ of energy at 298 K. The energy source used to convert CO_2 should be provided without producing more CO_2 such as solar energy. CO_2 photoreduction refers to the conversion of CO_2 to reduced C1 and C2 compounds using light-induced reactions. Since CO_2 does not absorb either visible or ultra-violet (UV) radiation in the wavelengths 200–900 nm, this process requires suitable photosensitizers. Several semiconductors have been utilized to absorb visible/UV radiation and transfer this energy to CO_2 [12].

Photocatalytic CO_2 reduction principles: The operating principle of semiconductor photocatalysts is illustrated in Figure 1. Absorption of light by a semiconductor electrode or particle causes the transition of an electron from the valence band (VB) to the conduction band (CB). Both an excited electron (e^-) and a hole (h^+) are generated concurrently in the CB and VB, respectively. The photo-generated excited e^- can potentially be used for CO_2 reduction. Because protons can also accept the excited elcctron, hydrogen evolution often competes with CO_2 reduction.

On the other hand, the photo-generated h^+ is quenched by electron injection from a reductant such as organic molecules or water. For the use of water as a reductant, the potential of the VB must be more positive than the oxidation potential of water [5, 13].

Products of CO_2 photoreduction: The CO_2 photoreduction to hydrocarbon fuel is very challenging process because it involves a multi-electron process and is made up of steps involving electron and proton transfer, C–O bond breaking and C–H bond formation. It is suggested that conversion of CO_2 to methane was a multistep reduction process as follows [14]:

$$H_2O + 2h^+ \rightarrow 1/2O_2 + 2H^+$$

$$CO_2 + 2H^+ + 2e^- \rightarrow HCOOH$$

$$HCOOH + 2H^+ + 2e^- \rightarrow HCHO + H_2O$$

$$HCHO + 2H^+ + 2e^- \rightarrow CH_3OH$$

$$CH_3OH + 2H^+ + 2e^- \rightarrow CH_4 + H_2O$$

It can be inferred from the above equations that photocatalytic CO_2 reduction can generate a wide range of products including the most commonly desirable products methanol and methane, but also oxalic and formic acids, carbon monoxide, formaldehyde, and even elemental carbon.

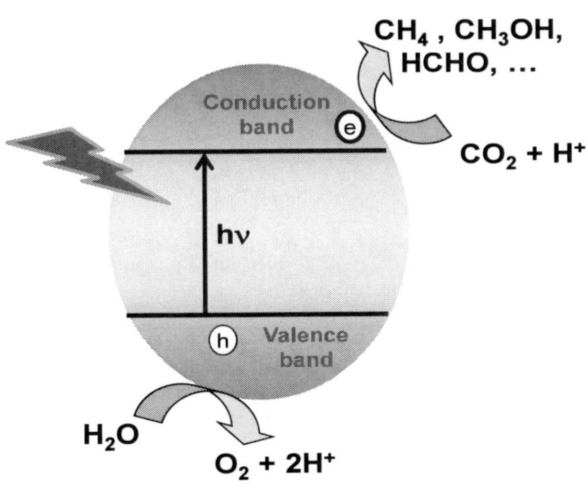

Figure 1. Illustration of the CO_2 photoreduction over semiconductor photocatalyst.

Generally, photocatalytic CO_2 reduction is not selective, and several products are simultaneously formed. Thus, separation of water-soluble products from CO_2 reduction and ways to recover them are also important problems when using H_2O as solvent that would limit the applicability of the process.

Photocatalysts for CO_2 reduction: The electronic band structure is the most important parameter of the photocatalysts for CO_2 reduction. To reduce CO_2 using water as reducing agent, the generated electrons and holes from photocatalysts must be able to reduce CO_2 and oxidize H_2O, respectively. This requires the semiconductor to have its CB electrons higher energy compared to the CO_2 reduction potential while the top of the VB must be more positive than the oxidation potential of water. Besides the electronic band structure, the resistance to chemical degradation by electrochemical corrosion, photo-corrosion and dissolution processes is another essential requirement to be met by the photocatalyst. A number of photocatalysts, especially metal sulfides, which exhibit suitable semiconducting properties for CO_2 photoreduction are not stable because the anions of these materials are more easily oxidized than water causing their degradation by oxidation of the material [5, 13]. Large band gap semiconductors are the most suitable photocatalysts for CO_2 reduction, because they are stable and provide sufficient negative and positive redox potentials in CBs and VBs, respectively. The disadvantage of using wide band gap semiconductors is the requirement for high energy input. For example, TiO_2 anatase, a stable photocatalyst with a large band gap energy (3.2 eV), is active only in the UV region of the solar spectrum.

3. Strategies toward Efficient Photocatalysts

A photocatalytic process involves many steps including light harvesting, charge generation, separation, transportation, and utilization. To design ideal photo-active materials for the efficient conversion of CO_2 using sunlight, increasing the light absorbance and improving the charge-carrier transportation are two of the main approaches [2, 3]. In addition, as a heterogeneous catalyst, a large surface area can also be the determining factor in photocatalytic reactions due to the fact that large amount of adsorbed reactant molecules promotes the reaction rate. In this regard, nano-scale materials with high surface area and unique properties that differ from those of corresponding bulk materials are particularly promising in achieving these goals and will attract considerable attention in the future development of the solar CO_2 photoreduction [13, 15-18]. In this part, strategies to improve the light absorption and charge separation of the photocatalysts including doping, coupling semiconductors, and using co-catalysts (see table 1) will be discussed.

3.1. Doping

Large band gap semiconductors such as metal oxides are the most suitable photocatalysts for CO_2 reduction because of their suitable band structure and good stability. However, their wide bandgaps make them active only in the UV region, which is associated with less than 5% of the total energy of the solar spectrum. Narrowing the bandgap of these semiconductor materials and shifting their absorption to the visible region is a potential strategy for increasing the efficiency of solar energy conversion. One effective method to obtain this goal is doping wide bandgap metal oxides with metal or non-metal elements.

Visible light absorption in doped large-bandgap semiconductors: Doping is well known to be an effective way to alter the optical and electronic properties of semiconductors for different applications. Technically, doping is the introduction of foreign elements into the parent materials without giving rise to a new crystallographic forms, phases or structures. In metal oxide semiconductors, the CB and VB usually consist of dominant metal 3d and oxygen 2p states, respectively [19]. Band gap narrowing of these materials requires either elevating the VB maximum or lowering the CB minimum. For example, when a metal (cation) or non-metal (anion) is used as a doping agent for TiO_2, its energy levels can be incorporated into the band gap of TiO_2. The new energy levels, also known as impurity states, can either accept electrons from the VB or donate electrons to the CB [20-24]. As shown in Figure 2, the energy gap between impurity states and CB or VB are narrower, in comparison with the energy gap between the VB and CB. Therefore, the doped TiO_2 may absorb visible light and generate electron-hole pairs.

Redox potentials in doped semiconductors: It is noted that the location of energy levels within the band-gap, which is highly dependent on the dopant used, is important when determining the reduction and oxidation strength of doped semiconductors. For instance, in the case of TiO_2, if electrons are being promoted from the VB of TiO_2 to the acceptor level (E_a, Figure 2), the oxidation power of TiO_2 is unchanged while the reduction power is decreased. In contrast, when electrons are being generated from the donor level (E_d, Figure 2), the oxidation power of TiO_2 is decreased whereas its reduction power is maintained.

Effect of dopant concentration: It is well known that doping can increase the photocatalytic activity of wide bandgap semiconductors in the visible region, which is resulted from a red-shift in the absorption spectrum. The extent to which the light absorption spectrum is red-shifted is dependent not only on the nature of dopant, but also on its concentration. A higher concentration of dopant often leads to a greater red-shift. However, the degree of red-shift is not always in line with photoactivity because in many cases doping ions can also act as trapping sites, and thus influence the lifetime of charge carriers [19, 25]. Increasing the dopant concentration may enhance the recombination of photogenerated electrons and holes, and therefore lower the reaction rate. Thus, there is an optimum metal content at which the efficiency is the highest.

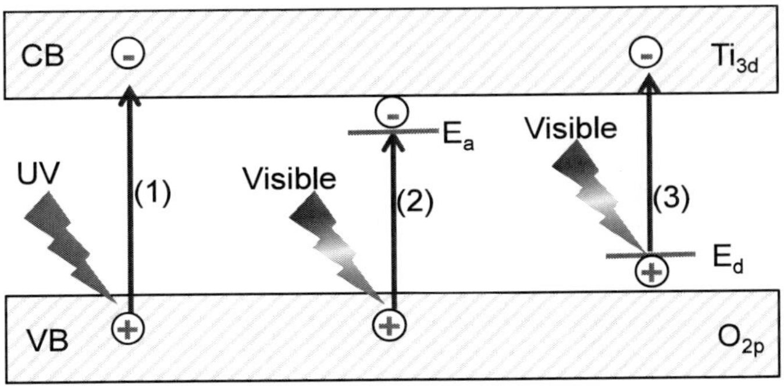

Figure 2. Three different transitions in TiO_2: (1) transition from VB to CB; (2) transition from VB to acceptor impurity level (E_a); and (3) transition from donor impurity level (E_d) to CB.

Effect of dopant distribution: Another important factor that affects the performance of dopant is its distribution in semiconductor matrix. It is obviously that a homogeneous distribution of dopant at the atomic level in the whole doped semiconductor matrix is needed to maximize the visible light absorption of doped semiconductors. However, the distribution of dopant is strongly affected by the doping method. For example, transition-metal doping, Cr-doped TiO_2 by impregnation has only a visible-light absorption shoulder, but Cr-doped TiO_2 by ion implantation has an impressive band-to-band absorption band edge up to ca. 550 nm. The reason is that the later method can produce Cr-doped TiO_2 in which Cr ions are evenly distributed throughout the TiO_2 matrix [26].

3.2. Coupling Semiconductors

Semiconductor/semiconductor heterojunction, formed by the direct contact of two semiconductors, represents an effective architecture for improving photocatalytic activity. When the two semiconductors with suitable band edge positions are combined, the charge transfer between them can increase the lifetime of the charge carriers, thus promoting the photocatalytic process. In addition, when the band gap of the coupled semiconductor is small, the energy range of photoexcitation for the system is also extended [27].

Types of semiconductor heterojunction: In general, the semiconductor heterojunction can be divided into three different types depending on the band alignments of semiconductors: straddling gap (type I), staggered gap (type II) or broken gap (type III) as shown in Figure 3. Among these three types of heterojunctions, type II band alignment with a staggered gap facilitates photocatalytic applications with promoted charge separation efficiencies due to the built-in-field formed at the junction interface [27, 28].

Charge transfer in type II heterojunction: The photocatalytic mechanism of type II semiconductor heterojunctions can be explained by the energetic diagrams shown in Figure 3. Assuming that both VB and CB of semiconductor B are more negative than those of semiconductor A, respectively. Photoexcited electrons can be transferred from CB(B) to CB (A) while photogenerated holes can be transferred from VB(A) to VB(B). Thus, whether the semiconductor A or B or both of them are excited by illumination, the electrons are located on CB(A) and the holes are located on VB(B), meaning they are spatially separated on the two semiconductors.

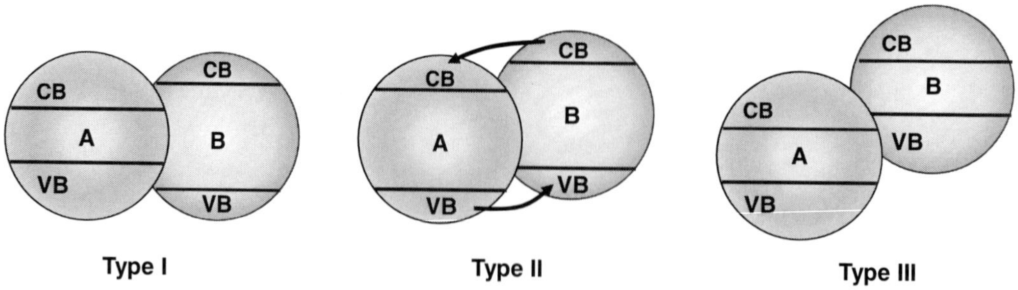

Figure 3. Different types of semiconductor heterojunction.

The result is the electron-hole recombination is diminished, leading to an enhancement of photocatalytic performance. Most of the examples of composite photocatalysts described in the literature are the type II heterojunction [27].

Effects of hetrojunction composition: There are two main factors that affect the performance of a heterojunction: composition and heterojunction interface. The composition of a semiconductor composite can affect its photocatalytic activity in two different ways. First, the composition strongly affects the light absorption of the composite. Due to the synergic absorption of the two semiconductors in the heterojunction, its light absorption spectra is a combination of those from both of them, and can absorb photons with energies larger than their respective band gaps. Therefore, the light absorption edge of heterojunction is mainly determined by the semiconductor with a narrower band gap. It means that the light absorption ability of the composite is governed by the amount of narrower band gap semiconductor in the composite. Second, the composition affects the catalytic process at the surface of the photocatalyst. As mentioned above, the type II heterojunction is the most frequently used in photocatalysis. In this type of heterojunction, the oxidation and reaction occur separately on the surface of each component. Thus, the specific surface area of each component in the composite which depends on their compositions, also influences the overall catalytic performance of the composite. Therefore, optimal compositions are usually needed for heterojunctions.

Effects of hetrojunction interface: Another factor that can greatly affect the photocatalytic efficiencies of heterojunctions is the interface. First, a large interface area would facilitate the charge separation between the two semiconductors. In addition, due to the lattice mismatch between semiconductors, defects could be formed as they are coupled to form heterojunctions, which would trap the photogenerated electrons and holes and thus prevent their diffusion. Therefore, the quality of the heterojunction interface, which depends on the synthesis method, also greatly affects the photocatalytic performance [27, 29].

3.3. Using Cocatalysts

Loading of cocatalysts have been widely used to improve the photocatalytic activity of photocatalysts, especially in the photocatalytic CO_2 reduction. These cocatalysts are typically very small particles of various metals such as Pt, Pd, Au, Ag, Ni, Cu, etc. [6, 50-64]. In general, metal cocatalysts can play two roles in enhancing the photocatalytic activity of a photocatalyst. First, they act to extend the lifetime of photogenerated charges that reach the surface of photocatalyst by enhancing the rates of electron–hole separation at the metal/ semiconductor interface. Second, metal cocatalysts can perform the catalytic function by providing chemically active sites where relevant chemical transformations can take place with lower activation barriers than on the semiconductor [6].

Cocatalysts for improving charge separation: When metal is in contact with the surface of a semiconductor, a Schottky junction can be formed. For example, in the metal/TiO_2 system, the Schottky junction allows the transfer of electrons from TiO_2 to the metal particles, but prevents electron transfer back across the Schottky barrier. Under band gap illumination, generated electron transfers from CB of TiO_2 to the metal cocatalysts, resulting in the spatial separation of electron and hole, which are located on metal particle and TiO_2 particle, respectively.

Table 1. Possible effects, influencing factors, and representative examples in different strategies toward efficient photocatalysts

Strategies	Possible effects	influencing factors	Representative examples
Doping 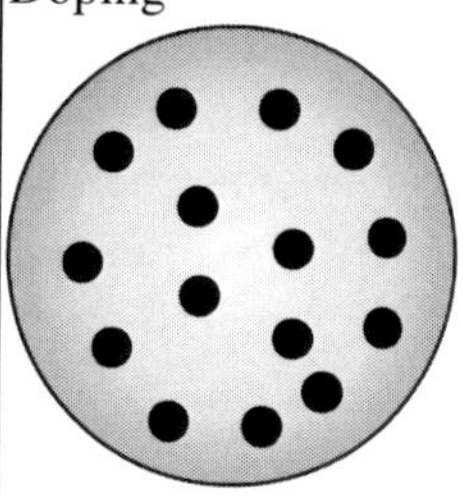	Improving light absorption Decrease charge separation Lowered redox potential of electrons or holes	Type of dopants Concentration of dopants Distribution of dopants	Metal doped TiO_2: Cr-TiO_2; [30] Fe-TiO_2; [31, 32] V-TiO_2 [33] Non-metal doped TiO_2: N-TiO_2; [34, 35] C-TiO_2; [36] F-TiO_2; [37] S-TiO_2 [38]
Coupling semiconductors 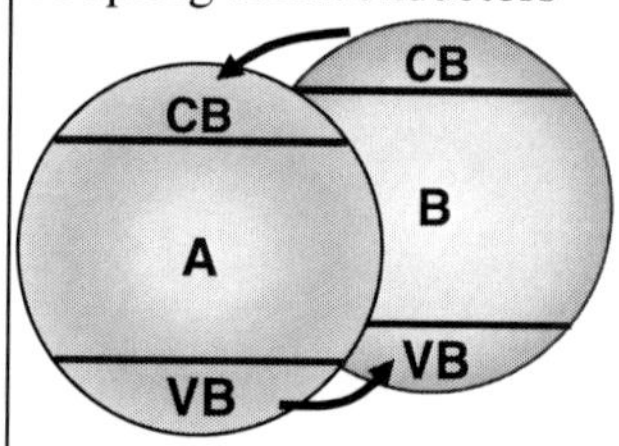	Improving charge separation Extending light absorption region Lowered redox potential of transferred electrons and hols	Electronic band structure Composition Interface	CdS/TiO_2; [17, 39-41] ZnO/TiO_2; [42-44] CuO/TiO_2; [45] CdSe/TiO_2; [46, 47] WO_3/TiO_2 [48]
Using cocatalysts 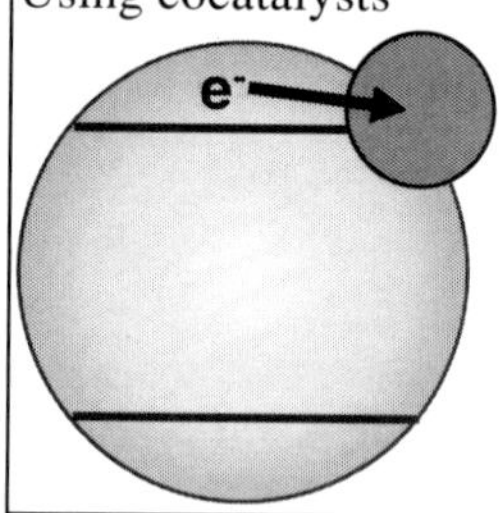	Improving charge separation Reducing overpotential for CO_2 reduction	Type of cocatalysts Size and distribution of cocatalysts Amount of cocatalysts	Pt, Pd, Au, Ag, Cu loaded TiO_2 [49-54];

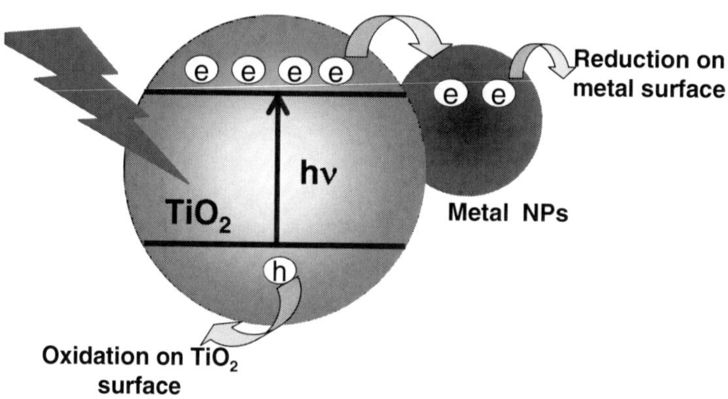

Figure 4. Illustration of electron transfer in metal-TiO_2 hybrids under UV illumination.

As a result, the hole is free to oxidize the organic compounds or water on the surface of TiO_2, while the electron can reduce water, CO_2 or oxygen on the surface of metal particles as illustrated in Figure 4 [65].

Cocatalysts as active sites: Another important role of metal cocatalysts is their own catalytic properties, which are very important for the photocatalytic hydrogen generation and photocatalytic reduction of CO_2. As a representative example, owing to a large overpotential for the reduction of CO_2 on the surface of TiO_2, the material alone becomes inactive. Loading TiO_2 with noble metals, such as Pt, Pd, Au, Rh, and Ag, which have a low overpotential CO_2 reduction, can make it easier to reduce CO_2 [56, 57, 61-63].

In addition, the cocatalysts can affect the selectivity of products in view of the formation of many different products during the process of CO_2 reduction. For example, deposition of a series of 2 wt% metals (Pt, Rh, Pd, Cu, Ru, and Au) on TiO_2 considerably accelerated photocatalytic reduction of CO_2 to CH_4 and/or acetic acid, and product distribution could be controlled by changing thy types of the metals deposited on TiO_2. CH_4 is the main product when TiO_2 was loaded with Pt and Pd while a considerable amount of acetic acid was produced using other metals loaded TiO_2 [60].

Effects of particle size and distribution: many factors such as the size of metal nanoparticles, metal dispersion and loading amount can affect the photocatalytic activity of metal/semiconductor nanocomposites.

Normally, at the same amount of metal loading, metal cocatalysts with smaller size have a larger surface area and more active sites, which result in higher catalytic activity [62]. In addition, the charge cariers are less likely to recombine in the bulk of small particles of a cocatalyst than in the large ones.

In some cases, smaller particles of metal cocatalysts may exhibit lower barriers for the interfacial charge transfer from semiconductors to the cocatalysts [66]. Compared with the metal particle size, more attention has been paid to the dispersion of metal nanoparticles.

A better metal dispersion provides a higher metal coverage and a larger metal/support interface, which can produce the same effect as increasing the amount of metal. Thus, the small size and high dispersion of metal nanoparticles as cocatalysts are crucial for the photocatalytic enhancement of metal/semiconductor composite.

Effects of cocatalyst loading: Another important factor influencing the photocatalytic activity of metal/semiconductor composite is the metal loading. There is an optimum loading value above which metal deposition has a detrimental effect on the photocatalytic activity [66]. As illustrated in Figure 5, there is a volcano-type trend between the loading of a given metal cocatalyst and the photocatalytic activity. Initially, the introduction of a metal onto a photocatalyst could gradually enhance the photocatalytic activity of system by facilitating charge separation and increasing the number of catalytic sites. When the loading of a metal cocatalyst on the semiconductor reaches the optimal value, the system achieves the highest activity. However, higher loading of the cocatalyst will drastically decrease the photocatalytic activity. This decrease is caused by the following factors: (i) high metal cocatalyst loading could result in covering the surface active sites of the semiconductor and hindering its contact with reactants.; (ii) excessive coverage of semiconductor catalyst limits the amount of light reaching the surface, reducing the number of photogenerated hole–electron and consequently lowering the photocatalytic activity; and (iii) once negatively charged, metal particles, especially for highly loaded samples, attract holes and subsequently recombine them with electrons. In this case, the metal deposits become recombination centers and decrease the photocatalytic activity [5, 66, 67].

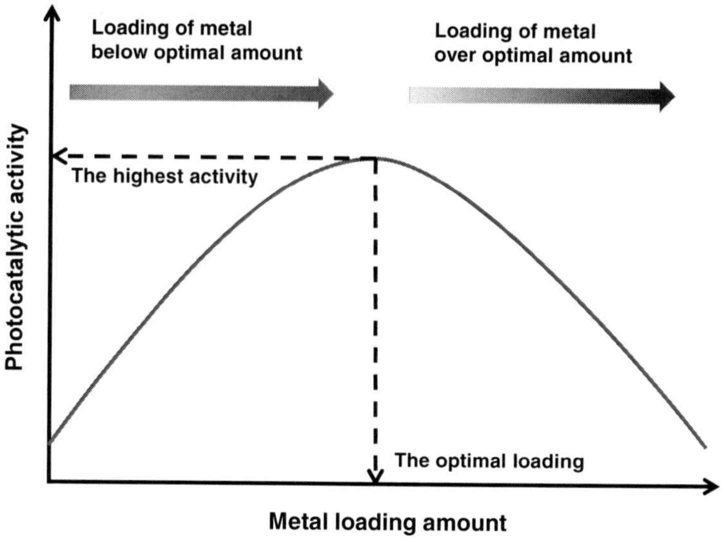

Figure 5. A volcano-type relationship between the loading amount of metal cocatalysts and the photocatalytic activity of the metal-loaded semiconductor photocatalyst.

4. High-Surface-Area Photocatalysts for CO_2 Reduction

The surface area plays important role in photocatalytic reaction which mainly occurs on the surface of the photocatalysts. A higher surface area usually results in a higher activity of the photocatalysts. Thus, several methods have been employed to increase the surface area of the photocatalysts including decreasing particle size or designing nanostructures such as mesoporous, nanotube arrays or hollow structures.

4.1. Nanoparticles

Particle size is an important parameter for photocatalytic efficiency, since it affects not only the total surface area of the photocatalysts, but also their physical and chemical properties. Since the photocatalytic reaction occurs at the surface of a semiconductor, increasing the surface area by decreasing the particle size is thus expected to improve the photocatalytic activity. However, the increase in surface area should not be accompanied by a decrease in crystallinity because it may be associated with large amounts of crystalline defects, which favor the recombination of electrons and holes leading to a poor photoactivity.

Another effect of particle size is the migration length of charge carriers. Once the electron and hole generated, they migrate to the surface of the semiconductor to react with adsorbed molecules. Thus, smaller particle sizes also mean shorter distance for photogenerated charge carriers to migrate before reaching surface reaction sites, potentially improving photocatalytic performance [13]. In addition, as the particle size decreases to the order of nanometers, the energy level may shift according to the quantum size effect Furthermore, when the size of the particle is too small, the electron hole separation is expected to be inefficient because the recombination rate is proportional to the probability at which an electron encounters a hole, then recombination will increase with the inverse of the particle volume [67].

Besides the size, the shape of the particle also affects the photocatalytic performance of a photocatalyst because it not only governs the adsorption of the reactant on the photocatalyst's surface, but also can influence the electronic band structure and charge separation of the semiconductor [68-70].

Nanosized photocatalyts for CO_2 reduction: A number of semiconductors with nanosize range and controlled shapes have been developed for the photocatalytic conversion of CO_2 (Table 2). As a representative example, Zhou et al. [71] synthesized ultrathin and uniform Bi_2WO_6 square nanoplates of ~9.5 nm thickness corresponding to six repeating cell units. The nanosheets exhibited a CH_4 production rate of 1.1μmol $g^{-1}h^{-1}$ which is much higher than that of Bi_2WO_6 prepared using solid state reaction (0.045 μmol $g^{-1}h^{-1}$) in the photocatalytic conversion of CO_2 under visible light illumination.

Controlling particle shape to improve the charge separation of the photocatalysts can help to improve the photocatalytic CO_2 reduction. For example, Zn_2GeO_4 nanoribbons (ca. 7 nm thickness, corresponding to five repeating cell unit) with ultralong and ultrathin geometry greatly improved photocatalytic activity toward reduction of CO_2 with H_2O into CH_4 relative to bulk Zn_2GeO_4 (Figure 7) [72]. The ultrathin geometry of the nanoribbons allows charge carriers to move rapidly from the interior to the surface to participate in the photoreduction reaction. The single crystalline hexagonal prism Zn_2GeO_4 nanorods (400 nm length and 50 nm width) also exhibited higher activity for photocatalytic reduction of CO_2 into CH_4 [73]. $In_2Ge_2O_7$ ultrathin nanowire (2–3 nm in diameter) proved to possess photocatalytic activity towards the conversion of CO_2 to CO and O_2 in the present water vapor under Xe arc lamp irradiation [74]. $NaNbO_3$ nanowires with a uniform diameter of about 100 nm and length of up to several tens micrometers exhibited about 20-time significant improvement for photocatalytic reduction of CO_2 into CH_4 under UV–vis light irradiation, relative to bulk $NaNbO_3$, due to superior crystallinity, the high surface area, and highly mobile charge carriers [75].

4.2. Mesoporous Materials

Mesoporous materials have been extensively used in photocatalysis due to their unique structural characteristics, such as high surface area and large pore. The continuity in mesoporous structure can be expected to make the electron transfer within the material easier, resulting in higher activity [76, 77].

However, synthesis of mesoporous semiconductor phtotocatalysts is still quite challenging. Generally, there are mainly two pathways to synthesize the mesoporous photocatalyts, one is soft-templating method and the other is hard-templating approach. Mesoporous structures of several semiconductors have been developed for the photoreduction of CO_2. For example, Ye and coworkers reported the mesoporous structure of N-doped Zn_2GeO_4(ZGON) [78], and an optimal balance between effective visible-light absorption and adequate redox potentials as well as improved crystallinity is achieved through adjusting the nitridation time. Compared with the ZGON synthesised by a solid-state reaction, the optimal mesoporous ZGON loaded with 1 wt% Pt exhibited enhanced activity in CO_2 photoreduction with H_2O vapour under visible light.

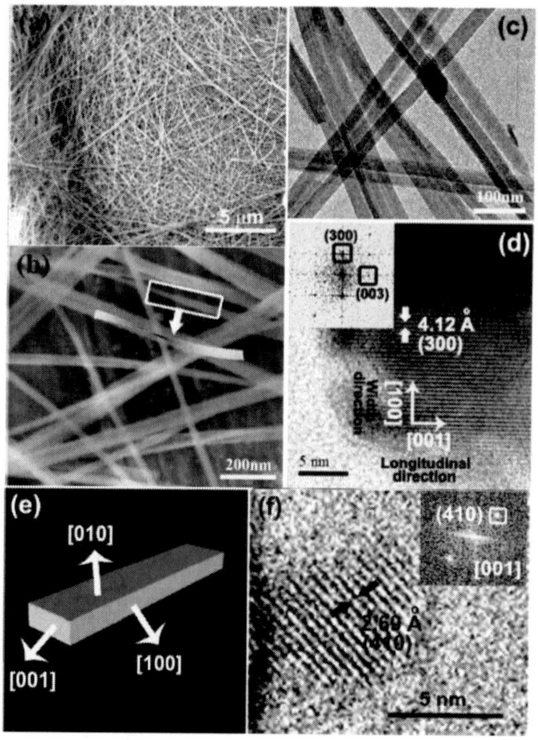

Figure 6. (a, b) FE-SEM, (c) TEM, and (d) HRTEM images of the Zn_2GeO_4 nanoribbons. The inset of (d) shows the FFT pattern obtained from the HRTEM image. (e) Structural model of a nanoribbon. (f) HRTEM image taken along [001] [72].

Guo et al. [79] synthesized $In(OH)_3$ with mesoporous structure which enhances activity in CO_2 photoreduction. The mesoporous $In(OH)_3$ loaded with co-catalysts could convert CO_2 into CH_4 with the rate of 0.8 μmol $g^{-1}h^{-1}$which is 20 times higher than that of $In(OH)_3$ without mesoporous structure under UV light irradiation. Xi et al. [80] reported the synthesis of hybrid TiO_2/ZnO mesoporous materials with high specific surface area, large pore volume, and nanojunction modified pore walls by a facile furfural alcohol-derived polymerization–oxidation route The homogeneously hybrid TiO_2/ZnO mesoporous materials exhibited a much higher photocatalytic activity of CO_2 reduction owing to their unusual hybrid mesoporous structure and high surface area compared to TiO_2/ZnO particles prepared by solid state reaction. The rate of CO_2 to CH_4 production obtained in the mesoporous hybrid TiO_2/ ZnO is at least fifty and six times higher than those obtained in the non-porous TiO_2/ZnO and commercial P25, respectively.

Nanoparticle photocatalysts supported on high-surface-area mesoporous support was also studied for the photocatalytic conversion of CO_2. For example, Ce-doped TiO_2 nanoparticles prepared by a simple sol-gel method were dispersed on SBA-15 to form a Ce-TiO_2/SBA-15 nanocomposite [81]. The photocatalysts were used to convert CO_2 and H_2O to fuels, mainly CO and CH_4, under UV-vis illumination. Compared with unsupported Ce-TiO_2, Ce-TiO_2 /SBA-15 remarkably enhanced the CO_2reduction rate, and Ce-TiO_2 /SBA-15 with a Ti : Si ratio of 1 : 4 demonstrated 8-fold enhancement in CO production and 115-fold enhancement in CH_4 production.

Table 2. Representatives of photocatalysts with different structures for CO_2 reduction (from 2009 to 2014)

Catalyst/Structure	Cocatalyst	Surface area (m^2/g)	Light source	Reaction medium	Main products	Activity	References (year)
TiO_2 / Nanocrystal	-	45	Xe lamp	CO_2 and H_2 vapor	CH_4	1.35 μmol h^{-1} g^{-1}	[94] (2014)
$NaNbO_3$ / Nanowire	Pt	12	Xe lamp	CO_2 and H_2 vapor	CH_4	653 ppm g^{-1} h^{-1}	[75] (2011)
$Na_2V_6O_{16}$ / Nanoribbon	Pt, RuO_2	-	Xe lamp, $\lambda \geq 420$ nm	CO_2 and H_2 vapor	CH_4	0.2 μmol g^{-1} h^{-1}	[95] (2013)
Bi_2WO_6 / Nanoplate	-	6	Xe lamp, $\lambda \geq 420$ nm	CO_2 and H_2 vapor	CH_4	1.1 μmol g^{-1} h^{-1}	[71] (2012)
$ZnGa_2O_4$ / Nanocube	RuO_2	10	Xe lamp	CO_2 and H_2 vapor	CH_4	26 μmol g^{-1} h^{-1}	[96] (2012)
Zn_2GeO_4 / Nanobelt	Pt, RuO_2	28	Xe lamp	CO_2 and H_2 vapor	CH_4	25 μmol g^{-1} h^{-1}	[72] (2010)
HNb_3O_8 / Nanobelt	-	39	Xe lamp	CO_2 and H_2 vapor	CH_4	3.58 μmol g^{-1} h^{-1}	[97] (2012)
TiO_2 / Nanosheet	-	-	Xe lamp	CO_2 and H_2 vapor	CH_4	5.8 ppm g^{-1} h^{-1}	[98] (2013)
TiO_2 / Nanorod	Pt	25	Xe lamp	CO_2 and H_2 vapor	CH_4	6 μmol g^{-1} h^{-1}	[99] (2011)
Zn_2GeO_4 / Mesoporous	Pt	90	Xe lamp	CO_2 and H_2 vapor	CH_4	28.9 ppm g^{-1} h^{-1}	[100] (2011)
$ZnGa_2O_4$ / Mesoporous	RuO_2	110	Xe lamp	CO_2 and H_2 vapor	CH_4	504 ppm g^{-1} h^{-1}	[101] (2010)
$In(OH)_3$ / Mesoporous	Pt	65	Xe lamp	CO_2 and H_2 vapor	CH_4	0.8 μmol g^{-1} h^{-1}	[79] (2013)
TiO_2/ZnO / Mesoporous	-	225	Xe lamp	CO_2 and H_2 vapor	CH_4	55 μmol g^{-1} h^{-1}	[80] (2011)
TiO_2/SiO_2 / Mesoporous	Cu	386	Xe lamp	CO_2 and H_2O vapor	CH_4	60 μmol g-cat^{-1} h^{-1}	[82] (2010)
TiO_2/Nanotube	Cu-Pt	-	Solar simulating light	CO_2 (0.998 % in N_2) and H_2O vapor	CH_4	3.55 mL g^{-1} h^{-1}	[86] (2012)
N-doped TiO_2/Nanotube	Cu, Pt	-	Natural sunlight	CO_2 and H_2O vapor	CH_4	160 μl g^{-1} h^{-1}	[63] (2009)
CdS (or Bi_2S_3)/TiO_2/Nanotube	-	100	Visible light, $\lambda \geq 400$ nm	CO_2 bubbled NaOH solution containing Na_2SO_3	CH_3OH	45 μmol g^{-1} h^{-1}	[102] (2012)
CuO-$TiO_{2-x}N_x$/Hollow nanocube	-	24	Solar simulating light	CO_2 and H_2O vapor	CH_4	41.3 ppm g^{-1} h^{-1}	[92] (2012)
Bi_2WO_6/Hollow sphere	-	24	Xe lamp, $\lambda \geq 420$ nm	CO_2 bubbled H_2O solution	CH_3OH	16.3 μmol g^{-1} h^{-1}	[93] (2012)
Graphene-$Ti_{0.91}O_2$/Hollow sphere	-	-	Xe lamp	CO_2 and H_2O vapor	CO	8.91 μmol g^{-1} h^{-1}	[91] (2012)

Enhanced CO_2 reduction rates were also observed on mesoporous silica supported Cu/ TiO_2 catalysts [82].

The peak production rates of CO and CH_4 is 60 and 10 μmol g^{-1} h^{-1}, respectively, for the 0.5% Cu/TiO_2-SiO_2 composite. The high surface area mesoporous silica substrate (>300 m^2/g) enhances the yield of CO_2 photoreduction possibly due to improved TiO_2 dispersion and increased adsorption of CO_2 and H_2O on the catalyst.

4.3. Nanotube Arrays

One-dimensional, vertically oriented, self-organized metal oxide nanotube arrays formed by electrochemical anodization are currently of great interest in photocatalysis because of their high aspect ratio and high surface area originated from hollow architecture. In addition, the self-assembled thin wall nanotube array geometry, vertically oriented from the substrate, promotes ready separation of photogenerated electron hole pairs and charge transport [83-85].

Among various materials, TiO_2 nanotube arrays (TiNT) have been mostly investigated in photocatalytic conversion of CO_2 due to its favorable electronic properties [63, 86]. For example, Vargheses and colleagues reported that nitrogen doping of anodically oxidized TiNT decorated with Cu and Pt nanoparticles shifts the band edge to nearly 540 nm, greatly increasing the visible light absorbance. Using outdoor sunlight 100 mW/cm^2 on a fixed bed reactor, a hydrocarbon production rate of 160 μL $g^{-1}h^{-1}$ is obtained using this TiNT [63] Titanium nanotube membranes are also being examined for photocatalytic reduction of CO_2 under solar irradiation. Zhang et al. [86] reported an approach that is able to achieve high-rate sunlight-driven conversion of diluted CO_2 to light hydrocarbons in which an optimized combination of a Cu-Pt coating and modulated-diameter TiO_2 nanotubes are used as the photocatalyst (Figure 7).

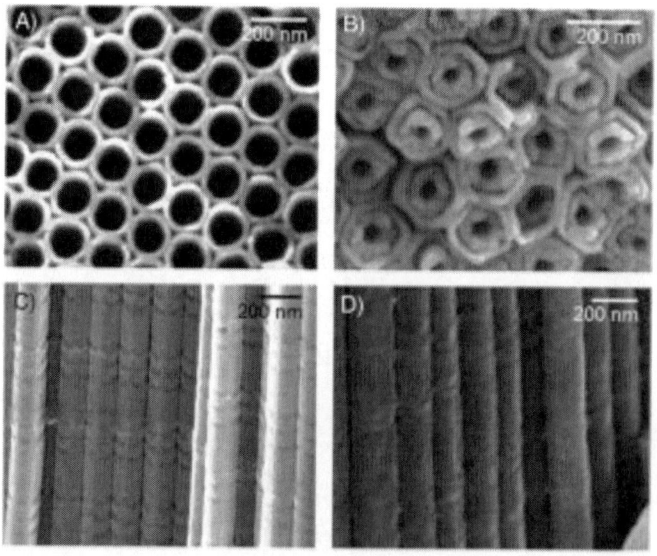

Figure 7. A,B) Plan-view and C,D) cross-sectional SEM images of the as-prepared PMTiNT platform (A,C) and Cu-Pt loaded nanotubes (B,D) [86].

The photocatalytic reaction occurs at room temperature and generates CH_4, C_2H_4, and C_2H_6 as reaction products. Under AM1.5 one-sun illumination, using 99.9% CO_2, a hydrocarbon production rate of 3.51 $mLg^{-1}h^{-1}$ was obtained. The photocatalytic conversion of CO_2 was further increased when dilute CO_2 was used as reactant. The periodic modulation of the diameters of the nanotube arrays found to increase the surface area and improved the utilization of light while the bimetallic coating increased catalyst activity and selectivity.

4.4. Hollow Nanostructures

Hollow structured materials have received much attention owing to their low density, high specific surface areas, and good photocatalytic activity [87, 88]. It was reported that such materials were expected to exhibit a high light-collection efficiency and a fast motion of charge carriers because of their hollow structures, closely packed interpenetrating networks, and large internal surface area [89, 90].

Tu et al. reported the synthesis of hollow spheres consisting of molecular-scale alternating titania ($Ti_{0.91}O_2$) nanosheets and graphene (G) nanosheets using a layer-by-layer assembly technique with polymer beads as sacrificial templates [91]. The designed hollow sphere photocatalysts exhibited a nine times increase in the photocatalytic activity compared to commercial P25-TiO_2 in the photocatalytic conversion of CO_2 in the presence of water. However, in this case, CO was the major product with a relatively small amount of CH_4. In et al. [92] reported the design of a new nanostructured material, hollow CuO nanocubes decorated with nanoparticles of titanium oxynitride ($TiO_{2-x}N_x$), that photocatalytically converts CO_2 into methane at ambient temperature and without platinum co-catalysts (Figure 8). The hollow hybrid was synthesized using uniform Cu_3N nanocubes as both morphological and compositional template for subsequent deposition of a partial TiO_2 shell, oxidation of the Cu_3N core to form hollow CuO nanocubes, and low-temperature (450 °C) incorporation of the expelled nitrogen into the TiO_2 framework to form crystalline TiO_{2-x} N_x. The CuO-$TiO_{2-x}N_x$ hollow nanocubes yield 41.3 ppm $g^{-1}h^{-1}$ of methane, which represents a production rate that is 2.5 times faster than that of Degussa P25-TiO_2 when measured under the same conditions (16.2 ppm $g^{-1}h^{-1}$). Hollow microspheres of Bi_2WO_6 were also prepared for the conversion of CO_2 to methanol using an anion exchange method [93]. The Bi_2WO_6 hollow microspheres, which possess large surface area and high CO_2 adsorption capacity, display efficient visible light CO_2 photocatalytic conversion into methanol in solution without any co-catalyst. In CO_2-saturated solution, the Bi_2WO_6 hollow microspheres gave rise to 32.6 μmol g^{-1} of CH_3OH, which is 25.5 times higher than that of Bi_2WO_6 prepared by solid-state reaction.

CONCLUSION

The catalytic conversion of CO_2 to liquid fuels is a critical goal that would positively impact the global carbon balance by recycling CO_2 into usable fuels. The challenges presented here are great, but the potential rewards are enormous.

Figure 8. (d-f) TEM images of CuO-TiO_{2-x} N_x hollow nanocubes; g) FESEM images of the CuO-$TiO_{2-x}N_x$ hollow nanocubes [92].

Photocatalytic conversion of CO_2 to useful fuels is a scientifically challenging problem, requiring efficient photocatalysts that effectively absorb sunlight.

To develop efficient photocatalysts, it is necessary to narrow the band gaps to harvest visible light in the longer-wavelength regions, and to enhance the photogenerated charge separation in photocatalysis. Thus, factors such as morphology, doping and dispersion of cocatalysts, all of which determine the photocatalytic activity of such materials, need to be elucidated in great detail. Coupling semiconductors is another promising approach to improve sunlight absorption and charge separation. Moreover, nanostructures such as mesoporous materials, nanotube arrays and hollow exhibiting high surface area and effective light absorption can be also integrated with other strategies for further enhancement of the CO_2 photocatalytic conversion efficiency.

REFERENCES

[1] Lewis, N. S.; Nocera, D. G. *Proc. Natl. Acad. Sci. US* 2006, 103, 15729-15735.
[2] Qu, Y.; Duan, X. *Chem. Soc. Rev.* 2013, 42, 2568-2580.
[3] Zhou, H.; Qu, Y.; Zeid, T.; Duan, X. *Energy Environ. Sci.* 2012, 5, 6732-6743.

[4] Chen, X.; Li, C.; Graetzel, M.; Kostecki, R.; Mao, S. S. *Chem. Soc. Rev.* 2012, 41, 7909-7937.

[5] Chen, X.; Shen, S.; Guo, L.; Mao, S. S. *Chem. Rev.* 2010, 110, 6503-6570.

[6] Roy, S. C.; Varghese, O. K.; Paulose, M.; Grimes, C. A. *ACS Nano* 2010, 4, 1259-1278.

[7] Koci, K.; Obalova, L.; Lacny, Z. *Chem. Pap.* 2008, 62, 1-9.

[8] Mori, K.; Yamashita, H.; Anpo, M. *Rsc Advances* 2012, 2, 3165-3172.

[9] Morris, A. J.; Meyer, G. J.; Fujita, E. *Acc. Chem. Res.* 2009, 42, 1983-1994.

[10] Fujishima, A.; Zhang, X. T. *C.R. Chim.* 2006, 9, 750-760.

[11] Habisreutinger, S. N.; Schmidt-Mende, L.; Stolarczyk, J. K. *Angew. Chem. Int. Ed.* 2013, 52, 7372-7408.

[12] Usubharatana, P.; McMartin, D.; Veawab, A.; Tontiwachwuthikul, P. *Industrial and Engineering Chemistry Research* 2006, 45, 2558-2568.

[13] Kudo, A.; Miseki, Y. *Chem. Soc. Rev.* 2009, 38, 253-278.

[14] Inoue, T.; Fujishima, A.; Konishi, S.; Honda, K. *Nature* 1979, 277, 637-638.

[15] Dinh, C.-T.; Nguyen, T.-D.; Kleitz, F.; Do, T.-O. *ACS Nano* 2009, 3, 3737-3743.

[16] Dinh, C.-T.; Nguyen, T.-D.; Kleitz, F.; Do, T.-O. *Chem. Commun.* 2011, 47, 7797-7799.

[17] Dinh, C.-T.; Pham, M.-H.; Kleitz, F.; Do, T.-O. *J. Mater. Chem. A* 2013, 1, 13308-13313.

[18] Kubacka, A.; Fernández-García, M.; Colón, G. *Chem. Rev.* 2011, 112, 1555-1614.

[19] Liu, G.; Wang, L.; Yang, H. G.; Cheng, H.-M.; Lu, G. Q. *J. Mater. Chem.* 2010, 20, 831-843.

[20] Nikolay, T.; Larina, L.; Shevaleevskiy, O.; Ahn, B. T. *Energy Environ. Sci.* 2011, 4, 1480-1486.

[21] Choi, W. Y.; Termin, A.; Hoffmann, M. R. *J. Phys. Chem.* 1994, 98, 13669-13679.

[22] George, S.; Pokhrel, S.; Ji, Z.; Henderson, B. L.; Xia, T.; Li, L.; Zink, J. I.; Nel, A. E.; Mädler, L. *J. Am. Chem. Soc.* 2011, 133, 11270-11278.

[23] Wang, J.; Tafen, D. N.; Lewis, J. P.; Hong, Z.; Manivannan, A.; Zhi, M.; Li, M.; Wu, N. *J. Am. Chem. Soc.* 2009, 131, 12290-12297.

[24] Park, J. H.; Kim, S.; Bard, A. J. *Nano Lett.* 2005, 6, 24-28.

[25] Parida, K. M.; Sahu, N. *J. Mol. Catal. A: Chem.* 2008, 287, 151-158.

[26] Anpo, M.; Takeuchi, M. *J. Catal.* 2003, 216, 505-516.

[27] Marschall, R. *Adv. Funct. Mater.* 2014, 24, 2421-2440.

[28] Wang, H.; Zhang, L.; Chen, Z.; Hu, J.; Li, S.; Wang, Z.; Liu, J.; Wang, X. *Chem. Soc. Rev.* 2014.

[29] Zong, X.; Yan, H.; Wu, G.; Ma, G.; Wen, F.; Wang, L.; Li, C. *J. Am. Chem. Soc.* 2008, 130, 7176-7177.

[30] Dholam, R.; Patel, N.; Adami, M.; Miotello, A. *Int. J. Hydrogen Energy* 2009, 34, 5337-5346.

[31] Khan, M. A.; Woo, S. I.; Yang, O. B. *Int. J. Hydrogen Energy* 2008, 33, 5345-5351.

[32] Umebayashi, T.; Yamaki, T.; Itoh, H.; Asai, K. *J. Phys. Chem. Solids* 2002, 63, 1909-1920.

[33] Liu, B.; Chen, H. M.; Liu, C.; Andrews, S. C.; Hahn, C.; Yang, P. *J. Am. Chem. Soc.* 2013, 135, 9995-9998.

[34] Liu, G.; Yang, H. G.; Wang, X.; Cheng, L.; Pan, J.; Lu, G. Q.; Cheng, H.-M. *J. Am. Chem. Soc.* 2009, 131, 12868-12869.

[35] Asahi, R.; Morikawa, T.; Ohwaki, T.; Aoki, K.; Taga, Y. *Science* 2001, 293, 269-271.
[36] Shen, M.; Wu, Z. Y.; Huang, H.; Du, Y. K.; Zou, Z. G.; Yang, P. *Mater. Lett.* 2006, 60, 693-697.
[37] Zhou, J. K.; Lv, L.; Yu, J.; Li, H. L.; Guo, P.-Z.; Sun, H.; Zhao, X. S. *J. Phys. Chem. C* 2008, 112, 5316-5321.
[38] Ohno, T.; Akiyoshi, M.; Umebayashi, T.; Asai, K.; Mitsui, T.; Matsumura, M. *Appl. Catal., A* 2004, 265, 115-121.
[39] Park, H.; Choi, W.; Hoffmann, M. R. *J. Mater. Chem.* 2008, 18, 2379-2385.
[40] Jang, J. S.; Ji, S. M.; Bae, S. W.; Son, H. C.; Lee, J. S. *J. Photochem. Photobiol., A* 2007, 188, 112-119.
[41] Jang, J. S.; Choi, S. H.; Kim, H. G.; Lee, J. S. *J. Phys. Chem. C* 2008, 112, 17200-17205.
[42] Agrawal, M.; Gupta, S.; Pich, A.; Zafeiropoulos, N. E.; Stamm, M. *Chem. Mater.* 2009, 21, 5343-5348.
[43] Liao, D. L.; Badour, C. A.; Liao, B. Q. *J. Photochem. Photobiol., A* 2008, 194, 11-19.
[44] Liao, S. J.; Huang, D. G.; Yu, D. H.; Su, Y. L.; Yuan, G. Q. *J. Photochem. Photobiol., A* 2004, 168, 7-13.
[45] Qin, S.; Xin, F.; Liu, Y.; Yin, X.; Ma, W. *J. Colloid Interface Sci.* 2011, 356, 257-261.
[46] Wang, C.; Thompson, R. L.; Baltrus, J.; Matranga, C. *J. Phys. Chem. Lett.* 2010, 1, 48-53.
[47] Robel, I.; Subramanian, V.; Kuno, M.; Kamat, P. V. *J. Am. Chem. Soc.* 2006, 128, 2385-2393.
[48] Song, K. Y.; Park, M. K.; Kwon, Y. T.; Lee, H. W.; Chung, W. J.; Lee, W. I. *Chem. Mater.* 2001, 13, 2349-2355.
[49] Yui, T.; Kan, A.; Saitoh, C.; Koike, K.; Ibusuki, T.; Ishitani, O. *ACS Appl. Mater. Interfaces* 2011, 3, 2594-2600.
[50] Shiraishi, Y.; Sakamoto, H.; Sugano, Y.; Ichikawa, S.; Hirai, T. *ACS Nano* 2013, 7, 9287-9297.
[51] Su, R.; Tiruvalam, R.; He, Q.; Dimitratos, N.; Kesavan, L.; Hammond, C.; Lopez-Sanchez, J. A.; Bechstein, R.; Kiely, C. J.; Hutchings, G. J.; Besenbacher, F. *ACS Nano* 2012, 6, 6284-6292.
[52] Su, R.; Tiruvalam, R.; Logsdail, A. J.; He, Q.; Downing, C. A.; Jensen, M. T.; Dimitratos, N.; Kesavan, L.; Wells, P. P.; Bechstein, R.; Jensen, H. H.; Wendt, S.; Catlow, C. R. A.; Kiely, C. J.; Hutchings, G. J.; Besenbacher, F. *ACS Nano* 2014, 8, 3490-3497.
[53] Hai, Z.; El Kolli, N.; Uribe, D. B.; Beaunier, P.; Jose-Yacaman, M.; Vigneron, J.; Etcheberry, A.; Sorgues, S.; Colbeau-Justin, C.; Chen, J.; Remita, H. *J. Mater. Chem. A* 2013, 1, 10829-10835.
[54] Foo, W. J.; Zhang, C.; Ho, G. W. *Nanoscale* 2013, 5, 759-764.
[55] Takai, A.; Kamat, P. V. *ACS Nano* 2011, 5, 7369-7376.
[56] Zhai, Q.; Xie, S.; Fan, W.; Zhang, Q.; Wang, Y.; Deng, W.; Wang, Y. *Angew. Chem. Int. Ed.* 2013, 52, 5776-5779.
[57] Zhang, X.; Han, F.; Shi, B.; Farsinezhad, S.; Dechaine, G. P.; Shankar, K. *Angew. Chem. Int. Ed.* 2012, 51, 12732-12735.
[58] Lin, H.-Y.; Yang, H.-C.; Wang, W.-L. *Catal. Today* 2011, 174, 106-113.

[59] Tian, H.; Zhang, X. L.; Scott, J.; Ng, C.; Amal, R. *J. Mater. Chem. A* 2014, 2, 6432-6438.
[60] Ishitani, O. *J. Photochem. Photobiol., A* 1993, 72, 269-271.
[61] Iizuka, K.; Wato, T.; Miseki, Y.; Saito, K.; Kudo, A. *J. Am. Chem. Soc.* 2011, 133, 20863-20868.
[62] Wang, W.-N.; An, W.-J.; Ramalingam, B.; Mukherjee, S.; Niedzwiedzki, D. M.; Gangopadhyay, S.; Biswas, P. *J. Am. Chem. Soc.* 2012, 134, 11276-11281.
[63] Varghese, O. K.; Paulose, M.; LaTempa, T. J.; Grimes, C. A. *Nano Lett.* 2009, 9, 731-737.
[64] Cargnello, M.; Diroll, B. T. *Nanoscale* 2014, 6, 97-105.
[65] Linsebigler, A. L.; Lu, G. Q.; Yates, J. T. *Chem. Rev.* 1995, 95, 735-758.
[66] Ran, J.; Zhang, J.; Yu, J.; Jaroniec, M.; Qiao, S. Z. *Chem. Soc. Rev.* 2014.
[67] Carp, O.; Huisman, C. L.; Reller, A. *Prog. Solid State Chem.* 2004, 32, 33-177.
[68] Cernuto, G.; Masciocchi, N.; Cervellino, A.; Colonna, G. M.; Guagliardi, A. *J. Am. Chem. Soc.* 2011, 133, 3114-3119.
[69] Roy, N.; Sohn, Y.; Pradhan, D. *ACS Nano* 2013, 7, 2532-2540.
[70] Pan, J.; Liu, G.; Lu, G. Q.; Cheng, H.-M. *Angew. Chem. Int. Ed.* 2011, 50, 2133-2137.
[71] Zhou, Y.; Tian, Z.; Zhao, Z.; Liu, Q.; Kou, J.; Chen, X.; Gao, J.; Yan, S.; Zou, Z. *ACS Appl. Mater. Interfaces* 2011, 3, 3594-3601.
[72] Liu, Q.; Zhou, Y.; Kou, J.; Chen, X.; Tian, Z.; Gao, J.; Yan, S.; Zou, Z. *J. Am. Chem. Soc.* 2010, 132, 14385-14387.
[73] Yan, S.; Wan, L.; Li, Z.; Zou, Z. *Chem. Commun.* 2011, 47, 5632-5634.
[74] Liu, Q.; Zhou, Y.; Ma, Y.; Zou, Z. *RSC Advances* 2012, 2, 3247-3250.
[75] Shi, H.; Wang, T.; Chen, J.; Zhu, C.; Ye, J.; Zou, Z. *Catal. Lett.* 2011, 141, 525-530.
[76] Zhang, R.; Elzatahry, A. A.; Al-Deyab, S. S.; Zhao, D. *Nano Today* 2012, 7, 344-366.
[77] Ismail, A. A.; Bahnemann, D. W. *J. Mater. Chem.* 2011, 21, 11686-11707.
[78] Zhang, N.; Ouyang, S.; Kako, T.; Ye, J. *Chem. Commun.* 2012, 48, 1269-1271.
[79] Guo, J.; Ouyang, S.; Kako, T.; Ye, J. *Appl. Surf. Sci.* 2013, 280, 418-423.
[80] Xi, G.; Ouyang, S.; Ye, J. *Chem. Eur. J.* 2011, 17, 9057-9061.
[81] Zhao, C.; Liu, L.; Zhang, Q.; Wang, J.; Li, Y. *Catal. Sci. Technol.* 2012, 2, 2558-2568.
[82] Li, Y.; Wang, W.-N.; Zhan, Z.; Woo, M.-H.; Wu, C.-Y.; Biswas, P. *Appl. Catal., B* 2010, 100, 386-392.
[83] Hayden, S. C.; Allam, N. K.; El-Sayed, M. A. *J. Am. Chem. Soc.* 2010, 132, 14406-14408.
[84] Paramasivam, I.; Jha, H.; Liu, N.; Schmuki, P. *Small* 2012, 8, 3073-3103.
[85] Ryu, J.; Lee, S. H.; Nam, D. H.; Park, C. B. *Adv. Mater.* 2011, 23, 1883-1888.
[86] Zhang, X.; Han, F.; Shi, B.; Farsinezhad, S.; Dechaine, G. P.; Shankar, K. *Angew. Chem.* 2012, 124, 12904-12907.
[87] Joo, J. B.; Dahl, M.; Li, N.; Zaera, F.; Yin, Y. *Energy Environ. Sci.* 2013, 6, 2082-2092.
[88] Li, H.; Bian, Z.; Zhu, J.; Zhang, D.; Li, G.; Huo, Y.; Li, H.; Lu, Y. *J. Am. Chem. Soc.* 2007, 129, 8406-8407.
[89] Dinh, C.-T.; Yen, H.; Kleitz, F.; Do, T.-O. *Angew. Chem. Int. Ed.* 2014, 53, 6618-6623.
[90] Hu, J.; Chen, M.; Fang, X.; Wu, L. *Chem. Soc. Rev.* 2011, 40, 5472-5491.
[91] Tu, W.; Zhou, Y.; Liu, Q.; Tian, Z.; Gao, J.; Chen, X.; Zhang, H.; Liu, J.; Zou, Z. *Adv. Funct. Mater.* 2012, 22, 1215-1221.
[92] In, S. I.; Vaughn, D. D.; Schaak, R. E. *Angew. Chem. Int. Ed.* 2012, 51, 3915-3918.

[93] Cheng, H. F.; Huang, B. B.; Liu, Y. Y.; Wang, Z. Y.; Qin, X. Y.; Zhang, X. Y.; Dai, Y. *Chem. Commun.* 2012, 48, 9729-9731.

[94] Yu, J.; Low, J.; Xiao, W.; Zhou, P.; Jaroniec, M. *J. Am. Chem. Soc.* 2014.

[95] Feng, S.; Chen, X.; Zhou, Y.; Tu, W.; Li, P.; Li, H.; Zou, Z. *Nanoscale* 2014, 6, 1896-1900.

[96] Yan, S.; Wang, J.; Gao, H.; Wang, N.; Yu, H.; Li, Z.; Zhou, Y.; Zou, Z. *Adv. Funct. Mater.* 2013, 23, 758-763.

[97] Li, X. K.; Pan, H. Q.; Li, W.; Zhuang, Z. J. *Appl. Catal., A* 2012, 413, 103-108.

[98] Xu, H.; Ouyang, S.; Li, P.; Kako, T.; Ye, J. *ACS Appl. Mater. Interfaces* 2013, 5, 1348-1354.

[99] Pan, J.; Wu, X.; Wang, L.; Liu, G.; Lu, G. Q.; Cheng, H.-M. *Chem. Commun.* 2011, 47, 8361-8363.

[100] Zhang, N.; Ouyang, S.; Li, P.; Zhang, Y.; Xi, G.; Kako, T.; Ye, J. *Chem. Commun.* 2011, 47, 2041-2043.

[101] Yan, S. C.; Ouyang, S. X.; Gao, J.; Yang, M.; Feng, J. Y.; Fan, X. X.; Wan, L. J.; Li, Z. S.; Ye, J. H.; Zhou, Y.; Zou, Z. G. *Angew. Chem. Int. Ed.* 2010, 49, 6400-6404.

[102] Li, X.; Liu, H.; Luo, D.; Li, J.; Huang, Y.; Li, H.; Fang, Y.; Xu, Y.; Zhu, L. *Chem. Eng. J.* 2012, 180, 151-158.

In: Comprehensive Guide for Mesoporous Materials, Volume 4 ISBN: 978-1-63482-092-9
Editor: Mahmood Aliofkhazraei

Chapter 5

FUNCTIONALIZED MESOPOROUS SILICA NANOPARTICLES FOR DRUG DELIVERY

Pablo Botella* and Carlos Muniesa
Instituto de Tecnología Química (UPV-CSIC), Valencia, Spain

ABSTRACT

Functionalized mesoporous silica nanoparticles (FMSNs) have been proposed as efficient drug delivery systems, able to target a large quantity of a biologically active molecule to specific tissues and cells with no effect on other cell types, avoiding side effects that many times limit their therapeutic application. In this work, we review the main features of a new generation of FMSN-based nanodrugs, with specific examples in preclinical research. Here, the control of particle size, morphology and surface properties is crucial to ensure the stability in biological fluids, as well as a progressive and sustained release of the delivered compound, allowing monitoring of the therapeutic agent diffusion. Furthermore, the incorporation of stimuli-responsive triggers in nanoparticles promotes specific intracellular release in target cells, which improves therapeutic activity and minimizes possible side effects.

1. INTRODUCTION

Since the discovery of surfactant-templated synthesis of mesoporous silica materials [1], a lot of work has been carried out regarding the functionalization and use of these materials for catalysis [2], separation [3] and sensors [4]. However, it was not until the beginning of the last decade that a MCM-41-type material was first reported as drug delivery system (DDS) [5]. In recent years, mesoporous silica materials have been considered to be excellent candidates for the preparation of DDSs. Firstly, their textural properties favor loading of significant quantities of therapeutic molecules within the pore channels. Secondly, it is possible to obtain nanoparticles of these materials, which results in an accurate control over

* E-mail: pbotella@itq.upv.es

the pore structure, particle shape and size [6, 7]. And finally, the silanol-containing surface can be easily functionalized, introducing additional features that allow imposing a stimuli-responsive controlled drug release [8]. Furthermore, biocompatibility test results have shown the potential of these materials in the preparation of pharmaceutical forms that might find application in theranostics and personalized medicine [9]. To date, many models of nanodrugs have been studied for adsorption and release in modified mesoporous silica-based DDSs, some of the most significant are reported in Table 1. In this review, we intend to provide an overview of the recent research progress on the preparation and application of novel mesoporous silica-based nanodrugs with abilities to target and track therapeutic molecules to the targeted cells or tissues in a tailored release manner to enhance drug efficiency and reduce toxicity.

Table 1. Examples of DDSs proposed models based on FMSNs

Nanoparticles	Drug	Loading	Release	Cell Line	Ref.
PEG-modified MCM-41	Doxorubicin	Adsorption	Diffusion (Fick)	HeLa	10
Phosphonate-modified MCM-41	Camptothecin	Adsorption	Diffusion (Fick)	PANC-1	11
Cyclodextrin-modified MCM-41	Doxorubicin	Adsorption	Diffusion/ Glutathione	A549	12
PEG-modified MCM-41	Methotrexate	Adsorption	Diffusion/ Glutathione	MCF-7	13
PEG-modified MCM-41	Doxorubicin	Adsorption	Diffusion/pH	BALB	14
PEG-dAC-modified MCM-41	Doxorubicin	Adsorption	Diffussion/Heat	NCr	15
Azobenzene-modified MCM-41	Doxorubicin	Adsorption	Photoactivation	HeLa	16
Glucidex-modified MCM-41	Doxorubicin	Gating	β-D-galactosidase	HeLa LLC-PK1	17
CdS-capped MCM-41	Vancomycin	Gating	DTT	Astrocytes	18
APTMS-modified MCM-41	Paclitaxel	Disulfide bond	Glutathione	HeLa	19
MPTES-modified MCM-41	Cysteine	Disulfide bond	Glutathione	HuH7	20
MPTES-modified SiO_2@MS	Camptothecin	Disulfide bond	Glutathione	HeLa	21
MPTES-modified MCM-41	Cysteine	Disulfide bond	NADH	HeLa	22
PEG-modified MCM-41	PdTPP	Amine bond	Photoactivation	MCF-7	23

PEG: polyethylene glycol; DTT: dithiothreitol; APTES: 3-aminopropyltriethoxysilane; MPTES: 3-mercaptopropyltriethoxysilane; dAC: diacrylate; MS: mesoporous silica; PdTPP: Pd-porphyrin.

2. Drug Delivery Systems

The use of mesoporous silica nanoparticles (MSNs) in drug delivery was formerly based on the physical and textural properties of the conventional materials, such as MCM-41, MCM-48 and SBA-15, with empty mesopore channels to absorb relatively large amounts of bioactive molecules. Here, several groups have systematically studied the influence of pore diameter, pore structure, surface area and pore volume on drug loading and release rate [24, 25]. It has been shown that the decrease in pore diameter leads to a decrease in drug loading quantity and release rate. At the same time, the pore structure type in terms of pore connectivity may condition the diffusion process and, in this sense, a one-dimensional pore structure with cage-like pores is the most promising pore geometry for providing high drug loading and slow drug release. Additionally, both pore volume and surface area favor the incorporation of drug molecules within the mesoporous structure.

The incorporation of drugs in MSNs takes place through non-covalent interactions, such as hydrogen bonding, physical adsorption, electrostatic interaction and p-p stacking [26]. Unfortunately, in most cases, these kinds of interactions are very weak, and some or total premature release of the cargo may occur before reaching the destination. The premature release problem not only limits the use of a DDS for effective therapy, but also poses a major challenge on possible side effects that may be related to the activity of the active principle outside the targeted cells or tissue. In this sense, surface functionalization of MSNs with appropriate organic groups allows for the incorporation of therapeutic molecules by more stable interacting forces, such as ionic bond and covalent bond. These functionalized mesoporous silica nanoparticles (FMSNs) are highly stable DDSs, able to deliver the drug with no leaking before reaching the designated site of cells or tissue. Furthermore, the surface of MSNs can be engineered with specific organic moieties for the development of stimuli-responsive systems, that is, delivery nanostructures that release their cargo under the action of a specific stimulus [27]. When used as drug carriers, these stimuli-responsive FMSNs are good candidates for strong therapeutic activity with no toxicity effects.

3. Drug Loading

The incorporation of the therapeutic molecules inside the pores and channels of the functionalized mesoporous materials may take place through different mechanisms, essentially, pore adsorption, encapsulation within the micelles of the structure directing agent (SDA), covalent bonding and gate-keeping.

3.1. Adsorption

The modification of the mesoporous silica surface through organic groups provides numerous possibilities to control drug adsorption and release. Mesoporous silica shows a high density of silanol groups that may be reacted by co-condensation of a tetraalkoxysilane with one or more organoalkoxysilane, obtaining an inorganic-organic hybrid network [10-15]. Alternatively, MSNs can be grafted by sililation on free and geminal silanol groups [28]. Organic moieties are able to link to drug molecules through intermolecular forces but especially through an ionic bond. A very well-known example of this is the adsorption of ibuprofen on amine functionalized MCM-41 and SBA-15 matrices [29, 30]. Figure 1 shows, schematically, drug molecule assemble on the organic-inorganic hybrid. The ionic interaction between the carboxyl groups in ibuprofen and the amino groups of the matrix surface allows the release rate of ibuprofen to be accurately controlled. However, the delivery of large bioactive molecules inside the pores of regular MSNs (~4 nm) is strongly hindered by the limited diameter of the internal channels and cavities. We overcame this issue by synthesizing MSNs with very high pore volume and very large pore entrance size (up to 17 nm) and cavity diameter (up to 20 nm) [7]. Figure 2 illustrates a small particle showing the local arrangement of the ultra large pores in a face-cubic centered (fcc) symmetry. After surface modification with aminopropyl groups, DNA molecules could be adsorbed by ionic interaction between the phosphate moieties of the nucleic acid and the amine groups of the modified silica mesophase. The stability of these DNA-MSN conjugates offered additional protection for enzymatic degradation during the intracellular trafficking, and imposed an efficient, controlled release of the nucleic acid.

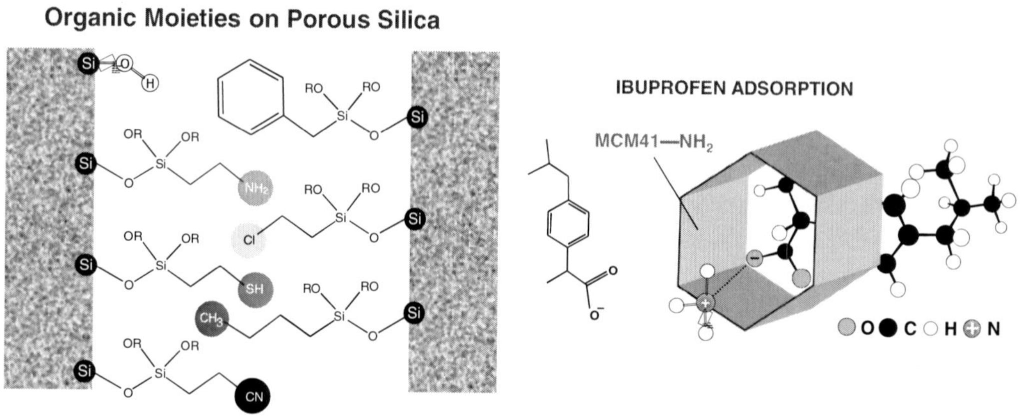

Figure 1. Artistic representation of mesoporous silica surface functionalization with organic moieties by reaction with alcoxysilane molecules. High yield ibuprofen adsorption can be carried out by ionic bonding on amine derivatized silica surface (MCM41-NH_2).

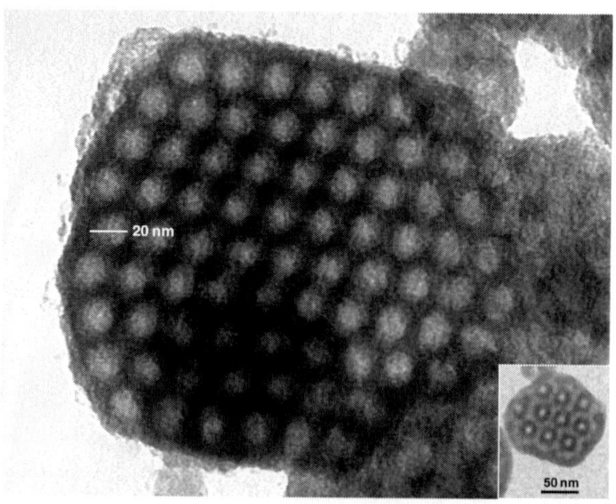

Figure 2. Detail of an ultra large pore mesoporous silica nanoparticle showing the dimensions of its ordered internal cavities; the inset shows the local arrangement of fcc symmetry in a small particle. Reprinted with permission from ref. 7. Copyright 2009, American Chemical Society.

3.2. Encapsulaption in Micelles of the Structure Directing Agent

A different approach to modulate the incorporation and release of a drug form in a mesoporous matrix under physiological conditions is to encapsulate the delivered molecule within the micelles of the organic SDA, and then make the inorganic structure grow over the loaded micelles. Then, the drug release process will be surfactant-assisted, as the discharge of the therapeutic compound depends on the dissolution of the surfactant micelles. Figure 3 depicts the nanocomposite synthesis process, drug incorporation and controlled release. A few reports describe this process for mesoporous silica materials prepared with the cationic surfactant cetyltrimethylammonium [31-34], although its application is limited by the high toxicity of this template [34, 35]. It is more feasible to use an anionic surfactant, as a phosphate monoester anionic surfactant, for the encapsulation of resveratrol in MSNs [36]. Also, we obtained good results by using non-toxic, biocompatible, non-ionic surfactants as alkyl-glucosides. After strictly tailoring the synthetic conditions, it was possible to obtain ordered mesostructures with some disaccharide surfactants, as *n*-decyl-β-D-maltoside and *n*-dodecyl-β-D-maltoside [37, 38]. The incorporation of ibuprofen within the template micelles in the synthesis gel and further condensation of silica led to a surfactant-assisted delivery system for small therapeutic molecules, which introduced improved kinetics in drug release [39].

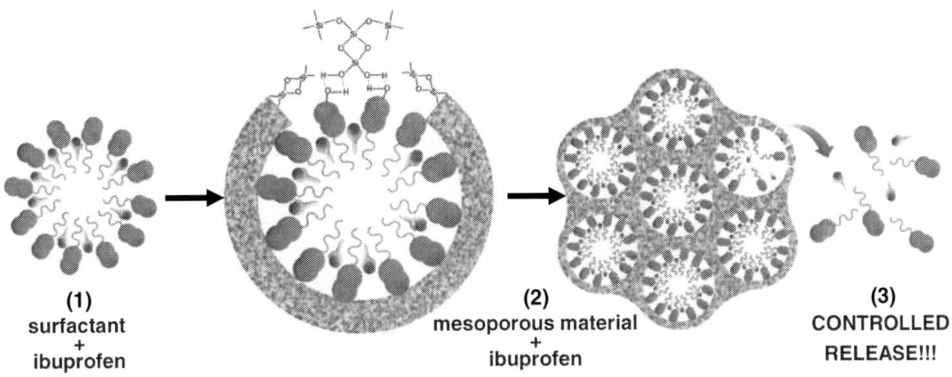

Figure 3. Schematic representation of the alkyl maltoside–silica composites synthesis process, ibuprofen incorporation within surfactant micelles and further drug release under dissolution of SDA micelles: (1) drug loading in surfactant micelles; (2) silica condensation over surfactant micelles and further growth of silica wall; (3) controlled release of ibuprofen after dissolution of SDA micelles. Reproduced by permission of The Royal Society of Chemistry from ref. 38.

3.3. Covalent Bond

Covalent linking of drug molecules on the external surface of MSNs is the most stable and safe way of delivery with no premature release. This is especially relevant in the case of the administration of cytotoxic drugs, like antitumorals. Unfortunately, the direct coupling of many of these molecules to an inorganic nanoparticle is circumvented by the reduced therapeutic activity of the structure-modified derivatives [40]. Therefore, it is desirable to develop covalent linking models that allow for the controllable release of the free drug after cell entry [41]. Secondly, it is necessary to state precisely when, where and how the pharmaceutical payload is released into the cells. In this sense, FMSNs have been proposed to incorporate drugs covalently linked by disulfide bridges [19-22], amide bonds [23] or ester bonds [42]. Most of these systems are sensitive to an external stimulus able to cleavage the covalent bond. Figure 4 presents a covalent linking model for the release of camptothecin (CPT) which can be activated by intracellular reducing compounds.

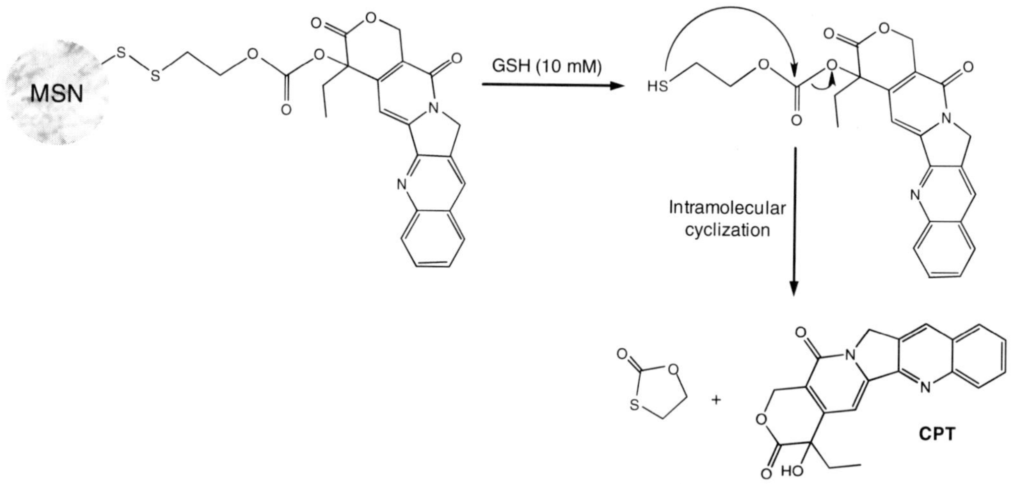

Figure 4. Redox sensitive release mechanism of camptothecin (CPT) incorporated on MSN surface under the action of intracellular glutathione (GSH).

3.4. Gate-Keeping

The regular and short range ordered porous systems of MSNs introduce a singular delivery property, as the active payload can be incorporated within the mesoporous channels before capping them with a variety of entities working as "gate-keepers". In this sense, confined guests release will be regulated by the uncapping process, usually mediated by an external stimulus. This type of DDSs with no premature release performance is particularly useful if the cargo to be delivered is toxic. Figure 5 represents a MSN loaded with guest molecules and end-capped with a general gate-keeper. The nanoparticles used as gate-keepers have been CdS, Fe_3O_4 or gold, which can be chemically attached to FMSNs through a disulfide linker. The activity of intracellular reducing agents unlocks the pores and allows release [18, 43]. Gating may also be carried out by grafting on the mesoporous wall organic moieties which are prone to suffer an oligomerization reversible process that may open/close the pores of MCM-41 [44]. A third option to gate-keep bioactive molecules inside mesoporous silica channels is to lock them with supramolecular assemblies. This combination has resulted in novel organic/inorganic hybrid materials with improved functionalities. A representative example is the system formed by a dendrimer tethered on FMSNs [17, 45]. Here, drug release may happen by structural changes provoked under certain stimuli. Next section describes the physical, chemical and biological stimuli that promote the release of the therapeutic molecule by breakage of its union to the inorganic nanoparticles.

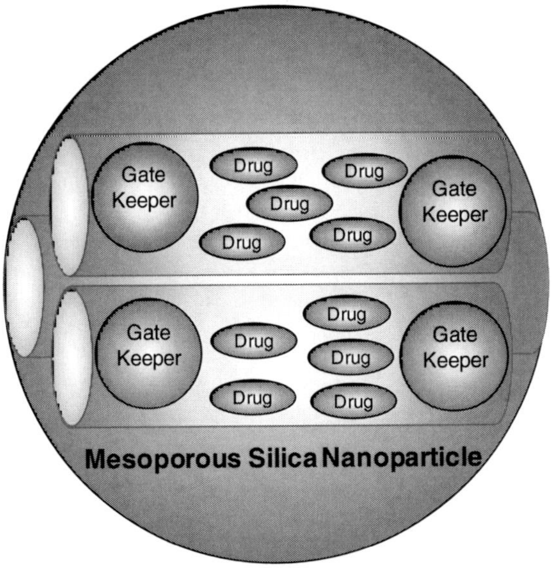

Figure 5. Schematic representation of the MSN loaded with guest molecules and end-capped with a general gate-keeper.

4. Stimuli-Responsive Systems

If the incorporation of the drug within the mesoporous channels determines the amount and stability of the cargo, the release mechanism is responsible for the discharge of the therapeutic molecule when reaching the targeted cells of tissues with no premature leaking. Among the different models proposed to achieve specific intracellular release, the most significant are those depending on pH change, redox reactions, vis/NIR light-induced bond cleavage and the activity of some cytosolic enzymes. Some of the most significant proposed stimuli-responsive systems are presented in Table 2. Nevertheless, other external stimuli may be useful to activate a specific release mechanism, e.g., ultrasounds, temperature and magnetic fields [66].

Table 2. FMSN-based stimuli responsive systems for intracellular drug delivery

Drug loading	Release System	Stimulus	Release Mechanism	Ref.
Gate-keeping	Rotaxanes and pseudorotaxanes	Lysosomal pH	Amines protonation and dis-sociation of cyclodextrins	46-50
Gate-keeping	Ionizable linear polyamines	Lysosomal pH	Amines protonation	17, 51, 52
Gate-keeping	Pseudorotaxanes	Lysosomal pH	Amines protonation and dis-sociation of cyclodextrins	53
Gate-keeping	Polyelectrolyte multilayers	Lysosomal pH	Sulfonate and amines protonation	54,55
Gate-keeping	CdS/Fe_3O_4 nanoparticles	Glutathione	Disulfide reduction and dissociation of nanoparticles	18,43
Gate-keeping	Poly(amido amine) dendrimer	Glutathione	Disulfide reduction and dissociation of dendrimers	45

Table 2. (Continued)

Drug loading	Release System	Stimulus	Release Mechanism	Ref.
Covalent bond	Self-immolative ligand	Glutathione	Disulfide reduction and intra-molecular cyclization	21
Gate-keeping	Cyclobutane coumarin dimmer	UV-light	Coumarin dimmerization	44
Gate-keeping	Azobenzene derivative ligand	UV-light	Trans-cis photoisomerization of azobenzene	16, 56-59
Gate-keeping	Avidin molecule	Trypsin	Hydrolysis of avidin molecules	60
Gate-keeping	Saccharide ligand	β-D-galactosidase	Saccharide hydrolysis	17, 61
Gate-keeping	Organosilica coating	Esterases	Ester bond hydrolysis	62
Gate-keeping	Peptide Fmoc-derivative	Thermolysis/ elastase	Degradation of peptide molecules	63
Gate-keeping	Cyclodextrin derivative	α-amylase	Degradation of cyclodextrins	64
Gate-keeping	Cucurbituril derivative	Lysine decarboxylase	Lysine decarboxylation and dissociation of cucurbituril	65

4.1. pH-Responsive Systems

Intracellular pH-responsive DDSs are based on the acidic conditions of lysosomes (pH~5-6.5), which may be used for the selective scission of drug-carrier linking bonds [67]. Furthermore, because the extracellular pH of tumor tissue is usually lower than that of normal tissues, it makes sense to design DDSs sensitive to local biological pH. Zink and co-workers developed several pH-responsive FMSNs by using pH-dependent pseudorotaxanes, rotaxanes and other analogues. In these systems, aromatic amine stalks were immobilized on the surface of MSNs, and β-cyclodextrin movable gates were introduced to encircle the stalks and control the flow of the drug loaded in the pore channels [46-50]. These β-cyclodextrin supramolecular gates encircled aromatic amine stalks as a result of non-covalent bonding interaction under neutral pH conditions, and effectively blocked the nanopore openings and trapped the included molecules. By decreasing the pH mildly, protonation of the aromatic amines took place followed by dissociation of β-cyclodextrin caps, opening of the nanovalve and drug release. Martinez-Mañez and co-workers modified pore outlets of MCM-41 by covalent bonding of ionizable linear polyamines [17, 51, 52], resulting in pH-controlled gate-like effects. At physiological pH, polyamines were mostly unprotonated and chemical doors remained closed due to hydrogen-bonding interactions between amine groups. However, at acidic pH, Coulombic repulsions in closely located polyammoniums caused the opening of these molecular gates, allowing for the diffusion of entrapped molecules. Similar models to those described have also been proposed by Kim et al. [53], and Shi et al. [54, 55].

4.2. Redox-Responsive Systems

Redox-sensitive, disulfide-based mechanisms have shown significant efficiency for selective intracellular release, based mainly on glutathione-driven cleavage [68, 69]. With a

10 mM cytosolic concentration, this tripeptide is able to cleavage disulfide bridges. Therefore, gate-keeping systems and structural derivatives of therapeutic molecules may be formulated in order to incorporate a disulfide linker in their structures. In the first case, the work of Lin and co-workers is remarkable, as they developed a series of redox-responsive FMSNs by using different gate-keepers to cap the pores, as nanoparticles (CdS, Fe_3O_4) [18, 43], or supramolecular assemblies (poly(amido amine) dendrimers) [45]. Briefly, MSNs surface modified with mercaptopropyl groups tethered the gate-keeping system at the pores mouth by a disulfide bridge-containing organic linker (e.g., 2-(propyldisulfanil) ethylamine). In the second case, the direct coupling of the drug molecule to the inorganic support must be carried out with no structural modification, as this would involve a serious drop of the pharmacological activity. To address these targets, we developed a novel nanoplatform of CPT based in a mercapto-functionalized silica hybrid, containing a non-porous core and a mesoporous shell [21]. A new synthetic strategy was implemented by reacting novel (pyridin-2-yldisulfanyl)alkyl carbonate derivatives of CPT with thiol groups of the inorganic support. This allowed us to attach the drug over the pores of the external coating by a cleavable disulfide linker, sensitive to thiols, such as glutathione, which was able to release the naked molecules after an intramolecular cyclization mechanism. Figure 4 showed this unique release mechanism under the action of glutathione.

4.3. Light-Sensitive Systems

Photoactivated FMSNs are able to release a drug under external control at a specific time and location, due to presence on the pores of a photoactivable moving component, sensitive to vis/NIR light. In this sense, Tanaka and co-workers designed a photo-triggered reversible DDS based on coumarin-modified MCM-41 [44]. Under UV irradiation of 310 nm, coumarin suffered a photodimerization reaction yielding cyclobutane coumarin dimmers that obstructed the entrances to the MCM-41 pores. These chemical doors could be reversibly opened if the wavelength of UV irradiation was changed to 250 nm. Furthermore, Brinker and co-workers grafted photoactivated moving parts based on the photoisomerization of azobenzene derivatives on mesoporous silica [16, 56-58]. Decrease in size of *trans* to *cis* isomers of azobenzene molecules attached to pore interiors was used to regulate the transport of molecules through the pores. Under continuous illumination at 413 nm, a constant trans–cis photoisomerization about the N—N bond caused dynamic back and forth wagging motion of the azobenzene derivatives, working as a molecular impeller that regulates the release of molecules through and out of the mesopores. By using this concept, Zink et al., showed that after loading the antitumoral drug CPT in these light-sensitive FMSNs, it was only released inside the cancer cells illuminated at the specific wavelength that activates the impellers (413 nm) [59]. Furthermore, very recently Shi et al., have made this system sensitive to non-hazardous NIR irradiation by protecting upconverting nanoparticles (e.g., NaYF4:TmYb) with a mesoporous silica coating containing the azobenzene derivatives [16]. Upon absorption of NIR light (980 nm), the nanoparticles emit photons in the UV/Vis region, which can be absorbed immediately by the photoresponsive azo molecules in the pore network of the mesoporous silica layer.

4.4. Enzyme-Sensitive Systems

Intracellular enzime-responsive models are the most representative among bioestimuli-responsive DDSs. Most of these releasing enzymes are located in endo-lysosomes and cytosol. Here, Bein et al., attached caps to the biotinylated colloid mesoporous silica surface, closing the pores [60]. The addition of protease trypsin caused the hydrolysis of the attached protein avidin and enabled the release of the entrapped guest. Also, Martinez-Mañez et al., functionalized the outlet of mesoporous silica pores with several saccharides [17, 61]. Then, the addition of β-D-galactosidase enzyme induced the uncapping of the mesopores through selective hydrolysis of the saccharides anchored on external surface, allowing for the discharge of transported molecules. In a different example, our group developed a hybrid organic–inorganic nanosphere composed of a purely organic internal liposomal phase in which bioactive molecules were encapsulated [62]. Covering this part, they constructed an external self-assembled organic–inorganic shell comprising covalently connected organic fragments and silica units. This organosilica shell could stabilize the internal liposomal phase and, consequently, isolate and protect the drug molecules. By properly selecting the organic counterpart of the organic–inorganic shell, the spheres responded to cytosolic esterases. Through this biological interaction, the organic moieties of the external organic–inorganic shell were broken, allowing the drug to be released. In other similar systems, peptides, cyclodextrins and cucurbituril attached to the functionalized mesoporous silica led to tight blocking of mesopores; introduction of the corresponding enzyme, resulted in the discharge of the guest molecules [63-65].

5. Multifunctional Materials

In certain applications, pure mesoporous silica carriers may be insufficient to achieve the expected efficiency in a specific diagnostic or therapy. At this point, some smart combinations of different functional materials with MSNs can lead to multifunctional medical nanoplatforms for simultaneous drug delivery, diagnosis and therapy. Here we propose two types of novel hybrid nanomaterials based on MSNs: magnetic nanoparticles and plasmonic nanoparticles. Some of the most representative examples of these multifunctional MSN-based materials are summarized in Table 3.

Table 3. Examples of hybrid multifunctional materials based on FMSNs

Material	Structure	Drug	Loading	Application	Ref.
Fe_3O_4@MSN	Core-Shell	None	None	MRI	70
Fe_3O_4@MSN	Core-Shell	Doxorubicin	Adsorption	MRI + FI + DDS	71
Fe_3O_4/MSN	Hybrid nanoparticles	Doxorubicin	Adsorption	MRI + DDS	72
Fe_3O_4@MSN	Rattle core@Shell	Docetaxel	Adsorption	MRI + DDS	73, 74
GNP@MSN	Nanocluster@Shell	Camptothecin	Adsorption	PT + DDS	75-77
GNR@MSN	Nanorod@Shell	Doxorubicin	Adsorption	PT + DDS	78

MRI: magnetic resonance imaging; FI: fluorescence imaging; GNP: gold nanoparticle; GNR: gold nanorod; PT: photothermal therapy.

5.1. Magnetic Nanoparticles

Magnetic targeting provides the ability to guide DDSs to the desired locations, hold them until the therapy is complete and remove them by means of an external magnetic field. This targeting ability is worthy since many drugs, especially anticancer drugs, have several toxic side effects. DDSs may carry the drugs and be guided to the targeted organs or locations inside the body, facilitating therapeutic efficiency and reducing the damage of normal organs due to drug toxicity [79]. In addition, superparamagnetic nanoparticles can deliver certain amount of heat under alternating magnetic fields, as a result of the Brownian rotation and Néel relaxation mechanisms. Resulting hyperthermia may be applied to cancer cells destruction [80]. Besides this, Fe_3O_4 nanocrystals/mesoporous silica hybrids are able to generate magnetic resonance contrast on transverse proton relaxation-time weighted sequence (T_1, T_2) due to the dipolar interaction of magnetic moments between nanoparticles and the proton in water [81]. Therefore, these multifunctional materials may also be used as magnetic resonance imaging (MRI) contrast agents.

Among different magnetic materials, Fe_2O_3 and Fe_3O_4 have been the most intensively studied. However, iron oxide nanoparticles show a strong trend to aggregate and collapse when exposed to a physiological medium. In this sense, the incorporation of iron oxide nanoparticules in a mesoporous silica matrix brings colloidal stability and some additional functionality, as the porous structure may also incorporate bioactive molecules. Here, several types of iron oxide-based magnetic mesoporus silica systems have been described. Firstly, core-shell structure composites can be obtained by depositing a mesoporous silica layer on iron nanoparticles. This architecture provides the magnetic composites with extremely high magnetization [70], whereas the outer mesoporous silica shell supplies enough surface area and pore volume for drug storage and release. Hyeon and co-workers synthesized discrete and monodisperse nanoparticles consisting of a single Fe_3O_4 nanocrystal core and a mesoporous silica shell [71]. This system showed integrated capability to be used as MRI and fluorescence imaging agent, along with potential use as drug delivery vehicle. The field dependent magnetism of hybrid particles at 300 K showed no hysteresis, indicating that they were superparamagnetic, which is a desirable characteristic for T_2 magnetic resonance contrast agents [82]. Moreover, by incorporation of doxorubicin (DOX) in the mesoporous coating, these nanoparticles showed high cytotoxic efficacy on SK-BR-3 cells.

Another possible hybrid model can be obtained by embedding iron oxide nanoparticles into mesoporous silica matrices. Lin et al., developed fibrous-structure mesoporous silica microspheres where 9 nm-Fe_3O_4 nanocrystals were trapped. This hybrid material showed a sustained DOX release profile, sufficient magnetic responsivity and redispersibility to the external magnetic field [72]. This system delivered DOX through the cell membrane and released the drug into cytosol, passing through the nucleus membrane and eventually assembling in the nucleus to kill cells.

Alternatively, magnetic nanoparticles can also be encapsulated as rattle cores inside hollow spheres with large interior spaces and mesoporous silica shell. Here, Shi and co-workers prepared a series of rattle-type hollow ellipsoids/spheres, with Fe_3O_4/Fe_2O_3 core and a single or double mesoporous silica shell [73, 74]. They proved that the saturation magnetization value of rattle-type hollow structure was clearly higher than that of the corresponding core-shell structure with an intact middle silica layer, due to the removal of the in-between silica wall. Thereby, the rattle-type structure was more suitable for drug delivery, as its loading ability of, for instance, DOX, improves about 20% due to the enhanced surface area and pore volume in the hollow interior. The magnetic Fe_3O_4 core could be used as MRI contrast (T_2) agent during the chemotherapeutic process, to monitor therapeutic efficiency and provide high-quality images for operators. Moreover, DOX-loaded material presented higher cytotoxicity than free drug over MCF-7 cell line, indicating that these iron oxide rattles-mesoporous silica capsules were excellent anticancer drug carriers [74].

5.2. Plasmonic Nanoparticles

Non-invasive photothermal therapy uses deep penetrating NIR lasers emitting at 650–900 nm, spectral range where hemoglobin and water, the major absorbers of visible and infrared light, have the lowest absorption coefficient [83]. Photothermia presents important benefits, such as its simplicity and ability to treat tumors that cannot be accessed by surgery. In this sense, if an active component capable of absorbing NIR light is located in the cancer cells, the photon energy absorbed efficiently turns into heat and, therefore, selective hyperthermia occurs only in the malign cells [84]. Given their compatibility, tunable light absorption and strong optical absorption in the vis/NIR region, Au nanostructures have been proposed as photothermal agents in cancer therapy [84, 85]. In this context, plasmonic materials with specific optical properties can be achieved by adjusting their shape and size, as well as controlling the assembly of gold nanoparticles [86, 87]. Here, the formation of nanoclusters of gold nanoparticles shifts the absorption to near-infrared, which can be efficiently used for photothermal cancer therapy, as a result of the collective effects associated with the assembled structures [88]. However, most systems based on naked gold nanoparticles fail during the *in vivo* application due to low stability and nonspecific aggregation in biological fluids, which makes it necessary to protect the nanoparticles or their clusters [89]. With this in mind, we developed a novel type of Au-mesoporous silica hybrid by encapsulation of gold nanoclusters in a biocompatible and multifunctional mesoporous silica scaffold. We controlled the assembly of gold nanoparticles with a homogeneous thin monolayer of amorphous silica into clusters of well defined size by tuning the pH of the colloid, e.g., through modification of the alkali concentration [75, 76]. Large hydrodynamic diameters and close packing of metal cores involved maximum red-shift absorbance and a constant tailing absorption in the NIR region due to optical coupling of their surface plasmon bands. However, due to the low stability of these nanoclusters in physiological fluids, it was convenient to protect them with a biocompatible and functionalizable mesoporous silica coating. As a consequence, these Au nanocluster/mesoporous silica particles were able to produce effective photothermolysis when internalized in 42-MG-BA human glioma cells and irradiated with a femtosecond pulse laser (790 nm) [77]. Moreover, the incorporation of a small quantity of CPT within the pores of the silica shell led to a multifunctional material showing combined effect between the cytotoxicity of the delivered drug and the photothermal activity of gold nanoclusters, providing almost double efficacy in the destruction of cancer cells than the single laser irradiation [76]. Similar effect has been reported for DOX-loaded mesoporous silica encapsulated gold nanorods (GNR@MSN) [78]. Significantly greater cell killing was observed when A549 cells incubated with drug-charged GNR@MSN were irradiated with NIR illumination, attributable to both GNR@MSN-mediated photothermal ablation and cytotoxicity of light-triggered DOX release.

6. *In Vivo* Studies

One of most significant issues in the therapeutics with nanodrugs is to achieve sufficiently high local concentration in the pathologic tissue, while sparing healthy tissue. The final therapeutic index is determined by biodistribution, metabolism and clearance of nanoparticulated DDS. In the specific case of cancer treatment, systematically administered FMSNs are usually quickly removed from plasma and delivered into all major organs, such as liver, lung and spleen, while drug accumulation in tumor is in comparison rather low. In this sense, hepatic, splenic and pulmonary accumulations are largely attributed to cells of the reticuloendothelial system (RES) found in these tissues [42, 90]. These cells are responsible for capturing blood circulating particles leading to a fast plasma clearance. In addition, the accumulation of nanoparticles in lung at shorter time points may be also due to initial aggregation of nanoparticles and their first-pass retention in lung capillaries, although dissociation occurs rapidly and particles subsequently redistribute mainly to liver and spleen. At this point, particles are expected to follow excretion via the hepatobiliary route without causing pathological damage to the targeted organs [91, 92]. Furthermore, renal clearance has been mentioned as a major route for excretion of small silica-based nanoparticles (lower than 50 nm) [93, 94], whereas larger particles are able to escape glomerular filtration and hold a longer plasma half-life, but are also prone to be uptaken by the RES system. Therefore, accumulation of FMSNs in the tumor compartment competes against uptake and clearance by monophagocytic cells. It has been long established that nanoparticles could selectively accumulate in tumor tissue via the enhanced permeation and retention (EPR) effect [95]. This passive targeting is due to the defective leaky vasculature and reduced lymphatic drainage systems of most solid tumors. Two main factors influence the EPR effect and tumor targeting. On the one hand, the physicochemical properties of nanoparticles such as particle size and surface charge, that can influence their pharmacokinetic behavior and their tumor accumulation [96]. On the other hand, the tumor characteristics, such as cancer type, stage of disease, tumor size, site of implantation and host species may also impact the tumor targeting properties of nanoparticles, as well as their tumor permeation [97]. Here, we carried out a biodistribution, therapeutic efficacy and biocompatibility evaluation after intravenous administration of CPT-loaded surface-modified silica nanoparticles (SNPs) in human colorectal cancer xenografts [96]. *In vivo* tumor-accumulation and whole-body tissue distribution were carried out based on the acquisition of fluorescence emission of a fluorophore (Cy5.5) conjugated to the CPT-loaded SNPs, as well as by HPLC quantification of tissue CPT levels. The results showed that, although these nanoparticles tended to accumulate in organs of the RES system, nanoparticles boost CPT concentration in tumor *vs* administration of the free drug. Accordingly, CPT-loaded SNP treatment delayed the growth of subcutaneous tumors while significantly reducing the systemic toxicity associated to CPT administration. Figure 6 illustrates the process of tumor growth delay under CPT-loaded SNP treatment.

To improve accumulation in tumor of CPT-loaded FMSNs, Zink and co-workers incorporated folic acid on the mesoporous surface as targeting molecule (F-FMSN) [98]. Mice bearing subcutaneous tumors were injected via the tail vein with fluorescein isothiocyanate-labeled FMSNs or F-FMSNs suspended in saline solution. After euthanasia, the tumor, liver, kidney, lung, heart, intestine, stomach, mesentery and spleen were collected and particle concentration in each organ was quantified by ICP-MS of Si content. Both FMSNs and F-FMSNs accumulated preferentially in tumor, but the gap, compared to other organs, was especially significant at 24 h in the case of the folic-targeted material. These results clearly demonstrated the preferential particle accumulation of FMSNs in tumor xenograft of mice, probably due to EPR effect, and folate-receptor-targeting conjugation further promoted this tumor-preferential accumulation. Figure 7 depicts the ICS-MS quantitative measurement of Si concentration in each organ of mice with xenograft tumors.

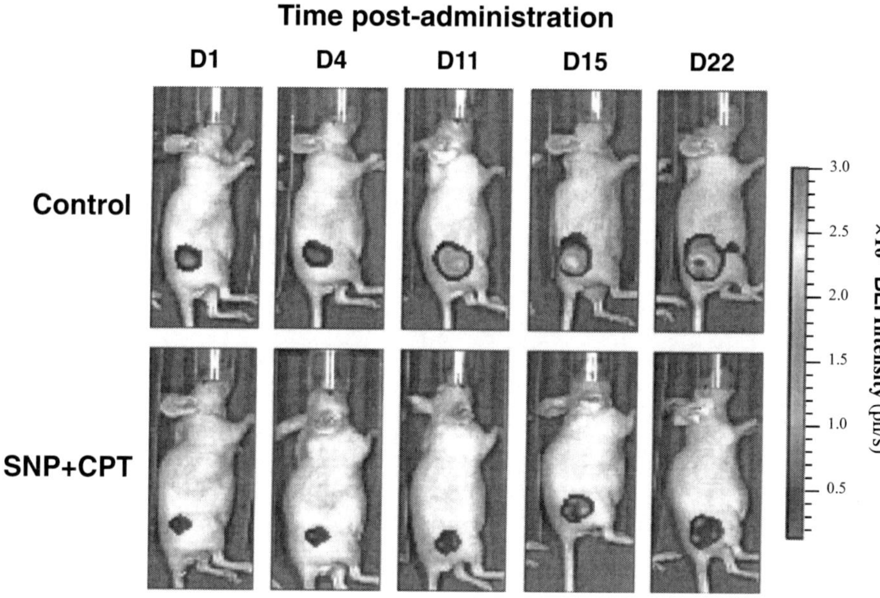

Figure 6. Tumor growth delay in athymic nude mice treated with CPT-loaded SNP in a 22-day experiment. Localized subcutaneous tumor growth (HT-29. Fluc colorectal cancer cells) can be observed by *in vivo* bioluminescence (BLI) monitoring of a representative mouse from each treatment group. Control group (receiving no drug) showed a significantly faster tumor growth from the beginning of the experiment.

Particle biodistribution and plasma clearance may also be enhanced significantly by PEGylation, which means surface modification with a poly(ethyleneglycol) (PEG), increasing blood circulation time and improving accumulation in pathologic tissue by EPR [94]. At this point, Shi and co-workers investigated the effects of particle size and PEGylation in MSNs on their biodistribution and urinary excretion in mice after tail-vein injection [99]. Both, naked MSNs and PEGylated MSNs of different particle sizes (80–360 nm) distributed mainly in the liver and spleen, a minority of them in the lungs, and a few in the kidney and heart. However, the PEGylated MSNs of smaller particle size escaped more easily from capture by RES tissues, show longer blood circulation lifetime, slower biodegradation, and lower excreted amount of degradation products in the urine.

Another FMSN aspect that may have significant impact in their *in vivo* application is the particle shape. Tang et al. [100], controlled FMSN fabrication to design particles of two different length/width ratios (1.5 and 5). After particle intravenous injection in mice Si content in blood and target organs was analyzed by ICP. Organ distributions show that intravenously administered FMSNs were mainly present in the liver, spleen and lung (>80%).

Short-rod FMSNs were easily trapped in the liver, whereas long-rod FMSNs were distributed in the spleen. Moreover, FMSNs were mainly excreted by urine and feces, and their clearance rate was primarily dependent on particle shape: short-rods had a more rapid clearance rate than long-rods in both excretion routes.

Although the *i.v.* is the most usual route to administer a stable suspension of FMSNs, other exposure routes may also be used depending on the delivered drug, including transdermal, hypodermic, intramuscular and oral administration [101]. Here, Tang and co-workers have carried out a systematic investigation of the absorption, distribution, excretion and toxicity of 110 nm silica nanoparticles (SNPs) in mice after intravenous, hypodermic, intramuscular injection and oral administration [102]. To further quantitatively analyze the biodistribution and clearance of particles, the Si content in blood and target organs was analyzed by ICP. The results showed that a fraction of the nanoparticles administered by intramuscular and hypodermic injection could cross different biological barriers into the liver but with a low absorption rate. Conversely, in the case of exposure *via* oral administration, particles were absorbed into the intestinal tract and persisted in the liver, whereas, as commented above, SNPs administered intravenously were mainly present in the liver and spleen. Finally, it was observed that particles were mainly excreted through urine and feces, regardless of the exposure route.

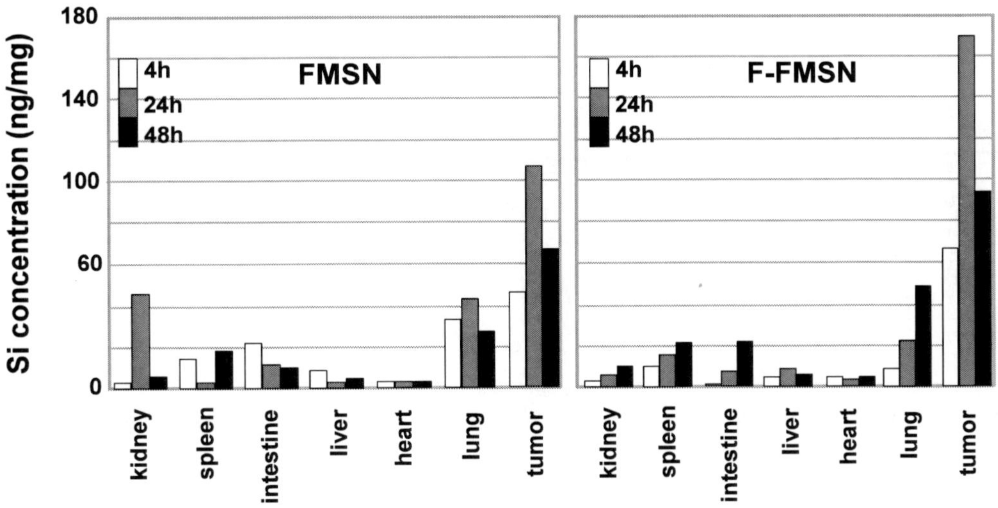

Figure 7. Quantitative measurement of Si concentrations as determined by ICP-MS in each organ of the mice with xenograft tumors. The results are shown as mean values±SD. Adapted with permission from ref. 97. Copyright 2010, Wiley.

Conclusion

FMSNs are excellent candidates for versatile DDSs, with unique advantages that may allow developing formulations for disease diagnosis and therapy. They show good biocompatibility and tolerance both in *in vitro* and *in vivo* studies, and may eventually be excreted from the body. They can be tailored in size, topology, pore diameter, shape and surface chemistry in order to achieve high drug loading and stimuli-responsive drug release profiles. The ease of combination with magnetic or plasmonic structures makes it possible to obtain multifunctional materials able to achieve, respectively, simultaneous bioimaging or photothermal performance together with drug delivery activity. However, these novel therapeutic nanoplatforms still have serious limitations to translate the formulations into clinical application, as they are prone to provoke a strong immune response in plasma and significant accumulation in RES cells-containing organs, mostly liver, spleen and lungs. In this sense, the outlook of future research on FMSN-based DDSs lies in studying such an immune reaction, optimizing the surface chemistry of nanoparticles in order to minimize this process, increasing plasma residence time, and favoring accumulation in targeted cells and tissues. Moreover, long term toxicity studies are needed in order to provide sufficient evidence regarding the safety of these materials in clinical use.

REFERENCES

[1] Kresge, C.T.; Leonowicz, M.E.; Roth, W.J.; Vartuli, J.C.; Beck, J.S. *Nature* 1992, 359, 710-712.

[2] Corma, A. *Chem. Rev.* 1997, 97, 2373-2419.

[3] Unger, K. K.; Kumar, D.; Grun, M.; Buchel, G.; Ludtke, S.; Adam, T.; Schumacher, K.; Renker, S. *J. Chromatogr. A* 2000, 892, 47-55.

[4] Huo, Q.; Feng, J.; Schueth, F.; Stucky, G.D. *Chem. Mater.* 1997, 9, 14-17.

[5] Vallet-Regí, M.; Rámila, A.; del Real, R. P.; Pérez-Pariente, J. *Chem. Mater.* 2001, 13, 308-311.

[6] Trewyn, B. G.; Whitman, C. M.; Lin, V. S.-Y. *Nano Lett.* 2004, 4, 2139-2143.

[7] Gao, F.; Botella, P.; Corma, A.; Blesa, J.; Dong, L. *J. Phys. Chem. B* 2009, 113, 1796-1804.

[8] Vivero-Escoto, J. L.; Slowing, I. I.; Trewyn, B. G.; Lin, V. S.-Y. *Small* 2010, 6, 1952-1967.

[9] Tarn, D.; Ashley, C. E.; Xue, M.; Carnes, E. C.; Zink, J. I.; Brinker, J. *Acc. Chem. Res.* 2013, 46, 792-801.

[10] Lin, Y.-S.; Abadeer, N.; Hurley, K. R.; Haynes, C. L. *J. Am. Chem. Soc.* 2011, 133, 20444-20457.

[11] Lu, J.; Liong, M.; Zink, J. I.; Tamanoi, F. *Small* 2007, 3, 1341-1346.

[12] Kim, H.; Kim, S.; Park, C.; Lee, H.; Park, H. J.; Kim, C. *Adv. Mater.* 2010, 22, 4280-4283.

[13] Cui, Y.; Dong, H.; Cai, X.; Wang, D.; Li, Y. *Appl. Mater. Interfaces* 2012, 4, 3177-3183.

[14] Chen, F.; Hong, H.; Zhang, Y.; Valdovinos, H. F.; Shi, S.; Kwon, G. S.; Theuer, C. P.; Barnhart, T. E.; Cai, W. *ACS Nano* 2013, 7, 9027-9039.

[15] Singh, N.; Karambelkar, A.; Gu, L.; Lin, K.; Miller, J. S.; Chen, C. S.; Sailor, M. J.; Bhatia, S. N. *J. Am. Chem. Soc.* 2011, 133, 19582-19585.

[16] Liu, J.; Bu, W.; Pan, L.; Shi, J. *Angew. Chem. Int. Ed.* 2013, 52, 4375-4379.

[17] Bernardos, A.; Mondragón, L.; Aznar, E.; Marcos, M. D.; Martínez-Máñez, R.; Sancenón, F.; Soto, J.; Barat, J. M.; Pérez-Payá, E.; Guillem, C.; Amorós, P. *ACS Nano* 2010, 4, 6353-6368.

[18] Lai, C.-Y.; Trewyn, B. G.; Jeftinija, D. M.; Jeftinija, K.; Xu, S.; Jeftinija, S.; Lin, V. S.-Y. *J. Am. Chem. Soc.* 2003, 125, 4451-4459.
[19] Yuan, L.; Chen, W.; Hu, J.; Zhang, J. Z.; Yang, D. *Langmuir* 2013, 29, 734-743.
[20] Sauer, A. M.; Schlossbauer, A.; Ruthardt, N.; Cauda, V.; Bein, T.; Bräuchle, C. *Nano Lett.* 2010, 10, 3684-3691.
[21] Muniesa, C.; Vicente, V.; Quesada, M.; Saez-Atienzar, S.; Blesa, J. R.; Abasolo, I.; Fernandez Y.; Botella, P. *RSC Adv.* 2013, 3, 15121-15131.
[22] Mortera, R.; Vivero-Escoto, J.; Slowing, I. I.; Garrone, E.; Onida, B.; Lin, V. S.-Y. *Chem. Commun.* 2009, 45, 3219–3221.
[23] Paula, A. J.; Araujo Júnior, R. T.; Martinez, D. S. T.; Paredes-Gamero, E. J.; Nader, H. B.; Durán, N.; Justo, G. Z.; Alves, O. L. *Appl. Mater. Interfaces* 2013, 5, 8387-8393.
[24] Vallet-Regí, M.; Balas, F.; Arcos, D. *Angew. Chem. Int. Ed.* 2007, 46, 7548-7558.
[25] Rosenholm, J. M.; Linden, M. *J. Control. Release* 2008, 128, 157-164.
[26] Tang, S.; Huang, X.; Chen, X.; Zheng, N. *Adv. Func. Mater.* 2010, 20, 2442-2447.
[27] Yang, P.; Gai, S.; Lin, J. *Chem. Soc. Rev.* 2012, 41, 3679-3698.
[28] Chen, H.-T.; Huh, S.; Lin, V. S.Y. In *Catalysis Preparation*; Regalbuto, J.; Ed.; CRC Press, 2007, pp 45-73.
[29] Muñoz, B.; Rámila, A.; Pérez-Pariente, J.; Díaz, I.; Vallet-Regí, M. *Chem. Mater.* 2003, 15, 500-503.
[30] Song. S.W.; Hidajat, K.; Kawi, S. *Langmuir* 2005, 21, 9568-9575.
[31] Corma, A.; Moliner, M.; Díaz-Cabañas, M. J.; Serna, P.; Femenía, B.; Primo, J.; García, H. *New. J. Chem.* 2008, 32, 1338-1345.
[32] Corma, A.; Díaz-Cabañas, M. J.; Moliner, M.; Rodriguez, G. *Chem. Commun.* 2006, 3137-3139.
[33] Clifford, N. W.; Iyer, K. S.; Raston, C. L. *J. Mater. Chem.* 2008, 18, 162-165.
[34] He, Q.; Shi, J.; Chen, F.; Xhu, M.; Zhang, L. *Biomaterials* 2010, 31, 3335-3346.
[35] Vlachy, N.; Touraud, D.; Heilmann, J.; Kunz, W. *Colloids Surf., B* 2009, 70, 278-280.
[36] Tsai, C.-H.; Vivero-Escoto, J. L.; Slowing, I. I.; Fang, I.-J.; Trewyn, B. G.; Lin, V. S.-Y. *Biomaterials* 2011, 32, 6234-6244.
[37] Rahman, M. S.; Rankin, S. E. *J. Colloid Interface Sci.* 2010, 342, 33-42.
[38] Lavrencic Stangar, U.; Huesing, N. *Silicon Chem.* 2003, 2, 157-165.
[39] Botella, P.; Corma, A.; Quesada, M. *J. Mater. Chem.* 2012, 22, 6394.
[40] Slowing, I.; Trewyn, B. G.; Lin, V. S.-Y. *J. Amer. Chem. Soc.* 2006, 128, 14792-14793.
[41] Feener, E. P.; Shen, W. C.; Ryser, H. J. P. *J. Biol. Chem.* 1990, 265, 18780-18785.
[42] Botella, P.; Abasolo, I.; Fernandez, Y.; Muniesa, C.; Miranda, S.; Quesada, M.; Ruiz, J.; Schwartz, S., Jr.; Corma, A. *J. Control. Release* 2011, 156, 246-257.
[43] Giri, S.; Trewyn, B. G.; Stellmaker, M. P.; Lin, V. S.-Y. *Angew. Chem. Int. Ed.* 2005, 44, 5038-5044.
[44] Mal, N. K.; Fujiwara, M.; Tanaka, Y. *Nature* 2003, 421, 350-353.
[45] Gruenhagen, J. A.; Lai, C.-Y.; Radu, D. R.; Lin, V. S.-Y. *Appl. Spectrosc.* 2005, 59, 424-431.
[46] Meng, H.; Xue, M.; Xia, T.; Zhao, Y.-L.; Tamanoi, F.; Stoddart, J. F.; Zink, J. I.; Nel, A. E. *J. Amer. Chem. Soc.* 2010, 132, 12690-12697.
[47] Patel, K.; Angelos, S.; Ditchel, W.R.; Coskun, A.; Yang, Y.-W.; Zink, J. I.; Stoddart, J. F. *J. Amer. Chem. Soc.* 2008, 130, 2382-2383.

[48] Du, L.; Liao, S.; Khatib, H. A.; Stoddart, J. F.; Zink, J. I. *J. Amer. Chem. Soc.* 2009, 131, 15136-15142.

[49] Angelos, S.; Khashab, N. M.; Yang, Y.-W.; Trabolsi, A.; Khatib, H. A.; Stoddart, J. F.; Zink, J. I. *J. Amer. Chem. Soc.* 2009, 131, 12912-12914.

[50] Ferris, D. P.; Zhao, Y.-L.; Khashab, N. M.; Khatib, H. A.; Stoddart, J. F.; Zink, J. I. *J. Amer. Chem. Soc.* 2009, 131, 1686-1688.

[51] Casasus, R.; Climent, E. D.; Marcos, M. D.; Martínez-Máñez, R.; Sancenón, F.; Soto, J.; Amorós, P.; Cano, J.; Ruiz, E. *J. Amer. Chem. Soc.* 2008, 130, 1903-1917.

[52] Coll, C.; Mondragón, L.; Martínez-Máñez, R.; Sancenón, F.; Marcos, M. D.; Soto, J.; Amorós, P.; Pérez-Payá, E. *Angew. Chem. Int. Ed.* 2011, 50, 2138-2140.

[53] Park, C.; Oh, K.; Lee, S. C.; Kim, C. *Angew. Chem. Int. Ed.* 2007, 46, 1455-1457.

[54] Zhu, Y.; Shi, J.; Shen, W.; Dong, X.; Feng, J.; Ruan, M.; Li, Y. *Angew. Chem. Int. Ed.* 2005, 44, 5083-5087.

[55] Zhu, Y.; Shi, J. *Microporous Mesoporous Mater.* 2007, 103, 243-249.

[56] Liu, N.; Yu, K.; Smarsly, B.; Dunphy, D. R.; Jiang, Y.-B.; Brinker, C. J. *J. Amer. Chem. Soc.* 2002, 124, 14450-14541.

[57] Liu, N.; Chen, Z.; Dunphy, D. R.; Jiang, Y.-B.; Assink, R. A.; Brinker, C. J. *Angew. Chem. Int. Ed.* 2003, 42, 1731-1734.

[58] Liu, N.; Dunphy, D. R.; Atanassov, P.; Bunge, S. D.; Chen, Z.; Lopez, G. P.; Boyle, T.; Brinker, C. J. *Nano Lett.* 2004, 4, 551-554.

[59] Lu, J.; Choi, E.; Tamanoi, F.; Zink, J. I. *Small*, 2008, 4, 421-426.

[60] Schlossbauer, A.; Kecht, J.; Bein, T. *Angew. Chem. Int. Ed.* 2009, 48, 3092-3095.

[61] Bernardos, A.; Aznar, E.; Marcos, M. D.; Martínez-Máñez, R.; Sancenón, F.; Soto, J.; Barat, J. M.; Amorós, P. *Angew. Chem. Int. Ed.* 2009, 48, 5884-5887.

[62] Corma, A.; Díaz, U.; Arrica, M.; Fernandez, E.; Ortega, Í. *Angew. Chem. Int. Ed.* 2009, 48, 6247-6250.

[63] Thornton, P. D.; Heise, A. *J. Amer. Chem. Soc.* 2010, 132, 2024-2028.

[64] Park, C.; Kim, H.; Kim, S.; Kim, C. *J. Amer. Chem. Soc.* 2009, 131, 16614-16615.

[65] Liu, J.; Du, X.; Zhang, X. *Chem. Eur. J.* 2011, 17, 810-815.

[66] Tang, F.; Li, L.; Chen, D. *Adv. Mater.* 2012, 24, 1504-1534.

[67] Giannotti, M. I.; Esteban, O.; Oliva, M.; Garcia-Parajo, M. F.; Sanz, F. *Biomacromolecules* 2011, 12, 2524-2533.

[68] Chen, R.; Feng, F.; Meng, F.; Deng, C.; Feijen, J.; Zhong, Z. *J. Controlled Release* 2011, 152, 2-12.

[69] Saito, G.; Swanson, J. A.; Lee, K.-D. *Adv. Drug Delivery Rev.* 2003, 55, 199-215.

[70] Huang, C.-C.; Tsai, C.-Y.; Sheu, H.-S.; Chuang, K.-Y.; Su, C.-H.; Jeng, U. S.; Cheng, F.-Y.; Su, C.-H.; Lei, H.-Y.; Yeh, C.-S. *ACS Nano* 2011, 5, 3905-3916.

[71] Kim, J.; Kim, H. S.; Lee, N.; Kim, T.; Kim, H.; Yu, T.; Song, I. C.; Moon, W. K.; Hyeon, T. *Angew. Chem. Int. Ed.* 2008, 47, 8438-8441.

[72] Gai, S.; Yang, P.; Ma, P. A.; Wang, D.; Li, C.; Li, X.; Niu, N.; Lin, J. *J. Mater. Chem.* 2011, 21, 16420-16426.

[73] Wu, H.; Liu, G.; Zhang, S.; Shi, J.; Zhang, Y.; Chen, Y.; Chen, F.; Chen, H. *J. Mater. Chem.* 2011, 21, 3037-3045.

[74] Chen, Y.; Chen, H.; Zeng, D.; Tian, Y.; Chen, F.; Feng, J.; Shi, J. *ACS Nano* 2010, 4, 6001-6013.

[75] Botella, P.; Corma, A.; Navarro, M. T. *Chem. Mater.* 2007, 19, 1979-1983.

[76] Botella, P.; Corma, A.; Navarro, M. T.; Quesada, M. *J. Mater. Chem.* 2009, 19, 3168-3175.
[77] Botella, P.; Ortega, I.; Quesada, M.; Madrigal, R. F.; Muniesa, C.; Fimia, A.; Fernández, E.; Corma, A. *Dalton T* 2012, 31, 9286-9296.
[78] Shen, S.; Tang, H.; Zhang, X.; Ren, J.; Pang, Z.; Wang, D.; Gao, H.; Qian, Y.; Jiang, X.; Yang, W. *Biomaterials* 2013, 34, 3150-3158.
[79] Akerman, M. E.; Chan, W. C. W.; Laakkonen, P.; Bhatia, S. N.; Ruoslahti, E. *Proc. Natl. Acad. Sci.* 2002, 99, 12617-12621.
[80] Néel, L.; Acad, C. R. *Science* 1949, 228, 664-666.
[81] Na, H. B.; Song, I. C.; Hyeon, T. *Adv. Mater.* 2009, 21, 2133-2148.
[82] Bulte, J. W.; Kraitchman, D. L. *NMR Biomed.* 2004, 17, 484-499.
[83] Weissleder, R. *Nat. Biotechnol.* 2001, 19, 316-317.
[84] Pissuwan, D.; Valenzuela, S. M.; Cortie, M. B. *Trends Biotechnol.* 2006, 24, 62-67.
[85] Hu, M.; Chen, J.; Li, Z.-Y.; Au, L.; Hartland, G. V.; Li, X.; Marquez, M.; Xia, Y. *Chem. Soc. Rev.* 2006, 35,1084-1094.
[86] Link, S.; El-Sayed, M. A. *J. Phys. Chem. B* 1999, 103, 8410-8426.
[87] Lu, W.; Melancon, M. P.; Xiong, C.; Huang, Q.; Elliot, A.; Song, S.; Zhang, R.; Flores, L. G., II; Gelovani, J. G.; Wang, L. V.; Ku, G.; Stafford, R. J.; Li, C. *Cancer Res.* 2011, 71, 6116-6121.
[88] Richardson, H. H.; Carlson, M. T.; Tandler, P. J.; Hernandez, P.; Govorov, A. O. *Nano Lett.* 2009, 9, 1139-1146.
[89] Dobrovolskaia, M. A.; McNeil, S. E. *Nat. Nanotechnol.* 2007, 2, 469-478.
[90] Lee, C.-H.; Cheng, S.-H.; Wang, Y.-J.; Chen, Y.-C.; Chen, N. T.; Souris, J.; Chen, C.-T.; Mou, C.-Y.; Yang, C.-S.; Lo, L.-W. *Adv. Func. Mater.* 2009, 19, 215-222.
[91] Kumar, R.; Roy, I.; Ohulchanskky, T.Y.; Vathy, L. A.; Bergey, E. J.; Sajjad, M.; Prasad, P. N. *ACS Nano* 2010, 4, 699-708.
[92] Souris, J. S.; Lee, C.H.; Cheng, S.H.; Chen, C.T.; Yang, C.S.; Ho, J.A.; Mou, C.Y.; Lo, L.W. *Biomaterials* 2010, 31, 5564-5574.
[93] Burns, A.A.; Vider, J.; Ow, H.; Herz, E.; Penate-Medina, O.; Baumgart, M.; Larson, S.M.; Wiesner, U.; Bradbury, M. *Nano Lett.* 2009, 9, 442-448.
[94] He, X.; Nie, H.; Wang, K.; Tan, W.; Wu, X.; Zhang, P. *Anal. Chem.* 2008, 80, 9597-9603.
[95] Maeda, H.; Wu, J.; Sawa, T.; Matsumura, Y.; Hori, K. *J. Control. Release* 2000, 65, 271-284.
[96] Perrault, S.D.; Walkey, C.; Jennings, T.; Fischer, H.C.; Chan, W.C. *Nano Lett.* 2009, 9, 1909-1915.
[97] Xiao, K.; Luo, J.; Li, Y.; Xiao, W.; Lee, J.S.; Gonik, A.M.; Lam, K.S. *Nanosci. Nanotech. Lett.* 2010, 2, 79-85.
[98] Lu, J.; Liong, M.; Li, Z.; Zink, J. I.; Tamanoi, F. *Small* 2010, 6, 1794-1805.
[99] He, Q.; Zhang, Z.; Gao, F.; Li, Y.; Shi, J. *Small* 2011, 7, 271-280.
[100] Huang, X.; Li, L.; Liu, T.; Hao, N.; Liu, H.; Chen, D.; Tang, F. *ACS Nano* 2011, 5, 5390-5399.
[101] Malfatti, M. A.; Palko, H. A.; Kuhn, E. A. *Nano Lett.* 2012, 12, 5532-5538.
[102] Fu, C.; Liu, T.; Li, L.; Liu, H.; Chen, D.; Tang, F. *Biomaterials* 2013, 34, 2565-2575.

In: Comprehensive Guide for Mesoporous Materials, Volume 4 ISBN: 978-1-63482-092-9
Editor: Mahmood Aliofkhazraei

Chapter 6

MESOPOROUS MATERIALS IN THE SYNTHESIS OF FRAGRANCES

Agnieszka Feliczak-Guzik** and Izabela Nowak
Adam Mickiewicz University in Poznan, Faculty of Chemistry, Poznan, Poland

ABSTRACT

The application of mesoporous molecular sieves, such as MCM-41, SBA-15, MSU-X and others in the synthesis of fragrances is discussed. Mesoporous materials by definition have pores of diameters from the range 2-50 nm [1]. The fragrance chemistry is an interesting blend of the chemistry of natural products, synthetic, physical and analytical chemistry to a certain degree open to original invention in composing new scents.

Especially terpenes that are a large class of organic compounds, are natural and sustainable feedstocks for the fine fragrance industry [2, 3]. These compounds can be used as intermediates and components of flavors, fragrances and pharmaceuticals. Transformation of several terpenes and terpenoids (geraniol, limonene, α-terpineol) by different chemical reactions over molecular sieves as catalysts is described, for example the epoxidation of terpenes is an attractive chemical transformation as such epoxides are versatile building units [4]. Products of hydrogenation, cyclization, oxidation or isomerization are important compounds in the flavor/fragrance industry, pharmaceutical industry and in general synthesis [5-7].

1. TERPENES AND TERPENOIDS

Hydrocarbons derived from plants of the general formula $(C_5H_8)_n$ and being oligomers of isoprene are called terpenes. In these compounds, the rests of the isoprene units are tail-head interconnected according to the Ruzički rule. Isoprene is 2-methyl-1,3-butadiene, classified as hemiterpene. This is a volatile liquid with a boiling point of 307 K, which is obtained by

* Corresponding author: Agnieszka Feliczak-Guzik, Adam Mickiewicz University in Poznań, Faculty of Chemistry, Umultowska 89b, PL61-614 Poznań, Poland. E-mail: agaguzik@amu.edu.pl.

dehydrogenation a mixture of isopentane and isopentene, or by dehydration of the condensation product of isobutene with formaldehyde [8]. The oxygen-containing derivatives of isoprene, such as isovaleric acid and prenol, are called hemiterpenoids [9-12].

The first to introduce the name terpenes for a mixture of $C_{10}H_{16}$ was Kekule, in 1880. His assistant Walach, in 1887, postulated hypothetically that terpenes are composed of two or more parts of isoprene. Thirty years later, Robinson modified the "isoprene rule" of Wallach – the parts of isoprene should be combined head to tail. Then Ruzicka modified the rule of Wallach by proving the head-tail connection of two or more parts of isoprene. However, this rule explained that terpenoids are separate from the chain precursors such as geraniol from monoterpenes, sesquiterpenes from farnesol. It should be mentioned that as many as three scientists dealing with issues of terpenes received the Nobel Prize over the span of 37 years [10, 13-16]. Table 1 shows scientists and their discoveries.

Table 1. Scientists and their discoveries [10, 13-16]

Scientist	Discovery
Kekule	Introduction the name of terpenes for a mixture of $C_{10}H_{16}$
Walach	He postulated hypothetically that terpenes are composed of two or more parts of isoprene
Robinson	Modification of Wallach's "isoprene rule" – the parts of isoprene should be combined head to tail.
Ruzička	Modification of Wallach's rule by proving the head-tail connection of two or more parts of isoprene.

Terpenes and terpenoids belong to the largest classes of natural chemicals. In plants, microorganisms and animals more than 30 000 terpenes have been isolated and identified [13-14]. These compounds play an important role as fragrances in perfumery, as environment-friendly enticing compounds to trap damaging insects, as constituents of aroma for spicing foods. Terpenes can be used as medicaments for the therapy of numerous diseases including cancer [3].

1.1. Occurrence of Terpenes and Their Role in Living Organisms

Both terpenes and their derivatives called terpenoids are widely distributed in nature. First of all, they are ingredients of fragrances, essential oils and resins of coniferous trees. Large amounts of these compounds are produced by marine organisms. Land animals produce small amounts of terpenes - mainly as pheromones and repellents [17].

These compounds play an important role in cellular metabolism. Terpenes affect the structure and function of cell membranes. They can be involved in many important biological functions; can participate in photosynthesis, respiration, hormonal regulation, regulation of growth and evolution, protection against attack by pathogens, intracellular signal conduction and vesicular transport into cells [18].

1.2. Biosynthesis of Terpenes

The straight forward substrate in the reactions of biosynthesis of terpenes is not isoprene, of whose units these compounds are formed, but its two equivalents: isopentyl pyrophosphate and dimethylallyl pyrophosphate [8], shown in Scheme 1.

For many years the only known pathway leading to the synthesis of the above described isoprene's precursors, has been that involving mevalonic acid. This pathway takes place in the kingdom of eukaryotes (in all mammals), the kingdom of archea, cytosol and in the mitochondria of plants. According to this pathway, the source of terpenes is mevalonic acid, is converted to isopentyl pyrophosphate or 3,3-dimethylallyl pyrophosphate. These compounds are transformed into the corresponding terpenes and other bioorganic compounds, which fit the isoprene rule.

The first step in this pathway involves the production of acetoacetyl-CoA from two acetyl-CoA molecules. Then the obtained acetoacetyl-CoA molecule is condensed with a third acetyl-CoA molecule, which leads to the formation of mevalonic acid. As a result of the phosphorylation and elimination of the acid formed, isopentyl pyrophosphate is formed at equilibrium with its dimethylallyl pyrophosphate [18-19]. This pathway is shown in Scheme 1.

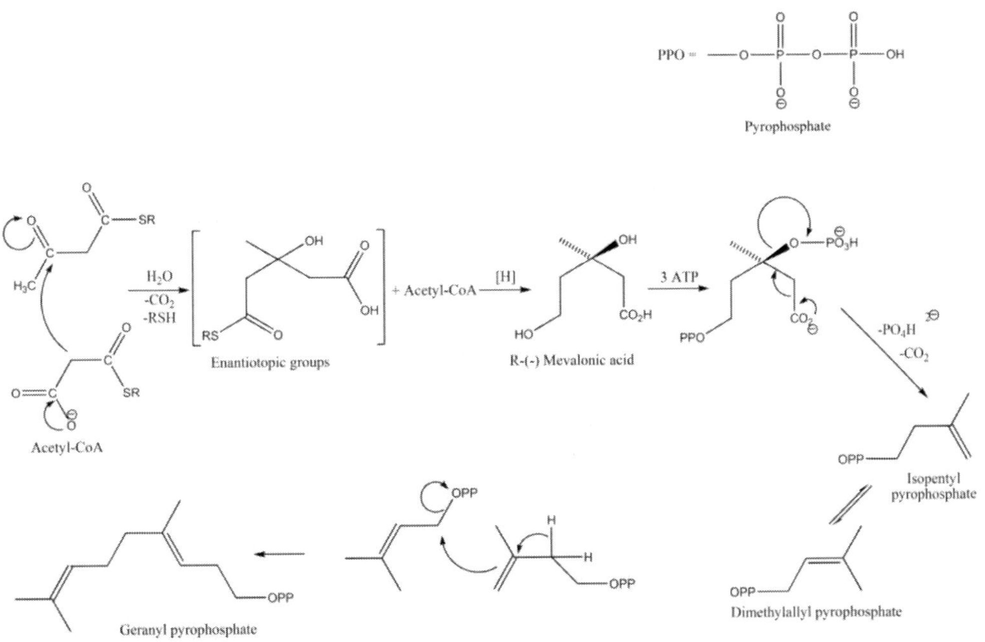

Scheme 1. Isoprenoid biosynthesis pathway [adapted from 8, 18-19].

Recently, a new pathway of isoprenoid biosynthesis has been discovered, which, unlike the above-described pathway, is independent of mevalonic acid. This route has been marked as DOXP-(1-deoxy-d-xylulose-5-phosphate). Isoprene precursors, in this pathway, are derived from pyruvic acid and glyceraldehyde 3-phosphate, and the intermediate products are the two compounds:

- DXP-1-deoxy-d-xylulose-5-phosphate
- MEP-2-metyloerytro-4-phosphate.

This alternative way of isoprenoid synthesis occurs in the chloroplasts of plants, algae, eubacteria and parasites. It consists of eight reactions which are catalyzed by nine enzymes [19-21]. This route is shown in Scheme 2.

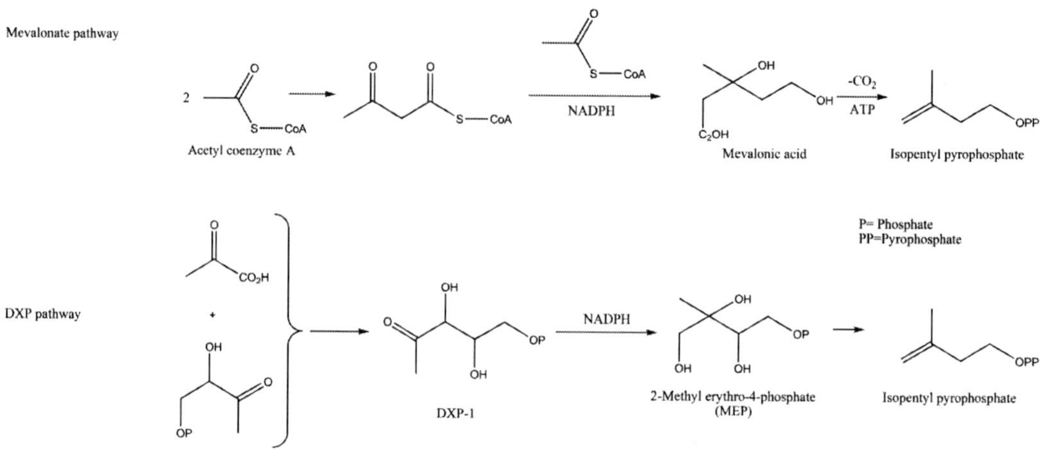

Scheme 2. Alternative pathways of isoprenoid biosynthesis [adapted from 19-21].

1.3. Classification of Terpenes

Table 2 shows the classification of terpenes according to the number of isoprene units in the molecule (n).

Table 2. Classification of terpenes

Groups of terpenes	Molecular formula	Number of isoprene units (n)
Monoterpenes	$C_{10}H_{16}$	*n=2*
Sesquiterpenes	$C_{15}H_{24}$	*n=3*
Diterpenes	$C_{20}H_{32}$	*n=4*
Sesterterpenes	$C_{25}H_{40}$	*n=5*
Triterpenes	$C_{30}H_{48}$	*n=6*
Tetraterpenes	$C_{40}H_{64}$	*n=8*
Polyterpenes	$(C_5H_8)_n$	n>8

2. Monoterpenes

Monoterpenes are widely used in aromatherapy, pharmaceutical and medical industries [22]. Furthermore, thanks to their taste and smell, they are also used in:

- perfume and cosmetic industry - as ingredients in many cosmetic fragrances, detergents, perfumes;
- food industry as a food flavors.

The high demand for these compounds is best illustrated by their annual consumption by humans, which is approximately 2.8 x 105 kg. Every year only in the perfume-cosmetic and food industries use as much as 30 000 tons of α-and ß-pinene [3].

2.1. Characterization of Selected Monoterpenes

Geraniol/nerol, limonene, pinene and citral are five of the most important terpenoids in the perfume industry. Especially the alcohols and their esters are of greatest importance. All of them are key starting compounds for other terpenoids. The ability to produce any of them opens up the potential to obtain all of them and thence a wide range of other terpenoids [3]. Geraniol is a valuable terpene, which is used widely because of its fresh flowery smell. Air oxidation of geraniol leads to formation of a mixture of the following products: geranial, neral, epoxygeraniol hydroxide, and the product of degradation of the secondary allylic alcohol [23]. These compounds are described below.

2.1.1. Geraniol – 3,7-Dimethyl-Trans-2,6-Octadien-1-ol

Geraniol [Figure 1] is a colorless liquid with a floral-rose fragrance. It is doubly unsaturated acyclic alcohol involved in a number of reactions, such as the rearrangement and cyclization. The rearrangement takes place in the presence of copper as a catalyst and leads to citronellal. Partial hydrogenation of geraniol leads to citronellol. Complete hydrogenation of double bonds yields tetrahydrogeraniol (3,7-dimethyloctan-1-ol). The oxidation of geraniol gives citral. Geranyl esters are obtained by esterification [24].

The parameters characterizing geraniol are collected in Table 3.

Table 3. Properties of geraniol

Formula	$C_{10}H_{18}O$	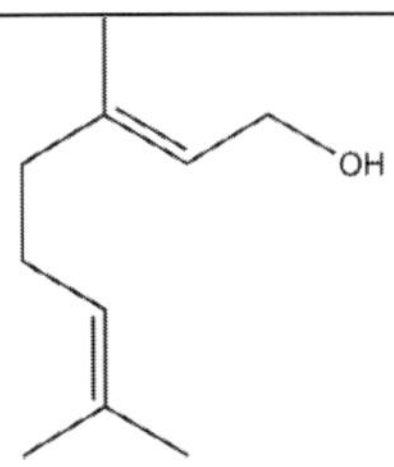
Molecular weight [g/mol]	154.25	
Boiling point [K]	503	
Density [g/cm^3]	0.8894	
Refractive index n^{20}_D	1.4777	Figure 1. Geraniol.

2.1.1.1. Production of Geraniol

The route for the production of geraniol consists of the hydrogenation of citral. Citral, the starting material, is used in large quantities as an intermediate in the synthesis of vitamin A. Currently, geraniol is obtained by isolation from essential oils, for example: from citronella oils and from Palmarosa oil. Fractional distillation of Java citronella oil yields a fraction containing 60% of geraniol, as well as citronellol and sesquiterpenes. Fractionation of Palmarosa oil after prior saponification of geranyl esters leads to a product with a higher content of geraniol and slightly different scent quality (application in fine fragrances).

Geraniol may be synthesized from β-pinene. Pyrolysis of β-pinene yields myrcene. Myrcene is converted into a mixture of mostly geranyl, neryl, linalyl chloride by addition of hydrogen chloride in the presence of small amounts of catalyst [24]. The mixture is reacted with sodium acetate in the presence of a nitrogen base after removal of the catalyst. Then this mixture is converted to geranyl acetate, neryl acetate and a small amount of linalyl acetate [Scheme 3] [24]. Geraniol is obtained after saponification and fractional distillation of the alcohols obtained.

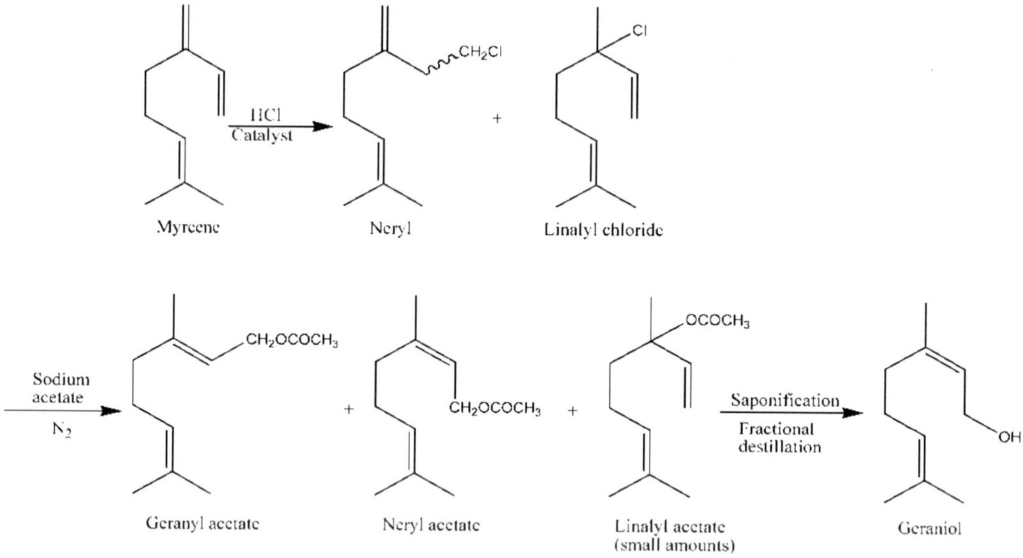

Scheme 3. Geraniol synthesis pathway [adapted from 23].

Pure synthetic geraniol is prepared by isomerization of linalool with orthovanadates used as catalysts to give a 90% yield of geraniol-nerol mixture [24]. Finally, highly pure geraniol is obtained by fractional distillation.

Geraniol is also produced by synthesis from citral. Citral has recently come to be produced petrochemically in very large quantities. Partial hydrogenation of citral has become a very economical route for the production of geraniol.

2.1.1.2. Application of Geraniol

Geraniol is one of the most widely used terpenoid in the fragrance industry. It can be used in all the flowery-roselike compositions and does not discolor soaps. In the fragrance compositions geraniol is used in small quantities, to enhance the citrus note. Geraniol is also an important intermediate in the manufacture of geranyl, citral and citronellol [24].

2.1.1.3. Reactions of Geraniol with the Participation of Mesoporous Materials

2.1.1.3.1. Oxidation of Geraniol

Selective oxidation of terpenes into oxygenated derivatives over heterogeneous catalysis is a key transformation in the synthesis of fine chemicals [25]. Geraniol is an interesting substrate in the reaction which leads to epoxy or diepoxygeraniol.

In geraniol oxidation in the liquid phase with hydrogen peroxide as the oxidant, mesoporous materials of MSU-X type (MSU – abbreviation for *Michigan State University*) were applied as catalysts. The results of oxidation reaction of this terpene show that 6,7-epoxygeraniol is the main products, followed by 2,3-epoxygeraniol and 2,3:6,7-diepoxygerniol [26]. Scheme 4 shows the geraniol oxidation pathway.

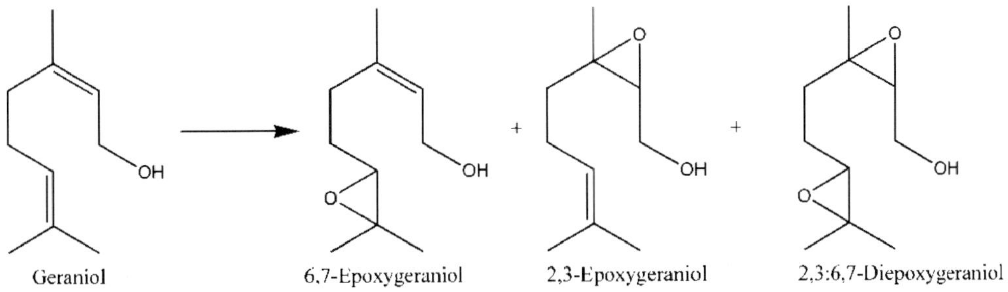

Scheme 4. Geraniol oxidation pathway [authors data].

Epoxidation occurs exclusively at the more nucleophilic double bond (C6–C7). This is a characteristic behavior of oxy species that react quicker with more electron-rich olefins. It is well known that the selectivity for epoxides could be understood through electronic structure calculation.

This property shows that the most nucleophilic group of each terpene is mainly epoxidised [27].

In the catalytic oxidation of geraniol by hydrogen peroxide also WO_3-SiO_2 mesoporous material was used [28].

This catalyst is selective towards epoxidation of geraniol and other alcohols in which the double bond is present. In such alcohols, the oxidation of C2-C3 bond is observed, leading to formation of an appropriate epoxyalcohol. This is a characteristic behavior of tungsten and titanium catalysts, in which the position of -OH group in the substrate determines the reaction product [29].

In 2011, Dapurkar et al. described the use of Cr-MCM-41 (abbreviation for *Mobil Composition of Matter No. 41*; hexagonally ordered mesoporous molecular sieve) as an efficient catalyst for the oxidation of geraniol with O_2 in super critical carbon dioxide as a solvent [30]. Marin-Astorga and co-workers [25] studied the selective oxidation of geraniol with hydrogen peroxide over lanthanum, titanium and niobium catalysts on mesoporous silica MCM-41. According to their work, the Nb-MCM-41 catalyst showed the highest activity and selectivity towards allylic epoxide. These results have shown that the differences in the acid sites, generated with different strengths, could influence the selectivity and activity. The main products, which can be obtained in the catalytic oxidation of geraniol using hydrogen peroxide as an oxidant are shown in Scheme 5 [25].

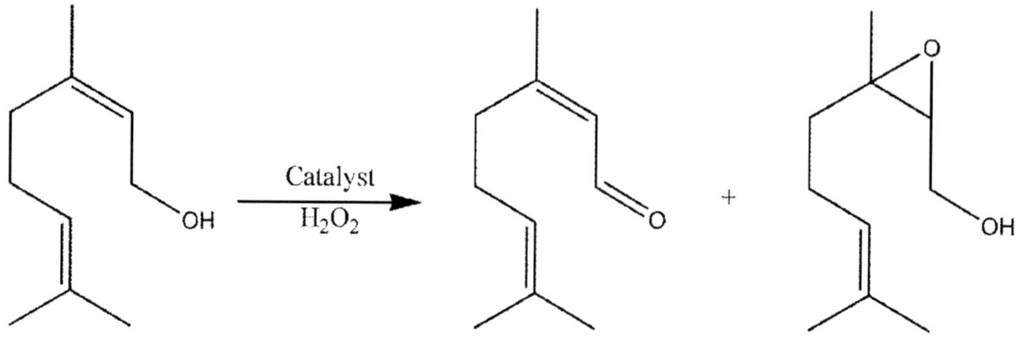

Geraniol Citral Allylic epoxide

Scheme 5. Catalytic oxidation of geraniol pathway [authors data].

Dorbes and co-workers [31] have used oxivanadium(IV) acetylacetonate immobilized onto CMK-3 (3D ordered mesoporous carbon) for the epoxidation of geraniol. These materials were tested in this reaction using *tert*-butyl hydroperoxide as an oxidant. They obtained substrate conversion higher than 98% and ~99% selectivity for 2,3-epoxygeraniol [31].

2.1.2. Citral – 3,7-dimethyl-2,6-octadien-1-al

Citral (Figure 2) occurs as trans and cis isomers (citral a – geranial and citral b – neral). These compounds are available in lemon grass oil (~85%), in *Litsea cubeba* oil (~75%) and in many other essential oils [32].

Citral a and citral b are colourless with a scent resembling of lemon. Properties of citrals are shown in Table 4 [24].

Table 4. Properties of citral

	Citral a Geranial	**Citral b Neral**	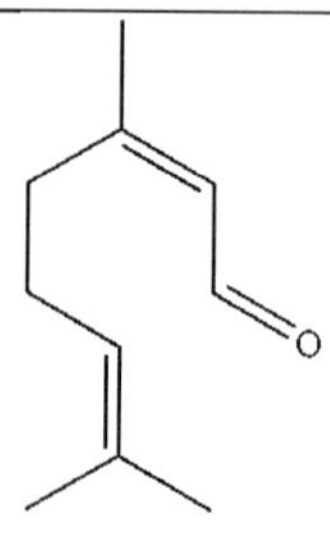 Figure 2. Citral
Formula	$C_{10}H_{16}O$		
Molecular weight [g/mol]	152.24		
Boiling point [K]	391-392	393	
Density [g/cm^3]	0.8888	0.8869	
Refractive index n^{20}_D	1.4898	1.4869	

Citral is industrially produced on a large scale because is used as a starting compound for the synthesis of Vitamin A [24]. 3,7-dimethyl-2,6-octadien-1-al is an α,β-unsaturated aldehyde with an additional double bond. This compound can be transformed into a number of addition materials.

2.1.2.1. Oxidation of Citral

Baeyer-Villiger oxidation of citral with hydrogen peroxide as oxidant over mesoporous materials containing tin has been described by Didillon et al. [33]. The main product of this reaction is melonal (2,6-dimethyl-5-hepten-1-al), which is used in melon and cucumber notes of fragrances [34]. Corma et al. have demonstrated the synthesis of melonal from citral. They have used a green oxidant in combination with heterogeneous materials, such as Sn-MCM-41 and Sn-beta (Sn-β, zeolite with a framework structure of BEA) as catalysts. The catalysts that contain tin are chemoselective and avoid the epoxidation of the double bond. Their selectivity to formate ester and melonal is around 95% [35].

2.1.2.2. Hydrogenation of Citral

Catalytic hydrogenation is one of the key reactions in the fine chemicals industry. Citral has three groups, which could be hydrogenated: a conjugate double bond (C=C), carbonyl group (C=O) and an isolated double bond (C=C) – Scheme 6 [36]. The number of hydrogenated products can be large, for example: hydrogenation of C=O (carbonyl group) leads to the unsaturated alcohols, such as: nerol and geraniol, hydrogenation of C=C (conjugate double bond) leads to isopulegol, menthol and citronellal and its derivatives and hydrogenation of C=C (isolated double bond) leads to 3,7-dimethylooctanol and 3,7-dimethyloctanal. Scheme 6 shows the products of citral hydrogenation [37].

As a result of selective hydrogenation of citral (carbonyl group) that can be performed with the use of many metals as promoters, Ni, Rh, Ru, Pt and Sn - used as an active center, it is possible to obtain nerol and geraniol. Selectivities around 95% to these products at 100% conversion can be obtained, however the final results depend on the tin/metal ratio. Also the procedure of the catalyst synthesis determines the metal-tin interaction and thus catalytic activity [38-39].

Other interesting products to obtain are isopulegol and menthol. Isopulegol is applied in perfumery as geranium note. It is an important intermediate in the synthesis of menthol showing cooling and refreshing properties. It is applied in cosmetics, toothpastes, sweets, chewing gum and pharmaceuticals [37].

Trasarti and co-workers [40] have described the selective one-pot synthesis of menthols from citral over Ni/Beta zeolite, Pd/Beta zeolite and Ni-MCM-41 used as catalysts. The best results, i.e., 90% yield to menthols, were obtained for Ni-MCM-41 [40]. MCM-41 has a large surface area and hexagonal array of uniform mesopores with a narrow pore size distribution. The catalytic processes with the use of MCM-41 are not disrupted by diffusion problems [41].

Citronellal can be produced synthetically by hydrogenation of citral and vapor-phase rearrangement of nerol and geraniol. These compounds can be cyclized into isopulegol, which is a precursor of menthol. Hydrogenation of isopulegol leads to eight stereoisomers of menthols, however only the (-)-menthol is interesting for industry, because this isomer has the peppermint scent and refreshing, cooling properties. Reaction of cyclization occurs in the presence of acid catalysts and more selective are Lewis acids [37].

The cyclization of citronellal to isopulegol under mild conditions has been described in numerous publications. Mesoporous materials, zeolites and supported Lewis acids on silica, alumina and titania have been also described. Shieh and co-workers [42] have shown the cyclization of citronellal over MCM-41, H-ZSM-5 (hydrogen form of MFI framework type zeolite), H-beta in toluene.

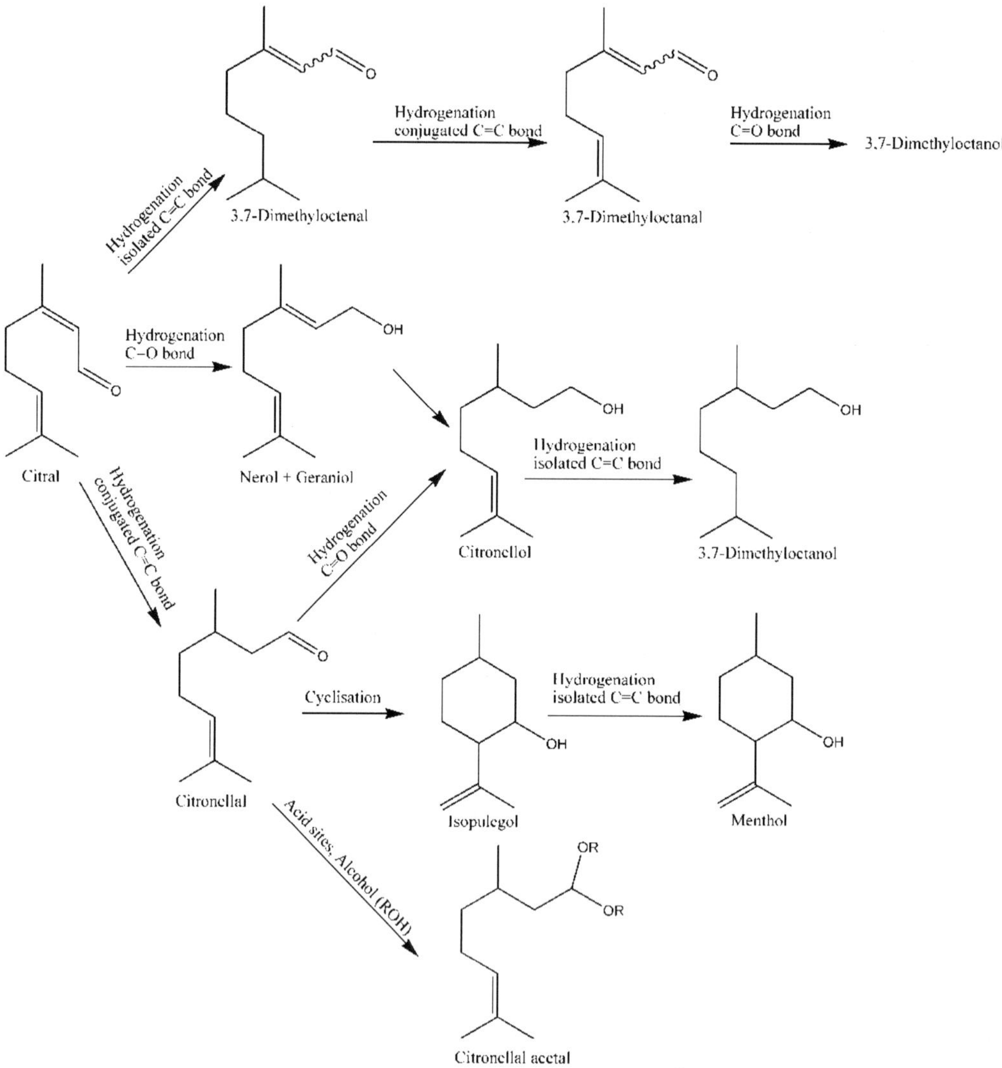

Scheme 6. The products of hydrogenation of citral [adapted from 36-37].

According to them the best catalyst is Al-MCM-41 and they obtained isopulegol with a selectivity of 62% and conversion around 92% after 2h at 333K [42].

Dmitry et al. [43] have studied the cyclization of citronellal over Al-MCM-22 (aluminum form of MWW framework type zeolite), Al-MCM-41, Al-ZSM-5 and Al-Beta and Al-Y aluminum form of FAU framework type zeolite) zeolites. They compared these catalysts with silica materials. The highest rates were obtained over mesoporous materials and zeolites having a 12-membered-ring (FAU, MFI). For these materials the selectivity to isopulegol was around 100%, while the conversion after 3h of reaction was around 63-68% [43]. Ekou and co-workers [44] studied the hydrogenation of citral over Ti-SBA-15 supported platinum nanocomposites. SBA-15 (abbreviation for *Santa Barbara Amorphous*) has a large surface area, large pore volume and adjustable pore size – from a few nanometers to more than 20 nanometers. They have prepared and characterized Pt catalysts on TiO_2 doped SBA-15 materials. The selective citral hydrogenation towards unsaturated aldehydes is realized over these materials. Pt/Ti-SBA-15 catalysts ensure a lower citral conversions than the catalyst made of platinum supported on bulk TiO_2, and increase the selectivity to unsaturated alcohol with the titanium loading [44].

The effect of the acid–base properties of the support on the performance of platinum catalysts in the partial hydrogenation of citral has been studied by Santiago-Pedro et al. [32]. They have shown the effect of mesoporous solid materials with different acid-base properties on the product distribution during the reaction. According to the same authors, basic materials were the most active catalysts (100% citral conversion after 1h of reaction). For basic materials, such as Pt/MgAl (hydrotalcite), the main product was citronellal, while nerol and geraniol were obtained on acid solids (Pt/SiO_2, Pt/SiO_2/TiO_2, Pt-SiO_2-ZrO_2) [32].

2.1.3. α-pinene – 2-pinene

Derivatives of pinane are among the main representatives of bicyclic monoterpenes. In nature, these terpenes, with a considerable dominance of α-pinene (70-90%), are in needles, cones and coniferous tree wood. They can also be met the oils of various species of pine, fir and cypress oils, myrtle and geraniums. The largest amounts of α-pinene can be found in turpentine oil. Both (+)-α-pinene, and (-)-α-pinene have optically active carbon atoms. Laevorotatory α-pinene is in turpentine oil, which is extracted from *Pinus pinaster*, while dextrorotatory variety is in the essential oil derived from *Pinus silvestris*. Pinenes have in their structure a ring of cyclohexane and ring of cyclobutane. The compound α-pinene commonly referred to as pinene (Figure 3), has a coupled system formed of a ring of cyclobutane and a double bond. The variety β-pinene (Figure 4), known as nopinene, differs from α-pinene in the location of the double bond. Properties α-pinene and β-pinene are shown in Table 5 [24, 45].

Table 5. Properties of α-pinene and β-pinene

	α-pinene		**β-pinene**	
Formula	$C_{10}H_{16}$	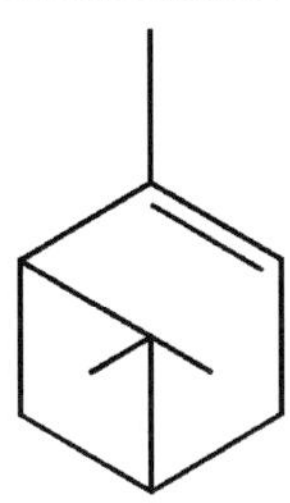	$C_{10}H_{16}$	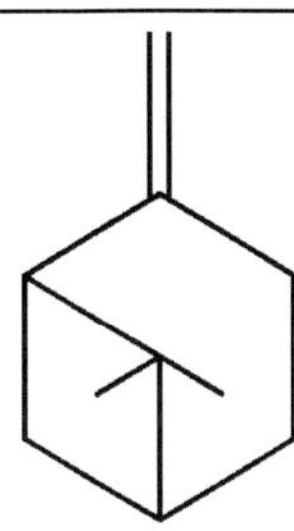
Molecular weight [g/mol]	136.24		136.24	
Boiling point [K]	429		437	
Density[g/cm^3]	0.8553		0.8712	
Reflactive index n^{20}_D	1.4662	Figure 3. α-pinene	1.4763	Figure 4. β-pinene
Optical rotation $[\alpha]_D^{20}$	+ or – 51.9°		+ or – 22.6o	

2.1.3.1. Application of α-Pinene

In chemical industry, α-pinene has been applied among others [46]:

- to produce of explosive materials - smokeless gunpowder, plastics;
- in perfumery - for the preparation of aromatic substances such as linalool, geraniol, nerol, citronellol, myrcene, camphene, ionones, isobornyl acetate, citral;
- as a flavoring agent - resin smell of turpentine;
- as a solvent for paints, varnishes and polishes.

2.1.4. Reaction of α-Pinene

2.1.4.1. Isomerization of α-Pinene

The reaction of isomerization in the presence of acid catalysts has been widely described. In the liquid-phase systems, the main products were: camphene and limonene. The selectivity and conversion of α-pinene depend on the nature, strength and number of acid sites of the catalysts [37]. Limonene is more reactive than camphene and it can be transformed into different compounds. Limonene is applied in cosmetic industry as a flavor and fragrance additive in cleaning products, food and pharmaceuticals. Camphene is an important compound for the synthesis of camphor. Natural camphor occurs in the wood of camphor tree, the synthetic camphor is made by isomerization of pinene [47].

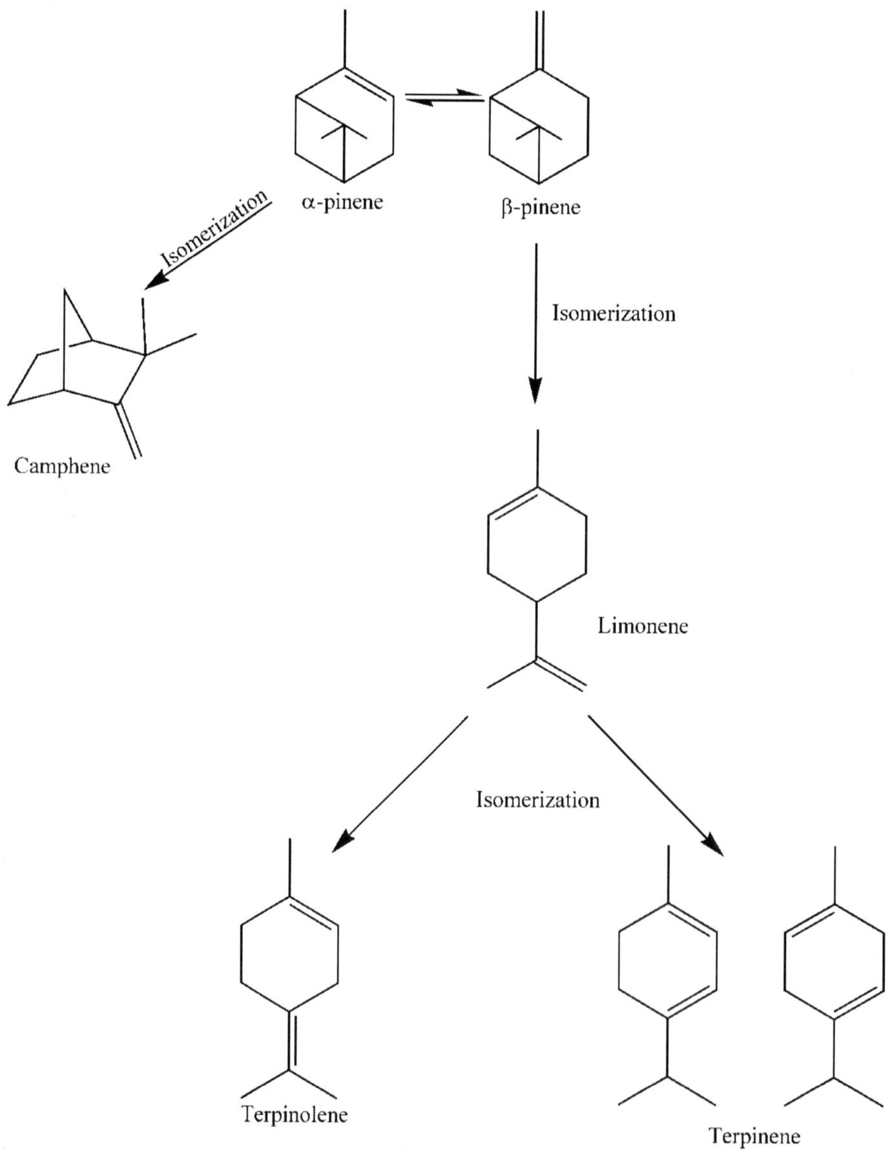

Scheme 7. Pathway of isomerization of α-pinene [adapted from 52].

Dimitrova and co-workers [48] reported on pinene isomerization using zeolites (ZSM-5, Beta and mordenite zeolites) and mesoporous material of MCM-41-type. In liquid phase at 373K, Beta zeolite gave 99% conversion and selectivity to camphene around 27% [48]. Yamamoto et al. [49] have studied the isomerization of pinene at 303 and 353K in the liquid phase over mesoporous silica of FSM-16-type (abbreviation for Folded Sheet Silica no. 16; material similar to that of MCM-41). This material has acid sites able to catalyze the isomerization of pinene. The activity of this catalyst depended on the reaction temperature. Conversion of pinene, at 353 K, was around 78%, while the selectivities to camphene and limonene were constant (around 41%) [49]. Another group of Ecormier and co-workers [50] has studied the isomerization of pinene over hexagonal mesoporous silica – HMS with different content of sulfated zirconia. They observed an increase in the activity of the hexagonal mesoporous silica with increasing Zr content. The selectivity to camphene or limonene depended on the strength of the acid catalyst [50].

The same reaction was studied by Stekrova and co-workers [51] over ceria supported mesoporous materials such as MCM-41 and SBA-15. This isomerization was carried out at 413K using N,N-dimethylacetamide as a solvent. The activity of such catalyst increased with increasing amount of Brønsted acid sites, and the most active catalyst was Ce-Si-MCM-41 [51].

Wang's group described the isomerization of α-pinene over MSU-S mesoporous materials. They used MSU-S(BEA) and MSU-S(Y) with various Si/Al ratio and prepared by direct assembly of nanoclustered zeolitic precursors, i.e., BEA and Y, respectively. The isomerization of pinene was carried out in the liquid phase and the catalytic behavior was compared with that of conventional zeolites and mesoporous materials. The catalyst MSU-S(BET) with Si/Al ratio of 67 showed the highest activity as 97% conversion of substrate and 91% yield to main products were obtained at 343K. Scheme 7 is shows the pathway of α-pinene isomerization [52]. Launay et al. [53] have studied the catalytic activity of mesoporous Ga-SBA-15 materials in α-pinene isomerization. The reaction was carried out at 353K for 1 h. The main products were camphene and limonene, which made more than 50% of the products formed. Conversion of pinene decreased with increasing Ga content [53].

2.1.4.2. Epoxidation of α-Pinene

Epoxides are obtained by oxidation of C=C bonds with the use of peroxides. These compounds are key crude materials in organic chemistry for the production of chemicals such as: alcohols, carbonyl compounds or ethers, etc. The epoxidation reactions with the use of heterogeneous catalysts such as metal complexes supported on microporous or mesoporous materials, titanium silicates, etc. have been widely studied in literature.

Corma et al. [54-55] have synthesized Ti-beta for the first time. As expected Ti-beta, which has larger pores than TS-1, is able to catalyze the epoxidation of larger olefin substrates and epoxidations with the more bulky *tert*-butyl hydroperoxide (TBHP). On the other hand titanium silicalite gave high activity and selectivity of oxidation reactions with a use of hydrogen peroxide [54-55]. The same group has obtained the mesoporous materials of MCM-41 containing titanium [54-55]. Chiker and co-workers [56-57] used mesoporous materials, such as MCM-41, SBA-15, HMS as supports of titanium oxide for the epoxidation of pinene with TBHP as an oxidant. Epoxide was produced with 100% selectivity and 91% conversion of pinene for Ti-SBA-15. The epoxidation of pinene with hydrogen peroxide as oxidant at 343K led to constant selectivity, however the conversion decreased to 35% [56-57].

2.1.4.3. Acetoxylation of α-Pinene

Machado et al. [58] have used SBA-15 with sulfonic acid groups for acetoxylation of pinene. The main products of acetoxylation of pinene are acetates, such as: α- terpinyl acetate, bornyl acetate or β-fenchyl acetate and hydrocarbons, such as: camphene, limonene, terpinolene or α-terpinene. The same authors have studied the influence of different parameters, for example: temperature, initial concentration of substrate, catalyst loading. In acetoxylation of pinene, the conversion increased with increasing surface area and pore volume of the catalysts [58].

2.1.5. Limonene – 1,8-p-Menthadiene

Limonene [Figure 5] is a derivative of menthane, it belongs to the group of monocyclic monoterpenes. This terpene occurs in two forms of optically active enantiomers, because limonene has an asymmetric carbon atom. The smell of the two enantiomers is defined as a pleasant lemon scent.

(+)-Limonene is found in the following oils: orange, lemon, cumin and celery, while (-)-limonene is a component of the essential oil of spruce, pine and fir. The racemic mixture of these limonenes called dipentene, is present in camphoric and bergamot oils. Dipentene can be obtained synthetically by: dimerization of isoprene at 553 K, thermal isomerization of α and β pinene and pyrolysis of natural rubber. Limonenes and dipentene are also used as solvents for paints and varnishes, to obtain p-cymene, terpin hydrate and terpineols. Anticancer activity of limonene has been well-documented [59-61]. The use of limonene in the aroma compositions is limited by the 7th Amendment to the Cosmetics Directive of the European Union, according to which the limonene is classified as a potential allergen. Properties of limonene are shown it Table 6.

Table 6. Properties of limonene

Formula	$\mathbf{C_{10}H_{16}}$
Molecular weight [g/mol]	136.24
Boiling point [K]	451
Density [g/cm^3]	0.8411
Reflective index n^{20}_D	1.4726
Optical rotation $[\alpha]_D^{20}$	+ or – 126.3°

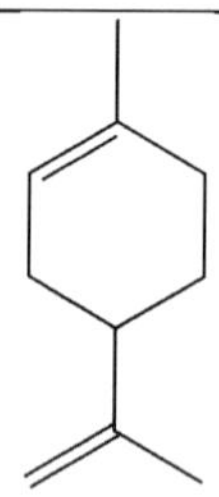

Figure 5. Limonene

2.1.5.1. Epoxidation of Limonene to Limonene Oxide

Limonene-1,2-epoxide occurs in natural sources and it is used in fragrances. Scheme 8 shows the potential products obtained during the epoxidation of limonene.

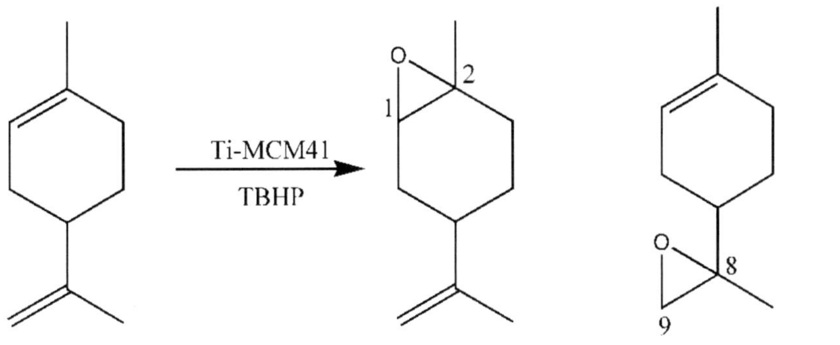

Scheme 8. The potential products obtained during the epoxidation of limonene.

Ti-zeolites and Ti-mesoporous materials have been used for epoxidation of limonene by Hutter and co-workers [62]. They have applied TBHP as an oxidant and dried silica-titania gel. The best result they have got was the yield of about 87%. Corma et al. [63] have studied the use of mesoporous material containing titanium (Ti-MCM-41) for the epoxidation of limonene with TBHP. They have achieved the yields of 1,2-limonene epoxide of 80%. Carniato and co-workers [64] have used Ti-POSS immobilized onto mesoporous silica for the epoxidation reaction. Polyhedral oligomeric silsesquioxane (POSS) containing titanium was anchored on the surface of SBA-15 and non-ordered amorphous mesoporous silica. The reaction was carried out at 358K in a glass batch reactor for 24h with TBHP as an oxidant. In the epoxidation of limonene all titanium-containing catalysts were active (Table 7). Conversion and selectivity values for Ti-POSS/SBA-15 and Ti-POSS/SiO_2 were comparable [64].

Table 7. Catalytic performance of the materials in epoxidation of limonene [adapted from 63, 64]

Catalyst	Conversion of limonene after 24h [%]	Selectivity to limonene monoepoxide after 24h [%]
Ti-MCM-41	94	98
Ti-POSS/SBA-15	25	88
Ti-POSS/ SiO_2	39	85
Ti/SBA-15	48	78
Ti/SiO_2	60	82
No catalyst	4	not determined
SiO_2	5	not determined

Oliviera et al. [65] have synthesized MCM-41 according to the method studied by Corma and co-workers [66]. Preparation of 3-hydroxypropyltrimethylsilyl functionalized MCM-41 was performed according to Scheme 9a.

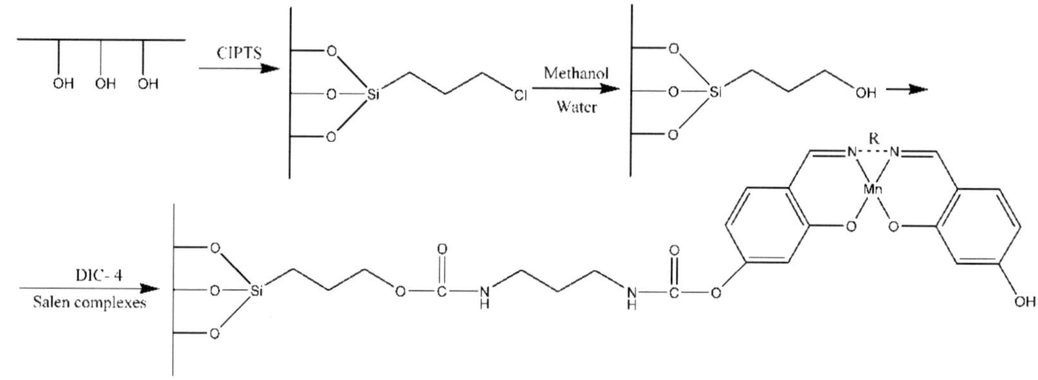

Scheme 9a. Preparation of 3-hydroxypropyltrimethylsilyl functionalized MCM-41 (CIPTS = 3-chloropropyltrimethoxysilane; DIC-4 = 1,4-diisocyanatobutane) [adapted from 65, 66].

Mesoporous material MCM-41 support was dehydrated and added to a 3-chloropropyltrimetoxysilane solution in anhydrous toluene, then this mixture was refluxed for 6h. Methanol was distilled off and the mixture was refluxed for 6h. The samples was filtered, washed and extracted (mixture of diethyl ether and dichloromethane). OHTPS@MCM-41 (3-hydroxypropyltrietoxysilyl functionalized materials) was synthesized by Oliveira's group [64] according to the synthesis proposed by Venkatesan and co-workers [67]. Scheme 9b presents the different manganese complexes prepared and anchored on the mesoporous material, MCM-41, surface by using 1,4-diisocyanatobutane (DIC-4) as a linking agent [65, 67].

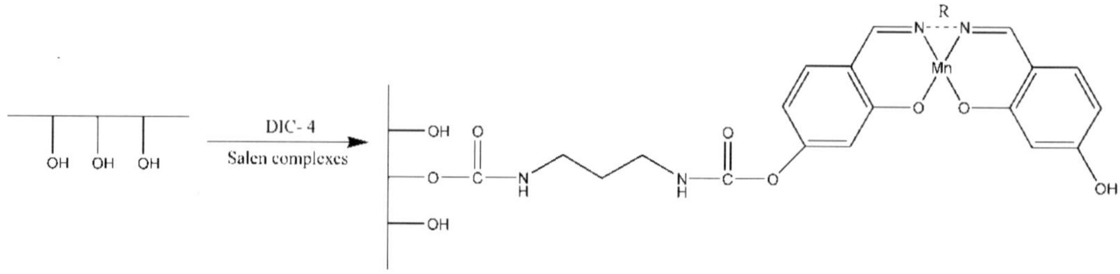

Scheme 9b. Preparation of 1,4-diisocyanatobutane functionalized MCM-41 [adapted from 65, 67]

MCM-41 anchored manganese salen complexes ([Mn(4-HOsalen)Cl], [Mn(4-HOsalhd)CH_3COO], [[Mn(4-HOsalophen)Cl]) were used as catalysts for limonene oxidation by Oliveira et al. [65].

They studied the limonene liquid phase oxidation with diluted TBHP (*tert*-butyl hydroperoxide) as an oxidant. Limonen oxidation was carried out at 333K, in a batch reactor, under atmospheric pressure. The main product was a polymer. Carveol, carvone and limonene oxide was obtained in small amounts. Table 8 shows the results of limonene oxidation over different catalysts.

Table 8. Limonene oxidation catalyzed by various salen catalysts [adapted from 65]

Catalyst	Conversion of limonene [%]	Selectivity to [%]				
		Polymer	Limonene oxide	Carveol	Carvone	Others
Mn(4-OHsalen)OHTPS-DIC@MCM-41	6.0	-	19.5	3.0	14.8	62.7
Mn(4-OHsalhd)OHTPS-DIC@MCM-41	80.5	63.8	2.2	0.4	3.3	30.3
Mn(4-OHsalophen)OHTPS-DIC@MCM-41	84.1	73.9	10.6	1.8	5.1	8.6
Mn(4-OHsalen) DIC@MCM-41	7.0	-	13.0	3.1	7.3	76.6
Mn(4-OHsalophen) DIC@MCM-41	51.4	72.0	3.8	1.2	1.0	21.0

Pirovano et al. [68], using the titanium containing silica catalysts, have synthesized limonene by rapid and straightforward method involving the reaction of oxidations. The catalysts Ti_4/SiO_2–D (Davisil silica) and Ti_4/SiO_2–N (Nippon-Kasei silica) were prepared by impregnation at room temperature of the tetranuclear ammonium citratoperoxotitanate(IV) complex $(NH_4)_8[Ti_4(C_6H_4O_7)_4(O_2)_4]x8H_2O$. These samples, after impregnation, were calcined at 823K for 3h under oxygen flow. The epoxidation test on (R)-(+)-limonene was carried out in a glass batch reactor at 355K using ethyl acetate as a solvent in the presence of the catalyst and TBHP as oxidant and mesitylene as internal standard. Table 9 shows the results for the catalytic limonene epoxidation with TBHP in the presence of catalysts.

Table 9. The results for the catalytic limonene epoxidation with TBHP in the presence of catalysts [adapted from 67]

Catalyst	Ti content [wt.%]	Conversion after 24h [%]	Selectivity to 1,2- monoepoxide of limonene [%]
Ti_4/SiO_2–D	1.26	65	85
Ti_4/SiO_2–N	1.19	65	82
Ti/MCM-41	1.88	68	62

The catalysts Ti_4/SiO_2–D and Ti_4/SiO_2–N ensured the highest conversion of limonene during the 24 h reaction. The catalyst Ti_4/SiO_2 displayed better selectivity to limonene epoxide than Ti-silica samples (reference sample) [68].

Guidotti et al. [69] have studied the use of Ti-MCM-41 in epoxidation of limonene. The catalyst was modified by post-synthesis silylation with hexamethyldisilazane. This reaction was carried out in a glass batch reactor at 363K using ethyl acetate as a solvent and TBHP as an oxidant. The conversion of limonene was around 90% for all the silylated samples. The main products of epoxidation of limonene were limonene mono-oxides [69].

Conclusion and Outlook

Mesoporous molecular sieves have been recently the objects of much attention owing to their scientific importance and great potential in practical application as catalysts. However, their applications often require incorporation of active sites in the walls of the mesopores.

A wide variety of terpenes are renewable, sustainable feedstock for the fine chemical industry, especially are an important source of intermediates for flavors and fragrances and vitamins A and E synthesis. Synthesis of fragrances by acid-base-redox reactions can be conducted by heterogeneous mesoporous catalysts. All researchers have the same view in the use of a heterogeneous catalyst, especially based on environmental safety considerations, simplifying methods in product separation, and its possibility to be re-used. In the fragrance synthesis, acid heterogeneous materials are commonly used compared with base heterogeneous catalysts. Mesoporous acid catalysts for preparation of fragrances are usually applied in acetalisation, esterification, isomerization, and rearrangement reactions, while heterogeneous base catalysts are applied in the condensation and isomerization reactions. The redox materials are active in oxidation reactions. The obtained fragrance chemicals can be used in cosmetic, pharmaceutical and chemical industry.

References

[1] Sing, K.S.W.; Everett, D.H.; Haul, R.A.W.; Moscou, L.; Pierotti, R.A.; Rouquerol, J.; Siemieniewska, T. *Pure Appl. Chem.* 1985, *57*, 603-619.

[2] Berger, R.G. *Flavours and Fragrances,* Springer: Berlin, 2007.

[3] Breitmaier, E. *Terpenes. Flavours, Fragrances, Pharmacia, Pheromones,* Wiley-VCH, 2006.

[4] Steffen, M.; Lucas, M.; Brandner, A.; Wollny, M.; Oldenburg, N.; Claus, P. *Chem. Eng. Technol.* 2007, *30*, 4, 481-486.

[5] Swift, K.A.D.; *Top Catal.* 2004, *27*, 143-155.

[6] Corma, A.; Renz, M.; Susarte, M. *Top Catal.* 2009, *52*, 1182-1189.

[7] Stolle, A.; Gallert, T.; Schmőger, Ch.; Ondruschka, B. *RSC Advances* 2013, *3*, 2112-2153.

[8] McMarry, J. E. *Organic chemistry,* 8th ed.;, Brooks, 2012.

[9] Maleew, A.; Stoyanov, S.; Neshev, G. *Rosanol-Bulgarian Rose Oil Pharma- Cologic and Clinical Investigation,* Pharmachim, Sofia, 1973.

[10] Torssell, K.; *Natural Products Chemistry: A Mechanistic, Biosynthetic and Ecological Approach,* Stockholm: Swedish Pharmaceutical Press, 1997.

[11] Evans, W.C.; *Trease and Evans' Pharmacognosy,* 15th ed.; Saunders: Londyn, 2002.

[12] Mann, J J.; Davidson, R.S.; Hobbs, J.B.; Banthorpe, D.V.; Harborne, J.B. *Natural Products: Their Chemistry and Biological Significance,* Londyn: Longman, 1994.

[13] Dewick, P.M. *Medicinal Natural Products: A Biosynthetic Approach,* 2nd ed.; Wiley: Chichester, 2002.

[14] Barton, D.H.R.; Meth-Cohn, O.; Nakanishi, K. *Comprehensive natural products chemistry. Isoprenoids Including Carotenoids and Steroids*, Pergamon: Amsterdam, 1999; Vol. 2.

[15] Dey, P.; Harborne, J.; Banthorpe, D. *Methods in Plant Biochemistry. Terpenoids,* Londyn, 1991; Vol. 7.
[16] Kołodziejczyk A. *Natural Organic Compounds,* PWN: Warsaw, 2006.
[17] Bansal, R.K. *A Textbook of Organic Chemistry*, 4th ed.; New Age International Publisher: India, Delhi, 2005.
[18] Wanke, M.; Skorupińska-Tudek, K.; Swiezewska, E. *Acta Biochim Pol.* 2011, *48*, 663-672.
[19] Hunter, W.N. *J. Biol. Chem.* 2007, *282*, 21573-21577.
[20] Hemmerlin, A. *Plant Science*. 2013, *203–204*, 41-54.
[21] Eisenreich, W.; Bacher, A.; Arigoni, D; Rohdich, F. *Cell. Mol. Life Sci.* 2004, *61*, 1401-1426.
[22] Casuscelli, S.G.; Eimer, G.A.; Canepe, A.; Heredia, A.C.; Poncio, C.E.; Crivello, M.E.; Perez, C.F.; Aguilar, A.; Herrero, E.R. *Cat. Today*. 2008, *133-135*, 678-683.
[23] Backtorp, C.; Hagvall L., Borje, A.; Karlberg, A.T.; Norrby, P.O.; Nyman, G. *JCTC,* 2008, *4*, 101-106.
[24] Surburg, H.; Panten, J. *Common Fragrance and Flavor Materials,* Viley-VCH, 2006.
[25] Marin-Astorga, N.; Martinez, J.J.; Borda, G.; Cubillos, J.; Suarez, D.N.; Rojas, H. *Top Catal.* 2012, *55*, 620-624.
[26] Feliczak-Guzik, A.; Nowak I. *Cat. Today.* 2009, *142*, 288-292.
[27] Silva, M. de S.; Vinhado, F.S.; Mandelli, D.; Schuchardt, U.; Rinaldi, R. *Journal of Molecular Catalysis* A: Chemical. 2006, *252*, 186-193.
[28] Somma, F.; Strukul, G. *J. Catal.* 2004, *227*, 344-351.
[29] Hoegaerts, D.; Sels, B.F.; de Vos, D.E.; Verpoort, F.; Jacobs, P.A. Cat. Today. 2000, 60, 209-218.
[30] Dapurkar, S.E.; Kawanami, H.; Chatterjee, M.; Rode, C.V.; Yokoyama; T.; Ikushima, Y. *Appl. Catal. A: Gen.* 2011, *394*, 209-214.
[31] Dorbes S.; Pereira, C.; Andrade, M.; Barros, D.; Pereira, A.M.; Rebelo, S.L.H.; Araújo, J.P.; Pires, J.; Carvalho, A.P.; Freire, C. Microporous Mesoporous Mater. 2012, *160*, 67–74.
[32] Santiago-Pedro, S.; Tamayo-Galvan, V.; Viveros-Garcia, T. Catal. Today 2013, *213*, 101-108.
[33] Didillon, B.; El Mansour, A.; Candy, J.P.; Basset, J.M.; Le Peletier, F.; Boitiaux, J.P. *Stud. Surf. Sci. Catal.* 1991, *59*, 137-143.
[34] Corma, A.; Iborra, S.; Mifsud, M.; Renz, M. *J. Catal.* 2005, *234*, 96-100.
[35] Boronat, M.; Corma, A.; Renz, M.; Viruela, P.M. *Chem. Eur. J.* 2006, *12*, 7067-7077.
[36] Zhu, J.; Li, M.; Lu, M.; Zhu, *J. Catal. Sci. Technol.* 2013, *3*, 737-744.
[37] Corma, A.; Iborra, S.; Velty A., *Chem. Rev.* 2007, *107*, 2411-2502.
[38] Vilella, I. M.J.; de Miguel, S.R.; de Lecea, C.S.M.; Linares-Solano, A.; Scelza, O.A. *Appl. Catal. A: General* 2005, *281*, 247-251.
[39] Singh, U.K.; Sysak, M.N.; Vannice, M.A. *J. Catal.* 2000, *191*, 181-191.
[40] Trasarti, A.F.; Marchi, A.J.; Apesteguia, C.R. *J. Catal.* 2004, *224*, 484-488.
[41] Coelho, J.V.; de Meireles, A.L.P.; M. da Silva, Rocha, K.A.; Pereira, M.C.; Oliveira, L.C.A.; Gusevskaya, E. V. *Appl. Catal. A: General* 2012, *443-444*, 125-132.
[42] Shieh, D.L.; Tsai, C.C.; Ko, A.N.; *React. Kinet. Catal. Lett.* 2003, *79*, 381-389.
[43] Maki-Arvela, P.; Kumar, N.; Nieminen, V.; Sjoholm, R.; Salmi, T.; Murzin, D.Y.; *J. Catal.* 2004, *225*, 155-169.

[44] Ekou, T.; Especel, C.; Royer, S.; *Cat. Today* 2011, *173*, 44-52.
[45] Singh G. *Chemistry Of Terpenoids And Carotenoids*, Discovery Publishing House, Vol. 3, 2007.
[46] Sell, Ch.S. *The Chemistry of Fragrances*, RSC Publishing, 2006.
[47] Solkinaa, Yu.S.; Reshetnikova, S.I., Estradab, M.; Simakovc, A.; Murzind, D. Yu.; Simakovaa, I.L. *Chem. Eng. J.* 2011, *176–177*, 42–48.
[48] Dimitrova, R.; Gunduz, G.; Spassova, M.; *J. Mol. Catal. A* 2006, *243*,17-23.
[49] Yamamoto, T.; Tanaka, T.; Funabiki, T.; Yoshida, S. *J. Phys. Chem. B* 1998, *102*, 5830-5839.
[50] Ecormier, M.A.; Lee, A.F.; Wilson, K. *Microporous Mesoporous Mater.* 2005, *80*,301-310.
[51] Stekrova, M.; Kumar, N.; Mäki-Arvela, P.; Ardashov, O.V.; Volcho, K. P.; Salakhutdinov, N. F.; Murzin, D. Yu. *Materials* 2013, *6*, 2103-2118.
[52] Wang, J.; Hua, W.; Yue, Y.; Gao, Z. *Bioresour. Technol.* 2010, *101*, 7224-7230.
[53] Launay, F.; Jarry, B.; Bonardet, J.L. *Appl. Catal. A: Gen.* 2009, *368*, 132-138.
[54] Corma, A.; Navarro, M. T.; Pariente, J. P. *J. Chem. Soc., Chem. Commun.* 1994, 147-148.
[55] Blasco, T.; Camblor, M. A.; Corma, A.; Esteve, P., Guil, J.; Martinez, M. A.; Perdigon-Melon, J. A.; Valencia, S. *J. Phys. Chem. B* 1998, *102*, 75-88.
[56] Chiker, F.; Nogier, J.P.; Launay, F.; Bonardet, J. L. *Appl. Catal. A: Gen.* 2003, *243*, 309-320.
[57] Chiker, F.; Launay, F.; Nogier, J. P.; Bonardet, J. L. *Green Chem.* 2003, 5, 318-322.
[58] Machado, J.; Castanheiro, J.E.; Matos, I.; Ramos, A.M.; Vital, J.; Fonseca, I.M.; *Microporous Mesoporous Mater.* 2012, *163*, 237-242.
[59] Bardon, S.; Foussard, V.; Fournel, S.; Loubat, A. Cancer Lett. 2002, *181*, 187-194.
[60] Gupta, A.; Myrdal, P.B.; *Int. J. Pharm.* 2004, *269*, 373-383.
[61] Stewart D. *The Chemistry of Essential Oils Made Simple: God's Love Manifest In Molecules*, Care publication, 2005.
[62] Hutter, R.; Mallat, T.; Baiker, A. *J. Catal.* 1995, *153*, 177-189.
[63] Corma, A.; Domine, M.; Gaona, J.A.; Jorda, J.L.; Navarro, M.T.; Rey, F.; Perez-Pariente, J.; Tsuji, J.; McCulloch, B.; Nemeth, L.T. *Chem. Commun.* 1998, *20*, 2211-2212.
[64] Carniato F.; Bisio, Ch.; Sordelli, L.; Gavrilova, E.; Guidotti, M.; *Inorg. Chim. Acta,* 2012, *380*, 244-251.
[65] Oliveira, P.; Machado, A.; Ramos, A.M.; Fonseca, I.; Braz Fernandes, F.M.;. Botelho do Rego, A.M.; Vital, J. *Catal. Commun.* 2007, *8*, 1366-1372 ; *Microporous Mesoporous Mater.* 2009, 120, 432-440.
[66] Corma, A.; Navarro, M.T.; Perez-Pariente, J.; Sanchez, F. *Stud. Surf. Sci. Catal.* 1994, *84*, 69-75.
[67] Venkatesan, C.; Singh, A.P. *J. Catal.* 2004, 227, 148-163.
[68] Pirovano, C.; Guidotti, M.; Dal Santo, V.; Psaro, R.; Kholdeeva, O.A.; Ivanchikova, I. D. *Catal. Today* 2012, *197*, 170-177.
[69] Guidotti, M.; Psaro, R.; Batonneau-Gener, I.; Gavrilova, E. *Chem. Eng. Technol.* 2011, *34(11),* 1924-1927.

In: Comprehensive Guide for Mesoporous Materials, Volume 4 ISBN: 978-1-63482-092-9
Editor: Mahmood Aliofkhazraei

Chapter 7

MESOPOROUS MATERIALS IN CHEMICAL ANALYSIS

Anna Olejnik, Joanna Goscianska and Izabela Nowak*
Adam Mickiewicz University in Poznań, Faculty of Chemistry, Poznań, Poland

ABSTRACT

Mesoporous materials are attractive adsorbents in analytical chemistry thanks to their specific properties such as large surface area, pore structure, possibility of surface functionalization to get hydrophobic/hydrophilic or positively or negatively charged surface. A combination of the size-exclusion effect of mesopores against large size molecules and the adsorption of small size analytes inside the mesopores actually mimics the effect of sample preparation before MS analysis. The use of mesoporous materials, such as ordered mesoporous silica and carbon, improves the efficiency and selectivity for profiling complicated samples, from human serum to cosmetic formulations. In this chapter a new applicability of mesoporous materials in sample analysis by using matrix assisted laser desorption/ionization mass spectrometry (MALDI-MS) is presented.

1. INTRODUCTION

Recent years have witnessed a rapid development of diverse techniques and methods in analytical chemistry. However, in order to achieve high quality of results a crucial step is still an appropriate sample preparation. High sensitivity and good specificity is especially important for complex and trace level analytes [1]. It is well known that reduction of sample complexity could improve detection of low-abundance compounds. This could be achieved by using mesoporous materials that have lately gained an increasing significance in sample preparation. The most studied mesoporous material is mesoporous silica, SiO_2. Other types of mesoporous materials are mesoporous titanium dioxide and mesoporous carbon. The mesoporous silica materials exhibit desirable properties such as highly ordered structure, large pore volume, large surface area and good mechanical stability [2].

* Corresponding author: Izabela Nowak. Adam Mickiewicz University in Poznań, Faculty of Chemistry, Umultowska 89b, PL61-614 Poznań, Poland. E-mail: nowakiza@amu.edu.pl.

Moreover, they possess tunable mesoporous channels with well-defined pore size distribution and controllable surface functionality [2, 3]. According to the IUPAC definition, mesoporous materials belong to the class of porous materials with the pore size between 2 and 50 nm [4]. The first mesoporous solid (M41S) was synthesized in 1992 by the Mobil Company [5, 6]. The best known representatives of M41S class are MCM-41, MCM-48 and MCM-50, which have different structures - hexagonal, cubic and lamellar, respectively. Other groups of mesoporous materials are SBA-family materials including SBA-1, SBA-3, SBA-15, SBA-16, and HMS and MSU. Ordered mesoporous materials are obtained by hydrothermal synthesis and a liquid templating mechanism [6-8]. So far, mesoporous materials have found various applications (Table 1) such as catalyst support [9], nano-templates [10], nanomaterials for advanced electronics [11], drug carriers [12] and also as good adsorbents [13-15].

Table 1. Some examples of mesoporous materials application

Mesoporous material	Application	References
SBA-15	Catalyst support	[9]
SBA-15	Nano-templates	[10]
Mesoporous carbon	Nanomaterials for advanced electronics	[11]
MCM-41	Drug carriers	[12]
SBA-3, SBA-15 SBA-16, KIT-6 C_{SBA-15}, C_{SBA-16} C_{KIT-6}, CMK-3, CMK-8	Adsorbents	[14-16]

The unique property of mesoporous materials such as size-controlled mesoporous structure, enables the size selective adsorption of small molecules but on the other hand the size-exclusion of larger molecules that permits the molecular weight cutoff in sample enrichment [3]. In addition, their large pore volumes and large surface areas are crucial and provide appropriate capacity for the adsorption of various compounds.

Furthermore, the silanol groups (Si-OH) located on the surface of mesoporous materials are responsible for binding with functional groups. Thus, the surface of the mesoporous materials can be modified by introduction of hydrophilic, hydrophobic, positively or negatively charged functional groups that can modify their adsorption properties. Therefore these materials have recently found applications in various analytical techniques including mass spectrometry. The chapter presents a new applicability of ordered mesoporous materials in sample analysis by using matrix assisted laser desorption/ionization mass spectrometry (MALDI MS). The main aim of this chapter is to review the recent achievements in this field.

2. MALDI Mass Spectrometry

Matrix assisted laser desorption/ionization mass spectrometry (MALDI-MS), known as the soft ionization MS method, was developed by Karas et al. [17] in 1985 and Tanaka et al. in 1988 [18]. Since that time this technique has become a powerful tool for analysis of a wide range of polymers and biomacromolecules such as peptides, proteins [19-21], oligonucleotides [22], oligosaccharides [23, 24], lipids [25] and synthetic macromolecules [26] (Table 2).

Table 2. Some examples of polymers analyzed by MALDI MS

Polymer	References
Proteins	[19-21]
Oligonucleotides	[22, 27, 28]
Oligosaccharides	[23, 24, 29]
Lipids	[25, 30, 31]
Macromolecules	[26, 32, 33]

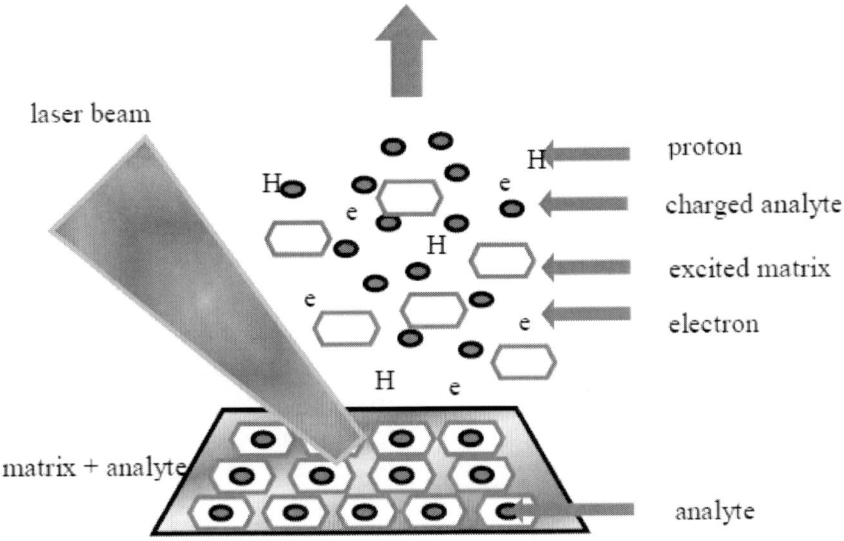

Figure 1. Desorption/ionization assisted by matrix.

In this method ions are desorbed from the solid phase. The sample analysed is initially dissolved in an appropriate solvent and then it is introduced to an excessive amount of a suitable matrix. Afterwards, the mixture is spotted on the MALDI target plate and air-dried [34]. Subsequently, the target plate containing the sample is directly inserted into the high vacuum of a mass spectrometer (below 10^{-6} mbar). The use of matrix enables the sample ionization, therefore the analyte is co-crystallized with matrix during the sample preparation. A UV laser beam (typically of a nitrogen laser of a wavelength of 337 nm) hits the sample-matrix crystal and excites the matrix that transfers the energy to the analytes (Figure 1). Subsequently, it leads to desorption and ionization of the sample analytes which are transformed into the gas phase [34]. Immediately after the ions have been formed, they are accelerated through the time-of-flight (TOF) mass analyzer, which is responsible for ion separation according to their m/z values. The best-known type of mass analyzer is an axial TOF. The ions fly down the time-of-flight tube and then they hit the detector. There are two types of TOF mass analyzers – a linear flight tube and a reflectron [34]. The choice of the appropriate matrix is crucial for successful MALDI experiments. However, there is no single matrix procedure and the matrix is usually selected through trial-and-error procedure.

The main role of the matrix in these experiments is to facilitate the analyte ionization. Therefore the typical matrices represent the group of UV absorbing organic compounds comprising an aromatic ring and carboxylic acid [35]. The matrix applied should:

- minimize the sample damage from the laser by absorbing the laser energy,
- mediate the energy and transfer protons to the analyte,
- be stable in high vacuum,
- not interact with the sample,
- be soluble in the solvent compatible with the analyte,
- have low molecular mass.

Table 3. Conventional matrixes applied in MALDI MS and their application

Matrix	Chemical structure	Application	Monoisotopic mass [Da]	References
α-cyano- 4-hydroxycinnmaic acid (CHCA)	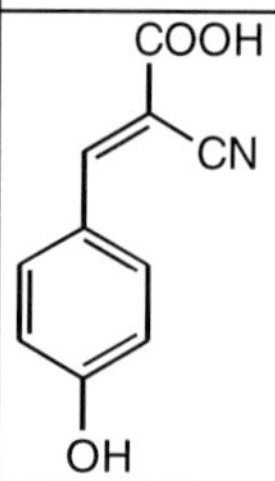	Peptides	189.04	[62, 63]
2, 5-dihydroxy benzonic acid (2,5 DHB)	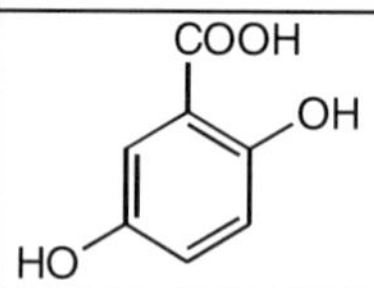	Proteins Peptides Carbohydrates Polymers	154.02	[64]
3,5-dimethoxy-4-hydroxycinnamic acid, sinapinic acid (SA)	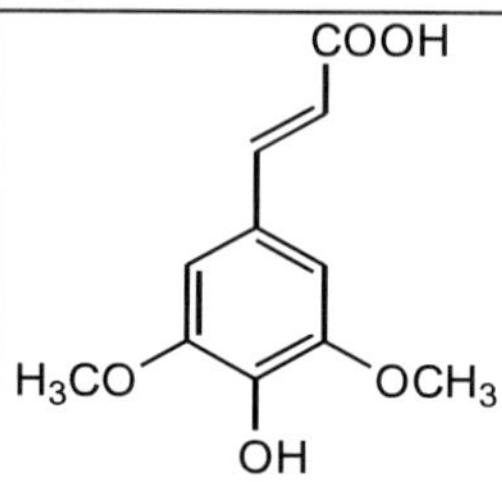	Proteins, Peptides Polymers	224.06	[65, 66]
5-chloro-2-mercapto-benzothiazole (CMBT)	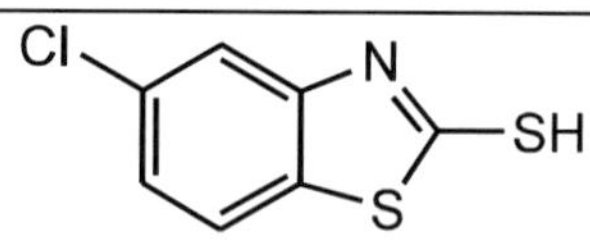	Protein, Peptides	201.69	[67]

It should be also mentioned that there are different sample preparation techniques in MALDI MS [36]. The most popular and highly applied is the dried droplet method, in which the analyte and the matrix are dissolved in a common solvent and the mixture is deposited on the MALDI target plate and is left for air drying until dry [37]. Another common preparation technique is known as the "thin layer" method. In this case the matrix solution is deposited on the MALDI target plate and is left to dry, so that a thin matrix layer is obtained. In the final step the analyte is applied on top [38]. Sometimes a "sandwich" method is applied, where an analyte layer is sandwiched between a matrix layer above and below. In some cases the solid/solid sample preparation is used where a solvent is not required. It means that the solid matrix and the solid sample are mixed, pulverized and subsequently a pastille is formed [39]. Other applied sample preparation methods are fast evaporation [40], vacuum drying [41] and two-layer [42] methods. The MALDI MS has been considered very convenient to analyze large molecules due to its high speed and throughput. Rapid and successful analysis of small molecules is also significant in medicinal chemistry and metabolomics [43].

This technique was unsuitable to determine the low-molecular-weight molecules (molecular weights below 500 Da), because of the interference of matrix-related peaks in the low-mass region [44, 45]. The use of traditional matrices such as α-cyano-4-hydroxycinnmaic acid and 2,5-dihydroxy benzoic acid (presented in Table 3) generates the strong background noise, which interferes with the target analytes. In order to overcome this problem various methods and alternative matrices have been introduced, such as inorganic materials [46], high mass molecules [47, 48], surfactant suppressed matrix [49], surface-assisted desorption/ionization [50, 51], desorption/ionization on silicon (matrix-free method) [52], carbon nanotubes [53], gold nanoparticles [54, 55], titanium dioxide nanoparticles [56], ionic liquids [57, 58], silicon nanoparticles [59, 60] and silicate [61]. The main purpose of using these materials was to reduce the peaks caused by the traditional matrices. An alternative matrix could be ordered mesoporous silica.

3. Application of the Ordered Mesoporous Materials in MALDI MS

There are many valuable attributes of mesoporous materials that can be useful in MALDI technique. Successful ionization/desorption can take place thanks to the correct and adjustable surface morphology and the proper pore size distribution of these materials. Moreover, the chemical activity of the surface can influence the adsorption capacity. Furthermore, the materials have large surface area and the surfaces that can contain active sites to which the analytes can be bound [68]. The utility of mesoporous materials as an additive matrix in MALDI-MS for analysis of compounds with low molecular weight has been described in several papers. Examples of the molecules analyzed by MALDI MS with the use of mesoporous materials are presented in Table 4 and Table 5.

It has been demonstrated that mesoporous materials are able to adsorb proteins and small bioactive compounds by the aperture adjustment and surface functionalization [69]. Mesoporous material MCM-41 with alkyl-diol group on the surface has been found useful for selective extraction of LMW proteins from complex biological samples [69]. Jun et al. [70] have applied MCM-41 type mesoporous silica as an additive matrix together with 2,5-dihydroxybenzoic acid (DHB) or α-cyano-4-hydroxycinnannic acid (CHCA) in MALDI-FT-ICR/MS for the analysis of peptides with different molecular weights such as Leu–Val, Phe–Val, Lue-Phe, bradykinin fragment 1–7, angiotensin II and ACTH fragment 18–39. The samples were prepared by dry-drop method. No peaks were detected when only the modified MCM-41 was used as a matrix. Therefore, it was applied as an additive matrix together with the conventional 2,5-dihydroxybenzoic acid (DHB). To reduce the number of ions of the matrix, the appropriate adjustment of laser power was much more important than the analyte to matrix ratio. Mesoporous materials are believed to have impact on peptide analysis because the laser-induced rapid temperature increase enhances the interaction of the analyte with Si-

OH located on the surface of MCM-41. While analyzing dipeptides, mesoporous materials were able to reduce DHB matrix peaks at low laser power. However, in the case of oligopeptides and polypeptides, no peaks of the molecular ions were found because the interaction between MCM-41 and analyte was too strong and the analyte could not desorb at low laser power. This could suggest that mesoporous materials suppress ionization of high molecular weight molecules such as oligopeptides and polypeptides. On the other hand, the use of mesoporous material as an additive matrix brings favorable effects for the detection of molecules below 500 Da with simple sample preparation [70].

A two-component matrix was also used to determine hexapeptide in cosmetic formulations, directly without initial processes such as separation or extraction of the component under determination. The matrix consisted of the classical matrix, DHB and supplementary material such as mesoporous SBA-15 (with mesopores diameter 4.3 nm and surface area of 800 m^2 g^{-1}). The use of additional matrix permitted a successful determination of hexapeptide by MALDI MS in such a complex sample like a cosmetic formulation [71] and intensive peaks of peptides were observed. A two-component matrix was also used to analyze the commercially available cosmetic peptides such as pentapeptide (Leuphasyl) and hexapeptide (Serilesine). A remarkable enrichment of low molecular peptides on the mesoporous silica based on their strong adsorption ability was reported. It should be mentioned that peptides adsorbed on the two-component solid matrix could be directly analyzed by using MALDI-TOF mass spectrometry. Because of the new sample preparation procedure, additional elution step has been eliminated and the risk of the potential sample mass loss is avoided [72].

The silica-based matrices have been also applied in the analysis of the low molecular weight molecules (<600 Da) such as two neurotransmitters, dopamine and serotonin [73], which are important in many regulatory pathways in the central nervous system. The new matrix was based on an ionic macro-complex between α-cyano-4-hydroxycinnmaic acid (CHCA, traditional matrix) and modified silica gel (silica gel Si-NH_3^+) or mesoporous silica (SBA-15-Si-NH_3^+). The aim of this study was to determine the relative utility of each matrix by taking into account the reduction of matrix peaks. It was found that the use of mesoporous silica ([SBA-15-Si-NH_3^+] [CHC^-]) in MALDI MS experiment has brought about minimal chemical noise and improved analyte signal intensity. The larger surface area of these mesoporous silica can explain this phenomenon. Moreover, large surface areas of these materials lead to more efficient laser desorption and ionization of the analyte [74, 75]. Therefore, the new matrix made of an ionic macro-complex based on mesoporous silica can be also attractive for the MALDI MS analysis of other low molecular compounds.

Similar studies have been carried out by Gao et al. [76]. An ionic macro-complex consisting of CHCA and SBA-15 was used for the analysis of melamine in milk by MALDI-TOF MS. A new matrix ([CHC^-] [NH^{3+}-Si-SBA-15-Si-NH^{3+}] [CHC^-]) was synthesized by covalently attaching SBA-15 to an ionic liquid comprising CHCA and 3-APTES (3-aminopropyltriethoxysilane). The performance of the new matrix was compared with that of the traditional CHCA matrix on the basis of two MALDI-TOF mass spectra of melamine. The mass spectrum obtained by using the modified SBA-15 matrix showed a more intensive peak assigned to melamine than that recorded with the use of the conventional matrix. Moreover, almost no background peaks in the low-mass region were detected.

The new matrix was also compared with the previously mentioned one ([SBA-15-Si-NH_3^+] [CHC^-] synthesized by Bach et al. [73]. The new matrix exhibited better reproducibility, higher homogeneity, required lower laser energy and minimized the effect of background interference ions [76]. The method's conditions were optimized and consequently a low detection limit of melamine (0.10 mg kg^{-1}) and high reproducibility of results were achieved.

Therefore, the new and high-throughput procedure to detect melamine in milk was developed. The new matrix has been suggested applicable for the analysis of amino acids and other low molecular weight compounds.

Another example of mesoporous materials application in the MALDI MS is the use of SBA-15 functionalized with 8-hydroxy quinoline group (SBA-15-8QSi) to determine saccharides, amino acids, metabolites and natural honey [77, 78]. The SBA-15 modified material was synthesized and then precisely characterized. The highly ordered 2D-hexagonal mesostructure and well-developed mesopores were confirmed by TEM. The surface area of SBA-15-QSi (668 m^2 g^{-1}) was smaller in comparison to that of unmodified SBA-15 (702 m^2 g^{-1}) because of the introduced functional groups (8QSi). SBA-15-8QSi exhibited less ion background noises and ensured better signal quality than those obtained when using the traditional 2,5-dihydroxybenzoic acid (DHB) matrix. Four different methods such as dried drop method, matrix-first two layer method, sample-first to later method and sandwich methods were tested to optimize the sample preparation. The sample-first two-layer method turned out to ensure the best performance in the analysis of low molecular compounds. In this method the SBA-15-8QSi matrix was placed as an upper layer on the analyzed sample thus the matrix could be dispersed on the sample surface. Therefore, a smoother and more homogeneous sample spot was obtained. In this method the laser first irradiated the matrix upper layer, increasing the local temperature and kinetic energy. Afterwards, the energy was transferred to the bottom layer allowing desorption and ionization of the molecules. The analytical conditions of the sample preparation were also optimized. Methanol turned out to be the ideal solvent and the analyte to matrix ratio of 3.5:10 was the most appropriate. After optimization of experimental conditions, the detection limit was 8 pmol with the RSD value below 29% [77]. It has been reported that the application of SBA-15-8QSi matrix for MALDI MS analysis of small molecules has a lot of advantages in comparison to the use of the traditional matrix such as better homogeneity and reproducibility of peak intensity and better desorption/ionization efficiency of the analytes [77]. SBA-15-8QSi has been successfully applied for determination of purity of natural honey containing low molecular saccharides such as xylose, glucose, sucrose and fructose [78].

The amount of peptides and proteins in biological samples may provide valuable information for clinical diagnosis [79, 80]. Therefore, the proper and selective analysis of peptides in the complex mixtures can be essential for better explanation and understanding the functions of endogenous peptides and for development of prognostic biomarkers [79]. The demand for reducing sample complexity to improve detection of low abundance peptides and proteins has stimulated development of various analytical techniques. MALDI-TOF MS could be a promising method for large-scale screening of body fluids (containing peptides and proteins) for detection of various human diseases.

Nanoporous silica particles have also found application in the selective binding of low-molecular weight compounds in human plasma [81]. The newly developed method allowed the detection of small peptides in the low-nanomolar range. Human plasma samples containing peptides and proteins were exposed to two types of silica with different pore sizes and various surface properties. A simple and fast sample preparation prior to MS analysis was developed.

The new procedure did not require the use of any column or other devices, therefore the risk of potential sample degradation was reduced. The silica particles were just suspended in human plasma and acted as sponges.

Table 4. Molecules analyzed by MALDI MS with the use of mesoporous silica materials

Type of MM	Molecules analyzed	MW	References
MCM-41	Leu–Val, Phe–Val, bradykinin fragment 1–7, angiotensin II ACTH fragment 18–39	230.3 264.3 756.4 1045.5 2464.2	[70]
SBA-15	Hexapeptide - ALA-ASP-LEU-LYS-PRO-THR	643.7	[71]
SBA-15	pentapeptide hexapeptide	569 615	[72]
[SBA-15-Si-NH_3^+][CHC^-] [silica gel-Si-NH_3^+] [CHC^-]	dopamine serotonin	153.08 176.09	[73]
SBA-15 SBA-15-8QSi	L-Arginine-HCl L-Histidine L-Lysine-HCl L-Phenylalanine L-Carnosine cytidine xylose sucrose	210.66 155.15 182.65 165.19 226.23 243.22 150.13 342.29	[77]
SBA-15 SBA-15-8QSi	xylose glucose sucrose fructose	150.13 180.16 342.29 180.16	[78]
([CHC^-] [NH^{3+}-Si-SBA-15-Si-NH^{3+}] [CHC^-])	melamine	126.12	[76]

Subsequently, the molecular species captured were profiled by MALDI-TOF MS. The obtained mass spectra of human plasma pretreated with silica nanoparticles showed a visible enrichment in the low-molecular weight proteome in comparison to those of the untreated plasma sample. It should be mentioned that the degree of enrichment depended on the silica type applied. When the silica pores were smaller, the lower molecular weight proteins were detected. However, as two different types of silica with two different classes of pore size were used, the proteins of a range of weight could be detected [81]. This phenomenon could be explained by different surface properties of silica and their ability to adsorb peptides and proteins.

The same research group continued their studies on developing the procedure based on mesoporous silica for selective binding and enrichment of low molecular weight plasma/ serum protein to be used for MALDI-TOF-MS [82]. In this paper they have presented a preparation protocol different than the previous one [81]. In this new protocol, mesoporous silica beads were incubated with the sample, washed and afterwards the suspension in CHCA was introduced on the MALDI target plate. The application of the new sample preparation protocol (suspension procedure) has brought a significant increase in the peaks intensity. Moreover, the background noises decreased. In addition, the sample spots obtained by two different procedures were analyzed by scanning electron microscopy (SEM). The results revealed a different behavior in matrix crystallization when the sample was introduced on the target plate by the suspension procedure. This could explain the better results of plasma analysis.

The functionalized mesoporous silica beads have also been used in MALDI-TOF MS for the profiling of human plasma and urine [83]. The adsorption properties of mesoporous silica were modified in order to modulate peptide enrichment. The surfaces of mesoporous materials were functionalized by introducing aminopropyl, N-(2-aminoethyly)-3-aminopropyl, and N,N,N'-tris-carboxymethyl ethylene diamine (TED). At first the urine and plasma peptides were extracted from modified mesoporous materials and then they were profiled by MALDI-TOF mass spectrometry. The results have proved that the functionalization of mesoporous silica with aminopropyl group increases specific plasma peptide subset recognition when compared to that of unmodified mesoporous materials [83]. It was found that the introduction of TED groups on the mesoporous silica increased the intensity of the peak assigned to urine peptides. These results proved that depending on the group introduced on the surface of mesoporous materials, different mass spectra of the same sample can be obtained.

The high-throughput screening MS-based platform for monitoring key peptidic-biomarkers for inflammatory lung and chronic respiratory diseases in induced sputum samples has been reported [84]. The sputum peptides were captured from both derivatized and non-derivatized mesoporous silica beads (MSBs) and then profiled by MALDI-TOF MS. Six m/z peaks emerged as potential diagnostic peptidic patterns able to differentiate these inflammatory airway diseases in the sputome range.

Another group of molecular sieve materials that have found application in MALDI-TOF MS are ordered mesoporous carbons discovered in 1999 by Jun et al. [85]. Ordered mesoporous carbon materials (OMC) are similar to mesoporous silica because they have mesoporous structure and large surface area. OMC are successfully applied in many fields due to their extremely well ordered pore structure, large pore volume and tunable pore diameters in the mesopore range [86]. These materials were used to obtain size-selective enrichment of N-linked glycans by MALDI-TOF MS [87]. The characterization of N-linked glycans is of great interests because it is informative in medical diagnostics [88, 89]. The effective and extensive determination of glycans is important to understand biological processes. A new method for their determination has been proposed by Zou et al. [87]. The mesoporous carbon (CMK-3) was synthesized and subsequently oxidized to O-CMK-3 by nitric acid [90, 91]. The oxidized mesoporous carbon could be well dispersed in aqueous solution while the unmodified material was poorly soluble. The obtained materials were characterized by various methods such as Fourier transform infrared spectroscopy (FT-IR), X-ray diffraction (XRD) and transmission electron microscopy (TEM). The results proved highly ordered mesoporous structure of CMK-3 with the pore size of ca. 3.4 nm. This pore size was appropriate to allow the intrapore capture of glycans. Moreover, the interference of the released protein residuals could be avoided as a result of the size-exclusion mechanism of mesopore to proteins [92]. In addition, it was reported that the large surface area (1576 m^2/g) of O-CMK-3 makes it promising enrichment adsorbent to capture glycans [87]. In this study the effective enrichment of the N-linked glycans from serum samples was obtained with the use of O-CMK-3. The N-linked glycans could be captured by the carbon material thanks to the large surface area, mesoporous structure and specific interaction between carbon and oligosaccharides in O-CMK-3. Twenty-four N-linked glycans derived from ovalbumin were successfully determined by MALDI-TOF MS with good repeatability, proper peak intensities and high signal-to-noise ratios. While analyzing the serum samples, 32 N-linked glycans were profiled and 5 were isolated from liver cancer and healthy samples [87]. The method presented above is promising for glycan analysis and can be potentially applied for early diagnosis of various diseases such as liver cancer. Taking into account of the advantage of the mesoporous structure and strong magnetic response, the new core-shell material synthesized in [93] was used to selectively extract low abundance endogenous peptides from human serum in the presence of high abundance of proteins. The mesoporous material was fabricated by two shells coated onto the Fe_3O_4 particles; the inner dense and thick silica shell could protect the magnetic core from harsh acidic solvents as well as induce the void between the core and the outer shell for the yolk–shell structure, while the outer organosilica shell was used as the template and carbon source for in situ preparation of a carbon shell of mesoporous structure [93]. Another mesoporous material that was applied in MALDI-TOF MS was ordered mesoporous tungsten titanium oxide (MTTO) [56]. This material was prepared by using triblock copolymer F127 as a structure-directing agent with the "acid-base pair" strategy. A highly ordered mesoporous material with two-dimensional hexagonal structure was successfully obtained. MTTO has a pore size of 9.8 nm, pore volume of 0.29 $cm^3\ g^{-1}$ and surface area of 150 $m^2\ g^{-1}$. Moreover, this material is able to absorb UV at 337 nm, therefore it was used as a matrix in MALDI-MS analysis to detect the cyclic decapeptide – Gramicidin S with a molecular weight 1140 Da. It is expected that MTTO might play a significant role while transferring energy from laser irradiation to the molecules of analyte. The mass spectra of Gramicidin S were obtained with a good signal-to –noise ratio when MTTO was used. In addition, the signal intensities of Gramicidin S with MTTO as a matrix were much stronger in comparison to the results obtained with nonporous tungsten titanium oxide. Therefore, the ordered MTTO is a promising material that can be applied in MALDI-MS to detect other biomolecules without using the conventional matrix, thanks to its ordered structured and the strong ultraviolet absorption ability. Chen and co-workers also took advantage of titania's capacity to absorb at 337 nm and carried out analyses of various molecules (200 Da-24 kDa) on titania sol-gel deposited on aluminum sheets, which formed mesoporous titania thin films [94]. The film was used as a matrix in MALDI MS.

Table 5. Molecules analyzed by MALDI MS with the use of other mesoporous materials

Type of MM	Analyzed molecule	References
ordered mesoporous carbons, O-CMK-3	N-linked glycans	[87]
magnetic mesoporous carbon microspheres with a yolk–shell structure, $Fe_2O_3@nSiO_2@mSiO_2$	low abundance peptides	[93]
ordered mesoporous tungsten titanium oxide, MTTO	cyclic decapeptide gramicidin S	[56]
mesoporous nanocrystalline titania sol–gel thin films	trypsinogen bradykinin	[94]

CONCLUSION

Mass spectrometry is one of the most powerful tools to analyze various complex samples. The extensive application of mass spectrometry is a result of its high sensitivity, high specificity and rapid analysis. So far, matrix assisted laser desorption/ionization mass spectrometry (MALDI MS) has been used to determine mostly the large non-volatile and labile molecules and its ability to detect low molecular compounds was limited due to matrix interference. The matrix signal can saturate the detector and as a consequence the detection efficiency of the small ions is decreased. The literature reviewed in this chapter revealed that mesoporous materials are attractive to be used in MALDI-TOF mass spectrometry. It has been reported that ordered mesoporous materials due to their stable mesoporous structure, large surface area and tailored size of mesopores are promising candidates for adsorption process. Moreover, due to the functionalization of the mesoporous materials surface with various organic and inorganic groups their properties can be changed. Thanks to these remarkable properties they can be used as an additive matrix together with the conventional matrix or even matrix-free in MALDI MS. The literature reviewed in this chapter has proved that a matrix consisting of mesoporous materials could exhibit better reproducibility, higher homogeneity and, what is the most important issue, could minimize the effect of background interference ions. The use of mesoporous materials enables facilitation of MS analysis. When analyzing biological samples, the use of mesoporous silica beads minimize the possible sample degradation. Recent studies on the synthesis and modification of mesoporous materials have revealed their great potential for protein and peptide enrichment. Mesoporous materials have the ability to harvest peptides and proteins from plasma and serum due to the existence of nanosized porous channels. Moreover, because of the adsorption properties of the ordered mesoporous materials, the complex samples could be analyzed directly without the need for time-consuming separation and extraction procedures. The new MALDI-TOF mass spectrometric detection method of small molecules can complement the existing analytical methods such chromatography, electrospray ionization mass spectrometry. In addition, the application of the ordered mesoporous materials in MALDI MS could be also useful for early diagnosis of various diseases.

REFERENCES

[1] Anderson, N. L., Anderson, N. G. *Mol. Cell. Proteomics* 2002, 1, 845.
[2] Ryoo, R., Joo, S., Kruk, M., Jaroniec, M. *Adv. Mater.* 2001, 13, 677.
[3] Zhao, L., Qin, H., Wu, R., Zou, H. *J. Chromatogr. A* 2013, 1228, 193.
[4] Sing, K. S. W., Everett, D. H., Haul, R. A. W., Moscou, L., Pierotti, R. A., Rouquerol, J., Siemieniewska, T. *Pure Appl. Chem.* 1985, 57, 603.
[5] Beck, J. S., Vartuli, J. C., Roth, W. J., Leonowicz, M. E., Kresge, C. T., Schmitt, K. D., Chu, C. T. W., Olson, D. H., Sheppard, E. W., McCullen, S. B., Higgins, J. B., Schlenker, J. L. *J. Am. Chem. Soc.* 1992, 114, 10834.
[6] Kresge, C. T., Leonowicz, M. E., Roth, W. J., Vartuli, J. C., Beck, J. S. *Nature* 1992, 359, 710.

[7] Monnier, A., Schuth, F., Huo, Q., Kumar, D., Margolese, D., Maxwell, R. S., Stucky, G. D., Krishnamurty, M., Petroff, P., Firoouzi, A., Janicke, M., Chemelka, B. F. *Science* 1993, 261, 1299.

[8] Karakassides, M. A., Bourlinos, A., Petridis, D., Coche-Guerente, L., Labbe, P. *J. Mater. Chem.* 2000, 10, 403.

[9] Huirache-Acuña, R., Nava, R., Peza-Ledesma, C. L., Lara-Romero, J., Alonso-Núñez, G., Pawelec, B., Rivera-Muñoz, E. M. *Materials* 2013, 6, 4139.

[10] Han, Y. J., Kim, J. M., Stucky, G. D. *Chem. Mater.* 2000, 12, 2068.

[11] Lee, J., Yoon, S., Oh, S. M., Shin, C. H., Heyon, T. *Adv. Mater.* 2000, 12, 359.

[12] Vallet-Regi, M., Ramila, A., del Real, R. P., Perez-Pariente. *Chem. Mater*, 2011, 13, 308.

[13] Liu, J., Shin, Y., Nie, Z., Chang, J. H., Wang, L. Q., Fryxell, G. E., Samuels, W. D., Exarhos, G. J. *J. Phys. Chem. A* 2000, 104, 8328.

[14] Goscianska, J., Olejnik, A., Pietrzak, R. *Mat. Chem. Phys.* 2013, 142, 586.

[15] Goscianska, J., Olejnik, A., Pietrzak, R. *Adsorption* 2013, 19, 581.

[16] Goscianska, J., Olejnik, A., Pietrzak, R. *J. Taiwan Inst. Chem. Eng.* 2014, 45, 347.

[17] Karas, M., Bachmann, D., Hillenkamp, F. *Anal. Chem.* 1985, 57, 2935.

[18] Tanaka, K., Waki, H., Ido, Y., Akita, S., Yoshida, Y., Yoshida, T. *Rapid Commun. Mass Spectrom.* 1988, 2, 151.

[19] Karas, M., Bahr, U., Ingendoh, A., Hillenkamp, F. *Angew. Chem. Int. Ed. Engl.* 1989, 28,760.

[20] Chan, T.-W. D., Colburn, A. W., Derrick, P. J. *Org. Mass Spectrom.*1992, 27, 53.

[21] Karas, M., Bahr, U., Ingendoh, A., Nordhoff, E., Stahl, B., Strupat, K., Hillenkamp, F. *Anal. Chim. Acta* 1990, 24, 175.

[22] Zagorevskii, D. V., Aldersley, M. F., Ferris, J. P. *J. Am. Soc. Mass Spectrom.* 2006, 17, 1265.

[23] Stahl, B., Steup, M., Karas, M., Hillenkamp, F. *Anal. Chem.* 1991, 63, 1463.

[24] Park, E., Yang, H., Kim, Y., Kim, J. *Food Chemistry*. 2012,134,1658.

[25] Balazy, M. *Prostag. Other Lipid Mediat.* 2004, 73, 173.

[26] Montaudo, G., Samperi, F., Montaudo, M. S. *Prog. Polym. Sci.* 2006, 31, 277.

[27] Sarracino, D., Richert, C. *Bioorg. Med. Chem. Lett.* 1996, 21, 2543.

[28] Streletskii, A. V., Kozlova, A. Yu., Esipov, D. S., Kayushin, A. L., Korosteleva, M. D., Esipov, S. E. *Russ. J. Bioorg.* 2005, 31, 139.

[29] Mechref, Y., Novotny, M. V., Krishnan, C. Anal. Chem. 2003, 75, 4895.

[30] Fuchs, B., Süß, R., Schiller, J. *Prog. Lipid. Res.* 2010, 49, 450.

[31] Fuchs, B., Schiller, J. *Eur. J. Lipid Sci. Technol.* 2009, 111, 83.

[32] Räder, H. J., Nguyen, T. T. T., Müllen, K. Macromolecules 2014, 47, 1240.

[33] Liu, J. S., Loewe, R. S., McCullough, R. D. *Macromolecules* 1999, 32, 5777.

[34] El-Aneed, A., Cohen, A., Banoub, J. *Appl. Spectrosc. Rev.* 2009, 44, 210.

[35] Bateson, H., Saleem, S., Loadman, P. M., Sutton, C. W. *J. Pharmacol. Toxicol.* 2011, 64, 197.

[36] Kussmann, M., Nordhoff, E., RahbekNielsen, H., Haebel, S., RosselLarsen, M., Jacobsen, L., Gobom, J., Mirgorodskaya, J., KrollKristensen, A., Palm, L., Roepstroff, P. *J. Mass Spec.* 1997, 32, 593.

[37] Karas, M., Hillenkamp, F. *Anal. Chem.* 1988, 60, 2299.

[38] Hanton, S. D., Cornelio Clark, P. A., Owens, K. G. *J. Am. Soc. Mass. Sepectrom.* 1999, 10, 104.
[39] Skelton, R., Dubois, F., Zenobi, R. *Anal. Chem.* 2000, 72, 1707.
[40] Vorm, O., Roepstroff, P., Mann, M. *Anal. Chem.* 1994, 66, 3281.
[41] Papac, D. I., Wong, A., Jones, A. J. S. *Anal. Chem.* 1996, 68, 3215.
[42] Dai, Y., Whittal, R. M., Li, L. *Anal. Chem.* 1999, 71, 1087.
[43] Shroff, R., Svatos, A. *Anal. Chem.* 2009, 81, 7954.
[44] Kinumi, T., Saisu, T., Takayama, M., Niwa, H. *J. Mass Spectrom.* 2000, 35, 417.
[45] McCombie, G., Knochenmuss, R. *Anal. Chem.* 2004, 76, 4990.
[46] Kinumi, T., Saisu, T., Takatyama, M., Niwa, H. *J. Mass. Spectrom.* 2000, 35, 417.
[47] Ayorinde, F. O., Garvin, K., Saeed, K. *Rapid Commun. Mass Spectrom.* 2000, 14, 608.
[48] Ayorinde, F. O., Hambright, P., Porter, T. N., Keith, Q. L. Jr. *Rapid Commun. Mass Spectrom.* 1999, 13, 2474.
[49] Guom, Z., Zhang, Q., Zou, H., Gou, B., Ni, J. *Anal. Chem.* 2002, 74, 1637.
[50] Finkel, N. H., Prevo, B. G., Velev, O. D., He, L. *Anal. Chem.* 2005, 77, 1088.
[51] Wu, J. Y., Chen, Y. C. *J. Mass Spectrom.* 2002, 37, 85–90.
[52] Trauger, S. A., Go, E. P., Shen, Z. J., Apon, V., Compton, B. J., Bouvier, E. S., Finn, M. G., Siuzdak, G. *Anal. Chem.* 2004, 76, 4484.
[53] Xu, S., Li, Y., Zou, H., Qiu, J., Guo, Z., Guo, B. *Anal. Chem.* 2003, 75, 6191.
[54] McLean, J. A., Stumpo, K. A., Russell, D. H. *J. Am. Chem. Soc.* 2005,127, 5304.
[55] Su, C.-L., Tseng, W.-L. *Anal. Chem.* 2007, 79, 1626.
[56] Yuan, M., Shan, Z., Tian, B., Tu, B., Yang, P., Zhao, D. *Micropor. Mesopor. Mat.* 2005, 78, 37.
[57] Li, Y. L., Gross, M. L., *J. Am. Soc. Mass Spectrom.* 2004, 15, 1833.
[58] Li, Y. L., Gross, M. L. *J. Am. Soc. Mass Spectrom.* 2005, 16, 679.
[59] Li, X. H., Wu, X. J., Kim, M., Kim, S. S., Jin, M. S., Li, D. H. *J. Am. Soc. Mass Spectrom.* 2009, 20, 2167.
[60] Wen, X., Dagan, S., Wysocki, V. H. *Anal. Chem.* 2007, 79, 434.
[61] Lee, C. S., Lee, J. H., Kang, K. K., Song, H. M., Kim, I. H., Rhee, H. K., Kim, B. G. *Biotechnol. Bioprocesses Eng.* 2007, 12, 174.
[62] Beavis, R. C. *Org. Mass Spectrom.* 1992, 27, 156.
[63] Crank, J. A., Amstrong, D. W. *J. Am. Soc. Mass Spectrom.* 2009, 20, 1790.
[64] Strupat, K., Karas, M., Hillenkamp, F. *Int. J. Mass Spectrom.* 1991, 111, 89.
[65] Beavis, R. C., Chait, B. T. *Mass Spectrom.* 1989, 3, 436.
[66] Beavis, R. C., Chait, B. Y. *Rapid Commun. Mass Spectrom.* 1989, 3, 432.
[67] Xu, N., Huang, Z.-H., Watson, J. T., Gage, D. A. *J. Am. Soc. Mass Spectrom.* 1997, 8, 116.
[68] Huang, L. C., Richman, E. K., Kirsch, B. L., Tolbert, S. H. *Micropor. Mesopor. Mat.* 2006, 96, 341.
[69] Qi, Y. X., Wei, J. Y., Wang, H., Zhang, Y. J., Xu, J., Qian, X. H., Guan, Y. F. *Talanta* 2009, 80, 703.
[70] Chen, J., Tianxi, H., Xuemei, F. *Talanta* 2012, 100, 419.
[71] Olejnik, A., Nowak, I., Schroeder, G., *Int. J. Pept. Res. Ther.* 2013, 19, 217.
[72] Olejnik, A., Schroeder, G., Nowak, I. *Extended abcstract, 22nd North American Meeting*, June 5-10, 2011.

[73] Mullens, C. P., Anugu, S. R., Gorski, W., Bach, S. B. H. *Int. J. Mass Spectrom.* 2011, 308, 311.

[74] Wei, J., Buriak, J. M., Siuzdak, G. *Nature* 1999, 399, 243.

[75] Canham, L. T., In: L. T. Canham (Ed.), *Properties of Porous Silicon*, Institution of Electrical Engineers, London, 1997, 83–88.

[76] Su, X., Zhou, H.-Y., Chen, F.-C., Gao, B.-X., Liu, Z.-W., Zhang, Y.-H., Liu, F., Liu, F., Li, Z.-R., Gao, Z.-X. *Int. J. Mass Spectrom.* 2013, 338, 39.

[77] Li, X., Wu, X., Kim, J. M., Kim, S. S., Jin, M. D., Li, D. *J. Am. Soc. Mass Spectrom.* 2009, 20, 2167.

[78] Li, X., Wu, X., Li, D. http://www.paper.edu.cn.

[79] Diamandis, E. P. *J. Proteome Res.* 2006, 5, 2079.

[80] Menschaert, G., Vandekerckhove, T. T. M., Baggerman, G., Schoofs, L., Luyten, W., Criekinge, W. V. *J. Proteome Res.* 2010, 9, 2051.

[81] Terracciano, R., Gaspari, M., Testa, F., Pasqua, L., Tagliaferri, P., Cheng, M. M. C., Nijdam, A. J., Petricoin, E. F., Liotta, L. A., Cuda, G., Ferrari, M., Venuta. *Proteomics* 2006, 6, 3243.

[82] Terracciano, R., Casadonte, F., Pasqua, L., Candeloro, P., Fabrizio, E. D., Urabni, A., Savino, R. *Talanta* 2010, 80, 1532.

[83] Terracciano, R., Pasqua, L., Casadonte, F., Frasca, S., Preiano, M., Falcone, D., Savino, R., *Bioconjugate Chem.*, 2009, 20, 913.

[84] Terracciano, R., Preiano, M., Palladino, G. P., Carpagnano, G. E., Barbaro, M. P., Pelaia, G., Savino, R., Maselli, R. *Proteomics*. 2011, 11(16) 3402.

[85] Ryoo, R., Joo, S. H., Jun, S. *J. Phys. Chem. B*. 1999, 103, 7743.

[86] Ndamanisha, J. C., Guo, L.-P. *Anal. Chim. Acta*. 2012, 747, 19.

[87] Qin, H., Zhao, L., Li, R., Wu, R., Wu, R., Zou, H. *Anal. Chem.* 2011, 83, 7721.

[88] Block, T. M., Comunale, M. A., Lowman, M., Steel, L. F., Romano, P. R., Fimmel, C., Tennant, B. C., London, W. T., Evans, A. A., Blumberg, B. S., Dwek, R. A., Mattu, T. S., Mehta, A. S., *Proc. Natl. Acad. Sci. US* 2005, 102, 779.

[89] Norton, P. A., Comunale, M. A., Krakover, J., Rodemich, L., Pirog, N., D'Amelio, A., Philip, R., Mehta, A. S., Block, T. M. *J. Cell. Biochem.* 2008, 104, 136.

[90] Jun, S., Joo, S. H., Ryoo, R., Kruk, M., Jaroniec, M., Liu, Z., Ohsuna, T., Terasaki, O. *J. Am. Chem. Soc.* 2000, 122, 10712.

[91] Bazula, P. A., Lu, A. H., Nitz, J. J., Schuth, F. *Micropor. Mesopor. Mat.* 2008, 108, 266.

[92] Tian, R., Zhang, H., Ye, Me., Jiang, X., Hu, L., Li, X., Bao, X., Zou, H. *Angew. Chem. Int. Ed.* 2007, 46, 962.

[93] Wan, H., Qin, H., Xiong, Z., Zhang, W., Zou, H. *Nanoscale*, 2013, 5, 10936.

[94] Chen, C. T., Chen, Y. C. *Rapid Comm. Mass Spec.* 2004, 18, 1956.

In: Comprehensive Guide for Mesoporous Materials, Volume 4 ISBN: 978-1-63482-092-9
Editor: Mahmood Aliofkhazraei

Chapter 8

Application of Mesoporous Materials for Drug Delivery

Grigoriy Sereda
University of South Dakota, Vermillion, SD, US

Abstract

This chapter addresses recent trends, advances, and challenges of the design and synthesis of mesoporous materials as drug delivery vehicles. Due to their ability to adsorb a drug or cocktail of drugs, be covalently linked to a drug delivering functionality, penetrate the cell membrane, and release the drug inside the cell upon changing pH, mesoporous materials are attracting increasing attention from both biomedicine and nanotechnology research communities. While structural, functional, and morphological versatility of silica nanoparticles have placed them into mainstream of research endeavors in the field and to the focus of the chapter, other mesoporous materials are also discussed as promising materials for drug delivery. This chapter overviews current synthetic strategies for assembling functional elements into an efficient drug delivery nanodevice and suggests potential directions for the field to grow in the future, which includes commercial applications.

1. Introduction

Biochemistry and molecular biology have always focused on the chemistry and properties of natural nanoparticles (proteins, peptides, nucleic acids), nanoassemblies (membranes, receptors), and microparticles (cells and their components). The expansion of nanotechnology over the last decade has resulted in a wide range of tools for biomedical applications with significant practical emphasis on drug delivery. A comprehensive review by Andrew Wang and coworkers published in 2012, summarizes utilization of a wide diversity of nanoparticles, ranging from metal nanoparticles to polynucleotides and synthetic block-polymers, for delivery of anti-cancer drugs [1]. A common drug delivering nanoassembly consists of a carrier nanoparticle (A), active drug, component responsible for affinity to the biological target (B), and surface modifiers controlling its permeability through biological barriers such as cell membranes. The drug delivery vehicle may also include a mechanism (C) for releasing the drug at the biological target (D) that can be triggered by the different chemical environment at the target, or by external stimuli such as illumination with light (Figure 1). For tracking and imaging purposes, a fluorescent label (usually the fluorescein derivative FITC) is often added to the system. Same components of the nanodevice may play multiple roles.

2. Morphology Control of Mesoporous Materials for Drug Delivery

Due to their unique morphology, mesoporous materials have shown significant advantages over other drug delivering nanoparticles. The high specific surface area of mesoporous materials allows for larger drug loads while the size-tunable pores can be utilized as controllable gates for the selective drug release. Limited diffusion rate between the particle and the environment may render unnecessary covalent conjugation of the nanoparticle with the drug, which alleviates concerns of its deactivation. The scope of available mesoporous materials includes mesoporous silica nanoparticles [2-6], hollow silica nanoparticles [7], titania nanoshells [8], mesoporous titania [9], and metal-organic frameworks (MOFs) [10]. Synthesis of these materials in the presence of organic templates allowing for the control over their pore size is extensively discussed in 1999 by Jackie Ying et al. [11] The pore size in a variety of mesoporous motifs (usually MCM-41, MCM-48, SBA-15) affects not only the drug loading capacity of the material, but also kinetics of drug loading and release [10]. In 2008, non-toxic and non-persistent in the body mesoporous hydroxyapatite nanoparticles were used as carriers of a model drug – *ibuprofen* adsorbed on their surface [12]. While most of the drug adsorbed on the outer surface was released into a neutral solution within one hour, it has taken 10h to release the remaining *ibuprofen* adsorbed inside the pores into solution [12].

By now, mesoporous silica nanoparticles have become one of the most common drug delivery vehicles with the capability of controlled drug release.

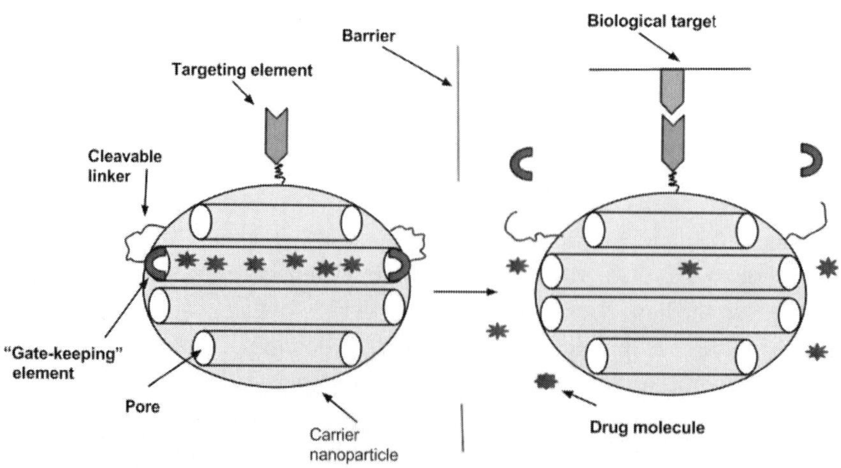

Figure 1. Components of a mesoporous drug delivery nanodevice.

In 2008, Victor Lin and coworkers have summarized reports showing that mesoporous silica nanoparticles are biocompatible, can be synthesized in various sizes and shapes, and easily functionalized with organic groups [13]. In 2001, a team of Spanish scientists discovered the ability of well ordered (MCM-41 type) mesoporous silica nanoparticles to function as a drug delivery system [14]. The rate of cellular uptake of spherical particles 100-200 nm by Human Fibroblasts and Chinese Hamster Ovarian Cancer cells is significantly higher, than for their tubular counterparts, and depends on the type of cells [15]. Shape and size of nanoparticles can be controlled by performing hydrolytic condensation of tetraethoxysilane in the presence of differently substituted triethoxysilanes. As a general rule, increased hydrophilicity of the silyl side chains favors formation of larger particles [16]. The reported ability of mesoporous silicas MCM-41 and SBA-15 to hold absorbed enzymes without their denaturation and loss of activity significantly contributes to their value as drug delivery vehicles [5]. Most common types of mesoporous drug carriers are listed in Table 1.

3. Biological Targets of Drug Delivery Nanoassemblies

Due to the urgency and importance of advancing novel approaches to treating different types of cancers, malignant cells and their components remain almost exclusive biological targets of drug delivery materials. The ability of mesoporous silica nanoparticles to penetrate the cellular membrane without the need of an affinity molecule such as antibody, was utilized to deliver a membrane-impermeable Cytochrome C inside Humane Cervical Cancer Cells [5]. This work expanded the use of such materials for transmembrane delivery of genes and other membrane-impermeable chemicals [17-20]. Particles of the size 100 nm or less are considered optimal for drug delivery purposes [1]. Therefore, silica-based mesoporous assemblies, which are readily available in this size range, show a great promise of shaping the field in the future because of their ability to deliver both biopolymers and small molecule drugs to cancer cells. Thus in 2012, mesoporous silica nanoparticles loaded with *paclitaxel* have been employed for targeting a variety of cancer cells expressing nucleolin, namely mouse fibroblast cells, breast cancer cells, and human glioma cells [2].

Table 1. Examples of mesoporous drug carriers and their preparation

Drug carrier	Technique for achieving mesoporocity	Delivered drug	References
Mesoporous silica nanoparticles	Organic template	*Variety of drugs*	[2-6, 10]
Hollow silica nanoparticles	Etching of calcium carbonate core with acid	*Model Brilliant Blue F*	[7]
Titania nanoshells	Etching of silica core with ammonia	*Camptothecin*	[8]
Mesoporous titania nanoparticles	Natural structural feature	*Doxorubicin*	[9]
Metal-organic frameworks	Natural structural feature	*Ibuprofen*	[10]
Hudroxyapatite	Organic template	*Ibuprofen*	[12]

The particles were delivered to the target by a covalently attached aptamer *AS1411* due to its affinity to nucleolin, which leads to inhibition of the binding of nucleolin to bcl-2 mRNA and induces apoptosis [2]. At the same time, the molecules of *paclitaxel* diffusing out of the particle, inhibited mitosis of cells by stabilizing of microtubules and, therefore, inhibiting their depolymerization back to tubulin [2]. Aptamers, which are short oligonucleotides by their chemical nature, are gaining popularity as a more stable, available, and modifiable alternative to more traditional antibodies. In this application, an aptamer was used both as a drug and a drug delivering component with a wide range of affinity allowing to simultaneously target several types of cancer cells. Simultaneous delivery of multiple drugs that act on different segments of the targeted process dramatically increases effectiveness of the therapy and allows for usage of lower dosages, and, therefore, minimizes unwanted side effects. Due to their large surface area and absorptive capability, mesoporous materials are uniquely suited as tools for carrying out this strategy. Modification of their surface with charged groups allows to fine-tune absorptive capacity of nanoparticles toward charged molecules of drugs. Thus, Huixin He and coworkers loaded *doxorubicin* into mesoporous silica nanoparticles functionalized with isocyanato-groups [4]. Next, the positive charge was imparted to the surface by covalent linking of isocyanato-groups to the generation 2 amino-terminated polyaminoamine dendrimer (PAMAM). Finally, the second drug (and at the same time the drug delivering component), negatively charged *Bcl-2 siRNA*, was electrostatically conjugated to the particle. The resulting material has shown high efficacy of internalization in the perinuclear region of multidrug-resistant human ovarian cancer cells, inducing cell death [4]. The *siRNA* delivered simultaneously with *doxorubicin*, suppressed m-RNA encoding the Bcl-2 protein responsible for nonpump resistance. As a result, doxorubicin bypasses pump resistance, which increases its efficacy by the factor of 132 [4].

A relatively new mesoporous material, titania hollow nanoshells, was introduced in 2012 by the group of Jyongsik Jang for intracellular delivery of *camptothecin* to human breast cancer cells [8]. The particles were prepared by growing porous titania shells around silica nanoparticles followed by dissolution of the templating silica by aqueous ammonia, and then functionalized with amino-groups by hydrolyzing an amino-derivative of trimethoxysilane on their surface. After conjugation of the surface amino-groups with an antibody *herceptin*, the particles were loaded with *camptothecin* to produce a material that showed high toxicity to cancer cells while maintaining low toxicity to normal cells [8]. While low solubility of silica in aqueous ammonia casts a doubt on the claim that the templating silica was actually removed from the titania shells, regardless, the mesoporous nature of silica has allowed for the drug absorption and, therefore, utilization of the material as a drug delivery vehicle.

Mesoporous materials are also explored for non-targeted drug delivery, or their time-controlled release. For instance, MCM-41 was found to slowly release *captopril* (a hypertension drug) [21] and *ibuprofen* [14] into aqueous medium. Adsorption and release of ibuprofen from mesoporous silica can be controlled by organic functionalization of the silica surface [22] as well as pore connectivity and geometry [23]. A more comprehensive list of such examples can be found in the review of Maria Vallet-Regi and co-workers [10]. Hydrogels containing mesoporous silica nanoparticles are finding growing application in regenerative medicine. A general strategy of simultaneous release of a cocktail of hydrophilic and hydrophobic drugs for wound healing was developed on the basis of mats containing mesoporous silica nanoparticles [24]. The composite hydrogels of chitosan and mesoporous silica nanoparticles were shown to simultaneously release model drugs (bovine serum albumin and gentamicin in a sustained manner [25]. Those composite hydrogels performed much better than chitosan hydrogels in supporting growth of chondrocytes and therefore can be useful for cartilage regeneration [25].

4. Controlled Drug Release

As opposed to non-functionalized non-mesoporous matrixes that release almost all of the absorbed drug within the first hour, non-functionalized mesoporous materials generally show a diffusion release that follows first-order kinetics, such as *alendronate* (osteoporosis drug) – loaded MCM-41 mesoporous silica [26]. The same drug loaded to amino-functionalized SBA-15 mesoporous silica followed a more desirable zero-order kinetics wherein the rate of release depends on time only [26]. This improvement is achieved due to stronger electrostatic attraction between positively charged amino-groups of the silica matrix and negatively charged phosphonato-groups of *alendronate*.

Although the mesoporous particle carrying a drug to its biological target significantly suppresses exposure of surrounding tissues to the drug, extensive research is performed to minimize release of the drug before it reaches its intended destination. Release of the drug is usually triggered by a different chemical composition of the medium at the biological target, however, sometimes it is induced by external stimuli such as illumination with light. The usual trend of higher acidity inside cancerous cells due to their higher metabolic rates makes the pH-value the simplest chemical stimulus that may trigger drug release. Silica is uniquely suited for this purpose due to the fact that its surface ζ-potential changes from -25.5 mV at pH 7.35 to -1.81 mV at pH 5.22 [5]. Due to such drastic drop of the ζ-potential absolute value, the electrostatic attraction between the loaded drug and the mesoporous matrix significantly weakens, which induces the drug release. This mechanism worked especially well for acid-triggered release of *cytochrome c* (that maintains its ζ-potential +15 and +17 mV between pH 5.5 and 8.0) inside human cervical cancer cells [5].

A more elaborate system consists of a cationic polyelectrolyte that obstructs pores of a drug-loaded SBA-15 mesoporous silica due to electrostatic attraction to ionized carboxy-groups introduced to the surface [27]. Lowering pH decreases the amount of ionized carboxy-groups, releases the polyelectrolyte from the pores, and dumps the drug to the solution. However, even after acidification to pH 5.0 – 6.0, a significant fraction of carboxy-groups (pK_a = 4.5-5.0) remains ionized, which still holds some "gate-keeping molecules" in the pore and necessitates development of alternative stimuli-responsive drug release mechanisms. Several of those systems, including thermo- and light-responsive assemblies (based on reversible photodimerization of coumarin [28] or azobenzene [29]), are described in the Maria Vallet-Regi [10] and Viktor Lin [13] reviews. Among the works cited in the review, it is especially worth mentioning utilization of CdS [19] and Au [30] nanoparticles as "gate keeping" agents inducing release of antibiotics, nucleic acids, genes, and factors of their expression in the presence of antioxidants cleaving disulfide bridges, that hold the particles at the "gate".

A rather interesting approach was taken by the group of Huixin He that employed *bcl-2 siRNA* as both a drug, drug-delivering component, and part of the gate-keeping system, releasing the second drug, *doxorubicin* inside the human ovarian cancer cell when exposed to glutathione, which is present in relatively high concentrations in many types of drug resistant cancer cells [4]. In this system, the RNA is electrostatically held by the PAMAM dendrimer covalently linked to the nanoparticle's surface. Interaction of polar glutathione with the resulting complex blocking the pore, opens it up and quantitatively releases the drug [4].

In 2011, a group of researchers from Spain has presented a proof of the concept that molecules of a peptide covalently attached to a mesoporous silica nanoparticle, can act as "gatekeepers" releasing a dye from the particle upon having been cleaved by a peptidase [3]. In that same year, the group of Kaiyong Cai has attached an extracellular matrix component (collagen) to a fluorescein-loaded mesoporous silica nanoparticle through a disulfide linker [31]. After being grafted with lactobionic acid, the nanoassembly was able to deliver the model fluorescein to *HepaG2* and endothelial cells and release it there because the disulfide linkage was reductively cleaved by glutathione present inside the cells. Interestingly, the grafted lactobionic acid has enabled cell-selective delivery of the drug to *HepaG2* (twice the amount delivered to endothelial cells) because the galactose group present in lactobionic acid is facilitating receptor-mediated endocytosis [32]. The redox-responsive nature of the drug release was confirmed by a control experiment with dithiothreitol as a release trigger [31].

Probably, among the most promising applications of controlled drug release is mimicking functions of endocrinal glands such as the pancreas. The group of Viktor Lin has constructed mesoporous silica nanoparticles functionalized with phenylboronic acid, loaded with cyclic adenosine monophosphate *(cAMP)* and having their pores sealed with gluconic acid-functionalized insulin *(G-ins)* bound to the boronic acid [33]. Those particles were internalized by rat pancreatic cells and were exposed to glucose inside the cells. Due to the higher affinity of boronic acid for glucose, than for noncyclic diols, glucose replaces gluconic acid-functionalized insulin and releases it into solution. In addition, release of *G-ins* opens up the pores and releases *cAMP*, which stimulates natural secretion of insulin by the pancreatic cell. In this example, *G-ins* acts as both drug and the "gate-keeping" element, releasing the second drug, which circumvents the problem of suppressing natural insulin production by its artificial sources [33].

A different approach to the pH-controlled drug release was taken by Korean researchers in 2011 [34]. Mesoporous silica nanoparticles modified with covalently anchored urease were loaded with *doxorubicin* and placed in an acidic solution, containing Ca^{2+}, $H_2PO_4^-$ and urea. The enzymatic hydrolysis of urea produces ammonia, which increases pH at the nanoparticle's surface and induces mineralization of pores with calcium phosphate. When the particles were internalized by breast cancer cells, calcium phosphate dissolved in a more acidic intracellular environment, releasing toxic *doxorubicin* [34].

Alternatively, the acid-soluble "gatekeeping" nanoparticles can be pre-synthesized and anchored to the drug delivering system. Thus, a Chinese research group has selectively functionalized the interior surface of mesoporous silica nanoparticles with amino-groups, the exterior surface – with carboxy-groups, and loaded with *doxorubicin* [35]. The pores were then sealed with amino-functionalized ZnO quantum dots via the amide-linkage with the surface carboxy-groups. The nanoassembly was internalized by human cervical cancer cells and released *doxorubicin* due to dissolution of the ZnO caps in the acidic intracellular environment. Presence of the amino-groups on the interior surface obviates adsorption of positively charged molecules of doxorubicin and facilitates their release [35].

An original method of targeted drug delivery not dependent on chemical affinity of the delivery system to its biological target, employs magnetic iron oxide nanoparticles attached to the pores of fluorescein-loaded mesoporous silica nanoparticles through a disulfide linker [18]. The particles were endocytosed by human ovarian cancer cells and released the model fluorescein due to the cleavage of the disulfide linker by intracellular antioxidants. The iron oxide nanoparticles can function not only as "gatekeeping" caps, but also as a handle for delivery of the assembly to the target destination in the body using an external magnet [18]. More research on the development of stimuli-controlled valves in the pores of mesoporous materials is underway, such as a base- or salt-triggered release system based on a rotaxan complex [36]. Some of those nanoscale devices can be potentially developed into novel drug delivery nanoassemblies.

5. Assembly of Targeting Drug Delivery Systems

Design of an efficient drug delivery nanoscale vehicle includes multiple components to impart the following desired properties to the final product:

- The drug itself must be stable and not interact with other drugs in case of multi-drug delivery systems.
- Ability to reversibly bind sufficient number of drug molecules.
- Biocompatibility and ease of internalization by cells.
- Selectivity for the biological target.
- Controlled drug release at the biological target.
- A fluorescent or other label for imaging, diagnostics, or research and development purposes.

Table 2. Design of mesoporous drug delivery systems

Mesoporous platform	Delivered drug	Biological target	Targeted delivery	Controlled release	References
Mesoporous silica nanoparticles	*Cytochrome C*	Human cervical cancer cells	-	-	[5]
Mesoporous silica nanoparticles	*Genes and other membrane-impermeable compounds*	Various targets	-	-	[17-20]
Mesoporous silica nanoparticles	*Paclitaxel, Aptamer AS1411*	Cancer cells expressing nucleolin	Aptamer AS1411	-	[2]
Mesoporous silica nanoparticles	*Doxorubicin, Bcl-2 siRNA*	Various drug resistant cancer cells	Bcl-2 siRNA	Disrupting of hydrogen bonding with glutathione	[4]
Titania hollow nanoshells	*Camptothecin*	Human breast cancer cells	Herceptin	-	[8]
Mesoporous silica	*Captopril, Ibuprofen, drugs for wound healing,*	-	-	Controlled kinetics by surface charge and morphology	[14, 21, 24]

Table 2. (Continued)

Mesoporous platform	Delivered drug	Biological target	Targeted delivery	Controlled release	References
Gels of chitosan and mesoporous silica nanoparticles	*BSA, gentamicin*	-	-	Controlled kinetics by surface charge and morphology	[25]
Mesoporous silica nanoparticles	*Alendronate*	-	-	Controlled kinetics by surface charge and morphology	[26]
Mesoporous silica	*Various drugs*	-	-	Surface charge change triggered by pH, releases a "gate-keeping" cationic polyelectrolyte from the pores	[27]
Mesoporous silica	*Cholestane, model rhodamine*	-	-	Photo- and thermo-induced configurational changes or dimerization of "gatekeeping" azobenzene and coumarin	[28, 29]
Mesoporous silica nanoparticles	*Antibiotics, nucleic acids, genes, factors of their expression*	-	-	Cleavage of a disulfide bridge tethering "gatekeeping" CdS or Au nanoparticles to the pores	[19, 30]
Mesoporous silica nanoparticles	*A model dye*	-	-	Cleavage of a "gatekeeping" peptide from the pores	[3]
Mesoporous silica nanoparticles	*Model fluorescein*	HepaG2 and endothelial cells	Lactobionic acid	Cleavage of a disulfide bridge tethering "gatekeeping" collagen	[31]
Mesoporous silica nanoparticles	*Insulin-gluconic acid conjugate, cAMP*	Rat pancreatic cells	-	Glucose triggers release of the insulin-gluconic acid conjugate (also the "gatekeeping" element) bound to phenylboronic acid-functionalized nanoparticle.	[33]
Mesoporous silica nanoparticles coated by hydroxyapatite	*Doxorubicin*	Breast cancer cells	-	The "gatekeeping" hydroxyapatite shell is dissolved at increased acidity	[34]
Mesoporous silica nanoparticles	*Doxorubicin*	HeLa cells	-	The "gatekeeping" ZnO nanoparticles are dissolved at increased acidity	[35]
Mesoporous silica nanoparticles	*Model fluorescein*	Ovarian cancer cells	Iron oxide nanoparticles driven by an external magnet	Cleavage of a disulfide bridge tethering "gatekeeping" iron oxide nanoparticles	[18]
Mesoporous carbon	*Acetaminophen*	-	-	-	[37]

According to the examples already discussed in detail, a drug delivery system may contain all or some of the listed features. Sometimes artistically designed structural components perform more than one of those functions.

While several comprehensive recent reviews [10, 11] discuss types and specific applications of drug delivery systems based on mesoporous materials, this section will focus on synthetic approaches and challenges of putting together the necessary components into a functioning nanoassembly. Each structural element (molecule, nanoparticle, or other material) added to the system must not adversely interact with already introduced elements and with the functional groups necessary for conjugation with the next element. The discussed synthetic strategies geared toward silica mesoporous materials as most commonly used for drug delivery applications, are based on six general methods.

a) Ionization of the nanoparticle's surface (–Si-OH groups in the case of silica) to fine-tune the ζ-potential of the silica surface.
b) Electrostatic modification of the surface using charged scaffolds.
c) Topological grafting of functional elements into the mesoporous structure.
d) Direct anchoring of functional components into the mesoporous structure
e) Anchoring of nucleophilic groups to the surface, which enables covalent conjugation with electrophilic functional elements.
f) Anchoring of electrophilic and other groups to the surface, which enables covalent conjugation with nucleophilic functional elements.

a) The simplest method to modify the surface of silica is protonation or deprotonation of the Si-OH groups in an aqueous medium. The surface ζ-potential, that can vary from below -30 mV up to more than +13 mV in the physiological pH range 5.5-8.0 [5], has a significant effect on internalization of nanoparticles into cells, their exit from cells, and interaction with the loaded drug. Due to the negative charge of most cellular membranes, positively charged nanoparticles are better internalized by cells, but exit them slower [32]. The surface potential of the drug delivery systems can be adjusted by changing pH of solution and fine-tuned by modification of the surface with positively or negatively charged functional groups. The surface potential plays a significant role in the electrostatic attraction between the nanoparticle and the loaded drug, may trigger drug release and affect stability of the drug itself. In this respect, mesoporous silica nanoparticles are well suited for delivery of proteins that usually do not denature inside the pores, and retain their activity [5].

b) As already discussed, the drug as the key element of a drug delivery system, is usually held inside the pores by electrostatic forces and may be released when the attraction weakens. When alternative drug release mechanisms are employed, the electrostatic attraction between a positively charged drug molecule such as *doxorubicin* with the surface is intentionally suppressed by its modification with positively charged amino-groups [35]. Electrostatic attraction was very effective at holding a negatively charged molecule of *Bcl-2 siRNA* at the positively charged PAMAM dendrimer scaffold covalently attached to the drug delivery system [4].

The electrostatic or even physisorptive link between a drug delivery system and its functional components may turn out to be well more common than it is usually believed to be. Sometimes, utilization of chemically flawed procedures for covalent conjugation, luckily results in a functional drug delivery system.

Thus, a reported functional drug delivery system contained an aptamer covalently attached to mesoporous silica [2]. The surface amino-groups converted into carboxy-groups, and finally activated by N-hydroxysuccinimide (NHS) were used as anchors for the attachment [2]. However, the overall experimental procedure is lacking the step of grafting amino-groups into the surface. Therefore, in the final nanoassembly, the aptamer was held at the surface by the electrostatic forces or physisorption rather than a covalent linker.

c) Topological grafting of functional components into the mesoporous structure is very characteristic for this type of materials. When the exit pores are occluded with a gate-keeping element, the drug molecules trapped inside may not be chemically bound to the surface, but cannot leave the particle simply because there is no way out. This type of "topological bond' is also known for catenane and rotaxane-type structures. Topological grafting was also employed for deposition of a "gate keeping" element – calcium phosphate into the pores of mesoporous silica nanoparticles [34]. The size and shape of calcium phosphate crystallized in the outer region of the branched system of mesopores does not match the exit pore, which holds the crystals inside the assembly.

d) Direct grafting of functional components into a mesoporous material works especially well for alkaline hydroxyapatite, which can be directly modified by carboxy-terminated functional components via formation of surface calcium carboxylates [12]. The surface of silica is routinely modified by functional moieties containing a trialkoxysilyl group. Hydrolysis of this group at the surface of a silica nanoparticle or in the process of its formation covalently grafts the group into silica due to formation of Si-O-Si bonds. For instance, hydrolysis of trialkoxysilyl derivatives of fluorescein [6] and coumarin [28] at the surface of mesoporous silica nanoparticles has been employed to introduce a fluorescent label or coumarin (a photoactivated "gate keeping" molecule) into the drug delivery system. Due to significant affinity of titania toward the phosphate group, mesoporous titania nanoparticles have been successfully modified by a phosphate-containing fluorescent derivative of a nucleotide [9] Similarly to their silica counterparts, mesoporous titania nanoparticles are highly biocompatible, which is demonstrated by a high LC_{50} value of 400 $\mu g\ mL^{-1}$ toward the human breast cancer cell line BT-20 [9]. More often, however, a reactive "anchoring" group is grafted to the nanoparticle first, and then covalently conjugated to the functional element.

e) The most commonly grafted "anchoring" group is nucleophilic amino-group, which is either coupled with an electrophilic derivative of the functional element, or first converted to the carboxy-group by reaction with succinic anhydride. For instance, amino-groups grafted to pre-prepared mesoporous silica nanoparticles were converted to carboxy-groups used for conjugation with the "gate keeping" macromolecules of collagen [31]. Alternatively, amino-functionalized mesoporous silica nanoparticles have been prepared by co-condensation of a trialkoxysilylalkylamine with tetraethylorthosilicate (TEOS) [16]. This method of grafting is more likely to produce higher surface density of the functional group, especially deep inside the pores, but may also affect morphology of the nanoparticles [16]. Presently, selectivity of grafting into different areas of the mesoporous structure remains a little explored field. So far, it has been shown that the outer area of mesoporous silica nanoparticles can by selectively functionalized by carboxy-groups (grafting of amino-groups followed by carboxylation with succinic anhydride) before the surfactant is removed from the pores. Next, after the surfactant is removed, the inner area of the particles was grafted with amino-groups [35]. Notably, trialkoxysilyl derivatives can be also used for grafting functional groups into various non-silica surfaces, for instance, grafting of amino-groups into titania nanoshells [8].

f) As an alternative to modification of the grafted surface amino-groups, it is a common practice to directly graft to the surface of silica electrophilic and other groups needed for immobilizing functional elements. The nanoparticles can be functionalized either at the step of their synthesis, or as a post-synthesis grafting step or sequence. High reactivity of the isocyanato-group with amines, and low reactivity with many other nucleophiles makes trialkoxysilylalkylisocyanates reagents of choice for grafting the isocyanato-group to a mesoporous particle after it is already loaded with the drug. Thus, mesoporous silica nanoparticles loaded with a ruthenium dye were grafted with either isocyanato- or azido-groups [3]. The introduced isocyanato-group was next used for amide coupling with a peptide, while the azido-group was employed as a "click-reagent" with an alkyne-functionalized peptide [3]. Another example demonstrates introduction of the mercapto-group to mesoporous silica nanoparticles at the step of their synthesis [19]. Due to their affinity to Cd, the HS-groups were employed to attach CdS quantum dots to the pores as "gate keeping" elements. Functionalization of non-oxide-based nanoparticle platforms for drug delivery requires different surface chemistry. In 2013, Thomas Rybolt et al. have reported that mesoporous carbons including carbon nanotubes are able to store acetaminophen and release it in a timed fashion [37]. Interestingly, carbon nanotubes besides of being a nanoparticle platform, are cytotoxic themselves, and can be additionally activated by near infrared or microwave irradiation. Among numerous functionalization techniques for carbonaceous materials, dipolar addition of N-glycine derivatives to graphene in the presence of formaldehyde [38] seems to be non-destructive and most universal method for introduction functional groups to mesoporous carbon nanoparticles.

6. Commercial Applications of Mesoporous Drug Delivery Systems

The title of a recent manuscript in *Science* "Nanoparticle Trojan horses gallop from the lab into the clinic" [39] well represents the current trend in this field of nanotechnology. A series of conjugates of an anticancer drug with protein or liposome nanoparticles are either approved for clinical use, or going through different phases of clinical trials. By 2007, more than 60 drugs based on protein as nanovectors have been approved by the US Food and drug Administration (FDA), and around 360 candidates to treat more than 200 diseases were under clinical studies [10].

The most common protein-based drug nanocarrier is albumin, which interacts with hydrophobic molecules through reversible non-covalent binding. The size of the binding cavity allows to formally consider this system as mesoporous. The first commercial product based on this system is *Abraxane*®, approved for the treatment of breast cancer [1]. *Genexol-PM* is a *paclitaxel*-loaded polymeric nanoparticle approved for clinical use, while *BIND-014* is its *docetaxel*-carrying analogue with controlled-release property still under clinical study [1]. Among the most popular natural platforms are liposome-based drug delivery systems that also can be classified as mesoporous. It is estimated, that liposomes and gold nanocarriers account for 45% of the total addressable market. Liposomes will offer the largest addressable market ($15 billion) in 2021 while gold nanocarriers will see the highest compound annual growth rate – 53.8% - in the next decade [43].

Table 3. Chemistry of drug delivery system assembly

Mesoporous platform	Grafted group	Chemistry of grafting	Mode of grafting	Conjugated component	Chemistry of conjugation	References
Mesoporous silica nanoparticles	-	-	-	Proteins	Electrostatic	[5]
Mesoporous silica nanoparticles	Isocyanate	Condensation of a trialkoxysilane	Post-synthesis	PAMAM dendrimer	Carbamate link	[4]
Mesoporous silica nanoparticles	Carboxyl	Conversion of amino-groups	Post-synthesis	Aptamer	Amide bond	[2]
Mesoporous silica nanoparticles	Hydroxyapatite nanocrystals	Topological	Post-synthesis	Doxorubicin	Topological	[34]
Hudroxyapatite	Ibuprofen	Carboxylate salts	Post-synthesis	-	-	[12]
Mesoporous silica nanoparticles	Fluorescein, coumarin	Condensation of a trialkoxysilane	Post-synthesis	-	-	[6, 28]
Mesoporous titania nanoparticles	Phosphorylated fluorescent nucleotide	Titanium phosphate salt	Post-synthesis	-	-	[9]
Mesoporous silica nanoparticles	Carboxyl	Conversion of amino-groups	Post-synthesis	Collagen	Amide bond	[31]
Mesoporous silica nanoparticles	Amine	Co-condensation of a trialkoxysilane	During synthesis	-	-	[16]
Mesoporous silica nanoparticles	Amine, carboxyl	Sequential condensation of trialkoxysilanes	Post-synthesis	Aminated ZnO nanoparticles	Amide bond	[35]
Titania nanoshells	Amine	Condensation of a trialkoxysilane	Post-synthesis	Herceptin	Amide bond	[8]
Drug-loaded mesoporous silica nanoparticles	Isocyanate	Condensation of a trialkoxysilane	Post-synthesis	Peptide	Carbamate link	[3]
Drug-loaded mesoporous silica nanoparticles	Azide	Condensation of a trialkoxysilane	Post-synthesis	Alkynyl-peptide	Cycloaddition	[3]
Mesoporous silica nanoparticles	Thiol	Co-condensation of a trialkoxysilane	During synthesis	CdS quantum dots	Cadmium sulfide salt	[19]

Today, there are at least three liposomal formulations of doxorubicin (*DOXIL, NPLD*, and *DaunoXome*) [1]. Pegylated liposomal doxorubicin (*DOXIL*) is approved for treating advance breast cancer. Pegylation slows down release of the drug from ~105 nm vesicles, which increases the half-life time of doxorubicin in the bloodstream from < 5 min to 2-3 days. The ~150 nm vesicles of *NPLD* present another alternative for treating breast cancer. Liposomal vesicles of daunorubicin (*DaunoXome*) are ~45 nm in diameter and have been approved for the treatment of AIDS-associated Kaposi's sarcoma [1]. The liposome-based drug delivery systems are not limited to treating cancers. They are proved to be efficient to deliver *cytarabine* for disseminated limphomatous meningitis (*DepoCyte*). The review of Wang, Langer, and Farokhzad [1] contains a comprehensive list of nanoformulations under clinical and preclinical studies, which includes delivery antibiotics, alkaloids, and *si*RNA.

Another comprehensive manuscript [40] discuss drug delivery by non-nanoparticulate implantable devices. However, such systems may or may not be classified as mesoporous.

Despite the significant promise of mesoporous silica nanoparticles in nanomedicine, the drug delivering systems based on this platform, are just entering the preclinical development stage [42]. In 2011, a silica nanoparticle-based melanoma-targeting system was FDA approved for stage 1 human clinical trial [42]. Many other mesoporous silica-based delivery systems, especially ordered mesoporous ones are on the pipeline. In the next decade, one should expect an avalanche of new approved nanoformulations taking advantage of the sophisticated targeting and controlled drug release mechanisms overviewed in this chapter. The total market for nanoscale drug delivery systems in 2010 was $ 1 billion, including $ 178 million for targeted drug delivery [44]. A recent review [44] lists more than two dozens of drug delivery system that have entered the market between 1995 and 2009. Out of that list, the following systems based on the liposomal, virosomal platforms or nanotubes, can be considered mesoporous: *Doxil, DaunoXome, AmBisome, EpaxalBerna, Inflexal V Berna, Depocyt, Miocet, Visudyne, Invivac, DepoDur, Somatuline Depot, and Octocogalfa.*

CONCLUSION

By 2012, anti-cancer drug delivery systems based on mesoporous nanoparticles have not yet been among numerous non-mesoporous counterparts approved for the clinical use or clinical studies [1]. Nevertheless, this booming field is quite ripe to produce a series of revolutionary effective therapeutics for cancer, diabetes, and other diseases at the molecular level. Extensive choice of tools for synthesis, modification, and characterization of mesoporous materials is now available, which warrants a vigorous development of more elaborated and artistically designed nanodevices for targeted drug delivery in the next decade. The field of mesoporous materials for drug delivery will definitely benefit from looking back and reassessing the employed synthetic procedures, including their chemical correctness. Surprisingly, the issue of regioselective functionalization of mesoporous structures remains very little explored. The expected development of the field along those lines is bound to result in prospective practical applications in nanomedicine. The ability of nanoparticles to be functionalized by targeted delivery functionalities makes them especially useful in the wake of the paradigm of the healthcare market shifting away from a ‘one-size fits all’ approach to a more personalized medicine based on an individual’s unique genome and immune response [43]. “The more scientists learn about the molecular causes for disease, the more targeted and effective nanotechnology-enabled drug delivery therapies will become” [43].

REFERENCES

[1] Wang, A.; Langer, R.; Farokhzad, O. Nanoparticle Delivery of Cancer Drugs. *Ann. Rev. Med.* 2012, 63, 185-198.

[2] Aravind, A.; Veeranarayanan, S.; Poulose, A.; Nair, R.; Nagaoka, Y.; Yoshida, Y.; Maekawa, T.; Kumar, D. Aptamer-Functionalized Silica Nanoparticles for Targeted Cancer Therapy. *BioNanoSci* 2012, 2, 1-8.

[3] Coll, C.; Mondragon, L.; Martinez-Manez, R.; Sancenon, F.; Marcos, M.; Soto, J.; Amoros, P.; Perez-Paya, E. Enzyme-Mediated Controlled Release Systems by Anchoring Peptide Sequences on Mesoporous Silica Supports. *Angew. Chem. Int. Ed.* 2011, 50, 2138-2140.

[4] Chen, A.; Zhang, M.; Wei, D.; Stueber, D.; Taratula, O.; Minko, T.; He, H. Co-delivery of Doxorubicin and Bcl-2 siRNA by Mesoporous Silica Nanoparticles Enhances the Efficacy of Chemotherapy in Multidrug-Resistant Cancer Cells. *Small* 2009, 5, 23, 2673-2677.

[5] Slowing, I.; Trewyn, B.; Lin, V. Mesoporous Silica Nanoparticles for Intracellular Delivery of Membrane-Impermeable Proteins. *J. Am. Chem. Soc.* 2007, 129, 8845-8849.

[6] Lin, Y.; Tsai, C.; Huang, H.; Kuo, C.; Hung, Y.; Huang, D.; Chen, Y.; Mou, C. Well-Ordered Mesoporous Silica Nanoparticles as Cell Markers. *Chem. Mater.* 2005, 17, 4570-4573

[7] Li, Z.; Wen, L.; Shao, L.; Chen, J. Fabrication of porous hollow silica nanoparticles and their applications in drug release control. *Journal of Controlled Release*. 2004, 98, 245-254.

[8] Kim, C.; Kim, S.; Oh, W.; Choi, M.; Jang, J. Efficient Intracellular Delivery of Camptothecin by Silica/Titania Hollow Nanoparticles. *Chem. Eur. J.* 2012, 18, 4902 – 4908.

[9] Wu, K.; Yamauchi, Y.; Hong, C.; Yang, Y.; Liang, Y.; Funatsu, T.; Tsunoda, M. Biocompatible, Surface Functionalized Mesoporous Titania Nanoparticles for Intracellular Imaging and Anticancer Drug Delivery. *Chem. Commun.* 2011, 47, 5232-5234.

[10] Vallet-Regi, M.; Balas, F.; Arcos, D. Mesoporous Materials for Drug Delivery. *Angew. Chem. Int. Ed.* 2007, 46, 7548-7558.

[11] Ying, J.; Mehnert, C.; Wong, M. Synthesis and Applications of Supramolecular-Templated Mesoporous Materials. *Angew. Chem. Int. Ed.* 1999, 38, 56-77.

[12] Yang, P.; Quan, Z.; Li, C.; Kang, X.; Lian, H.; Lin, J. Bioactive, Luminescent and Mesoporous Europium-Doped Hydroxyapatite as a Drug Carrier. *Biomaterials* 2008, 29, 4341-4347.

[13] Slowing, I.; Viverto-Escoto, J.; Wu, C.; Lin, V. Mesoporous Silica Nanoparticles as Controlled Release Drug Delivery and Gene Transfection Carriers. *Advanced Drug Delivery Reviews* 2008, 60, 1278-1288.

[14] Vallet-Regi, M.; Ramila, A.; Real, R.; Perez-Pariente, J. A New Property of MCM-41: Drug delivery System. *Chem. Mater.* 2001, 1, 308-311.

[15] Trewyn, B.; Nieweg, J.; Zhao, Y.; Lin, V. Biocompatible Mesoporous Silica Nanoparticles with Different Morphologies for Animal Cell Membrane Penetration. *Chemical Engineering Journal* 2008, 137, 23-29.

[16] Huh, S.; Wiench, J.; Yoo, J.; Pruski, M.; Lin, V. Organic Functionalization and Morphology Control of Mesoporous Silicas via a Co-Condensation Synthesis Method. *Chem. Mater.* 2003, 15, 4247-4256.

[17] Gruenhagen, J.; Lai, C.; Radu, D.; Lin, V.; Yeung, E. Real-Time Imaging of Tunable Adenosine 5-Triphosphate Release from an MCM-41-Type Mesoporous Silica Nanosphere-Based Delivery System. *Appl. Spectrosc.* 2005, 59, 424-431.

[18] Giri, S.; Trewyn, B.; Stellmaker, M.; Lin, V. Stimuli-responsive controlled-release delivery system based on mesoporous silica nanorods capped with magnetic nanoparticles. *Angew. Chem. Int. Ed.* 2005, 44, 5038-5044.
[19] Lai, C.; Trewyn, B.; Jeftinija, D.; Jeftinija, K.; Jeftinija, S.; Lin, V. A mesoporous silica nanosphere-based carrier system with chemically removable CdS nanoparticle caps for stimuli-responsive controlled release of neurotransmitters and drug molecules. *J. Am. Chem. Soc.* 2003, 125, 4451-4459.
[20] Radu, D. R.; Lai, C.; Jeftinija, K.; Rowe, E.; Jeftinija, S.; Lin, V. A Polyamidoamine Dendrimer-Capped Mesoporous Silica Nanosphere-Based Gene Transfection Reagent. *J. Am. Chem. Soc.* 2004, 126, 13216-13217.
[21] Qu, G.; Zhu, S.; Huang, S.; Li, S.; Qiu, S. Effective controlled release of captopril by silylation of mesoporous MCM-41. *ChemPhysChem* 2006, 7, 400-406.
[22] Munoz, B.; Ramila, A.; Perez-Pariente, J.; Diaz, I.; Vallet-Regi, M. MCM-41 Organic Modification as Drug delivery rate Regulator. *Chem. Mater.* 2003, 15, 500-503.
[23] Andersson, J.; Rosenholm, J.; Areva, S.; Linden, M. Influences of Material Characteristics on Ibuprofen Drug Loading and Release Profiles from Ordered Micro- and Mesoporous Silica Matrices. *Chem. Mater.* 2004, 16, 4160-4167.
[24] Song, B.; Wu, C.; Chang, J. Controllable Delivery of Hydrophilic and Hydrophobic Drugs from Electrospun Poly(lactic-co-glycolic acid)mesoporous Silica Nanoparticles Composite Mats. *J. Biomed. Mater. Res. Part B* 2012, 100B, 2178-2186.
[25] Zhu, M.; Zhu, Y.; Zhang, L.; Shi, J. Preparation of Chitosan/mesoporous Silica Nanoparticle Composite Hydrogels for Sustained co-Delivery of Biomacromolecules and Small Chemical Drugs. *Sci. Technol. Adv. Mater.* 2013, 14, 045005.
[26] Balas, F.; Manzano, M.; Horcajada, P.; Vallet-Regi, M. Confinement and Controlled Release of Bisphosphonates on Ordered Mesoporous Silica-Based Materials. *J. Am. Chem. Soc.* 2006, 128, 8116-8117.
[27] Yang, Q.; Wang, S.; Fan, P.; Wang, L.; Di, Y.; Lin, K.; Feng-Shou, X. pH-Responsive Carrier System Based on Carboxylic Acid Modified Mesoporous Silica and Polyelectrolyte for Drug Delivery. *Chem. Mater.* 2005, 17, 5999-6003.
[28] Mal, N.; Fujuwara, M.; Tanaka, Y. Photocontrolled Reversible Release of Guest Molecules from Coumarin-Modified Mesoporous Silica. *Nature* 2003, 421, 350-353.
[29] Lu, J.; Choi, E.; Tamanoi, F.; Zink, J. Light-Activated Nanoimpeller-Controlled Drug Release in Cancer Cells. *Small* 2008, 4, 4, 421-426.
[30] Torney, F.; Trewyn, B.; Lin, V.; Wang, K. Mesoporous Silica Nanoparticles Deliver DNA and Chemicals into Plants. *Nat. Nanotech.* 2007, 2, 295-300.
[31] Luo, Z.; Cai, K.; Hu, Y.; Zhao, L.; Liu, P.; Duan, L.; Yang, W. Mesoporous Silica Nanoparticles End-Capped with Collagen: Redox-Responsive Nanoreservoirs for Targeted Drug Delivery. *Angew. Chem. Int. Ed.* 2011, 50, 640-643.
[32] Slowing, I.; Trewyn, B.; Lin, V. Effect of Surface Functionalization of MCM-41-Type Mesoporous Silica Nanoparticles on the Endocytosis by Human Cancer Cells. *J. Am. Chem. Soc.* 2006, 128, 14792-14793.
[33] Zhao, Y.; Trewyn, B.; Slowing, I.; Lin, V. Mesoporous Silica Nanoparticle-Based Double Drug Delivery System for Glucose-Responsive Controlled Release of Insulin and Cyclic AMP. *J. Am. Chem. Soc.* 2009, 131, 8398-8400.

[34] Rim, H.; Min, K.; Lee, H.; Jeong, S.; Lee, S. pH-Tunable Calcium Phosphate Covered Mesoporous Silica Nanocontainers for Intracellular Controlled Release of Guest Drugs. *Angew. Chem. Int. Ed.* 2011, 50, 8853-8857.

[35] Muhammad, F.; Guo, M.; Qi, W.; Sun, F. Wang, A.; Guo, Y.; Zhu, G. pH-Triggered Controlled Drug Release from Mesoporous silica Nanoparticles via Intracellular Dissolution of ZnO Nanolids. *J. Am. Chem. Soc.* 2011, 133, 8778-8781.

[36] Leung, K.; Nguyen, T.; Stoddart, J.; Zink, J. Supramolecular Nanovalves Controlled by Proton Abstraction and Competitive Binding. *Chem. Mater.* 2006, 18, 5919-5928.

[37] McCary, S.; Rybolt, Th. Storage and Timed Release of Acetaminophen from Poous Carbonaceous Materials. *Open J. Phys. Chem.* 2013, 3, 76-88.

[38] Quintana, M.; Grzelczak, M.; Browne, W.; Rudolf, P.; Prato, M. Functionalization of Graphene via 1,3-Dipolar Cycloaddition. *ACS Nano* 2010, 4(6), 3527-3533.

[39] Service, R. F. Nanoparticle Trojan Horses Gallop from the Lab into the Clinic. *Science* 2010, 330, 314-15.

[40] Fournier, E.; Passirani, C.; Montero-Menei, C. N.; Benoit, J. P. Biocopatibilty of Implantable Synthetic Polymeric Drug Carriers: Focus on Brain Biocompatibility. *Biomaterials* 2003, 24, 3311-3331.

[41] Vallet-Regi, M. Mesoporous Silica Nanoparticles: Their Projection in Nanomedicine. *Mater. Sci.* 2012, 608548.

[42] Rosenhalm, J. M.; Mamaeva, V.; Sahlgren, S.; Linden, M. Nanoparticles in Targeted Cancer Therapy: Mesoporous Silica Nanoparticles Entering Preclinical Development Stage. *Nanomedicine* 2012, 7, 1, 111-120.

[43] *Report by Cientifica Ltd. Nanotechnology for Drug Delivery: Global Market for Nanocarriers 2012*, 129 p.

[44] Kulve, H.; Rip, A. Economic and Societal Dimensions of Nanotechnology-Enabled Drug Delivery. *Expert Opinion* 2013, 10, 5, 611-622.

In: Comprehensive Guide for Mesoporous Materials, Volume 4 ISBN: 978-1-63482-092-9
Editor: Mahmood Aliofkhazraei

Chapter 9

MESOPOROUS SILICA NANOPARTICLES AS AN INNOVATION IN CONTROLLED DRUG DELIVERY

***Tasnuva Tamanna*[1], *Jurgen Bulitta*[2] *and Aimin Yu*[1,*]**

[1]Faculty of Science, Engineering and Technology, Swinburne University of Technology, Hawthorn, Australia
[2]Faculty of Pharmacy and Pharmaceutical Sciences, Monash University, Parkville, Australia

ABSTRACT

Research on mesoporous silica nanoparticles (MSNs) has explored an immense opportunity to develop an innovative type of nano-drug vehicle. Both systemic delivery and implantable local delivery devices were reported to be achievable with mesoporous silica material.

Mesoporous silica shows itself advantageous for drug loading and controlled release, due to a tunable morphological parameter, surface modification/functionalization capability, and biocompatibility properties. Functionalization of mesoporous silica can alter the drug loading capacity and release profile by changing the pore size and surface property. In this chapter, research works on MSNs' properties influencing drug loading and drug release profile, and functionalization of mesoporous materials for applications in drug delivery systems are reviewed to illustrate the significance of MSNs as nano-drug carriers and how surface modification can improve the drug loading and controlled drug delivery.

[*] Corresponding author: Dr Aimin Yu, Email: aiminyu@swin.edu.au, Tel: +61 3 9214 8161

1. INTRODUCTION

The broadening appliance of nanotechnology in the vital fields of science including electronics, aerospace, and defense, medical, and dental applications drew global attention. The innovation of ordered mesoporous silica material, by Mobil Corporation in 1992, conveyed a breakthrough in wide application [1]. Modulation over the particle size, morphology, pore structure, pore geometry, pore connectivity, dimension, and surface property expanded the significance of mesoporous silica nanoparticles (MSNs) in diverse spheres, particularly biomedicine, catalysis, environmental protection, optics, etc. [2]. According to the International Union of Pure and Applied Chemistry (IUPAC), porous materials exhibiting pore diameter of 2-50 nm range are categorized as mesoprorous materials [3]. The biomedical application of MSNs alone covers their efficiency as diagnostic tools, biosensors, enzyme immobilizers, biocatalysts, vectors for gene delivery, and so on (Table-1).

In 2001, M. Vallet-Regi et al., explored the potential of mobile crystalline material-41 (MCM-41) type MSNs in controlled drug delivery systems [11]. A silica based drug delivery system is promising, due to the biocompatible nature [12]. In general, biocompatibility, high loading efficiency, zero premature release, regulated and site-specific drug release, cellular internalization, etc. are fundamental considerations in designing an effective, successful, and reliable drug delivery system.

The development of a liquid crystal templating mechanism (LCT), or sol-gel process, improved the synthesis of MSNs [13]. Briefly, the sol–gel synthetic process encompasses hydrolysis and condensation reactions, thereby producing two new phases namely "sol" and "gel". In this synthetic procedure, silica precursors such as tetramethylorthosilicate (TMOS), or tetraethoxysilane (TEOS), undergoes hydrolysis and condensation to form a solution called "sol".

$$Hydrolysis: OH^- + Si(OR)_4 \rightarrow Si(OR)_3OH + RO^-$$

$$Condensation: SiO^- + Si(OH)_4 \rightarrow Si-O-Si + OH^-$$

Table 1. Biomedical applications of different types of MSNs

Types of MSNs	Significant property	Biomedical application	Reference
Mobile crystalline material-41 (MCM-41)	Highly efficient MRI contrast agents	Diagnostic tool	[4]
Hexagonal mesoporous silica (HMS)	Electrochemistry of immobilized hemoglobin (Hb) on HMS	Biosensor	[5]
MSNs, FSM-16 and MCM-41	Enzyme immobilization	Biomolecule carrier	[6, 7]
MSNs	Intracellular delivery of DNA and siRNA	Gene delivery	[8]
MSNs	Encapsulation of siRNA	siRNA carrier	[9]
MSNs	Encapsulation of plasmid DNA	Gene delivery	[10]

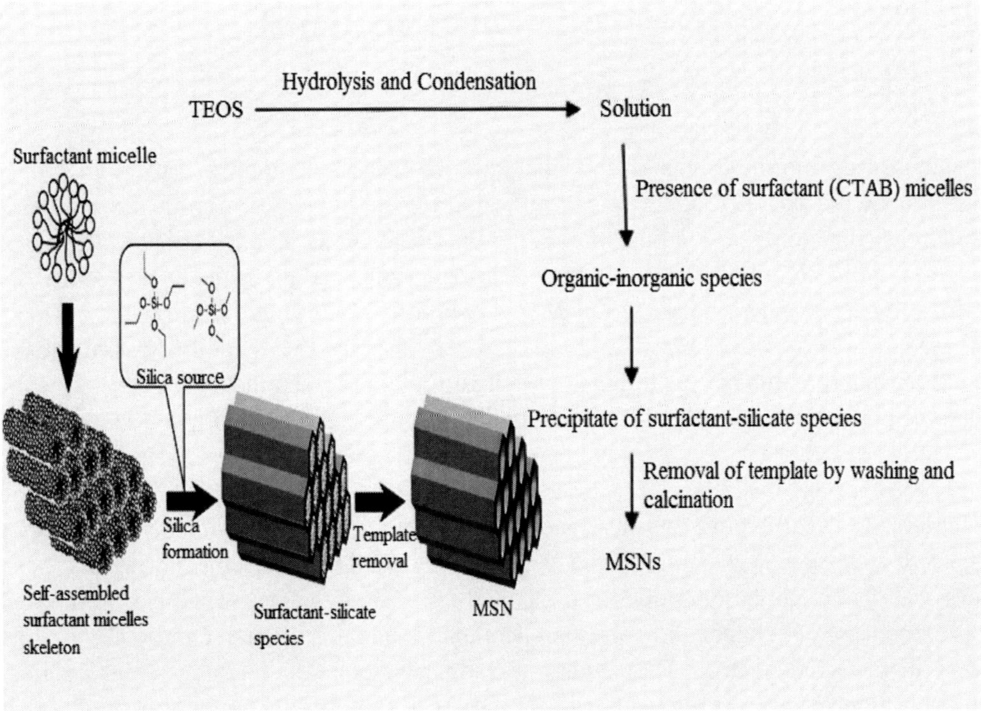

Figure 1. Schematic diagram of MSNs synthesis [17]. (Reproduced and slightly modified from Hoffmann, F. et al., *Angewandte Chemie International Edition* 2006, *45* (20), 3216-3251. With permission)

Due to the presence of a surfactant such as cetyltrimethylammonium bromide (CTAB) (most commonly used), hydrolysis and condensation of a silica precursor generates an organic-inorganic species, which is then polymerized and forms a "gel". Self-assembled of surfactant-silicate material, it eventually precipitates out of the solution (Figure 1). Finally, high temperature during calcination burns off the surfactants, eventually leaving MSNs [14]. Co-templating agents, for example, Pluronic F127 can also be used along with CTAB to alter the micelle packing behavior [15]. In addition, Pluronic F127 also helps to protect the particles from agglomeration and oligomerization. Furthermore, Pluronic F127 during MSNs synthesis serves as a particle size designer [16].

The inherent textural properties of MSNs display their worthiness to controlled drug delivery [18]. The unique feature includes

a. A homogeneous ordered pore network bestowing satisfactory regulation over drug entrapment and release kinetics
b. A high pore volume to accommodate sufficient drug quantity
c. A high surface area offering an extensive area for drug adsorption
d. A silanol-containing surface conveying a platform for functionalization to govern efficient control over drug loading and release

In this chapter we intend to cover MSNs' properties imparting drug loading and drug release profile, functionalization of MSNs surface, provoking themselves as effective drug delivery carriers, using different strategies to control drug delivery.

2. Drug Loading In MSNs

2.1. Approaches for Drug Loading

Inclusion of drug molecules in the host particles can be accomplished by in many ways. Drugs can be incorporated either after synthesis of drug carriers or during the synthesis of mesoporous silica (drug co-loading strategy). Sometimes, direct incorporation of drugs during the carrier synthesis minimizes steps of synthesizing and loading procedures. Nevertheless, the drug molecule must withstand the MSNs synthesis conditions to apply the co-loading strategy. However, when the condition applied for MSN synthesis facilitates drug loading, direct drug incorporation seems to be an attractive route. For example, Q. He and co-workers loaded the poorly water-soluble drug, doxorubicin, in MSN by means of a co-loading strategy [19].

They reported that the high temperature (80ºC) requirement of MSN synthesis accelerated the doxorubicin solubility instead of imparting significant damage. This method is associated with some problems. Likewise, the inserted drug molecule during the sol gel process can influence the final pore size of the drug-loaded vehicle, which can influence the release kinetics. P. P. Nampi and co-researchers incorporated gentamicin antibiotic via sol gel process in the silica matrix. They found variations in porosity features with variation in gentamicin concentration [20]. Although, the sol gel process is a unique method for drug incorporation, some common limitations include [21].

- The unlikely solubility of drug limiting the total amount drug loading amount
- Permanent entrapment of a fraction of drug in closed pores inhibiting to outside diffusion
- Compatible features of drug molecules to bear up with carrier synthesis condition and steps
- Difficulty in tuning the drug release profile due to interference of drug molecules towards porosity

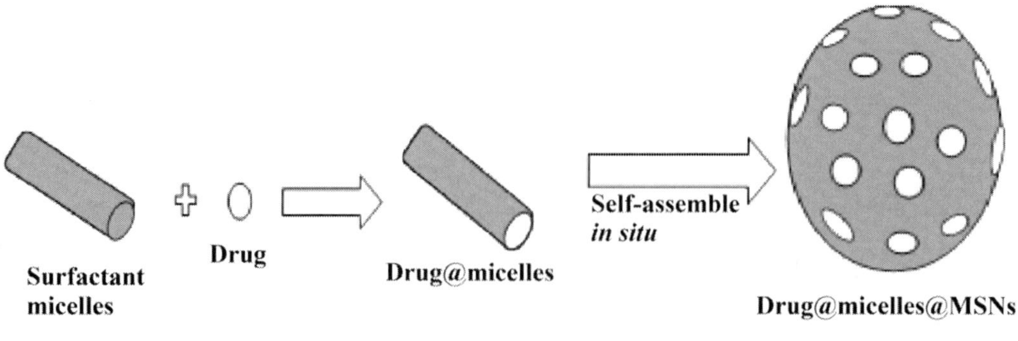

Figure 2. Drug loading in MSNs by co-drug loading strategy.

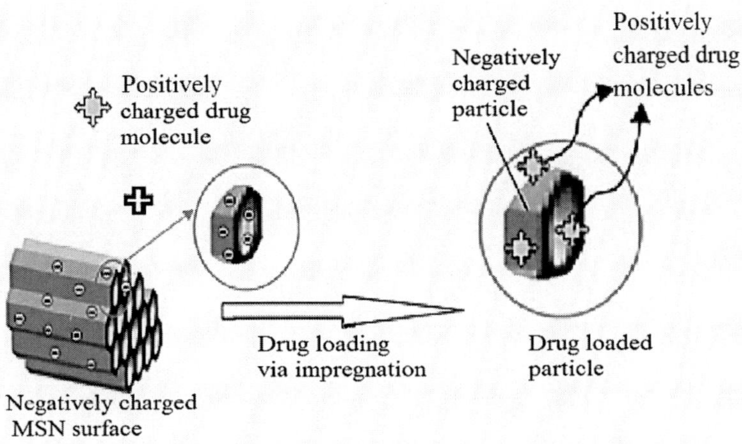

Figure 3. Drug loading in unmodified MSN by impregnation strategy.

The widely used drug loading method is impregnation or soaking. A wide number of research works involved this simple technique as a guest-loading tool [22]. Unmodified MSNs can be easily loaded with positively charged drug molecules via electrostatic interaction (Figure 3). Loading via post-impregnation can be improved by selecting optimum solvent system. C. Charnay and workers carried the ibuprofen loading through impregnation in different solvent systems [23]. Highly polar solvents, namely dimethylsulfoxide (DMSO), dimethylformamide (DMF), and dimethylacetamide (DMA), resulted in negligible adsorption of ibuprofen to both MSN-41 and aerosol while ethanol and hexane facilitated ibuprofen loading.

Based on this, they remain concerned about the solvent property while loading drugs via impregnation; hence, solvent properties show great influence.

2.2. Effect of MSNs' Textural Properties on Drug Loading

2.2.1. Pore Size

Hence the drug incorporation depends solely on the adsorptive properties of MSNs, thus, the pore size of MSNs highly limits the adsorption of guest molecules. Generally, soaking is the method of choice for drug inclusion; therefore, particles with numerous larger pore diameters than the drug molecule permit diffusion as well as confinement of the drug inside the pores. In addition, the pore diameter of mesoporous silica nanoparticles can be adjusted to the houseguest molecule by different approaches [24]. L. Jia et al., reported the influence of pore size on the loading capacity of anticancer agent paclitaxel in MSNs [25]. In their study, MSNs with pore size 3.03±0.13 nm, 5.63±0.25 nm, and 9.68±0.28 nm were synthesized and applied for paclitaxel loading. Drug-loading amounts were reported 12.1± 2.1%, 16.7±1.2%, and 21.6±1.7% for MSNs with pore size 3.03±0.13, 5.63±0.25, and 9.68±0.28 nm, respectively. Not only the drug loading capacity but rather the release of the drug was also reported to be dependent on the pore size [26]. The cumulative release of paclitaxel was found to be 23.1, 32.9, and 41.3% with correspondingly increasing pore diameter. Time required to gain maximum drug entrapment is also conditional to the pore size of the drug carriers.

2.2.2. Surface Area

The drug loading strategy is the adsorption of drug on the surface area, merely, surface-adsorption phenomenon.

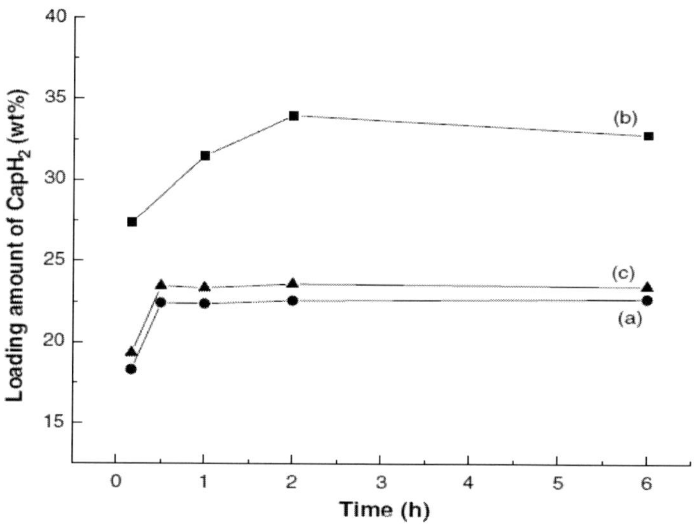

Figure 4. Increased drug loading with an increase in surface area where (a), (b), (c) refer SBA-15, MCM-41$_{16}$, and MCM-41$_{12}$ respectively. (Reproduced from Qu, F. et al., *Microporous and Mesoporous Materials* 2006, *92* (1-3), 1-9. With permission)

Therefore, the surface property of the host molecule can heavily influence the captivity of guest molecule. Usually, either physical or chemical adsorption between the drug molecule and the silanol group (Si—OH) of mesoporous silica keeps the drug attached in the carrier. Hence, a large surface area can be endowed with an increased quantity of Si—OH groups, which interact with a large extent of drug molecules, consequently leading to high drug loading. In fact, the large surface area (700–1500 m^2/g) of ordered mesoporous silica materials drew the attention of their use in a controlled drug delivery system [27]. A direct relation of drug loading and BET surface area was revealed by Qu F. et al. [28]. With three different types of mesoporous silica particles viz. SBA-15, MCM-41$_{16,}$ and MCM-41$_{12}$ the controlled release of the model drug captopril was measured. The highest drug loading 33.99% was found with the particle possessing largest surface area 1211 m^2g^{-1}. On the contrary, the lowest captopril-loading amount had the lowest surface area, 910 m^2g^{-1} (Figure 4).

The dependency of drug loading on the surface area was also confirmed by I. Barba et al. [29].

2.2.3. Pore Structure and Morphology

Along with other textural parameters, mesopore structure, pore connectivity, and pore geometry as well as morphology, can tune the drug loading capacity followed by controlling the release rate [30]. Variability in pore dimension such as 2-D hexagonal, 3-D hexagonal, cubic structure, 3-D bicontinuous structure, lamellar, cage like arrangement, etc. is achievable with different synthetic procedures by using different templating agents [31]. Interconnected pore systems accelerate the quick release compared with unconnected pore systems. Similarly, crisscrossed pore systems retain the drug molecule for relatively longer periods of time. By hindering the diffusion through altering pore connectivity, dimension, and geometry, a slow-release pattern can be accomplished. J. Andersson et al., proved how intensely the structural pore properties modulate the ibuprofen release from three different mesoporous silica carriers [32]. In this regard, the one dimensional hexagonal arrangement of cylindrical pores, i.e., MCM-41, SBA-3, were judged against a three dimensional cubic arrangement of spherical cavity linking via small pores, i.e., SBA-1. Comparing the release patterns, they concluded that the one-dimensional cage like pore-system is the most promising for slow release for a prolonged period of time. Based on ibuprofen and vancomycin loading in ordered mesoporous silica materials, V. Cauda et al., also reported a similar potential of pore morphology in both drug confinement and release profiles [33]. M. Stromme et al., mentioned the influence of pore connectivity and geometry as key parameters in controlling the release [34].

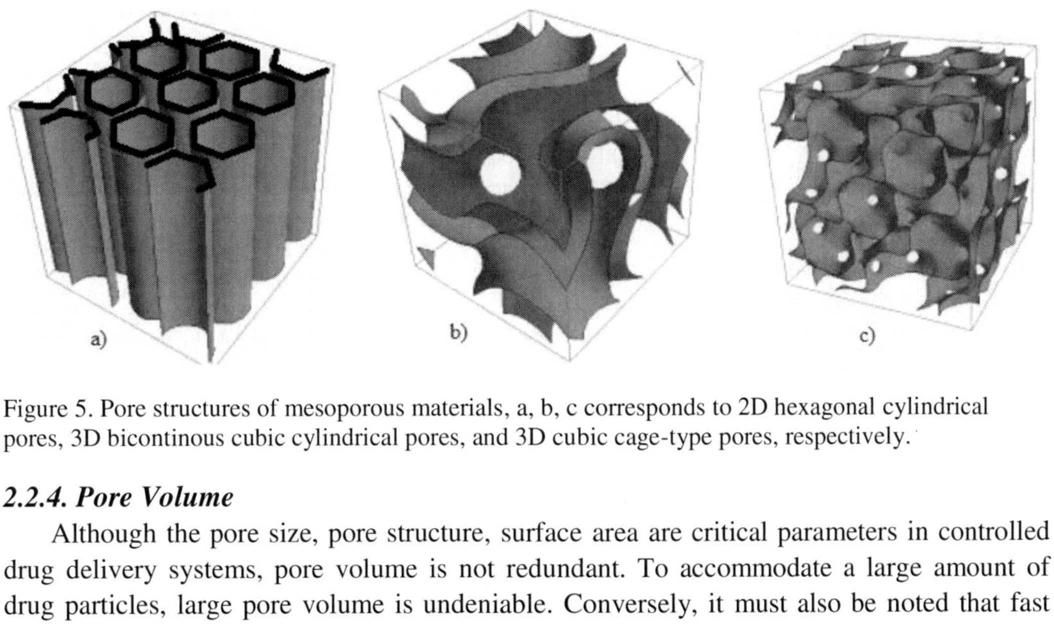

Figure 5. Pore structures of mesoporous materials, a, b, c corresponds to 2D hexagonal cylindrical pores, 3D bicontinous cubic cylindrical pores, and 3D cubic cage-type pores, respectively.

2.2.4. Pore Volume

Although the pore size, pore structure, surface area are critical parameters in controlled drug delivery systems, pore volume is not redundant. To accommodate a large amount of drug particles, large pore volume is undeniable. Conversely, it must also be noted that fast drug release is a limitation with high pore volume. Despite attaining high loading efficiency, high pore volume is a key factor that governs the loading capacity. Instead of having the most accessible mesopore network, favorable interconnecting 3-D pore geometry, wide pore size distribution, and a low pore volume characteristic could result in low drug (Ibuprofen) retention as observed by T. Heikkila et al., [35]. In order to load the poorly water-soluble drug, telmisartan, Y. Zhang et al., used mesoporous silica as a vehicle and revealed a similar effect of pore volume in drug loading efficacy [36].

Nevertheless, to enclose large molecules, such as large-size drug molecules or proteins, the pore volume is a vital aspect that can regulate the confinement of the desired molecule inside the mesoporous carrier [37].

3. Funtionalization of MSNs Carriers To Improve Drug Loading Efficiency

Functionalization of mesoporous host particles are usually carried out to improve adsorption capacity and for modulating the release properties for drugs and other biomolecules. In case of loading a negatively charged drug molecule in the mesoporous silica carrier, surface functionalization is done to anchor cationic functional groups, such as amino groups, to assist the interaction between the host and guest molecule. On the other hand, MSNs are also functionalized with anionic groups such as carboxylic groups to strengthen the interaction between the silica surface and drug molecules, eventually increasing the loading capacity. There are two general approaches, namely post-grating and co-condensation, towards the functionalization of mesoporous silica. T. Asefa and co-workers used both approaches to anchor vinyl, 3- mercaptopropyl, 3-aminopropyl, and secondary amine functional groups to facilitate higher adsorption of ibuprofen and rhodamine 6G followed by controlled release [38]. These two methods are illustrated briefly in the following sections.

3.1. Grafting Method to Anchor the Desired Functional Group on MSN Surface

Grafting is the most commonly used method to modify the surface of the MSN with intended functional groups. This technique is also referred to as post-synthetic or post-grafting method, and it involves two steps, such as synthesis of unmodified mesoporous silica, and modification of synthesized particles to anchor the desired functional group [39]. Grafting of functional groups is usually performed after the removal of surfactants. Functionalization is obtained by the silylation on free (≡Si– OH) and germinal silanol (=Si $(OH)_2$) groups present in the MSNs surfaces. It is worth mentioning that the associated silanol groups, which are bonded by hydrogen bonding, are less accessible for further modification. The surface chemistry of MSN is shown in Figure 6.

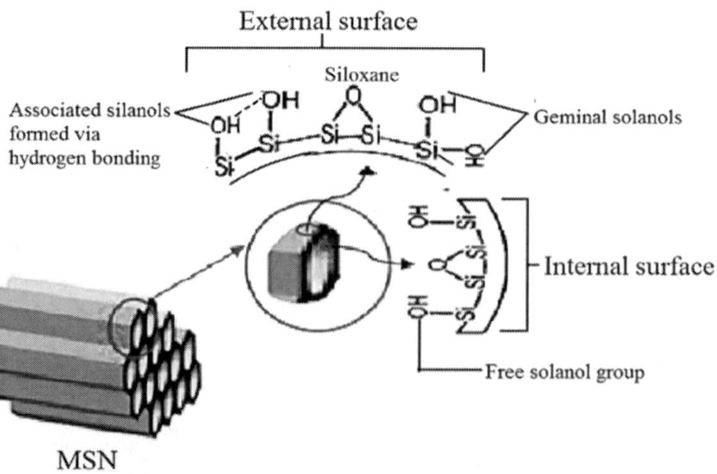

Figure 6. Surface chemistry of MSN.

$$\equiv Si{-}OH + Cl{-}SiR_3 \xrightarrow{\text{Base, } 25^{\circ}C} \equiv Si{-}O{-}SiR_3 + HCl.\,Base$$

$$\equiv Si{-}OH + RO{-}SiR_3 \xrightarrow{100^{\circ}C} \equiv Si{-}O{-}SiR_3 + HOR$$

$$2 \equiv Si{-}OH + HN(SiR_3)_2 \xrightarrow{25^{\circ}C} 2{\equiv}Si{-}O{-}SiR_3 + NH_3$$

X. S. Zhao and G. Q. Lu explained that the existence of hydrophilic networks, formed among the hydrogen-bonded silanol groups, made the associated silanol (hydrogen bonded silanol) groups less accessible for silylation [40]. Usually, silylation can be undertaken by one of the following schemes, using silylating agents such as chlorosilanes ($Cl-SiR_3$), alkoxysilanes ($RO-SiR_3$), and disilazane ($HN(SiR_3)_2$), as stated below [41].

For the large exposure of desired functional groups, the high density of silanol groups is mandatory after the removal of surfactant. Generally, surfactant removal is accomplished either by calcination or appropriate extraction methods. A matter of concern is that the high temperature, specifically 400-550°C, during the calcination treatment can destroy some unreacted silanol groups, which may be considered a limitation of this two-step functionalization method.

Anchoring of functional groups by the post-grafting method also relies on the solvent used for the reaction. In 2007, K. K. Sharma and co-workers studied how the solvent's properties, more precisely, polarity and dielectric constants of solvents, influence the concentration of grafted organic groups [42]. The presence of water during the grafting method can hamper the even distribution of grafted functional groups over the MSNs as well [43]. N. Gartmann et al., investigated the influence of water in the deposition of amino groups over MCM-41 [44]. APTES and toluene, containing various amounts of water, were used as functional group donors and solvent systems, respectively. They reported that the presence of water stimulated clustering of APTES in mesopores, resulting in higher grafting densities at the pore entrances, eventually lead to pore blocking and uneven distributions of functional groups.

The post-grafting technique has been performed by a large number of researchers to improve the controlled drug delivery system. Recently, in 2012, M. D. Popova and co-workers applied post-synthetic modification of MCM-41 and SBA-15 in order to study the adsorption capacity and release profile of sulfadiazine [45].

At first, both of the carriers were modified with amino groups for further anchoring of carboxylic groups. In case of the MCM-41 carrier, carboxylic groups grafted MCM-41 showed higher loading efficacy (49.78 wt.%) compared to non-modified MCM-41 (39.05 wt.%). Contrarily, carboxylic SBA-15, both modified and non-modified, resulted in similar adsorption efficiency for sulfadiazine (53.63 wt.% for SBA-15 and 52.40 wt.% for SBA-15COOH). As it was explained, the reason behind this phenomenon was that the loading in functionalized MCM-41 was governed only by functionality, while for SBA-15, the influence of pore blocking restricted the advantage of surface functionalization. However, both functionalized carriers, MCM-41-COOH and SBA-15-COOH, afforded a slower release of sulfadiazine because of relatively stronger electrostatic bonds between sulfadiazine and the carboxylic groups of the modified mesoporous samples rather than week hydrogen bonding between the sulfadiazine and silanol groups of the parent silica carriers.

There are many examples of post-grafting functionalization techniques implemented for sufficient drug loading and regulated release profiles. A few of them are summarized in Table-2.

Table 2. Post-grafting method to achieve improved MSNs as drug carriers

Type of MSNs	Grafted Functional group	Model drug	Types of drug	Reference
SBA-15	Amino group	Baicalein	Antibacterial, anti-viral, anti-inflammatory, and anti-pyretic bioactive molecule obtained from Chinese herbal medicine *Scutellaria baicalensis Georgi.*	[46]
SBA-15	Amino group	Captopril	Angiotensin-converting enzyme inhibitor	[47]
SBA-15	Amino group	Valproic acid	Anticonvulsant	[48]
SBA-16	Alcoxysilane (propyl-triethoxysilane)	Atenolol	Anti-hypertensive	[49]
SBA-15	Carboxylic acid	Vancomycin	Antibiotic	[39 (a)]

3.2. Co-condensation Method to Anchor Desired Functional Group on MSN Surface

Co-condensation is a widely used alternative way for synthesizing the functionalized mesoporous silica surface. Owing to some limitations with the post-synthesis grafting technique, a number of researchers suggest this direct functionalization technique. After the initial works carried out by the research group of S. Mann, D. J. Macquarrie, and A. Stein, one-pot synthesis has been used widely to produce organically functionalized silica [50]. In co-condensation reaction, phase separation of the precursors during the sol–gel reaction, and the cleavage of Si–C bonds during surfactant removal are circumvented [51]. Mainly, the reaction involves co-condensation of tetraalkoxysilane and one or more organoalkoxysilanes, to provide an expected functional group in the silica particle [52, 53].

Based on the undergoing reaction pathway, the condensation reaction for surface modification can be classified into the following categories, namely co-condensation by $S^{+}I^{-}$ pathway, co-condensation by $S^{+}X^{-}I^{+}$ pathway [54], co-condensation by $N^{0}I^{0}$ pathway [55], co-condensation by $S^{0}I^{0}$ pathway [56], and co-condensation by $S^{-}I^{+}$ pathway [57].

To aid the desired control over drug loading into, and drug release from, the carrier, functionalized MSN was preferred in earlier years. Besides the grafting technique, co-condensation is also a technique of choice to the researcher for anchoring the desired functional group on the mesoporous silica. Recently, M. Xie et al., used the condensation method for synthesizing carboxyl group functionalized MSN for drug delivery of anti-cancer drug doxorubicin (DOX) in 2013 [58].

They reported substantial drug loading and pH sensitive drug release profiles according to their expectation. A number of researchers have used this technique to improve the controlled drug delivery from MSNs (Table-3).

Table 3. Co-condensation method to achieve improved MSNs as drug carriers

Type of MSNs	Incorporated functional group	Model drug	Type of drug	Strategy of controlled drug release	Reference
MSNs	Trimethyl-ammonium (TA)	Sulfasalazine	Anti-inflammatory pro-drug	pH	[59]
MSNs	Trimethyl-ammonium (TA)	Orange II	Fluorescent tracing molecule	pH	[59]
MCM-41	Thiol groups with further grafting of 2-(pyridyl disulfanyl)ethyl-amine functionality	Adenosine triphosphate (ATP) and vancomycin	Neurotrans-mitter and antibiotic	Chemical reduction	[60]

Table 4. Examples of bifunctionalized MSNs

Type of MSN	Incorporated functional groups	Reagents Used
MSNs [61]	Ureidopropyl (UDP) groups, and 3-[2-(2-aminoethylamino)ethylamino]-propyl (AEP) groups	TEOS, 3-[2-(2-aminoethylamino)ethylamino]propyltrimethoxysilane; ureidopropyltrimethoxysilane
MCM-41 [62]	Amine and sulfonic acid groups	TEOS; 3-mercaptopropyltrimethoxysilane (MPTMS); and 3-tertbutyloxy carbonyl(aminopropyl) trimethoxysilane

Table 4. (Continued)

Type of MSN	Incorporated functional groups	Reagents Used
Periodic Mesoporous Organosilica [63]	Isocyanurate moieties (ICS) and thiol functional group; ICS and imidazole; ICS and ureidopropyl; ICS and tetra sulphide	TEOS and other mixed organosilanes: tris[3-(trimethoxysilyl)-propyl] isocyanurate; bis [3-(triethoxysilyl)propyl] tetrasulfide (4S); N-(3-triethoxysilylpropyl)4,5-dihydroimidazole (IM); and ureidopropyltrimethoxysilane (UR)
HMS [64]	Phenyl groups and cobalt (Co^{2+}) ions	TEOS; cobaltous acetate; phenyltriethoxysilanes
Highly ordered mesoporos silica, SBA-15, MSN [37, 65]	Amino and carboxyl groups	TEOS; 3-aminopropyl triethoxysilane; carboxyethylsilanetriol sodium
SBA-15 [66]	Amino and methyl groups	TEOS; 3-aminopropylmethyldiethoxysilane
SBA-15 [67]	Sulfonic acid and benzyl sulfonic acid and methyl	TEOS; mercaptopropyltrimethoxysilane and second organotrialkoxysilane: $(RO)_3$–Si–R´ with R´ = benzyl and methyl

Not only mono-functionalized MSN but also bi-functionalized or multi-functionalized MSN can be obtained by applying this one-pot synthesis technique. By co-condensation reaction between TEOS and two different organoalkoxysilanes, both acidic and basic groups can be incorporated in mesoporous silica (Table-4).

Bi-functionalized MSNs can be used as effective tools in numerous applications such as biomedicine, catalysis, sensing, biotechnology, furthermore, in environmental applications as a potential adsorbent [37, 63]

3.3. Comparison Between Grafting and Co-condensation Methods

Both of these functionalization techniques are enriched with some of their inherent advantages, although, holding some unavoidable limitations as well. The noticeable benefits and drawbacks of both of the functionalization techniques are briefly summarized in Table 5.

In 2005, W. Zeng et al., compared the effectiveness of these two functionalization processes in the context of a controlled drug delivery system using aspirin as a model drug [71]. Since different functionalization methods end up with vitiation in pore sizes, uniformity of functional group distributions as well as degrees of functionalization, they lead to variations in drug delivery properties. According to their findings, smooth distribution of amino groups along with higher orderliness was observed via the co-condensation method compared to the grafting method. Therefore, they concluded the former as the one developed for achieving controlled drug delivery. On the other hand, another research group preferred the post-grafting functionalization technique for controlled delivery of ibuprofen rather than co-condensation [72]. They used SBA-15 as host and subjected it to amino functionalization in order to control the release of ibuprofen. They justified that the amino-modified SBA-15 via post-grafting led to strong ionic interaction between the carboxylic group and the amino group resulting in more controlled release compared with pure SBA-15 and SBA-15 obtained via one-pot synthesis.

Table 5. Advantages and limitations of both post-grafting and co-condensation methods

Name of surface functionalization method	Advantages	Disadvantages
Post-grafting	i. Ensures hydrothermally stable particles after functionalization [68]. ii. Produces structurally well-defined particles after functionalization [68]. iii. Selective functionalization is possible [69].	i. Results non-homogeneity of grafted functional groups [68]. ii. Results in reduction of pore size and pore volume [70]. iii. Low density of reactive silanol group limits the loading of functional groups and promotes cross-linking between functional groups and silanol groups on surface [70]. iv. Tedious process as involving more than one step of preparation [70].
Co-condensation	i. Ensures even distribution of grafted groups [68]. ii. Ends up with higher density of functional group [68].	i. Results reduced degree of mesoscopic order of the mesostructure with increasing concentration of organosilane groups, which can lead to totally disordered mesostructures [17] ii. Incorporated organic functionality can be destroyed during removal of the surfactant [17].

4. MSNs and Controlled Drug Delivery

It is expected that an efficient and ideal drug carrier should deliver reserved molecules in a controlled manner rather than showing a premature burst release in a few hours. To fulfill the desire of controlled release, several strategies have been applied for different types of guest particles. Some research achieved the expected release profile by changing the morphology or physical properties whereas other studies promoted chemical modification along with the implementation of different techniques, such as stimuli-responsive release, photo-responsive drug release, ultrasound induced drug release, magnetic field stimulated release, and so on. Some favorable drug release approaches are briefly illustrated here.

4.1. pH-Dependent Drug Delivery

Several successful approaches have been attempted to attain pH dependent release profiles from MSNs. In this regard, research involved the coating over the MSNs to retard unwilling release as well as to facilitate the inclined pH-responsive release [73]. On the other hand, pH sensitive gate keeping was also favored by grafting specific pH liable linker groups on the MSN surface [74]. Organic groups can also be anchored on the mesoporous silica pore opening to introduce a pH controlled "molecular switch" [75]. Q. Gao et al., grafted propyldiethylenetriamine groups to mesoporous silica sphere. Hence, neutral propyldiethylenetriamine chains at higher pH attract each other through hydrogen bonding, thus, opening the pathway for drug release. In contrast, at acidic conditions, propyldiethylenetriamine becomes protonated and blocks the pore orifice, eventually hindering the drug release (Figure 7). pH controlled gate keeping tactics are also implemented by F. Muhammad and co-workers to impede premature doxorubicin release from MSNs host carriers [76]. They capped the mesopore with acid liable ZnO quantum dots (QDs) to permit a pH-triggered drug release mechanism. Reliance of drug release on pH was examined by using the release medium with a different pH. Fast release of doxorubicin from ZnO QDs capped MSNs was reported at an acidic environment (pH 5.0) compared with physiologic condition (pH 7.4). Acidic pH caused rapid dissolution of ZnO QDs nanolids, and thus promoted quick drug release.

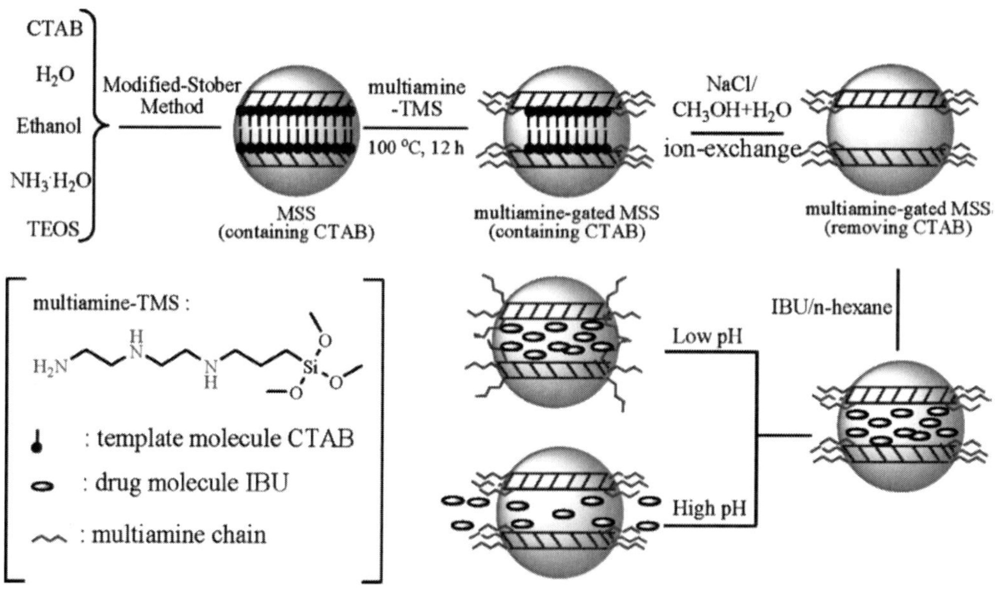

Figure 7. pH dependent ibuprofen release from mesoporous silica sphere (MSS). (Reproduced from Gao Q. et al., *Langmuir* 2010, *26* (22), 17133-17138. With permission).

Therefore, stable ZnO QDs nanolids at physiological pH inhibited rapid doxorubicin release, and ultimately demonstrated a pH-triggered controlled drug delivery system. Surface modification can result in pH-responsive drug release. By grafting pH-responsive poly (acrylic acid), controlled release of an anti-cancer drug was achieved by L. Yuan et al. [77]. Poly (acrylic acid) formed a covalent bond with the surface silanol groups of MSNs. Drug loading resulted due to electrostatic interaction between poly (acrylic acid) and doxorubicin. Reduction in pH value accelerated protonation of poly (acrylic acid) following the increased dissociation of doxorubicin from poly (acrylic acid) grafted MSNs. This phenomenon ended with a pH-controlled release of the anticancer drug at a higher pH that is physiological pH. Coating the polymer over the drug-loaded particle instead of gate keeping can also provide pH-sensitive drug release. For example, pH-sensitive polymer hydroxypropyl methylcellulose phthalate (HPMCP) was used as a coating material over the famotidine loaded mesoporous SBA-15 tablet to avoid drug release in simulated gastric fluid (SGF) [78]. HPMCP is insoluble at a lower pH; therefore, it formed a protective shell to delay the drug release in SGF. However, in SIF (simulated intestinal fluid, pH 7.4); the carboxylic group became ionized, thus boosting the electrostatic repulsion force between the polymer chains, finally resulting in quick dissolution of the coating material. Consequently, a pH-triggered release was achieved in SIF while drug release was sufficiently prohibited in SGF. Coating with different polyelectrolyte layers is another approach to establish a pH-dependent sustained release of guest molecules from the host reservoir. Y. Zhu et al. applied this technique by generating polyelectrolyte multilayers comprising of sodium polystyrene sulfonate (PSS) and poly allylamine hydrochloride (PAH) over the ibuprofen loaded hollow mesoporous silica [79].

4.2. Redox-Responsive Drug Release

Among different type of stimuli-triggered controlled drug delivery approaches, redox-responsive release is a discovery worth mentioning. In this type of controlled delivery scheme, drug release from the carriers is regulated by the presence of different reducing agents.

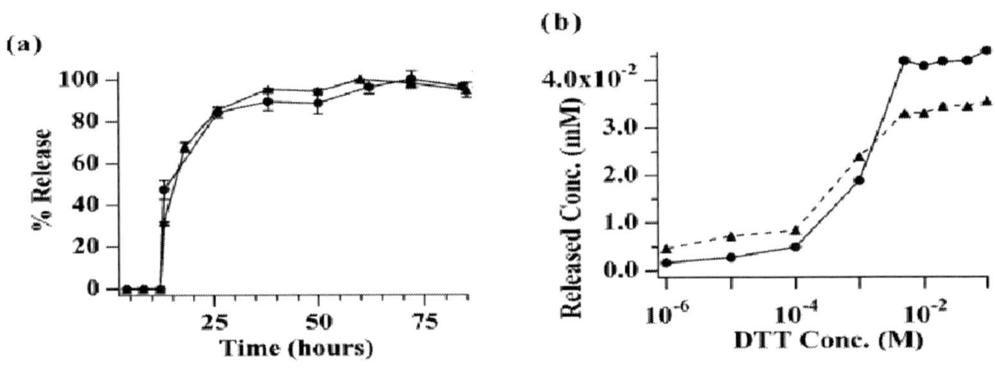

Figure 8. Redox based controlled delivery of vancomycin (—●—) and ATP (- -▲- -) from the CdS-capped MSN system: (a) represents % release over time and (b) represents DTT concentration vs. analyte concentrations after 24 h of the DTT additions. (Reproduced from Lai, C. et al. *Journal of the American Chemical Society* 2003, *125* (15), 4451-4459. With permission).

The principle of this approach is to cap the pore opening of the MSNs host particle and reopening the mesopore of these drug-loaded-capped particles in the presence of reducing agents. C. Lai and co-workers demonstrated a redox-sensitive drug delivery system via capping of the mesopore of MCM-41 with CdS nanoparticles [60]. To develop this drug delivery system, they synthesized the MCM-41 type of mesoporous silica with 2-

(propyldisulfanyl) ethylamine functionality and used vancomycin as well as adenosine triphosphate (ATP) as a model drug and biomolecule respectively. 2-(propyldisulfanyl) ethylamine functional groups on the MCM-41 surface formed a covalent bond with mercaptoacetic acid-derivatized cadmium sulphide (CdS) nanocrystals through amidation reaction.

This linkage between the MSNs and the CdS resulted in a capped drug reservoir, consequently hindering the leakage of stored molecules in the absence of any disulfide reducing agent. Formerly, exposure to two different disulfide reducing agents, namely dithiothreitol (DTT) and mercaptoethanol (ME), caused the cleavage of disulphide linkage between MSNs and CdS leading to reopening of host particles. That is how they generated redox-responsive drug/neurotransmitter release from MSNs carriers. A good number of research work has been done on redox-responsive drug release applications. R. Liu et al. developed a cross-linked polymeric network as gatekeeper based on response to redox signals [80]. They grafted poly(N-acryloxysuccinimide) polymer on mesoporous silica denoted as PNAS-MS. Rhodamine-B was loaded in PNAS-MS as a model drug then the pore orifice was blocked by reacting with a disulphide based bifunctional primary amine viz. cystamine. Reaction between cystamine and N-oxysuccimide groups along the polymer chain led across linking of polymeric chains. Eventually, release of the loaded material was restrained. The polymeric network revived by breaking down the disulfide bond of cystamine with the addition of disulphide reducing agents (DTT). As a result, it is possible for the release pattern to modulate with the redox signal. By using the same dye as a model drug, but differing in the capping molecule i.e., β-cyclodextrin, P. Nadrah and co-workers also developed a redox based drug delivery system [81]. Following this principle, they developed another redox based controlled drug delivery system by capping the mesopore of MSNs with Meijer poly(propylene imine) type dendrimers via reducible disulfide bond [82]. DTT was used as a reducing agent to regulate the drug release and they underlined that DTT performs in a manner similar to the naturally occurring reducing agent glutathione. They also suggested smaller dendrimers as better capping agents to obtain redox-responsive controlled drug delivery.

4.3. Photo-Induced Drug Delivery

Regulation over the uptake, storage, and release of organic molecules in MSNs via photo-controlled fashion conveyed a new era in controlled drug delivery systems. In 2003, N. K. Mal and co-workers first introduced a novel pathway to regulate the open-close behavior of the mesopore entrance of MCM-41 via photo-induced reversible intermolecular dimerization [83]. They grafted coumarin derivatives to the MCM-41 pore outlet, which served as "double doors" around the pore orifice, and the MCM-41 was filled with steroid cholestane. Without any photo-irradiation or after photo-cleavage of coumarin derivatives dimers, this double door became exposed and permitted entry into the pore interior. On the other hand, photo dimerization led coumarin to cyclobutane dimers across the pore entrance, thus restricting the pore access. The steroid cholestane release was controlled by irradiating with UV light at 250 nm wavelength that caused the cleavage of the cyclobutane ring of the coumarin dimer, consequently, opening the pore entrance. Based on the photo-controlled release pattern, many research works have been done to ensure "zero-premature release" of the toxic drug molecule.

In 2009, Vivero-Escoto JL et al. developed photo-induced intracellular controlled release of anticancer drug paclitaxel by synthesizing gold nanoparticle-capped MCM-41 [84]. To achieve their aim, a gold nanoparticle surface was functionalized with a photo-responsive linker following the capping by electrostatic interaction between negatively charged MCM-41 and positively charged photo-responsive linker associated Au particles. Due to executing photo-irradiation, the photo-liable linker group cleaved off, altering the surface charge property. As a result, a repulsion force formed between the negatively charged gold nanoparticles, and the negative charged MSN surfaces led to the uncapping of the porous pathway, consequently aiding the release of the loaded guest molecules. Moreover, photo-triggered regulated drug delivery from the MSN reservoir was also established by manipulating either the irradiating time or the wavelength intensity of external irradiation [85]. Recently, in 2012 H. Yan et al. reported a controlled drug delivery system based on the photo-thermal responsive achieved by functionalizing MSNs with rotaxen [86]. The ∞-CD ring of rotaxen was threaded with a photothermal-responsive azobenzene axle, which could undergo reversible *cis-trans* isomerization upon photo-irradiation. Upon UV light irradiation at 365 nm, *trans-to-cis* photo-isomerization of the azobenzene unit took place resulting in pore obstruction. Again, exposure to visible light or heating produced *cis-to-trans* isomerization of the azobenzene unit, and eventually the ∞-CD ring receded to the *trans*-azobenzene position. This photo-thermal induced back and forth movement formed the basis of controlled-drug delivery form MSNs. Based on the similar photo-isomerization of azobenzene, controlled delivery of doxorubicin was demonstrated by Q. Yuan et al. [87]. They incorporated the azobenzene in DNA double strands then used this complex as gatekeeper around the pore opening of MSNs. The core strategy lied on the photo-induced isomerization of azobenzene causing the dehybridization/hybridization of complementary DNA, finally, regulating uncapping/capping of pore gates of mesoporous silica.

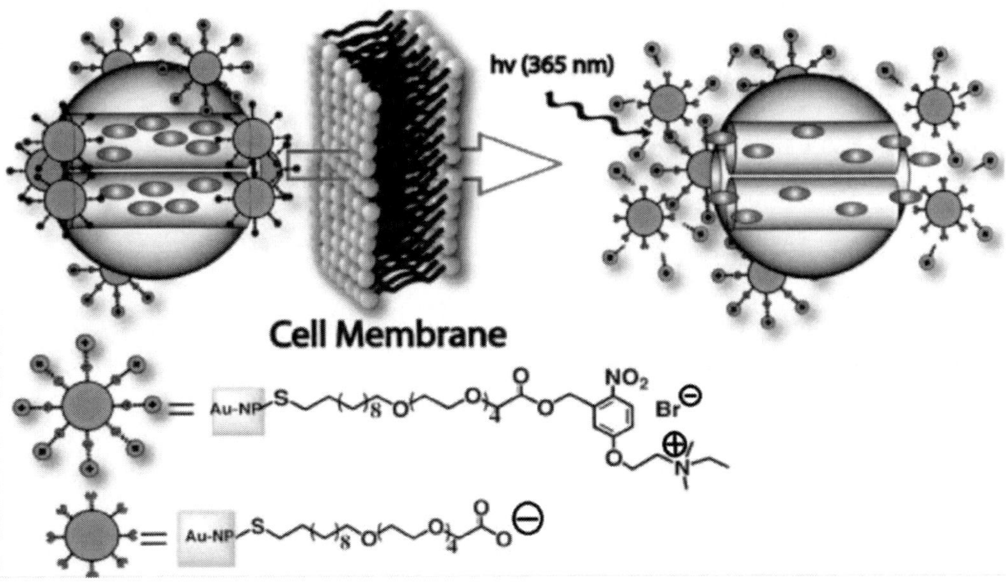

Figure 9. Photo-induced controlled release from MSNs carrier. (Reproduced from Vivero-Escoto, J. et al., *Journal of the American Chemical Society* 2009, *131* (10), 3462-3463. With permission.).

4.4. Magnetic Field Induced Drug Delivery

The magnetic field, as stimuli, drew the attention of researchers as an application in the field of controlled drug delivery. Targeted drug delivery can be accomplished by designing magnetism functionalized MSNs carriers for holding the drug until it reaches the desired location. However, the influence of a constant external magnetic field over the drug release can be omitted as reported by K. C. Souza et al. [88]. They designed a silica/magnetite nanocomposite of SBA-15 and Fe_3O_4 to generate controlled delivery of cisplatin and carboplatin. They investigated the influence of a low frequency alternating magnetic field dominating the rate of drug release. Actually, alternating magnetic fields aggravated the vibration of the magnetic particles and thus promoted quicker release of drug molecules. In addition, some research works combined the thermo regulated drug release behavior along with magnetic field implementation. For example, C. Liu and co-workers prepared a magnetic mesoporous silica vector comprising of Fe_3O_4 magnetic nanoparticles as core along with a thermo-responsive outer most polymer shell and sandwiched a mesoporous silica layer in between the two. They directed the temperature-triggered drug release by applying a magnetic field [89]. Similarly, A. Baeza et al. developed a drug delivery system involving both induced magnetic-fields, and temperature-controlled release profiles [90]. They prepared mesoporous silica nanoparticles with iron oxide and covered the surface with a thermo-responsive copolymer of poly(ethyleneimine)-b-poly(N-isopropylacrylamide) (PEI/NIPAM). At lower temperature (32°C), the poly(N-isopropylacrylamide) polymer became hydrated and blocked the mesopore entrance. Hence, the guest molecules became entrapped inside. Conversely, the increased temperature headed polymer shrinkage causing the reopening of pores. Thus, the trapped guest molecule could be released outside. The encapsulated iron oxide generated the required higher temperature in the presence of alternating magnetic fields. By applying a high-energy magnetic field, a thermo-responsive controlled release was accomplished. Based on a similar mechanism, C. R. Thomas et al. developed controlled delivery where they used nanovalves to obstruct drug leakage [91]. These nanovalves remain closed at physiologic temperature but unlocked upon heating. Application of oscillating magnetic induced hyperthermic effects of magnetic core silica nanoparticles consequently controlled the release profile.

4.5. Temperature-Regulated Drug Delivery

Although the thermo-responsive release pattern was implemented along with exposure to the magnetic field, many researchers designed temperature controlled drug release devices. In 2011, E. Aznar et al. reported paraffin gated mesoporous silica carriers to introduce a thermo responsive release profile [92]. MCM-41 was functionalized with octadecyltrimethoxysilane to act as carrier and afranine O was used as a model guest molecule. Octadecyltrimethoxysilane functionalized silica did not exhibit thermo-responsiveness but offered the platform to interact with paraffins via London forces. Owing to this, the hydrophobic layer of paraffin blocked the pores void as well as the constrained release of guest molecules. Application of heat higher than the paraffin's boiling point promoted the uncapping mechanism. A ten fold improvement in guest molecule retention was confirmed by treating thiol-functionalized micro-to-mesoporous silica nanoparticles (MSN) with poly(N-isopropylacrylamide) [93]. The guest uptake and release from this carrier was based on the thermo-responsive conformational change of the polymer. This thermo-responsive polymer exhibited soluble coil conformations below lower critical solution temperatures (LCST) while turning into insoluble globules or a collapsed state above the LCST. According to their experimental data, uptake and release of fluorescein was accomplished at room temperature, i.e., below the (LCST) of the polymer, and a low level of release was observed at 38°C i.e., above the LCST. They explained the uptake and release mechanism with the thermo-controlled phase state conversion of employed polymer. More precisely, the random coil conformation guided the diffusion through a porous channel; contrarily insoluble globule conformation provided a diffusion barrier.

4.6. Bio-Responsive Controlled Drug Delivery

Bio-responsive controlled drug release from MSNs is one of the remarkable developments in the field of controlled drug delivery. In 2013, Z. Chen and co-workers established a hyaluronic acid (HA) capped MSN nano-reservoir to demonstrate enzyme responsive, cell-specific, targeted, and controlled drug release at tumor sites [94]. CD44 is a major receptor for HA and interestingly over-expressed in many solid tumor cells. Again, HA is readily degradable in presence of lysosomal enzyme hyaluronidase-1 (Hyal-1). Hyal-1 is abundant in the tumor microenvironment, as well.

All these advantages were utilized to design this site-specific controlled drug delivery. To reconfirm the enzymatic degradation of HA as the mechanism of mesopores opening, they studied the release of rhodamine B in an acetate buffer, one with Hyal-1 and another one without Hyal-1. Result showed an 82% release in the presence of Hyal-1 whereas only about 9% without Hyal-1 for the same time of exposure. They also denatured the enzyme by applying heat and found zero release in that condition. Thereby, the enzymatic degradation of HA was confirmed as a pore reopening strategy. To evaluate the target specificity, NIH3T3 (normal cell) and MDA-MB-231 (CD44 over-expressed) were incubated with HA conjugated FITC doped MSNs (FITC-MSN-HA). Stronger fluorescence signals by MDA-MB-231, compared to weaker fluorescence signals by NIH3T3, indicated that HA-attached MSNs were more easily captured by the cancer cells rather than the normal cell line. This phenomenon happened due to high specific interaction of HA on MSNs with that of over expressed CD44 receptors on the plasma membrane of the MDA-MB-231 cells. Enzyme responsive controlled release is one of the noteworthy approaches to alter the release pattern according to desire. A. Bernardos and co-workers used a lactose derivative, comprising of β-D-galactose and β-D-glucose monosaccharides bonded via β1→4 glycosidic bond, as a capping element to provide a barrier at the pore entrance of the MCM-41 carrier [95]. Enzymatic hydrolysis resulted in uncapping of the pore allowing the cargo release. Hence, presence of the β-D-galactosidase enzyme induced hydrolysis of β1→4 glycosidic, giving disaccharide lactose into the monosaccharides, namely galactose and glucose. As a result, pore block and reopening was mainly triggered by enzymatic interaction. A. Popat et al. developed a novel enzyme triggered controlled drug delivery system by functionalizing the MCM-48 carrier surface with sulfasalazine [96]. Sulsalazine is a pro-drug, which is usually used to treat inflammatory bowel disease (IBM). The reduced form of salfasalazine, 5-aminosalicylic acid (5-ASA), produces the therapeutic effect in IBS. Once the pro-drug containing the MSN carrier reaches the colon, salfasalazine was reduced into 5-aminosalicylic acid (5-ASA) and sulfapyridine (SP) due to the action of the azo-reductase enzyme (produced by intestinal bacteria). Thus, the enzymatic reduction triggered the controlled as well as the site-specific drug release profile.

Aptamer based mesopore capping has also been used to design controlled drug delivery from MSN carrier. Aptamers can be defined as single-stranded oligonucleotides showing targets specifically binding to their high affinity and specificity. C. Zhu et al. designed stimuli-responsive controlled release from MSNs based on the aptamer-target interaction principle [97]. Immobilization of adenosine-5′-carboxylic acid (a derivative of ATP molecule) to mesoporous silica was done via amidation reaction, following the capping of pores with gold nanoparticles functionalized ATP aptamer (AuNP-aptamer). In the presence of ATP, a competitive displacement reaction to adenosine-aptamer interaction led to the dissociation of AuNP-aptamer caps from MSN, eventually triggering the release of the model drug. Thus, the uncapping mechanism was strongly reliant on high specificity and affinity of ATP aptamer towards its target, i.e., ATP. Using a similar principle, X. He and co-workers established controlled drug delivery from aptamer capped MSN [98].

In order to form the sandwich-type DNA structure, they hybridized the arm of single-stranded DNA1 (arm ssDNA1), and the arm of single-stranded DNA2 (arm ssDNA2) with ATP aptamer. Then, this sandwich-type DNA structure was anchored on the MSN surface through a click chemistry approach leading to the obstruction at the pore orifice. Exposure to ATP caused the cleavage of the ATP aptamer resulting in arm ssDNA1 and ssDNA2 only on the surface of MSN. As the single-stranded DNA is flexible in nature, the pore entrance reopened allowing the guest molecule to release.

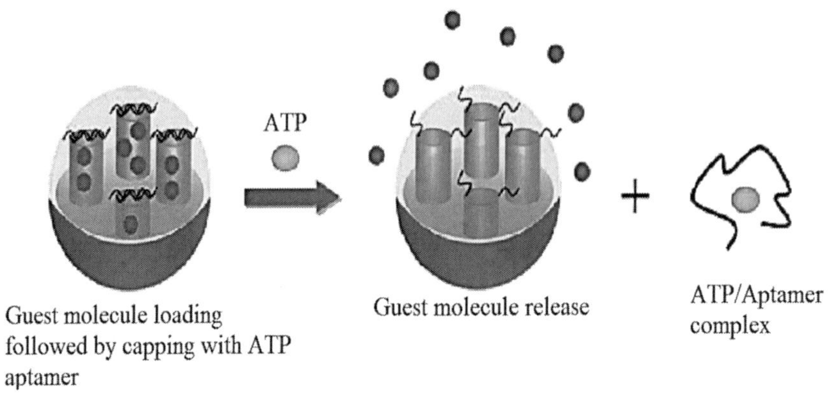

Figure 10. Schematic illustration of controlled drug delivery via aptamer based gate keeping. (Reproduced from He, X. et al. *Langmuir* 2012, *28* (35), 12909-12915. With permission).

Glucose responsive drug release from MSNs carriers was developed by Y. Zhao et al. [99]. To obtain the control over the release of both insulin and cAMP, gluconic acid-modified insulin (G-Ins) proteins were anchored to the external MSN surface as a capping element of mesopores. In this regard, boronic acid-functionalized MSN material (BA-MSN) and fluorescein isothiocyanate (FITC)-labeled G-Ins (FITC-G-Ins) were prepared. Then, void space of mesopores of BA-MSN (100 mg) was filled up with Camp and capped with FITC-G-Ins via reversible covalent bonding between phenyl boronic acid and the vicinal diols of FITC-G-Ins. The pore uncapping was attained by exposure to different saccharides. Hence, phenyl boronic acid forms much more stable cyclic esters with the adjacent diols of saccharides rather than with acyclic diols, thereby bonding between FITCG-Ins and BA-MSN cleaved off, allowing the release of stored materials.

Based on antigen-antibody interaction, a gate-keeping tactic was developed by E. Climent et al. to regulate guest outflow from MCM-41 type MSN nano-reservoir [100]. With the intention of demonstrating such a bio-responsive controlled release, MCM-41 was functionalized with hapten 4-(4-aminobenzenesulfonylamino) benzoic acid followed by capping via a polyclonal antibody. This antibody showed specific antibody-antigen reaction for sulfathiazole (respective antigen). To evaluate the antigen-antibody interaction dependent uncapping mechanism, the MSN carrier was loaded with dye $[Ru(bipy)_3]Cl_2$ and it revealed the release profile in the presence of sulfathiazole.

5. Drug Release Kinetics

Proper interpretation of drug release data is obligatory to comprehend the drug release profile. In this regard, different kinetic models are, for example, zero order, first order, Higuchi, Korsmeyer-Peppas, power law, Hixson-Crowell, square root mass, Weibull, linear probability, log – probability, reciprocal powered time, etc. are used to elucidate drug release data [101].

In case of MSNs as drug carriers, the drug release profile favorably follows the Higuchi relationship, which is

$$Q_t = k_{\mathrm{H}}a?\_t$$

where, Q_t is the amount of drug released in time t and k_{H} is the release rate constant for the Higuchi model [102]. The Higuchi release rate constant, $k_{\mathrm{H},}$ depends on the diffusivity and solubility of the drug in the release medium used, the tortuosity factor of the system, porosity of the matrix, A is the total amount of drug present in the matrix.

Usually, if the drug release profile is driven by the diffusion mechanism then it follows the conventional Higuchi Model. Typically, the drug release follows two steps, initial burst release followed by a simultaneous slow release. This model is suitable to interpret the release profile of drug molecules from MSNs with varying pore structure. For example, A. L. Doadrio and co-workers prepared SBA-15 type MSN for gentamicin loading. They obtained an initial prompt release continuing with a subsequent slow release profile that followed the Higuchi relationship [103]. Amino-functionalized MCM-41 type of mesoporous silica also follows the Higuchi Model as reported by A. Bernardos [104]. Riboflavin was used as model drug and the release kinetics showed good linearity up to 300 min, i.e., fitting the Higuchi model. After 300 min, a non-linearity in release kinetics reported indicating the presence of more than one step in the complete drug release profile. Drug release from MSNs with a cubic mesoporous structure, such as MCM-48, also adapts to the Higuchi model [105]. Ibuprofen and erythromycin were used as model drugs to elucidate the influence of pore size, and the Higuchi relationship in the drug release profile was observed in both cases.

Another popular mathematical model used to describe the kinetics of drug release from different type MSNs is the Korsmeyer–Peppas model, as follows.

$$\frac{M_t}{M_{a?\_}} = Kt^n$$

where, M_t, $M_{a?\_}$, K represent the cumulative mass of drug released at t time, cumulative mass of drug released at infinite time, and proportionality constant, respectively [102]. This proportionality constant confers the drug release rate as well. The release index is denoted by n and its value determines the mechanism of drug release. If n= 0.5, it indicates Fickian diffusion mechanism. When n> 0.5, it represents non-Fickian diffusion as the drug release mechanism. Many researchers used this model to reveal the first 60% of drug release from the porous silica carriers. For example, T. Heikkila et al. fitted the first 60% Ibuprofen release data from two different types of MSNs, such as TUD-1 and MCM-41 [106].

From the obtained K and n values they could conclude that drug release is slower from MSN-41 rather than TUD-1, as the former carrier showed a low k value compared with the latter. The n value determined the ibuprofen release mechanism of the TUD-1 material was governed diffusion mechanism rather than that of MCM-41. The Korsmeyer–Peppas model was also used by I. F. Alexa and co-workers to identify whether the Fickican diffusion mechanism, or the non-Fickian diffusion mechanism, governed the release of anti-hypertensive drug from unmodified SBA-15 and MgO modified SBA-15 [107].

CONCLUSION AND OUTLOOK

This chapter highlights some substantial breakthroughs in finding mesoporous silica nano-materials as groundwork for the development of controlled drug as well as biomolecule delivery. Different strategies have developed improvements in silica-based nano-carriers to load and release desired drug molecules in a controlled way. Some inherent properties of mesoporous silica influence the loading and release, as explained thoroughly.

Surface functionalization aids in high drug loading based on strengthening the interaction between drug and silica but commits a limitation of reduced pore size. One way to overcome this drawback is to employ MSNs with large pore sizes. Pore size of MSNs can be enlarged by several approaches. One of them could be using surfactant N, N-dimethyldecylamine followed by post hydrothermal treatment. This strategy helps to enlarge the pore size of the as-synthesized MSN. Over again, 1,3,5-trimethylbenzene (TMB) can also be used to broaden the pore diameter as well. Contrarily, without modifying the surface of MSN, a sustained release of guest molecule can be achieved by coating with different polymers namely PSS, PAH, poly(L-lysin), poly(L-glutamic acid), poly(ethylene glycol), poly(dimethylsiloxane), poly(D,L-lactide-co-glycolide), etc.

In addition, surface functionalization is an imperative tactic to strengthen the aim of the controlled drug release profile. In our recent research work, we demonstrated that both glutaraldehyde and glucose modified meso-structured cellular foam (MCF) showed significant control over the release of streptomycin sulphate [108]. The antibiotic loading was 5.25 and 7.01 wt % with pure MCF and amino modified MCF, respectively, whereas loading quantity increased to 25.43 and 25.5 wt % with glucose modified and glutaraldehyde modified MCF, correspondingly. This significant loading resulted due to the available hydroxyl groups, by glucose molecules, to the guest molecule streptomycin. Thus, hydrogen bonding formed between the host and guest molecule leading to high drug entrapment. On the other hand, formation of the double bond of Schiff base in between –CHO and streptomycin led to high drug loading in case glutaraldehyde modified MCF. The significant outcome of surface modification came across a with successful slow release pattern. Glucose modification successfully slowed down the streptomycin release but an even slower release was obtained from glutaraldehyde modified MCF. The underlying reason was the strong chemical interaction between the drug molecule and the drug carrier through the Schiff base double bond. This phenomenon illustrated the importance of surface functionalization as a tool to modulate drug release owing to an optimum, controlled, slow and sustained release over a long time.

Innovation of MSNs as drug and/or other biomolecule carrier has unwrapped a wide research scope for further development to commercialize such nanoparticles in the biomedical field. Moreover, it is exciting to look at how these further developments help establish MSNs as a commercial product in near future. Some other types of nanoparticles, for example, liposomes, protein nanoparticles, polymeric micelles, or polymeric nanoparticles are already being commercially used as drug or biomolecule delivering vehicles. There are some commercially available products in markets where liposomes are being used to deliver doxorubicin, estradiole, daunorubicin, morphine, verteporfin, and amphotericn B, under the trade names of Myocet, Estrasorb, Daunoxome, DepoDur, Visudyne, and AmBisome, respectively [109]. Therefore, the commercialization of MSNs can be a hot topic in the near future.

MSNs carriers will broaden their application in the biomedical field if they can be used to deliver two different guest molecules at a time. Bi-functionalization MSNs can be focused to supply both the drug and the drug synergist together, which will greatly affect the biomedical field. Moreover, this chapter reflects different ways to use MSNs as important vehicles for controlled drug delivery systems; furthermore, it highlights their prospects in biomedical applications.

REFERENCES

[1] Kresge, C. T.; Leonowicz, M. E.; Roth, W. J.; Vartuli, J. C.; Beck, J. S. *Nature,* 1992, 359 (6397), 710-712.

[2] (a) Chung, T. H.; Wu, S. H.; Yao, M.; Lu, C. W.; Lin, Y. S.; Hung, Y.; Mou, C. Y.; Chen, Y. C.; Huang, D. M. (b) Yang, P.; Quan, Z.; Lu, L.; Huang, S.; Lin, J. (c) Iwamoto, M.; Tanaka, Y.; Sawamura, N.; Namba, S. *Journal of the American Chemical Society,* 2003, 125 (43), 13032-13033; (d) Feng, X.; Fryxell, G. E.; Wang, L. Q.; Kim, A. Y.; Liu, J.; Kemner, K. M. *Science* 1997, 276 (5314), 923-926; (e) Lu, C.; Bai, H.; Su, F.; Chen, W.; Hwang, J. F.; Lee, H. H. *Journal of the Air and Waste Management Association* 2010, 60 (4), 489-496; (f) Huo, Q.; Zhao, D.; Feng, J.; Weston, K.; Buratto, S. K.; Stucky, G. D.; Schacht, S.; Schüth, F. *Advanced Materials* 1997, 9 (12), 974-978; (g) Tang, L.; Cheng, J. *Nano Today* 2013, 8 (3), 290-312; (h) Fang, X.; Zhao, X.; Fang, W.; Chen, C.; Zheng, N. *Nanoscale* 2013, 5 (6), 2205-2218.

[3] Rouquerol, J.; Avnir, D.; Fairbridge, C. W.; Everett, D. H.; Haynes, J. M.; Pernicone, N.; Ramsay, J. D. F.; Sing, K. S. W.; Unger, K. K. *Pure and Applied Chemistry,* 1994, 66 (8), 1739.

[4] (a) Taylor, K. M. L.; Kim, J. S.; Rieter, W. J.; An, H.; Lin, W.; Lin, W. *Journal of the American Chemical Society* 2008, 130 (7), 2154-2155.

[5] Dai, Z.; Liu, S.; Ju, H.; Chen, H. *Biosensors and Bioelectronics,* 2004, *19* (8), 861-867.

[6] Wang, Y.; Caruso, F. *Chemistry of Materials,* 2005, 17 (5), 953-961.

[7] Takahashi, H.; Li, B.; Sasaki, T.; Miyazaki, C.; Kajino, T.; Inagaki, S. *Microporous and Mesoporous Materials,* 2001, 44-45, 755-762.

[8] Xia, T.; Kovochich, M.; Liong, M.; Meng, H.; Kabehie, S.; George, S.; Zink, J. I.; Nel, A. E. *ACS Nano* 2009, 3 (10), 3273-3286.

[9] Hom, C.; Lu, J.; Liong, M.; Luo, H.; Li, Z.; Zink, J. I.; Tamanoi, F. *Small,* 2010, 6 (11), 1185-1190.

[10] M. H.; Na, H. K.; Kim, Y. K.; Ryoo, S. R.; Cho, H. S.; Lee, K. E.; Jeon, H.; Ryoo, R.; Min, D. H. *ACS Nano,* 2011, 5 (5), 3568-3576.

[11] Vallet-Regi, M.; Rámila, A.; Del Real, R. P.; Pérez-Pariente, *J. Chemistry of Materials* 2001, 13 (2), 308-311.

[12] (a) Lu, J.; Liong, M.; Li, Z.; Zink, J. I.; Tamanoi, F. *Small,* 2010, 6 (16), 1794-1805; (b) Li, X.; Zhang, L.; Dong, X.; Liang, J.; Shi, J. *Microporous and Mesoporous Materials,* 2007, 102 (1-3), 151-158; (c) Radin, S.; El-Bassyouni, G.; Vresilovic, E. J.; Schepers, E.; Ducheyne, P. *Biomaterials,* 2005, 26 (9), 1043-1052.

[13] (a) Beck, J. S.; Vartuli, J. C.; Roth, W. J.; Leonowicz, M. E.; Kresge, C. T.; Schmitt, K. D.; Chu, C. T. W.; Olson, D. H.; Sheppard, E. W.; McCullen, S. B.; Higgins, J. B.; Schlenker, J. L. *Journal of the American Chemical Society,* 1992, 114 (27), 10834-10843; (b) Wan, Y.; Zhao, D. *Chemical Reviews,* 2007, 107 (7), 2821-2860.

[14] Asefa, T.; Tao, Z. *Canadian Journal of Chemistry,* 2012, 90 (12), 1015-1031.

[15] Mai, W. X.; Meng, H. *Integrative Biology (United Kingdom)* 2013, 5 (1), 19-28.

[16] Kim, T. W.; Chung, P. W.; Lin, V. S. Y. *Chemistry of Materials,* 2010, 22 (17), 5093-5104.

[17] Hoffmann, F.; Cornelius, M.; Morell, J.; Fröba, M., *Angewandte Chemie International Edition,* 2006, 45 (20), 3216-3251.

[18] Vallet-Regí, M.; Balas, F.; Arcos, D. *Angewandte Chemie - International Edition,* 2007, 46 (40), 7548-7558.

[19] He, Q.; Gao, Y.; Zhang, L.; Zhang, Z.; Gao, F.; Ji, X.; Li, Y.; Shi, J. *Biomaterials,* 2011, 32 (30), 7711-7720.

[20] Nampi, P. P.; Mohan, V. S.; Sinha, A. K.; Varma, H. *Materials Research Bulletin,*2012, 47 (6), 1379-1384.

[21] Rosenholm, J. M.; Lindén, M. *Journal of Controlled Release,* 2008, 128 (2), 157-164.

[22] (a) Carino, I. S.; Pasqua, L.; Testa, F.; Aiello, R.; Puoci, F.; Iemma, F.; Picci, N. *Drug Delivery,* 2007, 14 (8), 491-495; (b) Gao, Q.; Xu, Y.; Wu, D.; Sun, Y.; Li, X. *Journal of Physical Chemistry, C* 2009, 113 (29), 12753-12758; (c) Andersson, J.; Rosenholm, J.; Linden, M. *Topics in Multifunctional Biomaterials & Devices*; (d) Sousa, A.; Souza, K. C.; Sousa, E. M. B. *Acta Biomaterialia,* 2008, 4 (3), 671-679; (e) Zhu, Y. F.; Shi, J. L.; Li, Y. S.; Chen, H. R.; Shen, W. H.; Dong, X. P. *Microporous and Mesoporous Materials,* 2005, 85 (1-2), 75-81; (f) Lopez, T.; Ortiz, E.; Alexander-Katz, R.; Basaldella, E.; Bokhimi, X. *Nanomedicine: Nanotechnology, Biology, and Medicine,*

2009, 5 (2), 170-177; (g) Demuth, P.; Hurley, M.; Wu, C.; Galanie, S.; Zachariah, M. R.; Deshong, P. *Microporous and Mesoporous Materials,* 2011, 141 (1-3), 128-134.

[23] Charnay, C.; Bégu, S.; Tourné-Péteilh, C.; Nicole, L.; Lerner, D. A.; Devoisselle, J. M. *European Journal of Pharmaceutics and Biopharmaceutics,* 2004, 57 (3), 533-540.

[24] (a) Huo, Q.; Margolese, D. I.; Stucky, G. D. *Chemistry of Materials,* 1996, 8 (5), 1147-1160; (b) Boissière, C.; Martines, M. A. U.; Tokumoto, M.; Larbot, A.; Prouzet, E. *Chemistry of Materials,* 2003, 15 (2), 509-515; (c) Kim, M. J.; Ryoo, R. *Chemistry of Materials,* 1999, 11 (2), 487-491.

[25] Jia, L.; Shen, J.; Li, Z.; Zhang, D.; Zhang, Q.; Duan, C.; Liu, G.; Zheng, D.; Liu, Y.; Tian, X. *International Journal of Pharmaceutics,* 2012, 439 (1-2), 81-91.

[26] (a) Van Speybroeck, M.; Mellaerts, R.; Mols, R.; Thi, T. D.; Martens, J. A.; Van Humbeeck, J.; Annaert, P.; Van den Mooter, G.; Augustijns, P. *European Journal of Pharmaceutical Sciences,* 2010, 41 (5), 623-630; (b) Marzouqa, D. M.; Zughul, M. B.; Taha, M. O.; Hodali, H. A. *Journal of Porous Materials,* 2012, 19 (5), 825-833.

[27] Trewyn, B. G.; Slowing, I. I.; Giri, S.; Chen, H. T.; Lin, V. S. Y. *Accounts of Chemical Research,* 2007, 40 (9), 846-853.

[28] Qu, F.; Zhu, G.; Huang, S.; Li, S.; Sun, J.; Zhang, D.; Qiu, S. *Microporous and Mesoporous Materials,* 2006, 92 (1-3), 1-9.

[29] (a) Xu, W.; Gao, Q.; Xu, Y.; Wu, D.; Sun, Y.; Shen, W.; Deng, F. *Powder Technology,* 2009, 191 (1-2), 13-20; (b) Izquierdo-Barba, I.; Colilla, M.; Vallet-Regi, M. *Journal of Nanomaterials,* 2008, 14.

[30] (a) Zhang, Y.; Jiang, T.; Zhang, Q.; Wang, S. *European Journal of Pharmaceutics and Biopharmaceutics,* 2010, 76 (1), 17-23; (b) Zhang, Y.; Zhang, J.; Jiang, T.; Wang, S. *International Journal of Pharmaceutics,* 2011, 410 (1-2), 118-124.

[31] (a) Zhao, D.; Huo, Q.; Feng, J.; Chmelka, B. F.; Stucky, G. D. *Journal of the American Chemical Society,* 1998, 120 (24), 6024-6036; (b) Liu, X.; Tian, B.; Yu, C.; Gao, F.; Xie, S.; Tu, B.; Che, R.; Peng, L. M.; Zhao, D. *Angewandte Chemie - International Edition* ,2002, 41 (20), 3876-3878; (c) Fan, J.; Yu, C.; Gao, F.; Lei, J.; Tian, B.; Wang, L.; Luo, Q.; Tu, B.; Zhou, W.; Zhao, D. *Angewandte Chemie - International Edition,* 2003, 42 (27), 3146-3150.

[32] Andersson, J.; Rosenholm, J.; Areva, S.; Lindén, M. *Chemistry of Materials,* 2004, 16 (21), 4160-4167.

[33] (a) Cauda, V.; Mühlstein, L.; Onida, B.; Bein, T. *Microporous and Mesoporous Materials,* 2009, 118 (1-3), 435-442; (b) Cauda, V.; Onida, B.; Platschek, B.; Muhlstein, L.; Bein, T. *Journal of Materials Chemistry,* 2008, *18* (48), 5888-5899.

[34] Strømme, M.; Brohede, U.; Atluri, R.; Garcia-Bennett, A. E. *Wiley Interdisciplinary Reviews: Nanomedicine and Nanobiotechnology,* 2009, 1 (1), 140-148.

[35] Heikkilä, T.; Salonen, J.; Tuura, J.; Kumar, N.; Salmi, T.; Murzin, D. Y.; Hamdy, M. S.; Mul, G.; Laitinen, L.; Kaukonen, A. M.; Hirvonen, J.; Lehto, V. P. *Drug Delivery,* 2007, 14 (6), 337-347.

[36] Zhang, Y.; Zhi, Z.; Jiang, T.; Zhang, J.; Wang, Z.; Wang, S. *Journal of Controlled Release,* 2010, 145 (3), 257-263.

[37] Colilla, M.; Izquierdo-Barba, I.; Sánchez-Salcedo, S.; Fierro, J. L. G.; Hueso, J. L.; Vallet-Regí, M. *Chemistry of Materials,* 2010, 22 (23), 6459-6466.

[38] Asefa, T.; Otuonye, A. N.; Wang, G.; Blair, E. A.; Vathyam, R.; Denton, K. *Adsorption,* 2009, 15 (3), 287-299.

[39] (a) Yang, Q.; Wang, S.; Fan, P.; Wang, L.; Di, Y.; Lin, K.; Xiao, F. S. *Chemistry of Materials* 2005, 17 (24), 5999-6003; (b) Angelos, S.; Liong, M.; Choi, E.; Zink, J. I. *Chemical Engineering Journal,* 2008, 137 (1), 4-13; (c) Basaldella, E. I.; Legnoverde, M. S. *Journal of Sol-Gel Science and Technology,* 2010, 56 (2), 191-196.

[40] Zhao, X. S.; Lu, G. Q. *Journal of Physical Chemistry, B* 1998, *102* (9), 1556-1561.

[41] Anwander, R.; Nagl, I.; Widenmeyer, M.; Engelhardt, G.; Groeger, O.; Palm, C.; Röser, T. *Journal of Physical Chemistry B* 2000, 104 (15), 3532-3544.

[42] Sharma, K. K.; Anan, A.; Buckley, R. P.; Ouellette, W.; Asefa, T. *Journal of the American Chemical Society,* 2008, 130 (1), 218-228.

[43] Ritter, H.; Nieminen, M.; Karppinen, M.; Brühwiler, D. *Microporous and Mesoporous Materials,* 2009, 121 (1-3), 79-83.

[44] Gartmann, N.; Schütze, C.; Ritter, H.; Brühwiler, D. *Journal of Physical Chemistry Letters,* 2010, 1 (1), 379-382.

[45] Popova, M. D.; Szegedi, A.; Kolev, I. N.; Mihaly, J.; Tzankov, B. S.; Momekov, G. T.; Lambov, N. G.; Yoncheva, K. P. *International Journal of Pharmaceutics,* 2012, 436 (1-2), 778-785.

[46] Wang, H.; Gao, X.; Wang, Y.; Wang, J.; Niu, X.; Deng, X. *Ceramics International,* 2012, 38 (8), 6931-6935.

[47] Popovici, R. F.; Seftel, E. M.; Mihai, G. D.; Popovici, E.; Voicu, V. A. *Journal of Pharmaceutical Sciences,* 2011, 100 (2), 704-714.

[48] Giraldo, L.; Moreno-Piraján, J. C. *Journal of Chemistry,* 2013.

[49] Andrade, G. F.; Soares, D. C. F.; Almeida, R. K. D. S.; Sousa, E. M. B. *Journal of Nanomaterials,* 2012, 2012.

[50] (a) Burkett, S. L.; Sims, S. D.; Mann, S. *Chemical Communications* 1996, (11), 1367-1368; (b) Macquarrie, D. J. *Chemical Communications* 1996, (16), 1961-1962; (c) Lim, M. H.; Blanford, C. F.; Stein, A. *Journal of the American Chemical Society* 1997, 119 (17), 4090-4091.

[51] Slowing, I. I.; Vivero-Escoto, J. L.; Wu, C. W.; Lin, V. S. Y. *Advanced Drug Delivery Reviews,* 2008, 60 (11), 1278-1288.

[52] Chong, A. S. M.; Zhao, X. S.; Kustedjo, A. T.; Qiao, S. Z. *Microporous and Mesoporous Materials,* 2004, 72 (1-3), 33-42.

[53] Huh, S.; Wiench, J. W.; Yoo, J. C.; Pruski, M.; Lin, V. S. Y. *Chemistry of Materials,* 2003, 15 (22), 4247-4256.

[54] Stein, A.; Melde, B. J.; Schroden, R. C. *Advanced Materials,* 2000, *12* (19), 1403-1419.

[55] Mercier, L.; Pinnavaia, T. J. *Chemistry of Materials,* 2000, 12 (1), 188-196.

[56] Richer, R.; Mercier, L. *Chemistry of Materials,* 2001, 13 (9), 2999-3008.

[57] Yokoi, T.; Yoshitake, H.; Tatsumi, T. *Chemistry of Materials,* 2003, 15 (24), 4536-4538.

[58] Xie, M.; Shi, H.; Ma, K.; Shen, H.; Li, B.; Shen, S.; Wang, X.; Jin, Y. *Journal of Colloid and Interface Science,* 2013, 395 (1), 306-314.

[59] Lee, C. H.; Lo, L. W.; Mou, C. Y.; Yang, C. S. *Advanced Functional Materials,* 2008, 18 (20), 3283-3292.

[60] Lai, C. Y.; Trewyn, B. G.; Jeftinija, D. M.; Jeftinija, K.; Xu, S.; Jeftinija, S.; Lin, V. S. Y. *Journal of the American Chemical Society,* 2003, 125 (15), 4451-4459.

[61] Huh, S.; Chen, H. T.; Wiench, J. W.; Pruski, M.; Lin, V. S. Y. *Angewandte Chemie - International Edition,* 2005, 44 (12), 1826-1830.

[62] Shang, F.; Sun, J.; Wu, S.; Yang, Y.; Kan, Q.; Guan, J. *Microporous and Mesoporous Materials,* 2010, 134 (1-3), 44-50.

[63] Olkhovyk, O.; Jaroniec, M. *Industrial and Engineering Chemistry Research,* 2007, 46 (6), 1745-1751.

[64] Chen, C.; Jie, X.; Qiaohong, Z.; Hong, M.; Hong, M.; Lipeng, Z. *Journal of Physical Chemistry, C* 2009, 113 (7), 2855-2860.

[65] (a) Han, L.; Ruan, J.; Li, Y.; Terasaki, O.; Che, S. *Chemistry of Materials,* 2007, *19* (11), 2860-2867; (b) Ma, Y.; Xing, L.; Zheng, H.; Che, S., Anionic-cationic switchable amphoteric monodisperse mesoporous silica nanoparticles. *Langmuir,* 2011, 27 (2), 517-520.

[66] Tang, T.; Li, X. A.; Xu, Y.; Wu, D.; Sun, Y. H.; Xu, J.; Deng, F. *Colloids and Surfaces B-Biointerfaces,* 2011, 84 (2), 571-578.

[67] Margolese, D.; Melero, J. A.; Christiansen, S. C.; Chmelka, B. F.; Stucky, G. D. *Chemistry of Materials,* 2000, 12 (8), 2448-2459.

[68] Lim, M. H.; Stein, A. *Chemistry of Materials,* 1999, 11 (11), 3285-3295.

[69] Gartmann, N.; Bruhwiler, D. *Angewandte Chemie-International Edition,* 2009, 48 (34), 6354-6356.

[70] Rahmat, N.; Abdullah, A. Z.; Mohamed, A. R. *American Journal of Applied Sciences,* 2010, *7* (12), 1579-1586.

[71] Zeng, W.; Qian, X. F.; Yin, J.; Zhu, Z. K. *Materials Chemistry and Physics,* 2006, 97 (2-3), 437-441.

[72] Song, S. W.; Hidajat, K.; Kawi, S. *Langmuir,* 2005, 21 (21), 9568-9575.

[73] Sun, J. T.; Hong, C. Y.; Pan, C. Y. *Journal of Physical Chemistry, C* 2010, 114 (29), 12481-12486.

[74] Liu, R.; Zhang, Y.; Zhao, X.; Agarwal, A.; Mueller, L. J.; Feng, P. *Journal of the American Chemical Society,* 2010, 132 (5), 1500-1501.

[75] Gao, Q.; Xu, Y.; Wu, D.; Shen, W.; Deng, F. *Langmuir* 2010, 26 (22), 17133-17138.

[76] Muhammad, F.; Guo, M.; Qi, W.; Sun, F.; Wang, A.; Guo, Y.; Zhu, G. *Journal of the American Chemical Society,* 2011, 133 (23), 8778-8781.

[77] Yuan, L.; Tang, Q.; Yang, D.; Zhang, J. Z.; Zhang, F.; Hu, J. *Journal of Physical Chemistry, C* 2011, 115 (20), 9926-9932.

[78] Xu, W.; Gao, Q.; Xu, Y.; Wu, D.; Sun, Y. *Materials Research Bulletin,* 2009, 44 (3), 606-612.

[79] Zhu, Y. F.; Shi, J. L.; Shen, W. H.; Dong, X. P.; Feng, J. W.; Ruan, M. L.; Li, Y. S. *Angewandte Chemie-International Edition,* 2005, 44 (32), 5083-5087.

[80] Liu, R.; Zhao, X.; Wu, T.; Feng, P. *Journal of the American Chemical Society,* 2008, 130 (44), 14418-14419.

[81] Nadrah, P.; Maver, U.; Jemec, A.; Tišler, T.; Bele, M.; Dražić, G.; Benčina, M.; Pintar, A.; Planinšek, O.; Gaberščček, M. *ACS Applied Materials and Interfaces,* 2013, 5 (9), 3908-3915.

[82] Nadrah, P.; Porta, F.; Planinsek, O.; Kros, A.; Gaberscek, M. *Physical Chemistry Chemical Physics,* 2013, 15 (26), 10740-10748.

[83] Mal, N. K.; Fujiwara, M.; Tanaka, Y. *Nature,* 2003, 421 (6921), 350-353.

[84] Vivero-Escoto, J. L.; Slowing, I. I.; Wu, C. W.; Lin, V. S. Y. *Journal of the American Chemical Society,* 2009, 131 (10), 3462-3463.

[85] Lin, Q.; Huang, Q.; Li, C.; Bao, C.; Liu, Z.; Li, F.; Zhu, L. *Journal of the American Chemical Society,* 2010, 132 (31), 10645-10647.

[86] Yan, H.; Teh, C.; Sreejith, S.; Zhu, L.; Kwok, A.; Fang, W.; Ma, X.; Nguyen, K. T.; Korzh, V.; Zhao, Y. *Angewandte Chemie - International Edition,* 2012, 51 (33), 8373-8377.

[87] Yuan, Q.; Zhang, Y.; Chen, T.; Lu, D.; Zhao, Z.; Zhang, X.; Li, Z.; Yan, C. H.; Tan, W. *ACS Nano,* 2012, 6 (7), 6337-6344.

[88] Souza, K. C.; Ardisson, J. D.; Sousa, E. M. B. *Journal of Materials Science-Materials in Medicine,* 2009, 20 (2), 507-512.

[89] Liu, C.; Guo, J.; Yang, W.; Hu, J.; Wang, C.; Fu, S. *Journal of Materials Chemistry,* 2009, 19 (27), 4764-4770.

[90] Baeza, A.; Guisasola, E.; Ruiz-Hernández, E.; Vallet-Regí, M. *Chemistry of Materials,* 2012, 24 (3), 517-524.

[91] Thomas, C. R.; Ferris, D. P.; Lee, J. H.; Choi, E.; Cho, M. H.; Kim, E. S.; Stoddart, J. F.; Shin, J. S.; Cheon, J.; Zink, J. I. *Journal of the American Chemical Society,* 2010, 132 (31), 10623-10625.

[92] Aznar, E.; Mondragõn, L.; Ros-Lis, J. V.; Sancenõn, F.; Marcos, M. D.; Martínez-Máñez, R.; Soto, J.; Pérez-Payá, E.; Amorõs, P. *Angewandte Chemie - International Edition,* 2011, 50 (47), 11172-11175.

[93] You, Y. Z.; Kalebaila, K. K.; Brock, S. L.; Oupický, D. *Chemistry of Materials,* 2008, 20 (10), 3354-3359.

[94] Chen, Z.; Li, Z.; Lin, Y.; Yin, M.; Ren, J.; Qu, X. *Chemistry - A European Journal,* 2013, 19 (5), 1778-1783.

[95] Bernardos, A.; Mondragón, L.; Aznar, E.; Marcos, M. D.; Martínez-Máñez, R.; Sancenón, F.; Soto, J.; Barat, J. M.; Pérez-Payá, E.; Guillem, C.; Amorós, P. *ACS Nano,* 2010, 4 (11), 6353-6368.

[96] Popat, A.; Ross, B. P.; Liu, J.; Jambhrunkar, S.; Kleitz, F.; Qiao, S. Z. *Angewandte Chemie - International Edition,* 2012, 51 (50), 12486-12489.

[97] Zhu, C. L.; Lu, C. H.; Song, X. Y.; Yang, H. H.; Wang, X. R. *Journal of the American Chemical Society,* 2011, 133 (5), 1278-1281.

[98] He, X.; Zhao, Y.; He, D.; Wang, K.; Xu, F.; Tang, J. *Langmuir,* 2012, 28 (35), 12909-12915.

[99] Zhao, Y.; Trewyn, B. G.; Slowing, I. I.; Lin, V. S. Y. *Journal of the American Chemical Society,* 2009, 131 (24), 8398-8400.

[100] Climent, E.; Bernardos, A.; Martínez-Máñez, R.; Maquieira, A.; Marcos, M. D.; Pastor-Navarro, N.; Puchades, R.; Sancenón, F.; Soto, J.; Amorós, P. *Journal of the American Chemical Society,* 2009, 131 (39), 14075-14080.

[101] Barzegar-Jalali, M.; Adibkia, K.; Valizadeh, H.; Shadbad, M. R. S.; Nokhodchi, A.; Omidi, Y.; Mohammadi, G.; Nezhadi, S. H.; Hasan, M. *Journal of Pharmacy and Pharmaceutical Sciences,* 2008, 11 (1), 167-177.

[102] Wang, S. *Microporous and Mesoporous Materials,* 2009, 117 (1–2), 1-9.

[103] Doadrio, A. L.; Sousa, E. M. B.; Doadrio, J. C.; Pérez Pariente, J.; Izquierdo-Barba, I.; Vallet-Regí, M. *Journal of Controlled Release,* 2004, 97 (1), 125-132.

[104] Bernardos, A.; Aznar, E.; Coll, C.; Martínez-Mañez, R.; Barat, J. M.; Marcos, M. D.; Sancenón, F.; Benito, A.; Soto, J. *Journal of Controlled Release,* 2008, 131 (3), 181-189.

[105] Izquierdo-Barba, I.; Martinez, Á.; Doadrio, A. L.; Pérez-Pariente, J.; Vallet-Regí, M. *European Journal of Pharmaceutical Sciences,* 2005, 26 (5), 365-373.
[106] Heikkilä, T.; Salonen, J.; Tuura, J.; Hamdy, M. S.; Mul, G.; Kumar, N.; Salmi, T.; Murzin, D. Y.; Laitinen, L.; Kaukonen, A. M.; Hirvonen, J.; Lehto, V. P. *International Journal of Pharmaceutics,* 2007, 331 (1), 133-138.
[107] Alexa, I. F.; Ignat, M.; Popovici, R. F.; Timpu, D.; Popovici, E. *International Journal of Pharmaceutics,* 2012, 436 (1-2), 111-119.
[108] Xu, S.; Yong, J.; Lai, G.; Zhang, H.; Wu, Y.; Yu, A. *Chemistry Letters,* 2013, 42 (3), 235-237.
[109] Ochekpe, N. A.; Olorunfemi, P. O.; Ngwuluka, N. C., *Tropical Journal of Pharmaceutical Research,* 2009, 8 (3).

In: Comprehensive Guide for Mesoporous Materials, Volume 4 ISBN: 978-1-63482-092-9
Editor: Mahmood Aliofkhazraei

Chapter 10

PERIODIC MESOPOROUS ORGANOSILICAS: HYBRID POROUS MATERIALS WITH EXCITING APPLICATIONS

***Els De Canck*[1], *Dolores Esquivel*[1], *Francisco J. Romero-Salguero*[2] *and Pascal Van Der Voort*[1]**

[1]Department of Inorganic and Physical Chemistry, Center for Ordered Materials, Organometallics and Catalysis (COMOC), Ghent University, Ghent, Belgium
[2]Department of Organic Chemistry, Nanochemistry and Fine Chemistry Research Institute (IUIQFN), Faculty of Sciences, University of Córdoba, Córdoba, Spain

ABSTRACT

Periodic Mesoporous Organosilicas or commonly named "PMOs" constitute of a new family of functional hybrid materials where organic groups are covalently integrated within the pore walls of a silica framework. With well-defined pore structures and unique properties, they combine the structural characteristics of ordered mesoporous silicas with the chemical functionalization possibilities of organic groups. These advantages convert these materials into promising candidates in areas such as catalysis, adsorption, nanoelectronics, and chromatography, among others. In this chapter, we will give an overview of all these applications where PMOs with specific functional groups have been successfully reported.

1. INTRODUCTION

The research field of nanoporous materials is an interesting and exciting topic with countless researchers involved worldwide.

One type of hybrid solid, combining organic and inorganic properties, was developed during the latest decade and is called *Periodic Mesoporous Organosilica* or *PMO*.

The first reports on this type of nanomaterial appeared in 1999, published by three research groups, namely Stein et al. [1], Inagaki et al. [2], and Ozin et al. [3].

This development occurred very shortly after the first account on the now well-known SBA-type materials [4, 5] and after the development of the famous ordered mesoporous silicas (MCM materials) by the Mobil Oil company [6]. Both materials, SBA and MCM, are considered to be the "ancestors" of PMOs as similarities can be found between the syntheses.

PMOs are prepared by the hydrolysis of a bis-silane and subsequent polycondensation around a template (Figure 1). The bis-silane is represented by $(OR')_3$-Si-R-Si-$(OR')_3$ with R the organic functionality and OR' a hydrolysable alkoxy moiety, mostly methoxy or ethoxy groups. The template or surfactant will aggregate in the aqueous environment and form micelles around which the bis-silane will polycondense. A variety of silanes and surfactants exists and the final properties can be altered by carefully choosing the reaction conditions. Naturally, this opens up many opportunities for the preparation of very different materials. After aging, surfactant removal is performed to open up the porous structure of the PMO material. Figure 1 presents a schematic overview of a general synthesis procedure with the template Pluronic® P123 ($EO_{20}PO_{70}EO_{20}$) and a generic bis-silane.

The functions inside the walls are not accessible and thus not available for any chemical modification. However, these organic groups are beneficial for the overall structural properties of the PMO material due to the addition of rigidity or flexibility.

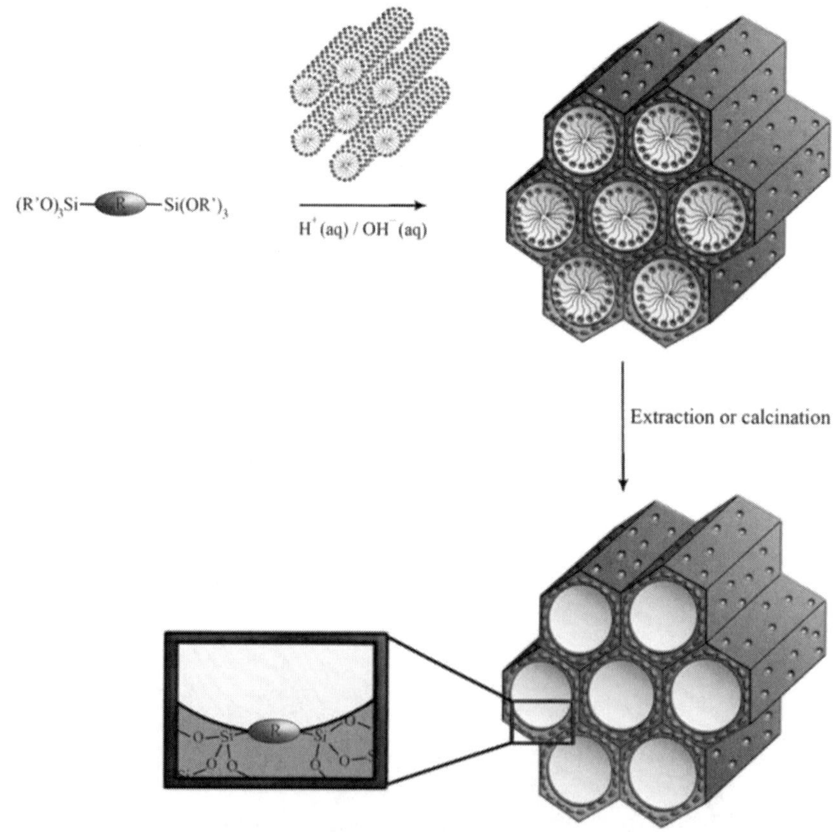

Reproduced from Reference [7] with permission from the PCCP Owner Societies.

Figure 1. Schematic representation of the synthesis of a Periodic Mesoporous Organosilica. R represents an organic functionality and OR' the alkoxy groups.

In contrast, the organic bridges on the surface of the PMO material are reachable and can be chemically modified.

A brief overview of the most commonly used bis-silanes and templates are given in Table 1 and Table 2, respectively. This first only shows a small amount of the numerous bis-silanes that exist. More complex and advanced structures are reported including multi-organic bridged precursors. This variation in silanes provides materials for all different kinds of applications including chromatography, sensing, heterogeneous catalysis, adsorption, micro-electronics and biomedical applications.

Several reviews have appeared on the subject of Periodic Mesoporous Organosilicas [8-18], including the very recent publication of a comprehensive review in *Chemical Society Reviews* [19].

Table 1. List of the most common silanes used in PMO synthesis

Abbreviation	Full name	Structural formula
BTME	1,2-Bis(trimethoxysilyl)ethane	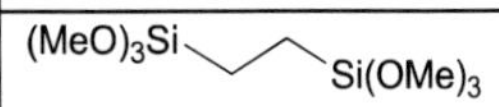
BTEE	1,2-Bis(triethoxysilyl)ethane	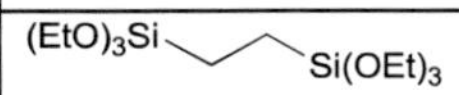
BTEENE	1,2-Bis(triethoxysilyl)ethylene	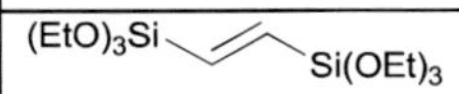
BTEM	1,2-Bis(triethoxysilyl)methylene	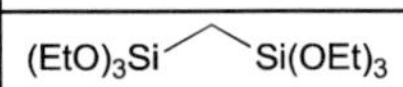
BTEB	1,4-Bis(triethoxysilyl)benzene	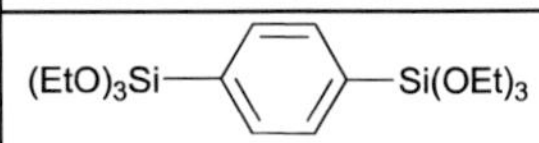
BTEEB	1,4-Bis(triethoxysilylethyl)benzene	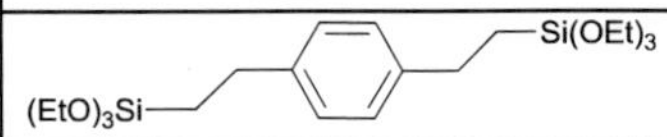
BTEVB	1,4-Bis((E)-2-(triethoxysilyl)vinyl)benzene	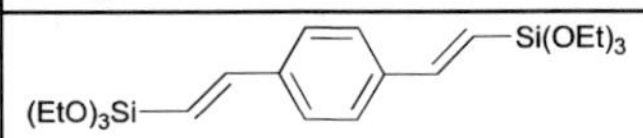
BTEVS	1,4-Bis((E)-2-(triethoxysilyl)vinyl)stilbene	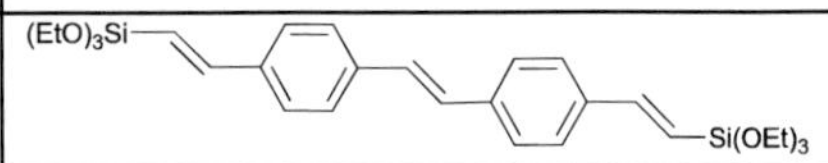
BTET	2,5-Bis-(triethoxysilyl)thiophene	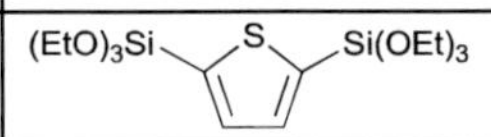
BTEPUE	Bis[3-(triethoxysilyl)propyl urethane] ethane	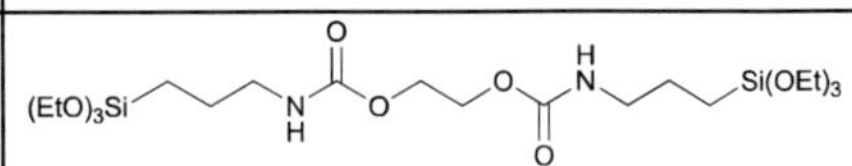
BTEPDS	Bis[3-(triethoxysilyl)propyl] disulfide	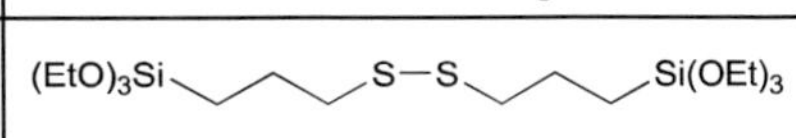
BTEPTS	1,4-Bis(triethoxysilyl)propane tetrasulfide	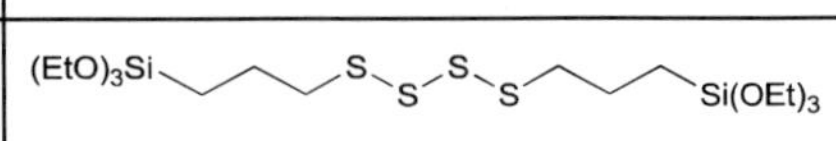
ICS	Tris [3-(trimethoxysilyl)propyl]isocyanurate	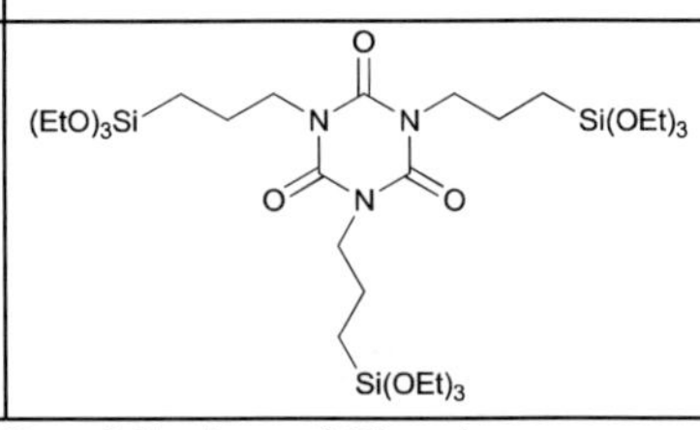

Reproduced from Reference [19] with permission from The Royal Society of Chemistry.

Table 2. List of the most common templates used in PMO synthesis

Abbreviation	Full name	Structural formula
CTAC/CTAB	cetyltrimethylammonium chloride/bromide	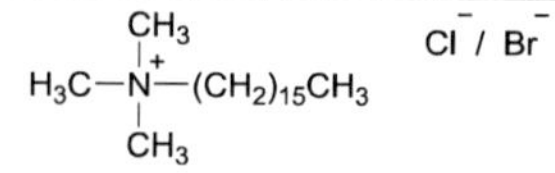
OTAC	octadecyltrimethylammonium chloride	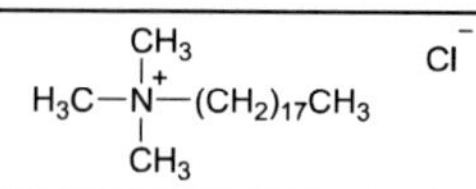
C_nTMACl/Br	alkyltrimethylammonium chloride/bromide	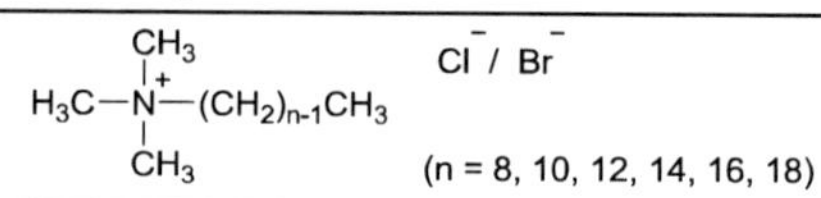
CPCl	cetylpyridinium chloride	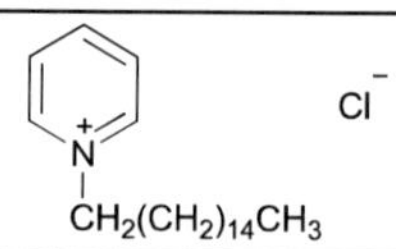
FC4	fluorocarbon surfactant	
Brij-30	polyoxyethylene (4) lauryl ether	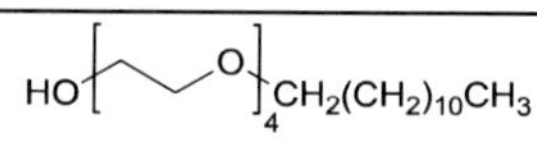
Brij-56	polyoxyethylene (10) cetyl ether	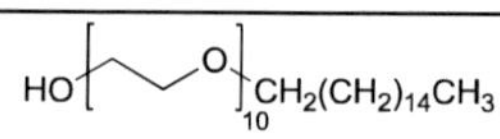
Brij-76	polyoxyethylene (10) stearyl ether	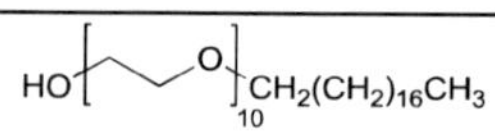
Triton-X100	polyoxyethylene (10) octylphenyl ether	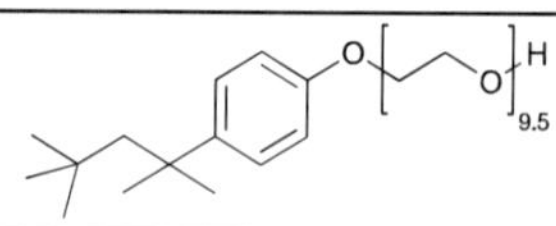
P123	Pluronic P123 poly(ethylene glycol)-poly(propylene glycol)-poly(ethylene glycol)	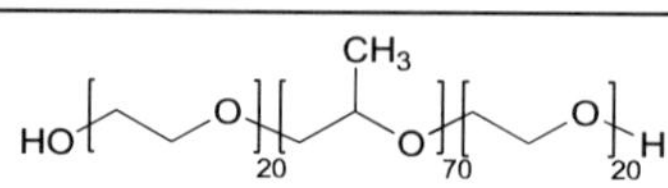
F127	Pluronic F127 poly(ethylene glycol)-poly(propylene glycol)-poly(ethylene glycol)	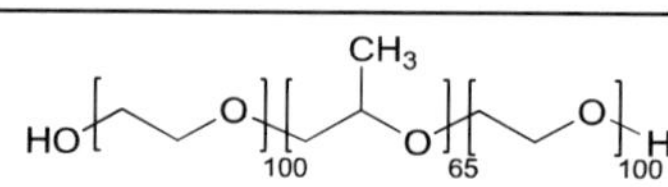
B50-6600	poly(ethylene oxide)–poly(butylene oxide)–poly(ethylene oxide)	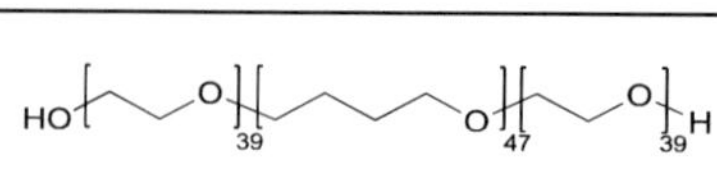
PEO-GLGA-PEO	poly(ethylene oxide)–poly(lactic acid-co-glycolic acid)–poly(ethylene oxide)	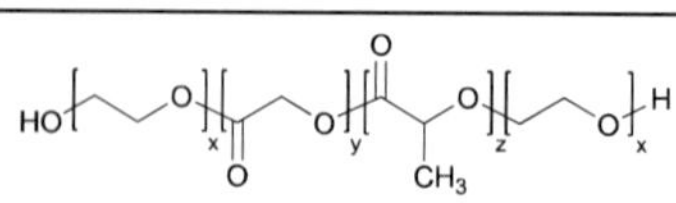
$C_{n\text{-}s\text{-}m}$	Divalent and *gemini* surfactants	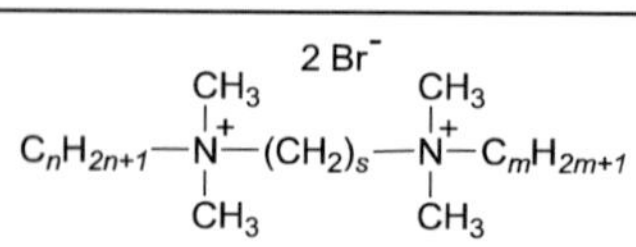

Reproduced from Reference [19] with permission from The Royal Society of Chemistry.

2. APPLICATIONS OF PMOs

Periodic Mesoporous Organosilicas are applied in many different fields due to their versatility as their structural and chemical properties can be easily altered. A large amount of research is performed to broaden their usability [19]. A concise overview is presented below.

2.1. Ultra Stable PMOs As Super Insulating Low-*K* Film

PMO materials with simple organic bridges such as methane, ethane and benzene possess an intrinsic quality which is of great importance in some applications, *i.e.*, the hydrophobicity induced by the organic moieties. This property is especially useful in chromatography or electronic devices when the PMO is applied as chromatographic phase or as low-*k* film, respectively.

The use of PMOs for chromatographic packing materials will be the subject of another chapter in this book. PMOs can also be applied as insulating interlayer or low-*k* material. These materials with a low dielectric constant (*k*) reduce the RC-delay induced between wires in close proximity. The *k*-value is preferably below 2. The intrinsic properties of PMOs make them very useful in micro-electronic devices. Their porosity or plainly stated, cavities filled with air, reduces the *k*-value significantly as $k_{air} \approx 1$. Furthermore the organic bridges (methane and ethane) are less polarisable then silicon dioxide ($k = 3.9$), which diminishes the *k*-value even more [8].

In order to function as an interlayer, PMO thin films must be prepared. This can be achieved by two methods: (1) the deposition on a substrate using the well-known Evaporation Induced Self Assembly or EISA method [20] or (2) the preparation by hydrothermal synthesis [21].

It was Lu et al. [22] who reported the first PMO thin films for this application, although the majority of the work was performed by the research groups of Ozin and Van Der Voort. An overview of the reported materials is presented in Table 3 together with the porosity and observed *k*-value.

Ozin et al. elaborately described the preparation of thin films consisting of methane-, ethane-, ethene- or ring-PMOs, the latter being a multi-organic bridged organosilica made out of interconnected $[Si(CH_2)]_3$ rings. The group thoroughly investigated their dielectric, mechanical and hydrophobic properties [23, 24, 26, 27].

Especially the hydrophobicity of the ring-PMO is a key factor as less silanol groups are present and the adsorption of water is reduced. The presence of water can significantly increase the dielectric constant as $k_{water} \approx 80$.

The group of Van Der Voort extended the applicability of the ring-PMO as low-*k* film (Figure 2) [28]. By employing Brij-76 instead of the surfactant CTAB, they improved the chemical stability of the thin films. This was explained by the presence of thicker pore walls in combination with the $-CH_2-$ moieties. The ring-PMO thin films sustained a treatment with 1 M NaOH. Moreover, they recently developed a method to modify the top surface of a low-*k* film in order to seal the pores [25, 30].

By applying these procedures, the *k*-value was lowered and the mechanical and chemical stability was improved. But most importantly, it also lowers the leak current [30].

Table 3. Overview of different types of PMO films and their dielectric constants

PMO	Surfactant	Porosity %	*k*-value	References
Methane-PMO	CTAC	n.a.	1.8	[23]
	CTAC	53	1.6	[24]
Ethane-PMO	Brij-56	56	1.98	[22]
	CTAC	n.a.	1.9	[23]
	CTAC	45	1.67	[24]
	Brij-76	39.5	2.15	[25]
	CTAC	47	1.7	[26]
Ethene-PMO	CTAC	44	1.7	[24]
Ring-PMO[a]	CTAC	n.a.	2.0	[27]
	CTAC	50	1.63	[24]
	Brij-76	55	1.8	[28]
POSS-PMO[b]	CTAC	39	1.7	[29]

[a]Prepared with the precursor 1,1,3,3,5,5-hexaethoxy-1,3,5-trisilacyclohexane. [b]Prepared with a polyhedral oligomeric silane.

Reproduced from Reference [19] with permission from The Royal Society of Chemistry.

Adapted from Reference [28] with permission from The Royal Society of Chemistry.

Figure 2. SEM and TEM image of a ring-PMO thin film, prepared *via* spin-coating with Brij-76 and ethanol in acidic medium. The length of the scale bar on the TEM image represents 20 nm.

Another type of interesting PMO material for low-*k* applications is presented by Ozin et al. [29] A thin film was spin-coated by using a polyhedral oligomeric silsesquioxane or POSS precursor. The resulting PMO possesses air pockets in the pore walls which is beneficial for the dielectric constant. A *k*-value of 1.7 was achieved.

This overview presented above clearly indicates that PMOs can be applied as low-*k* materials in microelectronic devices. However, their performance should still be improved by more extended research on for example the stability and adhesion to other components.

2.2. Functional PMOs As Catalysts

The use as catalysts is undoubtedly one of the main applications of PMOs. In order to rationalize the catalytic applications of PMOs, we have divided them according to the main catalytic function.

2.2.1. Acid and Base Catalysis by PMOs

Brönsted acid sites have been incorporated in PMOs almost exclusively as sulfonic acid groups. Figure 3 and Table 4 summarize the main sulfonated motifs found in these materials and their main catalytic applications. Different structures of sulfonated PMOs obtained by direct sulfonation (**a**,**b**), co-condensation (a silica precursor is occasionally included) (**c**-**i**), post-modification of the bridges (**j**-**m**) and grafting without (**n**) or with post-modification (**o**-**r**) are shown.

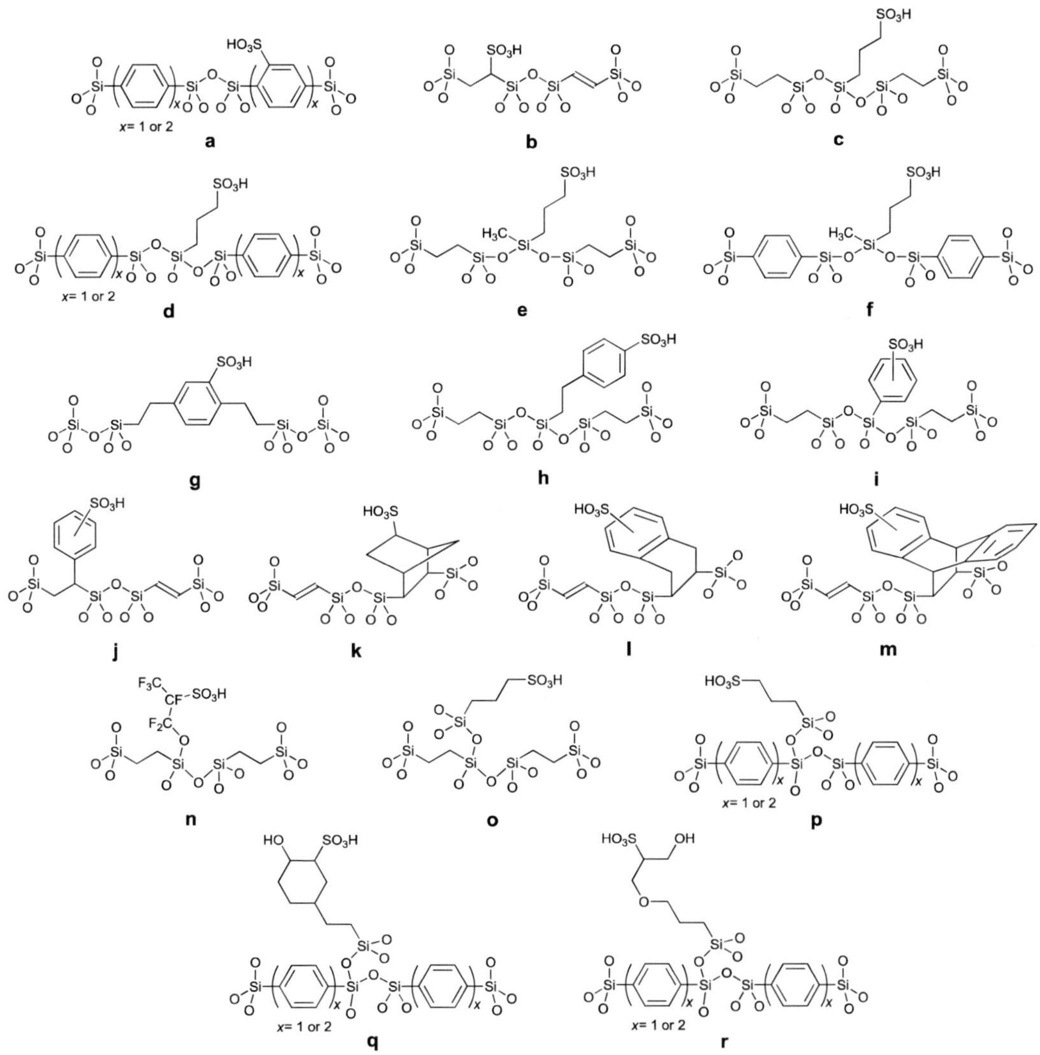

Reproduced from Reference [19] with permission from The Royal Society of Chemistry.

Figure 3. Types of sulfonated PMO precursors.

Table 4. Overview of reactions catalyzed by sulfonated PMOs

Reaction type	Structure
Acetalization of heptanal by 1-butanol	c[36], h[36], n[43]
Alkylation of isobutane with 1-butene	n[44]
Claisen-Schmidt condensation between benzaldehyde and acetophenone	c[35], i[33]
Self-condensation of heptanal to crotonaldehyde	j[45], n[43]
Dehydration of 1-butanol to butyl ether	c[46], h[47]
Dimerization of 2-phenylpropene	a[32], d[32]
Esterification Transesterification (biodiesel production)	a[42], b[48], c[49], d[50, 51], g[39], k[52], l[38, 52], m[52], p[50, 51], q[51], r[51] e[53], f[53]
Etherification	a[41], c[40], h[40]
Friedel-Crafts reactions Alkylation Acylation and Fries rearrangement Hydroxyalkylation	 a[32], c[54], d[32] a[32, 55], d[32] a[41], c[56], d[56], i[33], o[56], p[51, 56], q[51], r[51]
Hydration of propylene oxide to propylene glycol	c[46]
Hydrolysis of cyclohexylacetate of sugars	 c[49] c[31], o[31], p[31]
Rearrangement Beckmann rearrangement Pinacol-pinacolone rearrangement Rearrangement-aromatization of ketoisophorone	 l[37] l[37] a[32], d[32]

Reproduced from Reference [19] with permission from The Royal Society of Chemistry.

The maximum values of acidity reported for sulfonated PMOs used as catalysts are ca. 1.1, 1.8 and 1.4 mmol g^{-1}, which correspond, respectively, to those obtained by direct sulfonation, co-condensation and post-modification or grafting methods. In general, no clear relationship can be established between the catalytic activity and the amount of acid sites [31-33]. Many other factors such as structure, hydrophobicity, accessibility, acid strength and stability influence the activity.

Most of the literature published so far on catalytic applications of PMOs emphasizes the higher activity in organic reactions where water is involved [34], compared to their silica counterparts or even some zeolites and commercial resins, at least per acid site [35-38]. This increased activity is usually ascribed to the enhanced hydrophobicity near the active site, which facilitates diffusion of water out of the pores and organic compounds into the pores [36, 39-42].

The hydrophobicity of sulfonated PMOs also plays a key role in their stability and so they present a higher resistance to deactivation than their silica counterparts. Thus, sulfonated PMOs (type **c**) can be used even in a flow reactor for the alkylation of phenol with 2-propanol.

Many sulfonated PMOs have been reported to be stable for several runs without significant loss of activity [31, 38, 46]. Materials synthesized by co-condensation (type **d**) tend to be more stable than those obtained by grafting (type **p**) [50]. The manifest differences in stability described by different authors should be related to the variety of catalyst structures, organic processes and reaction conditions which are involved in the catalytic applications of these materials.

At present, only few examples of base catalysis by PMOs have been reported. Kapoor et al. [57] described the synthesis of dendritic silica hybrids and a phenylene-PMO grafted with 3-aminopropyltrimethoxysilane. These materials were applied as catalysts for the Knoevenagel condensation between benzaldehyde and malononitrile. PMOs were more active than mesoporous silicas. The same reaction was tested on crystal-like phenylene-PMO subjected to a two step amination process [58]. Very recently, Hesemann et al. [59] reported the synthesis of tris(3-(trimethoxysilyl)propyl)amine and its use as precursor for the synthesis of mesoporous organosilica. The tertiary amines present in the pore walls were highly active as catalysts for the Knoevenagel condensation between benzaldehyde and malononitrile and for the Henry reaction of benzaldehyde and nitroethane. This latter catalytic test was checked in highly ordered ethane-PMO with up to 40% APTMS in the initial mixture [60].

A co-condensation route was reported by Park et al. [61] to prepare a melamine-bridged PMO starting from a mixture of tris-silylated melamine (up to 10 mol%) and sodium metasilicate. This catalyst, which could be recycled, afforded a 40% conversion in the reaction between propylene oxide and CO_2 to the corresponding cyclic carbonate in DMF solvent.

Mono-silylated N-phenylthiazolium salts were grafted on benzene-PMOs, ethane-PMOs, MCM-41 and silylated-MCM-41 by Thiel et al. [62] The resulting benzene-PMO gave rise to the higher yields and better stability both in the benzoin condensation of benzaldehyde and in the cross-coupling of aldehydes and acyl imines.

2.2.2. Catalysis by Metal Sites in PMOs

Most examples of catalysis by metal sites incorporated in PMOs deal with redox processes (Figure 4).

Among the different metals, Ti, Sn, V and Nb have been successfully incorporated into the framework of PMOs, particularly ethane-PMOs, as isolated tetrahedral sites. Ti-PMOs were found to be moderately active toward the selective epoxidation of α-pinene [63] and 1-octene [64] with 30% H_2O_2 and TBHP, respectively. Also, Sn [65] and V-substituted [66] ethane-PMOs and Nb-substituted ethane- and octane-PMOs [67] were used as catalysts in the epoxidation of different olefins with H_2O_2 as oxidant. In addition, Ti-PMOs and Nb-PMOs were excellent catalysts for the ammoximation of cyclic ketones [68] and the oxidation of stearic acid [67], respectively. Cr-PMOs have also been prepared by using either BTEE or BTEE/TEOS mixtures with cationic surfactants under basic conditions [69], In general, the more hydrophobic character (*i.e.*, organic content), the higher activity per metal center, and so metal-substituted PMOs are usually more active and selective than their silica counterparts. For the same reason, the hybrid materials were usually more stable, even though metal leaching (particularly, V and Cr) was unavoidable in these catalysts.

Al-containing ethane-PMOs were used as acid catalysts for the alkylation of 2,4-di-*tert*-butylphenol with cinnamyl alcohol [70]. Interestingly, the hybrid materials gave higher conversion and selectivity to the cinnamylphenol than a similar Al-MCM-41.

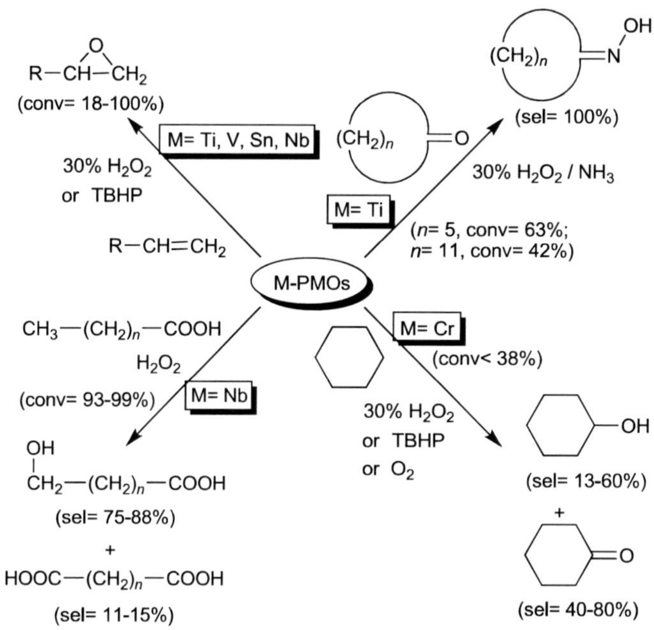

Reproduced from Reference [19] with permission from The Royal Society of Chemistry.

Figure 4. Redox processes catalyzed by metal sites incorporated in PMOs.

Very recently, Zr has been incorporated into the framework of phenylene-bridged PMOs to be used in biodiesel production by methanolysis of crude palm oil [71]. Moreover, Pd-supported PMO have shown to be active in the Suzuki cross-coupling reaction between bromobenzene and phenylboronic acid [72].

2.2.3. Catalysis by Metal Complexes in PMOs

The heterogeneization of organometallic compounds is a hot topic that offers significant advantages for "green catalysis", particularly the recovery of the usually expensive and/or toxic catalyst, the avoidance of product contamination and the possibility of operating under continuous flow. With this aim, various metal complexes have been incorporated as bridges in PMOs and used as catalysts (Table 5).

Catalytic PMOs containing Pd complexes are active for different coupling reactions and redox processes. However, they often undergo a significant reduction of their catalytic activity after successive reuses due to the degradation of their mesostructures [73] and/or the decomposition of their bridges with the concomitant formation of Pd nanoparticles, which act as homogeneous catalysts [73].

Thus, the main role of the PMO was to recapture and stabilize the Pd nanoparticles. Other bridges, such as bis(8-quinolinolato)dioxomolybdenum(VI) complexes [74] and metalloporphyrins [75] were stable and even more active than the related homogeneous complexes. No leaching was observed for vanadyl Schiff base complexes [76]. Also, bridges consisting of Ni(II) α-diimine complexes showed high activity in the ethylene polymerization at a wider range of temperatures than the homogenous complex. In addition, it gave a well-defined product morphology, unlike the homogeneous one [77]. Several PMOs containing different organophosphine complexes as bridges have been used as catalysts for a great variety of organic reactions (Figure 5). Thus, different bis- and tris-silylated phosphine complex precursors with different metals, *i.e.*, Pd(II), Rh(I), Ru(II) and Au(I), were co-condensed with TEOS, BTEB and/or 4,4'-bis(triethoxysilyl)biphenyl [78-81]. These materials exhibited similar activities to the homogeneous complexes and could be recycled repeatedly due to their resistence to leaching. Moreover, they were usually more active than similar grafted materials since the grafting method could lead to pore blockage.

2.2.4. Catalysis by Bifunctional PMOs

Although different PMOs with two or even more different functions have been previously described, here we will refer to bifunctional PMOs catalysts as those bearing two catalytic groups. At the moment, there are only few examples of such systems. In fact, Mehdi et al. [93] were the first to discuss this concept in an early work where they described the synthesis of bifunctional PMOs with acidic groups ($-SO_3H$) in the framework and basic groups ($-NH_2$) in the channel pores. No catalytic application was reported.

Thiel et al. [94] obtained a benzene-PMO with crystal-like pore walls bearing sulfonic acid (1.10 mmol g^{-1}) and amino groups (1.79 mmol g^{-1}) attached to the hydrophobic phenylene layers and to the hydrophilic silica layers, respectively. Remarkably, this material allowed the one pot transformation of benzaldehyde dimethyl acetal and nitromethane into *trans*-β-nitrostyrene (yield = 97.5%), which involves a deacetalization followed by a nitroaldol condensation by acid and basic groups, respectively. A similar approach was accomplished by Inagaki et al. [95] but using an ethene-PMO. By co-condensation and post-functionalization, they synthesized two materials with sulfonic acid groups and amino groups with these groups interchanged between the hydrophobic and the hydrophilic layers. Regardless of the material, the amount of $-SO_3H$ and $-NH_2$ groups were around 1.1 and 1.3 mmol g^{-1}, respectively. A similar approach was also used by Polarz et al. [96] to investigate the distribution of amino and carboxylic acid groups attached to phenylene bridges.

A final example of bifunctional catalytic PMOs has been recently reported by Li et al. [97]. This material was prepared by co-condensation of BTEE and MPTMS and was later impregnated with $HAuCl_4$. After drying, the Au species were in the metallic state whereas the S species were partially oxidized (the SH/SO_3H molar ratio was 15). The Au nanoparticles, which were coordinated to unreacted thiol groups, had an average size of 1-2 nm. The cooperative effect of the Au particles and the acid groups was demonstrated in the hydration of alkynes in the absence of H_2SO_4 in aqueous solution. Compared to SBA-15, the PMO support promoted the formation of smaller Au particles and provided higher activity due to its enhanced hydrophobicity which prevents the aggregation of the Au nanoparticles and facilitates the diffusion of the reactants in a variety of organic reactions under solvent-free or aqueous conditions.

2.2.5. Enantioselective Catalysis by PMOs

The preparation of efficient heterogeneous catalysts for the enantioselective synthesis of chiral compounds is of a paramount importance, particularly for the pharmaceutical industry. An overview is presented in Table 6.

An early work by Corma et al. [76, 98] described the introduction of chiral vanadyl Schiff base complexes into the walls of mesostructurated materials. Although they exhibited good activity in the cyanosilylation of benzaldehyde, they gave an enantiomeric excess (ee) comparable to the complex anchored on MCM-41 but significantly lower than similar complexes under homogeneous conditions.

Table 5. Overview of catalytic processes in PMOs with metal complexes as bridges

Reaction	Precursor	Catalytic Perf.	References
Cyanosilylation of aldehydes  R= C_6H_5, p-F-C_6H_4, p-(CH_3O)-C_6H_4, n-C_5H_{11}	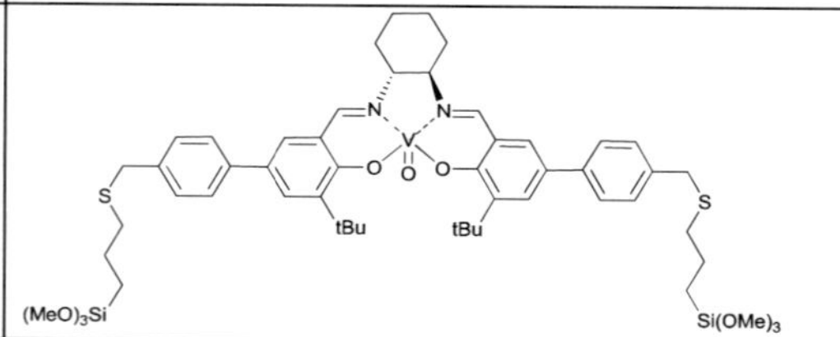	Conv=71-86%, Sel > 98%	[76]
Suzuki-Miyaura coupling 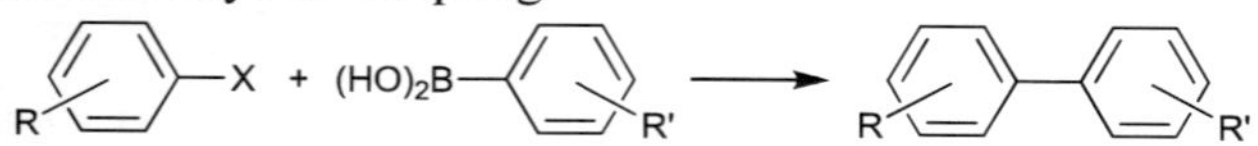 X=Cl, Br	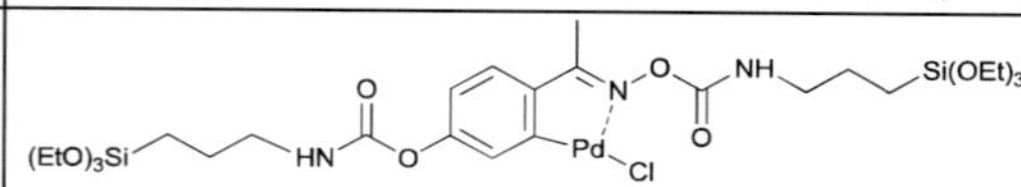	Y>99% (X=Br) Y=5-12% (X=Cl)	[73]
	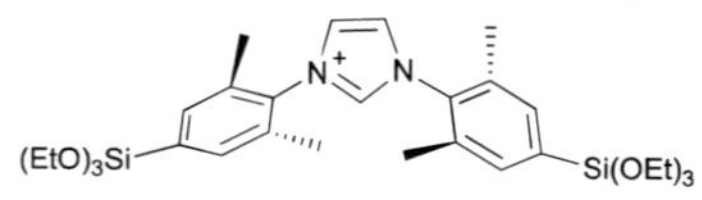 + Pd(acac)2	Y=81-93 (X=Br) Y=50-100% (X=Cl)	[82-84]
	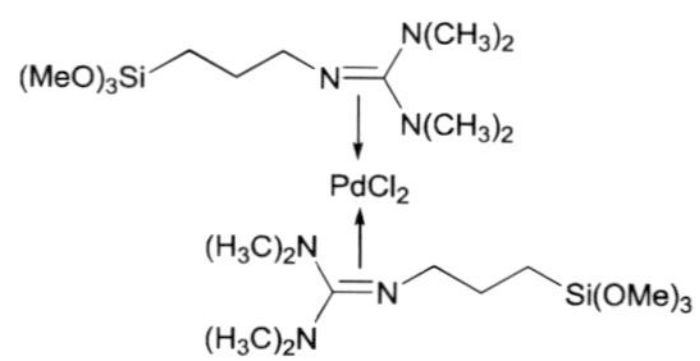	Y=76-99% (X=Br)	[85]
Coupling of benzylic/allylic chlorides and arylboronic acids 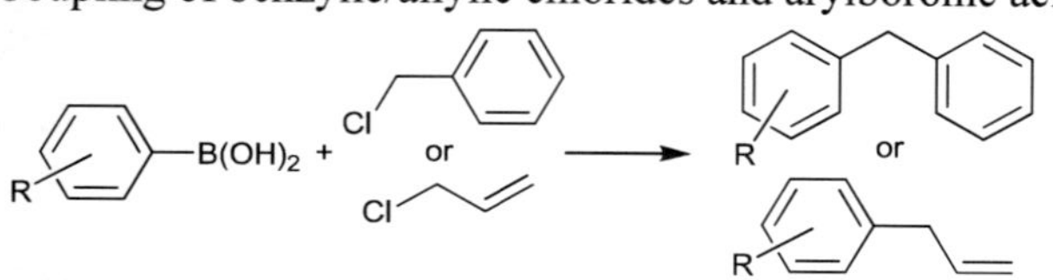	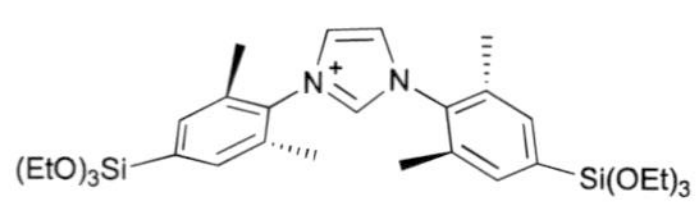 + Pd(acac)2	Y=61-95%	[82, 83]

Reaction	Precursor	Catalytic Perf.	References
Aerobic oxidation of benzylic alcohols 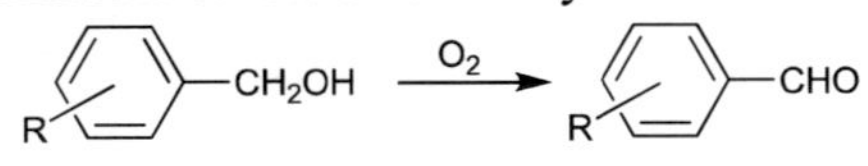 R= H, CH_3, CH_3O	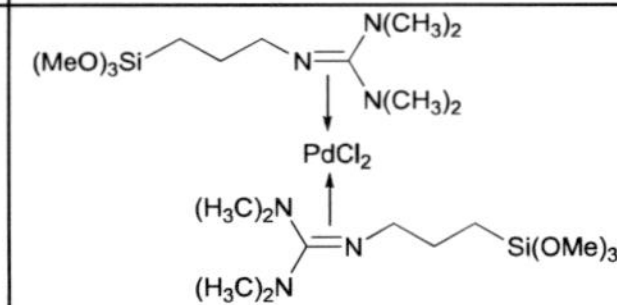	Y≈98%	[85]
Epoxidation of alkenes 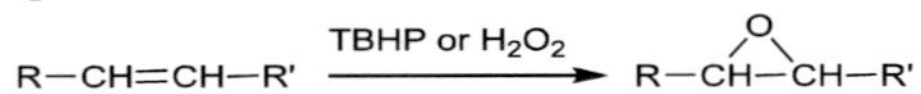	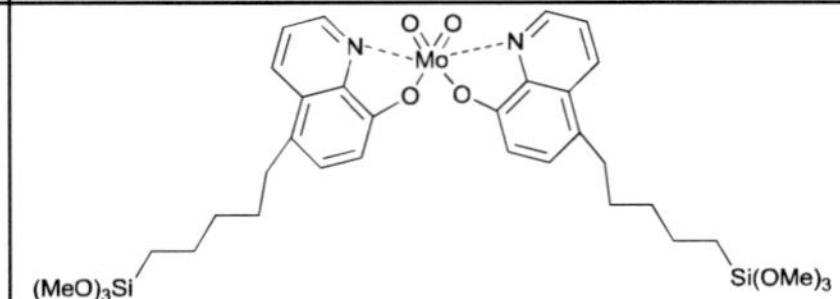	Y~49%	[74]
	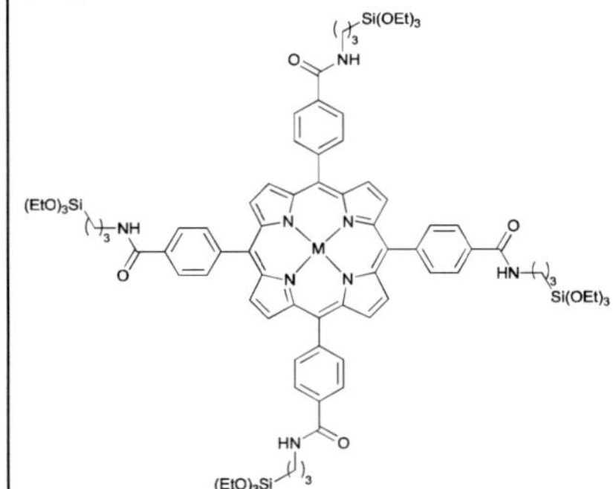	Conv=41%	[86]
	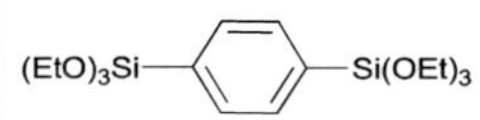 + M(CO)6 (M=Mo, Cr)	Conv=37-100%,	[87, 88]
	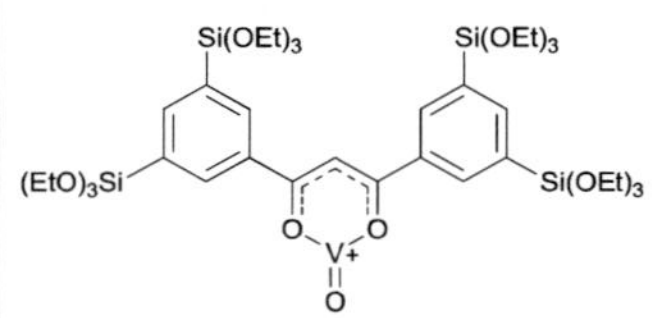	Sel=100% Conv=55%	[89]

Table 5. (Continued)

Reaction	Precursor	Catalytic Perf.	References
Aerobic Baeyer-Villiger oxidation 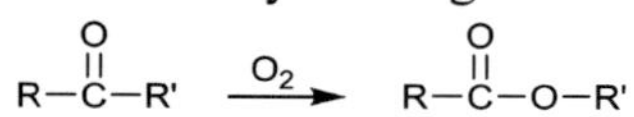	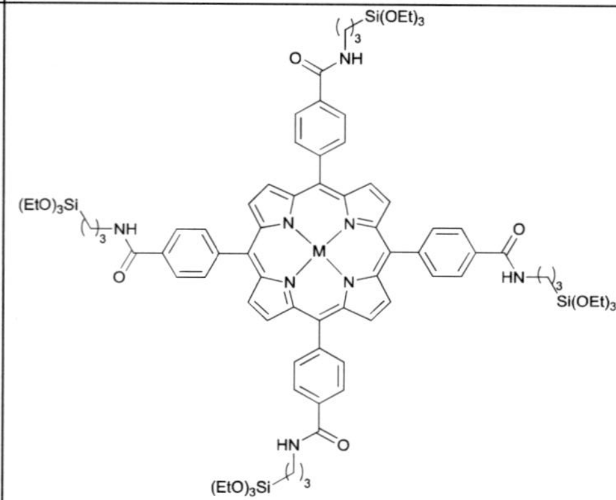	Y=50-100%	[75]
Hydroxylation of benzene 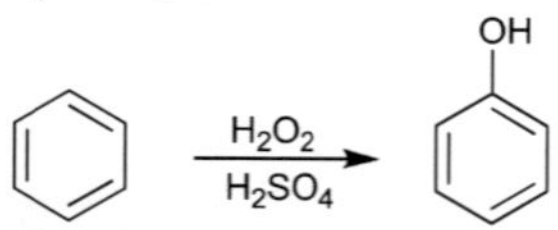	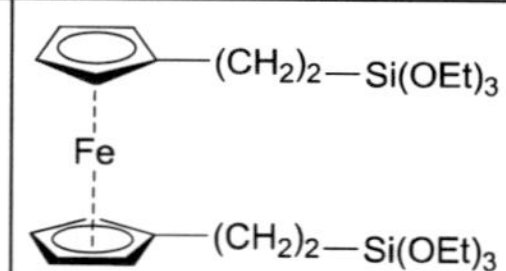	Y~20%, Sel~65%	[90]
Hydrosilylation of 1-hexyne 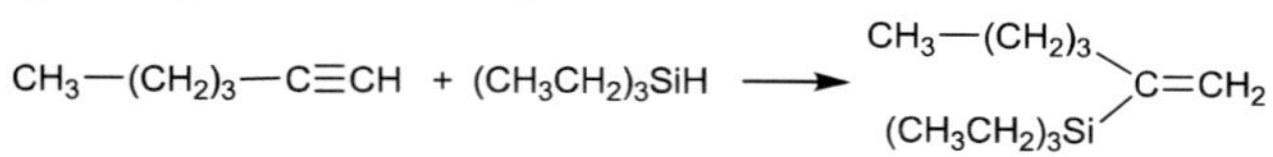	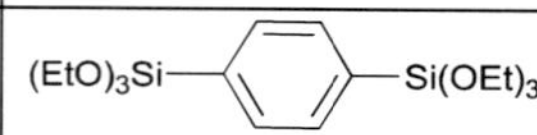 + [(CH3CN)3RuCp]PF6	Conv≈28%	[91]
Methylenation of cinnamaldehyde and Heck reaction	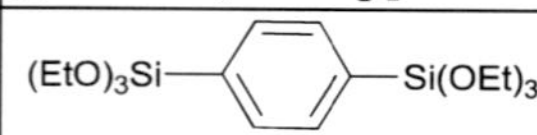 + 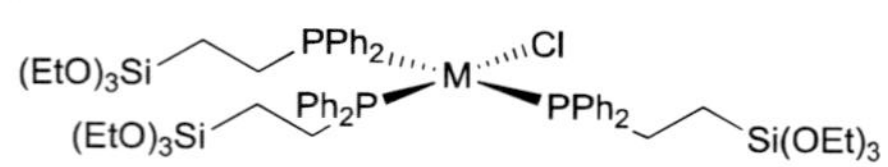 +	Sel=61-71% Conv=66-92%	[92]

Reaction	Precursor	Catalytic Perf.	References
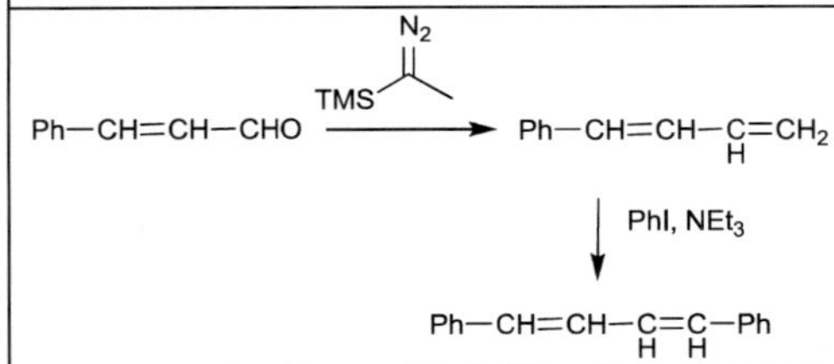	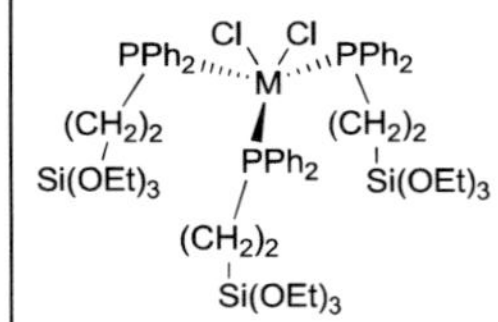 (M=Rh, Pd)	Sel=61-71% Conv=66-92%	[92]

Conv, conversion; Sel, selectivity; Y, yield: TMSCN, trimethylsilyl cyanide.
See references for details. Adapted from Reference [19] with permission from The Royal Society of Chemistry.

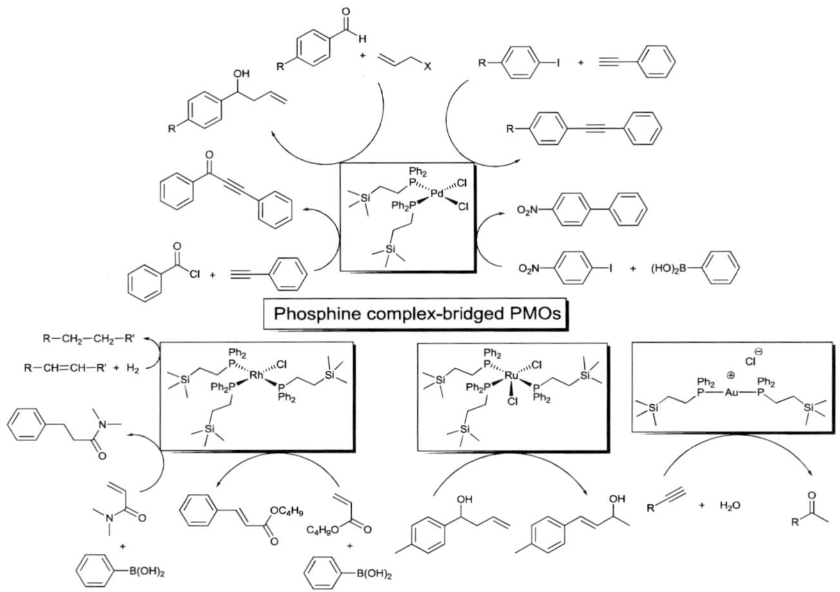

Reproduced from Reference [19] with permission from The Royal Society of Chemistry.

Figure 5. Catalytic processes on PMOs with metal organophosphine complex as bridges.

Table 6. Overview of catalytic processes in PMOs with chiral bridges

Reaction	Bridge	Catalytic performance	References
Cyanosilylation of benzaldehyde 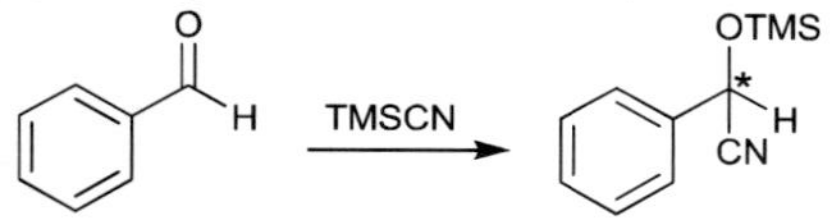	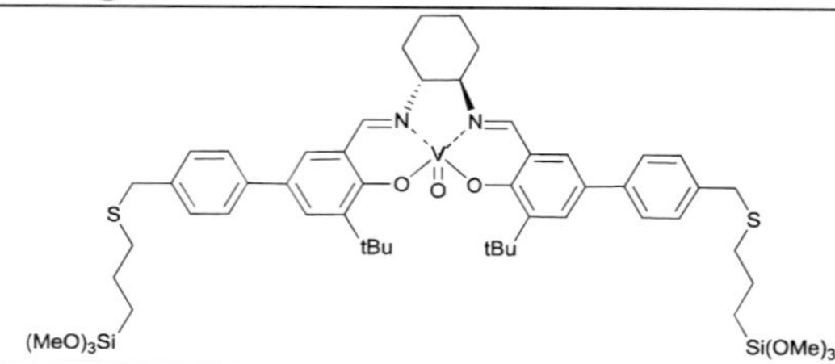	Conv=80%; ee=30%	[76, 98]
	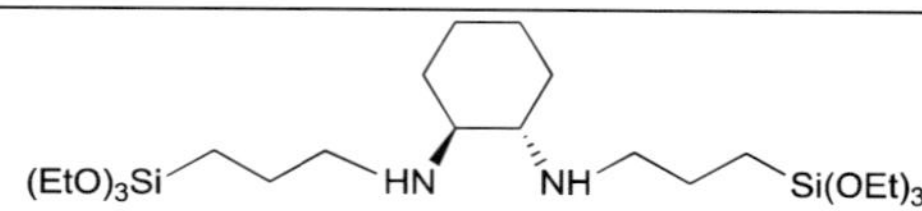 + $[Rh(cod)Cl]_2$	Conv=16-36%, ee=8%	[100]
Transfer hydrogenation of aromatic ketones 	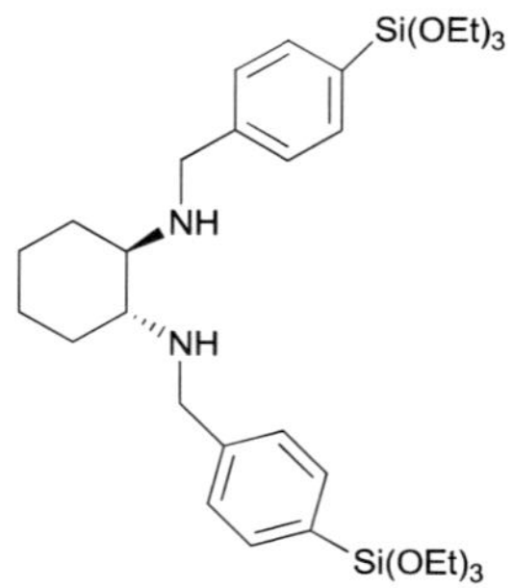 + $[Rh(cod)Cl]_2$	Conv=39-98%, ee=22-41%	[100, 101]
Sulfoxidation of methyl phenyl sulfide 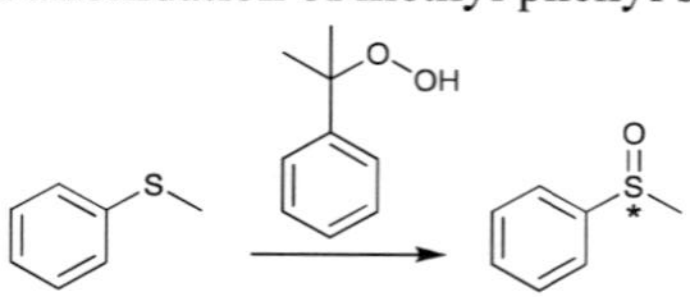	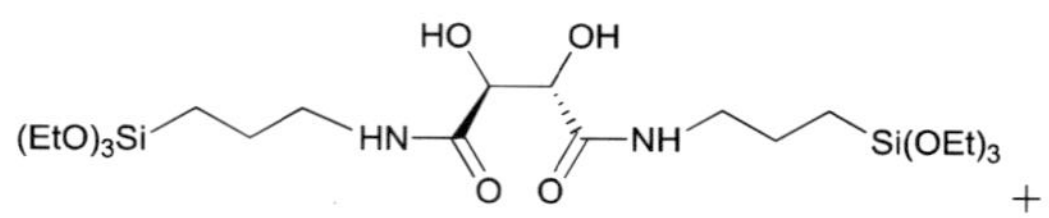 + $Ti(i\text{-}PrO)_4$	Conv=70%, ee=30%	[102]

Reaction	Bridge	Catalytic performance	References
Epoxidation of allyl alcohol 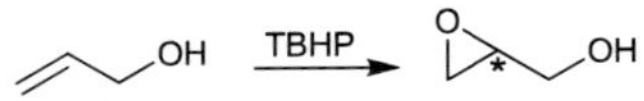	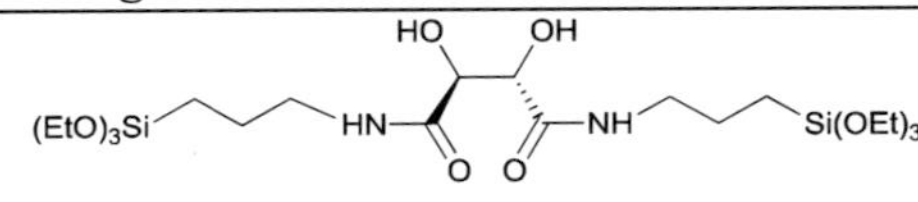 + Ti(i-PrO)$_4$	Conv=46%, ee=20%	[103]
Addition of diethylzinc to aldehydes 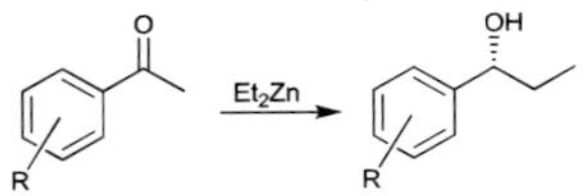	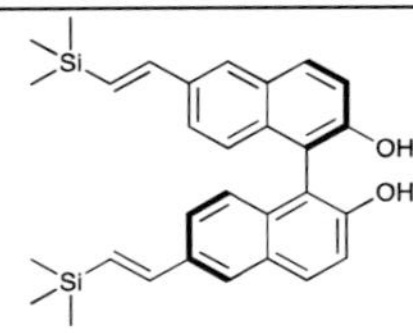 + Ti(i-PrO)$_4$	Conv=98% (R=H), ee=40%	[105]
	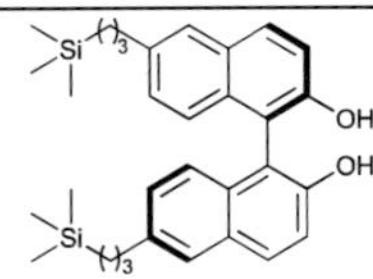 + Ti(i-PrO)$_4$	Conv=40-99%, ee=68-93%	[106]
Hydrogenation of β-keto esters 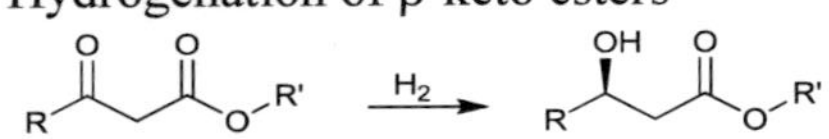	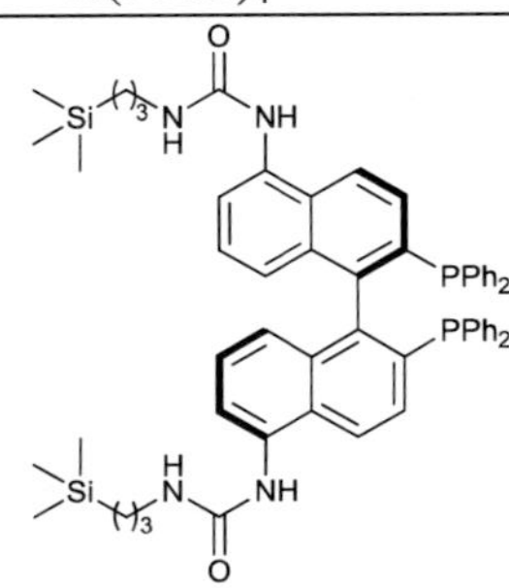 + [RuCl$_2$(benzene)]$_2$	Conv=40-99%, ee=92-99%	[112]
Carbonyl ene reaction 	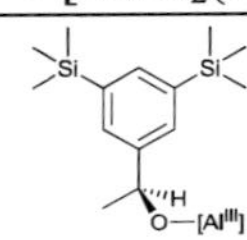	Y=60-80%, ee=0-80%	[111]

Reproduced from Reference [19] with permission from The Royal Society of Chemistry.

Different bis-silylated *trans*-(1*R*,2*R*)-cyclohexadiamine derivatives are among the most common chiral precursors to construct PMOs [99]. Soon afterwards, similar bridges complexed with $[Rh(cod)Cl]_2$ were used as catalysts [100, 101]. However, all these materials were deactivated after the first use probably due to the leaching of Rh metal.

Various bis-silylated tartrate derivatives have also been incorporated as bridges in PMOs [102, 103]. Once coordinated to titanium, they were used as oxidation catalysts in the presence of hydroperoxides. Garcia et al. [102] reported a higher catalytic performance of these materials than a similar grafted MCM-41. The catalyst could be recycled but the sulfoxide yield decreased considerably. A similar material was synthesized by in situ transamidation between (N-methylaminopropyl)trimethoxysilane and dimethyl 2,3-O-isopropylidene-L-tartrate to form the chiral bridge (up to 1.62 mmol tartramide per gram) and co-condensation with BTEE [104]. However, the yield and ee were similar to those reported previously and therefore with a much lower ee than the homogeneous catalyst (94% ee). Yang et al. [105-107] synthesized various bis-silylated R-(+)-BINOL derivatives which were later co-condensed with either TMOS or BTME, the latter leading to more active and stable catalysts. The same bridge was incorporated to PMO nanospheres with different pore structures by co-condensation with TEOS [107]. Both ee and activity increased with the structural order, *i.e.,* from worm like to *P6mm* mesostructures. The nanospheres with radiating channels having a highly ordered 2D-hexagonal structure exhibited an outstanding catalytic performance for the addition of diethylzinc to a variety of aromatic aldehydes. An ee value of 94% (at 99% conversion) was obtained for benzaldehyde, indicating that PMO is even more enantioselective than the homogeneous catalyst. These results highlighted the key role of the framework rigidity to favor that the bite angle of BINOL is correctly arranged. BINAP bridges incorporated in the wall of a mesoporous silica resulted in an excellent PMO catalyst after complexation with ruthenium, exhibiting very high enantioselectivities, close to that of its homogeneous counterpart (99.9% ee).

Some other strategies commonly used for silicas have been successfully applied to PMOs. Thus, hollow nanospheres with ethylene bridges have been functionalized with L-prolinamide by a co-condensation route [108] and by acid-base interaction of a chiral diamine with sulfonic acid groups incorporated previously in their wall [109]. Both materials were used as catalysts in the asymmetric aldol reaction of aromatic aldehydes with cyclohexanone with excellent conversion and ee. Hollow benzene-PMO spheres synthesized by co-condensation of BTEB and 3-azidopropyltrimethoxysilane were used for the immobilization of an imidazolidinone derivative (MacMillan catalyst) through click chemistry post-modification [110]. Its catalytic activity was evaluated in the Diels-Alder reaction between cyclopentadiene and *trans*-cynnamaldehyde in water where it exhibited a very high efficiency (98% yield, 81% ee). In this cases, the organosilica matrix can provide a higher hydrophobicity and as a consequence an increased hydrothermal stability and a facilitated diffusion of reactants.

A very interesting example has been recently reported by Polarz et al., [111] which have addressed the construction of PMOs by using exclusively a bis-silylated chiral precursor. Although at room temperature the ee was zero, it became much higher at lower temperatures. Thus, the ee at -55 °C was 72% for the PMO, whereas it was 17% for the homogeneous complex. As a consequence, steric constraints at the metal center were considered to be decisive.

This fact was further corroborated by the enhanced enantioselectivity of the PMO when restricting the pore size of the PMO and, particularly, when functionalizing its free silanol groups with SiR_3 groups of increasing size.

2.3. PMOs As Adsorbents

A number of studies have been carried out on the environmental applications of non-ordered and ordered mesoporous silicas, particularly addressing the adsorption of heavy metals, organic species and gasses. Excellent reviews on this matter have been published [113, 114]. Since the discovery of PMOs, the possibility of using them as adsorbents for pollutants has been explored and the main challenge is to develop PMO materials that can outperform the classical functionalised silica materials as well as commercial adsorbents such as activated carbon.

2.3.1. Metal Ion Adsorption

The adsorption of metal ions is one of the most investigated research areas because of its importance in waste water treatment. Various PMO materials are used as adsorbents. Some of them are reviewed by Walcarius et al. [113] in 2010, including ordered mesoporous silicas. A concise literature overview of PMO adsorbents for metals is presented in Table 7. Three different strategies are used to develop an adsorbent.

The first method consists of the introduction of one type of functionality, *e.g.*, urea, disulfide or pyridine containing bis-silanes for the adsorption of iron(III) [115], mercury(II) [116] and copper(II) [117], respectively. As these silanes are sometimes too flexible to obtain rigid PMO structures, a co-condensation with TEOS is performed.

Secondly, it is also very interesting to incorporate multiple functionalities in one PMO material. This is achieved by combining several silsesquioxanes and organotrialkoxysilanes during the synthesis. An example of this is presented in Figure 6 where mercury adsorbents are prepared through the combination of a large organic group such as isocyanurate and a simple functionality such as a tetrasulfide group [118].

The incorporation of a secondary group negatively affected the structural ordering of the material but this strongly depends on the type of functionality. Incorporation of the tetrasulfide unit leads to a decrease in pore size while the decrease for the ureido and imidazole functionality was less noticeable. The combination of an isocyanurate and tetrasulfide moiety resulted in a high amount of Hg(II) adsorbed on the PMO (1.37 mmol g^{-1}).

A third very remarkable approach to produce adsorbents has been reported by Burleigh et al. [119] and is known as *molecular imprinting*. The authors used the metal salts of nickel, copper or zinc during the synthesis of the PMO together with BTEE, N-[3-(trimethoxysilyl) propyl]ethylenediamine. The metal ion forms a complex with the amines during the synthesis and after preparation of the materials, the Ni(II), Cu(II) or Zn(II) is removed with HNO_3. This results in a material possessing cavities shaped as the metal with identical chemical environment. When using this material in an adsorption process, the metal ions of the same type which were used during synthesis, will selectively bind to the cavity, especially with solutions containing high concentrations in the range of 10^{-3} mol L^{-1}. Furthermore the materials prepared *via* the imprinting technique exhibited a higher metal binding capacity then the non-imprinted material.

Table 7. Overview of PMO materials investigated for metal (cation) adsorption

Bis-silylated precursor	Surface functionality	Metal	Metal capacity mmol g-1	References
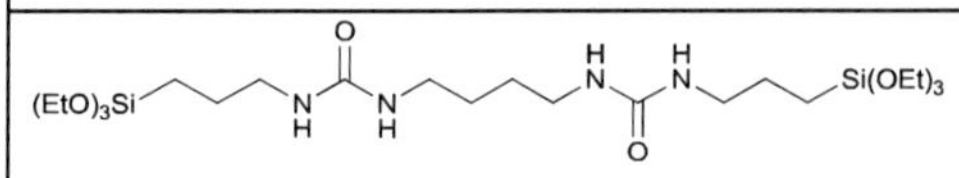 TEOS	-NH-C=O-NH-	Fe(III)	0.19	[115]
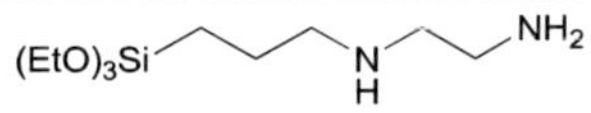 BTEE	$-NH_2$, -NH- $-CH_2CH_2-$	Cu(II) Ni(II) Zn(II)	-	[119]
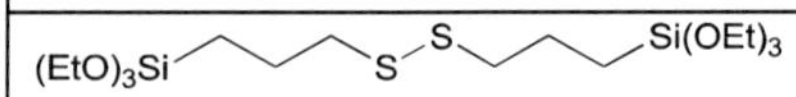 TEOS	-S-S-	Hg(II)	3.57	[116]
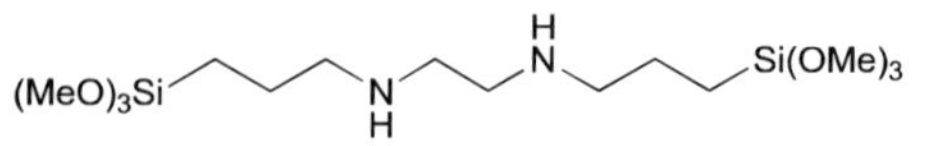 TEOS	-NH-	Cu(II) Ni(II) Zn(II)	0.020 - 0.034 $2.3 \cdot 10^{-3}$ - $1.5 \cdot 10^{-3}$ $1.1 \cdot 10^{-3}$ - $0.9 \cdot 10^{-3}$	[120]
BTEPTS,[a] TMOS	-S-S-S-S-	Hg(II)	0.92 - 7.48	[121]
ICS,[a] TEOS	-N-C=O-N-	Hg(II)	9.7	[122]
ICS,[a] TEOS 	-N-C=O-N- -SH	Hg(II)	4.23 - 5.64	[123]
ICS,[a] BTEPTS,[a] 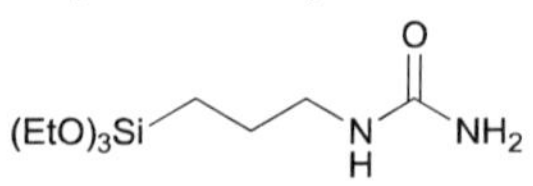 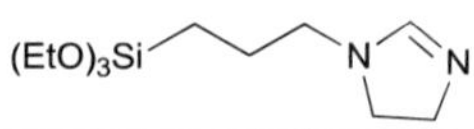	-N-C=O-N- -S-S-S-S- $-NH-C=O-NH_2$ $-N-C_3H_5-N$	Hg(II)	0.40 - 1.37	[118]
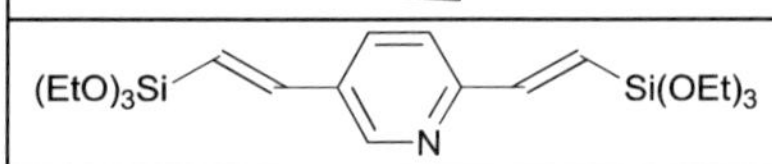	$-NC_5H_3-$	Cu(II)	2.4	[117]
BTEE, TMOS 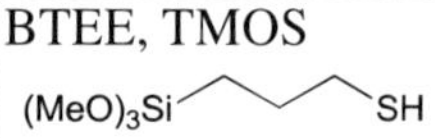	$-CH_2CH_2-$ -SH	Hg(II)	1.42 - 2.31	[124]
BTEE 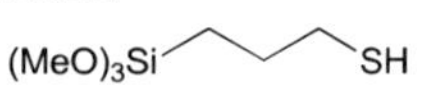	$-CH_2CH_2-$ -SH	Hg(II) Zn(II) Cu(II) Cr(III) Cd(II)	0.76 - 1.75 0.37 0.98 - 1.29 2.04 - 3.27 1.14 - 1.62	[125]
BTEPTS,[a] TEOS	-S-S-S-S-	Hg(II)	3.13 - 13.51	[126]

[a]See Table 1.

Adapted from References [19] with permission from The Royal Society of Chemistry.

Only a few reports have appeared on the very challenging topic of the adsorption of anionic metal species, *i.e.*, the adsorption of oxyanions such as ReO_4^- molecules [127, 128] and the adsorption of AsO_4^{3-} [129]. A key element in the investigation of PMOs as adsorbents should be the stability of the materials during the adsorption process, but also the reusability is of major importance, which is not discussed in many papers.

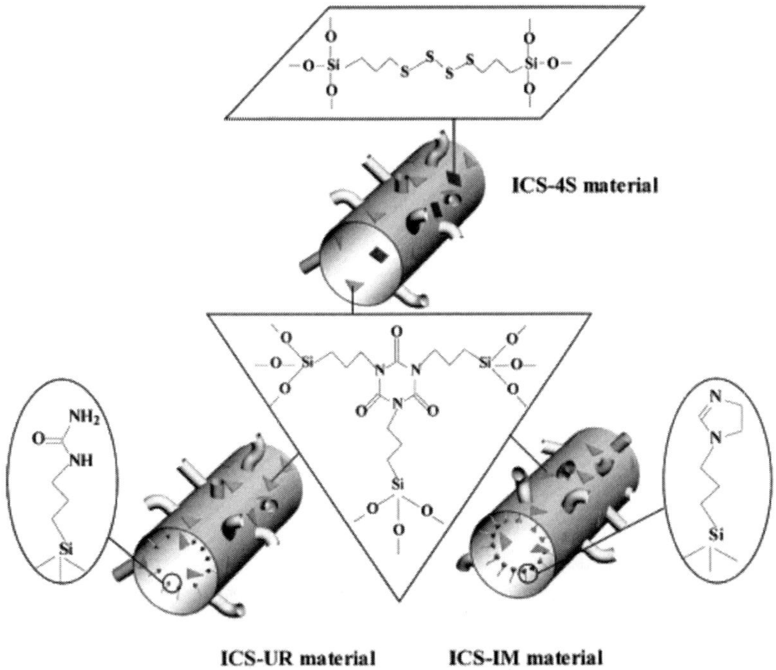

Reprinted with permission from references 118. Copyright (2007), American Chemical Society.

Figure 6. PMO materials synthesized *via* the combination of different precursors, *i.e.*, BTEPTS (top), ICS (middle), ureidopropyltrimethoxysilane (left) and N-(3-triethoxysilylpropyl)4,5-dihydroimidazole (right).

A PMO material can only compete with functionalised silica materials and commercial materials if it is reusable.

2.3.2. Gas Adsorption

Although less reports have been published on the adsorption of gases and volatile organic compounds (VOCs) on PMOs, some promising and interesting results have already been obtained. An overview is presented in Table 8.

The adsorption of VOCs, such as acetaldehyde, was firstly studied in 2006 by comparing benzene-PMOs with 2D-hexagonal and 3D-cubic symmetries as well as the corresponding sulfonated materials [130]. The 3D-structured materials outperformed the 2D-structured and the attachment of sulfonic acid groups even improved the adsorption. The same conclusion was obtained comparing PMOs with polyamidoamine dendritic species inside their pores and the non-functionalised material [57].

The interaction of different VOCs and gases is more examined in detail by Sozanni et al. [131] using a benzene-PMO material., They concluded that the interactions between adsorbate and adsorbent occurred *via* CH---π bonds and not *via* π---π bonds as one would initially expect.

Also, the benzene-PMO appeared to be an excellent adsorbent for CO_2 storage with an adsorption of 90 w% without further functionalisation. A few years later, a theoretical study appeared on the adsorption of CO, CO_2 and CH_4 on benzene-PMOs [135]. Additionally, the adsorption of benzene, toluene and hexane was studied on PMOs with different aromatic bridges [132].

Table 8. Overview of PMO materials investigated for gas adsorption

Bis-silylated precursor	Surface functionality	Gas/VOC	References
1,4-bis(triallylsilyl)phenylene	$-C_6H_4-$ $-C_6H_3(SO_3H)-$	acetaldehyde	[130]
1,4-bis(triethoxysilyl)benzene	$-C_6H_4-$	benzene, hexafluorobenzene, cyclohexane, ethanol, xenon, methane, carbon dioxide	[131]
1,4-bis(triethoxysilyl)benzene 4,4'-bis(triethoxysilyl)-1,1'-biphenyl 1,4-bis(trimethoxysilylethyl)benzene	$-C_6H_4-$ $-C_6H_3-C_6H_3-$ $-C_2H_4-C_6H_4-C_2H_4-$ porphyrin	benzene, toluene, hexane	[132]
3-aminopropyltrimethoxysilane, 3-isocyanatopropyl-triethoxysilane, 1,2-bis(trimethoxysilyl)ethane	$-CH_2CH_2-$ $-(CH_2)_3NH_2$ $-(CH_2)_3NCO$	sulfur dioxide, cyanogen chloride, ammonia, octane	[133]
1,2-bis(triethoxysilyl)ethane 1,2-bis(triethoxysilyl)ethylene 1,4-bis(triethoxysilyl)benzene 4,4'-bis(triethoxysilyl)-1,1'-biphenyl	$-CH_2CH_2-$ $-CH=CH-$ $-C_6H_4-$ $-C_6H_3-C_6H_3-$	hydrogen	[134]
1,4-bis(triethoxysilyl)benzene	$-C_6H_4-$	CO, CO_2, CH_4	[135]
bis(trimethoxysilylpropyl)amine	$-(CH_2)_3-NH-(CH_2)_3-$	CO_2	[136]
bis((triethoxysilyl)propyl)-ethylenediamine	$-(CH_2)_3-NH-(CH_2)_2-NH-(CH_2)_3-$	CO_2	[52, 53]

The CO_2 adsorption ability of PMOs can be improved by the introduction of nitrogen functionalities. This has been shown for PMO materials by the use of the amine containing precursors bis((triethoxysilyl)propyl)-ethylenediamine [137, 138] and bis (trimethoxysilylpropyl) amine [136]. To obtain a rigid and highly porous structure, these precursors were co-condensed with TEOS [137] or BTME [136], respectively. Promising results were already obtained with these hybrid materials.

Recently, Melde et al. [133] reported the adsorption of other gasses such as sulfur dioxide, cyanogen chloride, ammonia and octane on an ethane-PMO as well as the effect of humidity on its adsorption behavior and the influence of post-functionalisation with an aminopropyl and isocyanatopropyl functionality.

The adsorption of hydrogen gas and its interaction with the pore walls of ethenylene-, ethylene-, phenylene- and biphenylene-bridged PMOs has recently been compared [134]. PMOs containing π-electrons adsorbed more hydrogen than the ethane-PMO. But also the stacking of the aromatic rings in the benzene-materials had an influence on the interaction. Out of the tested materials, the ethene-PMO showed the highest amount of hydrogen stored with 2.23 molecules nm^{-2} (0.62 wt%).

2.3.3. Adsorption of Organic Compounds

Only a few studies have appeared on the use of PMOs as adsorbents for organic compounds in liquid phase.

Herein, we present a brief overview of some representative studies on the adsorption of organics on PMOs in Table 9. Especially materials containing aromatic bridges are used and investigated. The presence of benzene moieties is particularly beneficial by the adsorption of aromatic pollutants due to π---π interactions between adsorbate and adsorbent.

This interaction was clearly observed when researchers synthesized a phenylene derived PMO based on BTEEB and compared it with an ethane-PMO [139]. Although the phenylene derived material was disordered and less porous than the ethylene-bridged PMO, it showed a high uptake of phenol derivatives. A comparison between phenylene derived material and activated carbon showed a similar adsorption. However, a large difference in regeneration was observed. The PMO was regenerated with ethanol and ~98% of the phenolic compounds were removed from the material. Only 1-2% could be removed from the activated carbon after washing with EtOH. More investigation was performed on the adsorption of aromatic compounds *via* an in-depth kinetic study with aromatic compounds on phenylene-bridged PMOs [140]. The material was used in batch experiments but also in a column set-up. The adsorption capacity of BTX decreased in the following sequence: benzene > o-xylene > p-xylene > toluene.

The removal of other organic compounds from aqueous solutions with nitro functionalities or a more complex structure is also explored. A recent publication focused on the use of PMOs prepared with BTEB and BTEEB for the preconcentration of trinitrotoluene and the importance of imprinting similar to the method used for metal adsorption [141]. Furthermore, the adsorption of more complicated organic compounds was reported where the adsorption of polycyclic aromatic hydrocarbons and nicotine was described [142, 143].

The limited amount of reports published on the adsorption of organics in liquid phase emphasizes the need for further investigation. Issues such as the influence of introducing functionalities on the materials, batch and column set-ups and especially the adsorption of larger complex molecules are only vaguely explored for PMOs.

2.4. As Emerging Biological/Biomedical Supports

Up to date, the implication of PMOs in bio- and medical processes is an emergent area of application. PMOs with high surface areas and pore volumes, regular porous structure and tunable morphology are promising candidates in processes related to adsorption and release of biological molecules or substances. An overview of the most important biomedical applications of PMOs is presented in Table 10.

The first investigations on the application of PMOs in the biological/biochemical/ biomedical field involved the immobilization of cytochrome c [145, 146] and lysozyme [147-149]. The adsorption of cytochrome c in ethane-PMO was similar to that in SBA-15 since it was mainly controlled by electrostatic forces [145]. However, both electrostatic and hydrophobic interactions affected the adsorption of lysozyme.

Moreover PMOs have also been used as supports for the immobilization of lipase, a biocatalyst of great relevance for industrial applications [150-152] or other important biomolecules – proteins [153, 154], enzymes [155-157] and amino acids [158] -. Not only pure PMOs have been employed as supports for enzymes. For instance, Na et al. [159] synthesized bifunctional PMOs by co-condensation of BTEE with 3-glycidoxypropyltrimethoxylsilane in the presence of P123.

Table 9. Overview of PMO adsorbents used for organic compounds

Bis-silylated precursor	Organic compound	References
BTEB, BTEE	4-chlorophenol	[144]
BTEE, BTMEB[a]	4-nitrophenol, 4-chlorophenol, 4-methylphenol	[139]
BTEB, BTEMB	Trinitrotoluene	[141]
BTEB	benzene, toluene, o-xylene, and p-xylene	[140]
BTEB, 1-phenyltriethoxysilane	acenaphtene, fluoranthene, fluorene, naphthalene, pyrene	[142]
BTEB, BTES-biphenyl[c], BTMS-amine[d]	Nicotine	[143]

[a]1,4-bis(trimethoxysilylethyl) benzene. [b]4,4-bis(triethoxysilyl)biphenyl (BTES-biphenyl). [c]bis[3-(trimethoxysilyl)propyl]amine.

Adapted from Reference [19] with permission from The Royal Society of Chemistry.

In this case, the functional groups favored the immobilization of papain through interactions between epoxy and amino groups present on PMOs and papain, respectively. Besides their role as supports, ethane-PMO successfully participated in a new strategy for protein refolding. Its mesoporous channels could entrap the denaturated protein (hen egg white lysozyme), and promote its refolding, hereby increasing the release rate [160].

Concerning to biomedical applications, several types of drugs have been successfully inmovilized on PMOs. In 2009, Qiao et al. [161] reported the adsorption and release of an antibiotic *(tetracycline)* on PMOs. They utilized hollow and solid spheres of ethane-PMO as well as solid spheres of silica. Hollow spheres were more suitable to adsorb and store drugs in their empty cores than solid spheres. Moreover, the hydrophobic interactions between the ethane bridges and the drug molecules facilitated higher loading capacities and slower release rates than those in silica materials. Similar results were obtained recently by Asefa et al. [162] in the adsorption of an anticancer drug (*cisplatin*) on ethane-PMOs.

Parambadath et al. [163] used a bis-silylated precursor bearing sulfonamide and urea units for preparing hexagonally ordered PMOs. These materials were promising candidates for the adsorption/desorption of two drugs, *i.e.*, *captopril and 5-fluorouracil.* For the same purpose, this group also synthesized a PMO from a precursor bearing ureylene and piperazine units [164]. Ha et al. [165] recently succeeded to incorporate a high loading of amidoxime within PMO walls by co-condensation of 1,2-diureylenemaleonitrile and TEOS. The amphoteric character of these organic moieties allowed their application as drug delivery systems of hydrophobic (*ibuprofen*) and hydrophilic (*5-fluorouracil*) drugs. This same year, Wu et al. [166, 167] accomplished the synthesis of bifunctionalized ethane- or benzene-bridged PMOs with exceptionally high loadings of pendant carboxylic acids groups. The presence of negative charges owing to these carboxylic acids groups allowed their use in the loading and release of a positively charged, hydrophobic anticancer drug, *doxorubicin.*

Table 10. Bio-molecules and drugs immobilized onto PMOs

Type	Adsorbed molecule	Support or carrier (surface functionality)	Structure/Morphology	References
Protein	Cytochrome c *bovine heart* *horse heart*	 $-CH_2-CH_2-$ $-C_6H_4-$ / $-(C_6H_4)_2-$	 2D-hexagonal/rodlike disordered/ hollow spherical particles	 [145] [146]
	Serum albumin *bovine*	 $-CH_2-CH_2-$ $-CH_2-CH_2-$ functionalized with $-NH_2$ or $-COOH$	 cubic/hollow particles hollow spheres 2D-hexagonal/rodlike	 [149] [154] [153]
	Hemoglobin	$-CH_2-CH_2-$ functionalized with $-NH_2$ or $-COOH$	2D-hexagonal/rodlike	[153]
Enzyme	Lysozyme *hen egg white*	 $-CH_2-CH_2-$ $-(CH_2)_3-NH-(CH_2)_3-$ / $-C_6H_4-$ / $-(C_6H_4)_2-$ $-CH_2-CH_2-$	 2D-hexagonal/rodlike hollow spherical particles cubic/hollow particles	 [147] [148] [149]
	Lipase *rhizopus oryzae* *candida Antarctica fraction B* *thermomyces lanuginosus*	 $-CH_2-CH_2-$ $-CH_2-CH_2-$ $-CH_2-CH_2-$ / $-CH=CH-$ / $-C_6H_4-$	 n.a. 2D-hexagonal/rodlike -/mesocellular cagelike	 [150] [151] [152]
	Papain	$-CH_2-CH_2-$ functionalized with $-(CH_2)_3-O-CH_2-(CH)_2O$	2D-hexagonal	[159]
	Chloroperoxidase Horseradish peroxidase	$-(CH_2)_3-NH-(CH_2)_3-$ $-(CH_2)_3-NH-CO-NH-(CH_2)_3-$ / $-(CH_2)_3-NH-CO-S-(CH_2)_3-$ $-(CH_2)_3-S-CH_2-CHOH-CH_2-O-(CH_2)_3-$	disordered 2D-hexagonal 2D-hexagonal	[155] [156] [157]
Amino-acids	Glycine, L-lysine, isoleucine	$-C_6H_4-$ $-(C_6H_4)_2-$ / $-(CH_2)_3-NH-(CH_2)_3-$	disordered/ hollow spherical worm-like/hollow spherical	[158] [158]
Drug	Tetracycline	$-CH_2-CH_2-$	2D-hexagonal/spherical particles	[161]
	Cisplatin Captopril	$-CH_2-CH_2-$ $-(CH_2)_3-NH-CO-NH-C_6H_4-SO_2-NH-CO-NH-(CH_2)_3-$ $-(CH_2)_3-NH-CO-N((CH_2)_2)_2N-CO-NH-(CH_2)_3-$	2D-hexagonal 2D-hexagonal/rodlike 2D-hexagonal/ polyhedral bipyramidal crystals	[162] [163] [164]
	5-Fluorouracil	$-(CH_2)_3-NH-CO-NH-C_6H_4-SO_2-NH-CO-NH-(CH_2)_3-$ $-(CH_2)_3-NH-CO-N((CH_2)_2)_2N-CO-NH-(CH_2)_3-$ $-(CH_2)_3-NH-CO-NH-CCN=CCN-NH-CO-NH-(CH_2)_3-$	2D-hexagonal/rodlike 2D-hexagonal/polyhedral bipyramidal crystals 2D-hexagonal/shaped tubular-twisted columns-micro-elongated particles	[163] [164] [165]
	Ibuprofen	$-(CH_2)_3-NH-CO-NH-CCN=CCN-NH-CO-NH-(CH_2)_3-$	2D-hexagonal/shaped tubular-twisted columns-micro-elongated particles	[165]
	Doxorubicin	$-CH_2-CH_2-$ or $-(C_6H_4)-$ functionalized with $-COOH$	2D-hexagonal/spherical particles	[166, 167]

Very recently, in 2013, Ha et al. [168] reported the construction of a multifunctional PMOs composed of a heteroatomic pyridine ring and two N-H sites. This material showed a great potential for multidisciplinary applications, such as solid chemosensor, stimuli-responsive carrier of a range of therapeutic agents and as an efficient adsorbent for the removal of metal ions.

Conclusion

PMOs have demonstrated to be very useful for a variety of applications and many others can be envisaged in the near future. Numerous studies reveal their superior performance over analogous silicas as a result of their higher hydrophobicity and stability. Eventually, they could replace silicas as low-*k* materials. However, their use as adsorbents has to be further optimized. As catalysts, PMOs have sometimes shown to be more active than commercial resins and the first results on bifunctional and chiral reactions are promising.

Nevertheless, additional experiments to elucidate structure-activity relationships should be conducted. In general, the advantages of employing bridging instead of pendant organic moieties have to be fundamentally investigated for most applications.

Surely, the current results and the forthcoming advances in PMOs will make them the materials of choice for some high-tech applications in strong competition with other highly porous solids, such as functionalized silicas, metal organic frameworks and mesoporous polymers, besides other traditional materials, *i.e.*, zeolites, porous carbons and oxides.

Acknowledgments

The authors acknowledge the financial support from Ghent University BOF Grant (Nr. 01P02911T), GOA grant (Nr. 01G00710) and the Research Foundation-Flanders (FWO-Vlaanderen, project G.006813.N).

References

[1] Melde, B. J.; Holland, B. T.; Blanford, C. F.; Stein, A., *Chem. Mater.* 1999, 11 (11), 3302-3308.

[2] Inagaki, S.; Guan, S.; Fukushima, Y.; Ohsuna, T.; Terasaki, O., *J. Am. Chem. Soc.* 1999, 121 (41), 9611-9614.

[3] Asefa, T.; MacLachan, M. J.; Coombs, N.; Ozin, G. A., *Nature* 1999, 402 (6764), 867-871.

[4] Zhao, D. Y.; Feng, J. L.; Huo, Q. S.; Melosh, N.; Fredrickson, G. H.; Chmelka, B. F.; Stucky, G. D., *Science* 1998, 279 (5350), 548-552.

[5] Zhao, D. Y.; Huo, Q. S.; Feng, J. L.; Chmelka, B. F.; Stucky, G. D., *J. Am. Chem. Soc.* 1998, 120 (24), 6024-6036.

[6] Beck, J. S.; Vartuli, J. C.; Roth, W. J.; Leonowicz, M. E.; Kresge, C. T.; Schmitt, K. D.; Chu, C. T. W.; Olson, D. H.; Sheppard, E. W.; et al., *Journal of the American Chemical Society* 1992, 114 (27), 10834-10843.

[7] Van der Voort, P.; Vercaemst, C.; Schaubroeck, D.; Verpoort, F., *Phys. Chem. Chem. Phys.* 2008, 10 (3), 347-360.

[8] Hoffmann, F.; Cornelius, M.; Morell, J.; Froba, M., *Angew. Chem. Int. Edit.* 2006, 45 (20), 3216-3251.

[9] Hoffmann, F.; Cornelius, M.; Morell, J.; Froba, M., *J. Nanosci. Nanotechno.* 2006, 6 (2), 265-288.

[10] Hoffmann, F.; Froba, M., *Chem. Soc. Rev.* 2011, 40 (2), 608-620.

[11] Hoffmann, F.; Fröba, M., *The Supramolecular Chemistry of Organic-Inorganic Hybrid Materials* 2010, 39-102.

[12] Ford, D. M.; Simanek, E. E.; Shantz, D. F., *Nanotechnology* 2005, 16 (7), S458-S475.

[13] Hatton, B.; Landskron, K.; Whitnall, W.; Perovic, D.; Ozin, G. A., *Accounts of Chemical Research* 2005, 38 (4), 305-312.

[14] Hunks, W. J.; Ozin, G. A., *J. Mater. Chem.* 2005, 15 (35-36), 3716-3724.

[15] Jaroniec, M., *Nature* 2006, 442 (7103), 638-640.

[16] Kapoor, M. P.; Inagaki, S., *B Chem. Soc. Jpn.* 2006, 79 (10), 1463-1475.

[17] Mizoshita, N.; Tani, T.; Inagaki, S., *Chem. Soc. Rev.* 2011, 40 (2), 789-800.

[18] Xia, H. S.; Zhou, C. H.; Tong, D. S.; Lin, C. X., *J. Porous Mat.* 2010, 17 (2), 225-252.

[19] Van Der Voort, P.; Esquivel, D.; De Canck, E.; Goethals, F.; Van Driessche, I.; Romero-Salguero, F. J., *Chem. Soc. Rev.* 2013, 42 (9), 3913-3955.

[20] Wahab, M. A.; He, C. B., *Langmuir* 2009, 25 (2), 832-838.

[21] Park, S. S.; Park, D. H.; Ha, C. S., *Chem. Mater.* 2007, 19 (11), 2709-2711.

[22] Lu, Y. F.; Fan, H. Y.; Doke, N.; Loy, D. A.; Assink, R. A.; LaVan, D. A.; Brinker, C. J., *J. Am. Chem. Soc.* 2000, 122 (22), 5258-5261.

[23] Hatton, B. D.; Landskron, K.; Whitnall, W.; Perovic, D. D.; Ozin, G. A., *Advanced Functional Materials* 2005, 15 (5), 823-829.

[24] Wang, W. D.; Grozea, D.; Kohli, S.; Perovic, D. D.; Ozin, G. A., *Acs Nano* 2011, 5 (2), 1267-1275.

[25] Goethals, F.; Baklanov, M. R.; Ciofi, I.; Detavernier, C.; Van Der Voort, P.; Van Driessche, I., *Chem. Commun.* 2012, 48 (22), 2797-2799.

[26] Wang, W. D.; Grozea, D.; Kim, A.; Perovic, D. D.; Ozin, G. A., *Advanced Materials* 2010, 22 (1), 99-102.

[27] Landskron, K.; Hatton, B. D.; Perovic, D. D.; Ozin, G. A., *Science* 2003, 302 (5643), 266-269.

[28] Goethals, F.; Ciofi, I.; Madia, O.; Vanstreels, K.; Baklanov, M. R.; Detavernier, C.; Van der Voort, P.; Van Driessche, I., *J. Mater. Chem.* 2012, 22 (17), 8281-8286.

[29] Seino, M.; Wang, W. D.; Lofgreen, J. E.; Puzzo, D. P.; Manabe, T.; Ozin, G. A., *J. Am. Chem. Soc.* 2011, 133 (45), 18082-18085.

[30] Goethals, F.; Levrau, E.; Pollefeyt, G.; Baklanov, M. R.; Ciofi, I.; Vanstreels, K.; Detavernier, C.; Van Driessche, I.; Van der Voort, P., *Journal of Materials Chemistry C* 2013, 1 (25), 3961-3966.

[31] Dhepe, P. L.; Ohashi, M.; Inagaki, S.; Ichikawa, M.; Fukuoka, A., *Catal. Lett.* 2005, 102 (3-4), 163-169.

[32] Rac, B.; Hegyes, P.; Forgo, P.; Molnar, A., *Appl. Catal. a-Gen.* 2006, 299, 193-201.

[33] Yang, J.; Yang, Q. H.; Wang, G.; Feng, Z. C.; Liu, J., *J. Mol. Catal. a-Chem.* 2006, 256 (1-2), 122-129.

[34] Kang, C. M.; Huang, J. L.; He, W. H.; Zhang, F., *J. Organomet. Chem.* 2010, 695 (1), 120-127.

[35] Shylesh, S.; Samuel, P. P.; Srilakshmi, C.; Parischa, R.; Singh, A. P., *J. Mol. Catal. a-Chem.* 2007, 274 (1-2), 153-158.

[36] Rat, M.; Zahedi-Niaki, M. H.; Kaliaguine, S.; Do, T. O., *Micropor. Mesopor. Mater.* 2008, 112 (1-3), 26-31.

[37] Nakajima, K.; Tomita, I.; Hara, M.; Hayashi, S.; Domen, K.; Kondo, J. N., *Catal. Today* 2006, 116 (2), 151-156.

[38] Nakajima, K.; Tomita, I.; Hara, M.; Hayashi, S.; Domen, K.; Kondo, J. N., *Adv. Mater.* 2005, 17 (15), 1839-1842.

[39] Li, C. M.; Yang, H.; Shi, X.; Liu, R.; Yang, Q. H., *Micropor. Mesopor. Mater.* 2007, 98 (1-3), 220-226.

[40] Morales, G.; Athens, G.; Chmelka, B. F.; van Grieken, R.; Melero, J. A., *J. Catal.* 2008, 254 (2), 205-217.

[41] Karam, A.; Alonso, J. C.; Gerganova, T. I.; Ferreira, P.; Bion, N.; Barrault, J.; Jerome, F., *Chem. Commun.* 2009, (45), 7000-7002.

[42] Esquivel, D.; Jimenez-Sanchidrian, C.; Romero-Salguero, F. J., *J. Mater. Chem.* 2011, 21 (3), 724-733.

[43] Dube, D.; Rat, M.; Shen, W.; Beland, F.; Kaliaguine, S., *J. Mater. Sci.* 2009, 44 (24), 6683-6692.

[44] Shen, W.; Dube, D.; Kaliaguine, S., *Catal. Commun.* 2008, 10 (3), 291-294.

[45] Dube, D.; Rat, M.; Beland, F.; Kaliaguine, S., *Micropor. Mesopor. Mater.* 2008, 111 (1-3), 596-603.

[46] Liu, J.; Yang, J.; Li, C. M.; Yang, Q. H., *J. Porous Mat.* 2009, 16 (3), 273-281.

[47] Sow, B.; Hamoudi, S.; Zahedi-Niaki, M. H.; Kaliaguine, S., *Micropor. Mesopor. Mater.* 2005, 79 (1-3), 129-136.

[48] De Canck, E.; Vercaemst, C.; Verpoort, F.; Van der Voort, P., *Scientific Bases for the Preparation of Heterogeneous Catalysts: Proceedings of the 10th International Symposium* 2010, 175, 365-368.

[49] Yang, Q. H.; Kapoor, M. P.; Shirokura, N.; Ohashi, M.; Inagaki, S.; Kondo, J. N.; Domen, K., *J. Mater. Chem.* 2005, 15 (6), 666-673.

[50] Yang, Q. H.; Kapoor, M. P.; Inagaki, S.; Shirokura, N.; Kondo, J. N.; Domen, K., *J. Mol. Catal. a-Chem.* 2005, 230 (1-2), 85-89.

[51] Kapoor, M. P.; Fujii, W.; Kasama, Y.; Yanagi, M.; Nanbu, H.; Juneja, L. R., *J. Mater. Chem.* 2008, 18 (39), 4683-4691.

[52] Esquivel, D.; De Canck, E.; Jimenez-Sanchidrian, C.; Van Der Voort, P.; Romero-Salguero, F. J., *J. Mater. Chem.* 2011, 21 (29), 10990-10998.

[53] Karimi, B.; Mirzaei, H. M.; Mobaraki, A., *Catal. Sci. Technol.* 2012, 2 (4), 828-834.

[54] Yuan, X. D.; Lee, H. I.; Kim, J. W.; Yie, J. E.; Kim, J. M., *Chem. Lett.* 2003, 32 (7), 650-651.

[55] Kapoor, M. P.; Kasama, Y.; Yanagi, M.; Yokoyama, T.; Inagaki, S.; Shimada, T.; Nanbu, H.; Juneja, L. R., *Micropor. Mesopor. Mater.* 2007, 101 (1-2), 231-239.

[56] Yang, Q. H.; Liu, J.; Yang, J.; Kapoor, M. P.; Inagaki, S.; Li, C., *J. Catal.* 2004, 228 (2), 265-272.

[57] Kapoor, M. P.; Kasama, Y.; Yokoyama, T.; Yanagi, M.; Inagaki, S.; Hironobu, N.; Juneja, L. R., *J. Mater. Chem.* 2006, 16 (48), 4714-4722.

[58] Ohashi, M.; Kapoor, M. P.; Inagaki, S., *Chem. Commun.* 2008, (7), 841-843.

[59] El Hankari, S.; Motos-Perez, B.; Hesemann, P.; Bouhaouss, A.; Moreau, J. J. E., *J. Mater. Chem.* 2011, 21 (19), 6948-6955.

[60] Zhang, L.; Liu, J.; Yang, J.; Yang, Q. H.; Li, C., *Micropor. Mesopor. Mater.* 2008, 109 (1-3), 172-183.

[61] Prasetyanto, E. A.; Ansari, M. B.; Min, B. H.; Park, S. E., *Catal. Today* 2010, 158 (3-4), 252-257.

[62] Shylesh, S.; Zhou, Z.; Meng, Q. G.; Wagener, A.; Seifert, A.; Ernst, S.; Thiel, W. R., *J. Mol. Catal. a-Chem.* 2010, 332 (1-2), 65-69.

[63] Kapoor, M. P.; Bhaumik, A.; Inagaki, S.; Kuraoka, K.; Yazawa, T., *J. Mater. Chem.* 2002, 12 (10), 3078-3083.

[64] Melero, J. A.; Iglesias, J.; Arsuaga, J. M.; Sainz-Pardo, J.; de Frutos, P.; Blazquez, S., *J. Mater. Chem.* 2007, 17 (4), 377-385.

[65] Sisodiya, S.; Shylesh, S.; Singh, A. P., *Catal. Commun.* 2011, 12 (7), 629-633.

[66] Shylesh, S.; Singh, A. P., *Micropor. Mesopor. Mater.* 2006, 94 (1-3), 127-138.

[67] Feliczak, A.; Walczak, K.; Wawrzynczak, A.; Nowak, I., *Catal. Today* 2009, 140 (1-2), 23-29.

[68] Bhaumik, A.; Kapoor, M. P.; Inagaki, S., *Chem. Commun.* 2003, (4), 470-471.

[69] Shylesh, S.; Samuel, P. P.; Singh, A. P., *Micropor. Mesopor. Mater.* 2007, 100 (1-3), 250-258.

[70] Yang, Q. H.; Li, Y.; Zhang, L.; Yang, J.; Liu, J.; Li, C., *J. Phys. Chem. B* 2004, 108 (23), 7934-7937.

[71] Sanchez-Vazquez, R.; Pirez, C.; Iglesias, J.; Wilson, K.; Lee, A. F.; Melero, J. A., *Chemcatchem.* 2013, 5 (4), 994-1001.

[72] Corral, J. A.; Lopez, M. I.; Esquivel, D.; Mora, M.; Jimenez-Sanchidrian, C.; Romero-Salguero, F. J., *Materials* 2013, 6 (4), 1554-1565.

[73] Corma, A.; Das, D.; Garcia, H.; Leyva, A., *J. Catal.* 2005, 229 (2), 322-331.

[74] Yang, Y.; Zhang, Y.; Hao, S. J.; Kan, Q. B., *J. Colloid Interf. Sci.* 2011, 362 (1), 157-163.

[75] Jeong, E. Y.; Ansari, M. B.; Park, S. E., *Acs Catal.* 2011, 1 (8), 855-863.

[76] Baleizao, C.; Gigante, B.; Das, D.; Alvaro, M.; Garcia, H.; Corma, A., *J. Catal.* 2004, 223 (1), 106-113.

[77] Bahuleyan, B. K.; Jermy, B. R.; Ahn, I. Y.; Suh, H.; Park, D. W.; Ha, C. S.; Kim, I., *Catal. Commun.* 2009, 11 (4), 252-256.

[78] Zhang, F.; Kang, C. M.; Wei, Y. Y.; Li, H. X., *Adv. Funct. Mater.* 2011, 21 (16), 3189-3197.

[79] Huang, J. L.; Zhu, F. X.; He, W. H.; Zhang, F.; Wang, W.; Li, H. X., *J. Am. Chem. Soc.* 2010, 132 (5), 1492-1493.

[80] Dufaud, V.; Beauchesne, F.; Bonneviot, L., *Angew. Chem. Int. Edit.* 2005, 44 (22), 3475-3477.

[81] Yang, X. S.; Zhu, F. X.; Huang, J. L.; Zhang, F.; Li, H. X., *Chem. Mater.* 2009, 21 (20), 4925-4933.

[82] Yang, H. Q.; Li, G. A.; Ma, Z. C.; Chao, J. B.; Guo, Z. Q., *J. Catal.* 2010, 276 (1), 123-133.

[83] Yang, H. Q.; Han, X. J.; Li, G. A.; Ma, Z. C.; Hao, Y. J., *J. Phys. Chem. C* 2010, 114 (50), 22221-22229.

[84] Li, G.; Yang, H. Q.; Li, W.; Zhang, G. L., *Green Chem.* 2011, 13 (10), 2939-2947.

[85] Yang, H. Q.; Ma, Z. C.; Qing, Y.; Xie, G. Y.; Gao, J.; Zhang, L.; Gao, J. H.; Du, L., *Appl. Catal. a-Gen.* 2010, 382 (2), 312-321.

[86] Jeong, E. Y.; Burri, A.; Lee, S. Y.; Park, S. E., *J. Mater. Chem.* 2010, 20 (48), 10869-10875.

[87] Coelho, A. C.; Balula, S. S.; Bruno, S. M.; Alonso, J. C.; Bion, N.; Ferreira, P.; Pillinger, M.; Valente, A. A.; Rocha, J.; Goncalves, I. S., *Adv. Synth. Catal.* 2010, 352 (10), 1759-1769.

[88] Coelho, A. C.; Balula, S. S.; Antunes, M. M.; Gerganova, T. I.; Bion, N.; Ferreira, P.; Pillinger, M.; Valente, A. A.; Rocha, J.; Goncalves, I. S., *J. Mol. Catal. a-Chem.* 2010, 332 (1-2), 13-18.

[89] Kuschel, A.; Luka, M.; Wessig, M.; Drescher, M.; Fonin, M.; Kiliani, G.; Polarz, S., *Adv. Funct. Mater.* 2010, 20 (7), 1133-1143.

[90] Zhang, T. M.; Gao, C. G.; Yang, H. Q.; Zhao, Y. X., *J. Porous Mat.* 2010, 17 (5), 643-649.

[91] Kamegawa, T.; Saito, M.; Watanabe, T.; Uchihara, K.; Kondo, M.; Matsuoka, M.; Anpo, M., *J. Mater. Chem.* 2011, 21 (33), 12228-12231.

[92] Huang, J. L.; Zhang, F.; Li, H. X., *Appl. Catal. a-Gen.* 2012, 431, 95-103.

[93] Alauzun, J.; Mehdi, A.; Reye, C.; Corriu, R. J. P., *J. Am. Chem. Soc.* 2006, 128 (27), 8718-8719.

[94] Shylesh, S.; Wagener, A.; Seifert, A.; Ernst, S.; Thiel, W. R., *Angew. Chem. Int. Edit.* 2010, 49 (1), 184-187.

[95] Sasidharan, M.; Fujita, S.; Ohashi, M.; Goto, Y.; Nakashima, K.; Inagaki, S., *Chem. Commun.* 2011, 47 (37), 10422-10424.

[96] Kuschel, A.; Drescher, M.; Kuschel, T.; Polarz, S., *Chem. Mater.* 2010, 22 (4), 1472-1482.

[97] Zhu, F. X.; Wang, W.; Li, H. X., *J. Am. Chem. Soc.* 2011, 133 (30), 11632-11640.

[98] Baleizao, C.; Gigante, B.; Das, D.; Alvaro, M.; Garcia, H.; Corma, A., *Chem. Commun.* 2003, (15), 1860-1861.

[99] Benitez, M.; Bringmann, G.; Dreyer, M.; Garcia, H.; Ihmels, H.; Waidelich, M.; Wissel, K., *J. Org. Chem.* 2005, 70 (6), 2315-2321.

[100] Jiang, D. M.; Yang, Q. H.; Wang, H.; Zhu, G. R.; Yang, J.; Li, C., *J. Catal.* 2006, 239 (1), 65-73.

[101] Jiang, D. M.; Gao, J. S.; Li, J.; Yang, Q. H.; Li, C., *Micropor. Mesopor. Mater.* 2008, 113 (1-3), 385-392.

[102] Garcia, R. A.; van Grieken, R.; Iglesias, J.; Morales, V.; Gordillo, D., *Chem. Mater.* 2008, 20 (9), 2964-2971.

[103] Zhang, L.; Liu, J.; Yang, J.; Yang, Q. H.; Li, C., *Chem.-Asian J.* 2008, 3 (10), 1842-1849.

[104] Garcia, R. A.; van Grieken, R.; Iglesias, J.; Morales, V.; Villajos, N., *J. Catal.* 2010, 274 (2), 221-227.

[105] Wang, P. Y.; Yang, J.; Liu, J.; Zhang, L.; Yang, Q. H., *Micropor. Mesopor. Mater.* 2009, 117 (1-2), 91-97.

[106] Liu, X.; Wang, P. Y.; Yang, Y.; Wang, P.; Yang, Q. H., *Chem.-Asian J.* 2010, 5 (5), 1232-1239.
[107] Liu, X. A.; Wang, P. Y.; Zhang, L.; Yang, J.; Li, C.; Yang, Q. H., *Chem.-Eur. J.* 2010, 16 (42), 12727-12735.
[108] Gao, J. S.; Liu, J.; Tang, J. T.; Jiang, D. M.; Li, B.; Yang, Q. H., *Chem.-Eur. J.* 2010, 16 (26), 7852-7858.
[109] Gao, J. S.; Liu, J.; Bai, S. Y.; Wang, P. Y.; Zhong, H.; Yang, Q. H.; Li, C., *J. Mater. Chem.* 2009, 19 (45), 8580-8588.
[110] Shi, J. Y.; Wang, C. A.; Li, Z. J.; Wang, Q.; Zhang, Y.; Wang, W., *Chem.-Eur. J.* 2011, 17 (22), 6206-6213.
[111] Kuschel, A.; Polarz, S., *J. Am. Chem. Soc.* 2010, 132 (18), 6558-6565.
[112] Wang, P. Y.; Liu, X.; Yang, J.; Yang, Y.; Zhang, L.; Yang, Q. H.; Li, C., *J. Mater. Chem.* 2009, 19 (42), 8009-8014.
[113] Walcarius, A.; Mercier, L., *J. Mater. Chem.* 2010, 20 (22), 4478-4511.
[114] Jal, P. K.; Patel, S.; Mishra, B., *Talanta* 2004, 62 (5), 1005-1028.
[115] Benitez, M.; Das, D.; Ferreira, R.; Pischel, U.; Garcia, H., *Chem. Mater.* 2006, 18 (23), 5597-5603.
[116] Hao, N.; Han, L.; Yang, Y. X.; Wang, H. T.; Webley, P. A.; Zhao, D. Y., *Appl. Surf. Sci.* 2010, 256 (17), 5334-5342.
[117] Waki, M.; Mizoshita, N.; Ohsuna, T.; Tani, T.; Inagaki, S., *Chem. Commun.* 2010, 46 (43), 8163-8165.
[118] Olkhovyk, O.; Jaroniec, M., *Ind. Eng. Chem. Res.* 2007, 46 (6), 1745-1751.
[119] Burleigh, M. C.; Dai, S.; Hagaman, E. W.; Lin, J. S., *Chem. Mater.* 2001, 13 (8), 2537-2546.
[120] Hossain, K. Z.; Mercier, L., *Adv. Mater.* 2002, 14 (15), 1053-1056.
[121] Liu, J.; Yang, J.; Yang, Q. H.; Wang, G.; Li, Y., *Adv. Funct. Mater.* 2005, 15 (8), 1297-1302.
[122] Olkhovyk, O.; Jaroniec, M., *J. Am. Chem. Soc.* 2005, 127 (1), 60-61.
[123] Olkhovyk, O.; Pikus, S.; Jaroniec, M., *J. Mater. Chem.* 2005, 15 (15), 1517-1519.
[124] Wu, H. Y.; Liao, C. H.; Pan, Y. C.; Yeh, C. L.; Kao, H. M., *Micropor. Mesopor. Mater.* 2009, 119 (1-3), 109-116.
[125] Yang, L. M.; Wang, Y. J.; Luo, G. S.; Dai, Y. Y., *Microporous and Mesoporous Materials* 2005, 84 (1-3), 275-282.
[126] Zhang, L. X.; Zhang, W. H.; Shi, J. L.; Hua, Z.; Li, Y. S.; Yan, J., *Chem. Commun.* 2003, (2), 210-211.
[127] Lee, B.; Bao, L. L.; Im, H. J.; Dai, S.; Hagaman, E. W.; Lin, J. S., *Langmuir* 2003, 19 (10), 4246-4252.
[128] Lee, B.; Im, H. J.; Luo, H. M.; Hagaman, E. W.; Dai, S., *Langmuir* 2005, 21 (12), 5372-5376.
[129] Nakai, K.; Oumi, Y.; Horie, H.; Sano, T.; Yoshitake, H., *Micropor. Mesopor. Mater.* 2007, 100 (1-3), 328-339.
[130] Kapoor, M. P.; Yanagi, M.; Kasama, Y.; Yokoyama, T.; Inagaki, S.; Shimada, T.; Nanbu, H.; Juneja, L. R., *J. Mater. Chem.* 2006, 16 (32), 3305-3311.
[131] Comotti, A.; Bracco, S.; Valsesia, P.; Ferretti, L.; Sozzani, P., *J. Am. Chem. Soc.* 2007, 129 (27), 8566-8576.

[132] Johnson, B. J.; Anderson, N. E.; Charles, P. T.; Malanoski, A. P.; Melde, B. J.; Nasir, M.; Deschamps, J. R., *Sensors-Basel* 2011, 11 (1), 886-904.
[133] Johnson, B. J.; Melde, B. J.; Peterson, G. W.; Schindler, B. J.; Jones, P., *Chem. Eng. Sci.* 2012, 68 (1), 376-382.
[134] Kubo, M.; Ishiyama, K.; Shimojima, A.; Okubo, T., *Micropor. Mesopor. Mater.* 2012, 147 (1), 194-199.
[135] Martinez, U.; Pacchioni, G., *Microporous and Mesoporous Materials* 2010, 129 (1-2), 62-67.
[136] Bai, S. Y.; Liu, J.; Gao, J. S.; Yang, Q. H.; Li, C., *Microporous and Mesoporous Materials* 2012, 151, 474-480.
[137] Tang, Y. D.; Landskron, K., *Journal of Physical Chemistry C* 2010, 114 (6), 2494-2498.
[138] Qi, G.; Fu, L.; Duan, X.; Choi, B. H.; Abraham, M.; Giannelis, E. P., *Greenhouse Gases: Science and Technology* 2011, 1 (3), 278-284.
[139] Burleigh, M. C.; Markowitz, M. A.; Spector, M. S.; Gaber, B. P., *Environ. Sci. Technol.* 2002, 36 (11), 2515-2518.
[140] Moura, C. P.; Vidal, C. B.; Barros, A. L.; Costa, L. S.; Vasconcellos, L. C. G.; Dias, F. S.; Nascimento, R. F., *J. Colloid Interf. Sci.* 2011, 363 (2), 626-634.
[141] Trammell, S. A.; Zeinali, M.; Melde, B. J.; Charles, P. T.; Velez, F. L.; Dinderman, M. A.; Kusterbeck, A.; Markowitz, M. A., *Anal. Chem.* 2008, 80 (12), 4627-4633.
[142] Vidal, C. B.; Barros, A. L.; Moura, C. P.; de Lima, A. C. A.; Dias, F. S.; Vasconcellos, L. C. G.; Fechine, P. B. A.; Nascimento, R. F., *J. Colloid Interf. Sci.* 2011, 357 (2), 466-473.
[143] Shin, J. H.; Park, S. S.; Ha, C. S., *Colloid Surface B* 2011, 84 (2), 579-584.
[144] Burleigh, M. C.; Jayasundera, S.; Spector, M. S.; Thomas, C. W.; Markowitz, M. A.; Gaber, B. P., *Chem. Mater.* 2004, 16 (1), 3-5.
[145] Qiao, S. Z.; Yu, C. Z.; Xing, W.; Hu, Q. H.; Djojoputro, H.; Lu, G. Q., *Chem. Mater.* 2005, 17 (24), 6172-6176.
[146] Park, M.; Park, S. S.; Selvaraj, M.; Kim, I.; Ha, C. S., *J. Porous Mat.* 2011, 18 (2), 217-223.
[147] Qiao, S. Z.; Djojoputro, H.; Hu, Q. H.; Lu, G. Q., *Prog. Solid State Ch.* 2006, 34 (2-4), 249-256.
[148] Park, M.; Park, S. S.; Selvaraj, M.; Zhao, D. Y.; Ha, C. S., *Micropor. Mesopor. Mater.* 2009, 124 (1-3), 76-83.
[149] Li, N.; Wang, J. G.; Zhou, H. J.; Sun, P. C.; Chen, T. H., *Chem. Mater.* 2011, 23 (18), 4241-4249.
[150] Shakeri, M.; Kawakami, K., *Catal. Commun.* 2008, 10 (2), 165-168.
[151] Serra, E.; Diez, E.; Diaz, I.; Blanco, R. M., *Micropor. Mesopor. Mater.* 2010, 132 (3), 487-493.
[152] Zhou, Z.; Taylor, R. N. K.; Kullmann, S.; Bao, H. X.; Hartmann, M., *Adv. Mater.* 2011, *23* (22-23), 2627-2632.
[153] Zhu, L.; Liu, X.; Chen, T.; Xu, Z.; Yan, W.; Zhang, H., *Appl. Surf. Sci.* 2012, 258 (18), 7126-7134.
[154] Hao, N.; Wang, H. T.; Webley, P. A.; Zhao, D. Y., *Micropor. Mesopor. Mater.* 2010, 132 (3), 543-551.

[155] Hudson, S.; Cooney, J.; Hodnett, B. K.; Magner, E., *Chem. Mater.* 2007, 19 (8), 2049-2055.
[156] Lin, N.; Gao, L.; Chen, Z.; Zhu, J. H., *New J. Chem.* 2011, 35 (9), 1867-1875.
[157] Wan, M. M.; Gao, L.; Chen, Z.; Wang, Y. K.; Wang, Y.; Zhu, J. H., *Micropor. Mesopor. Mater.* 2012, 155 (0), 24-33.
[158] Shin, J. H.; Park, S. S.; Selvaraj, M.; Ha, C. S., *J. Porous Mat.* 2012, 19 (1), 29-35.
[159] Na, W.; Wei, Q.; Lan, J. N.; Nie, Z. R.; Sun, H.; Li, Q. Y., *Micropor. Mesopor. Mater.* 2010, 134 (1-3), 72-78.
[160] Wang, X. Q.; Lu, D. N.; Austin, R.; Agarwal, A.; Mueller, L. J.; Liu, Z.; Wu, J. Z.; Feng, P. Y., *Langmuir* 2007, 23 (10), 5735-5739.
[161] Lin, C. X.; Qiao, S. Z.; Yu, C. Z.; Ismadji, S.; Lu, G. Q., *Micropor. Mesopor. Mater.* 2009, 117 (1-2), 213-219.
[162] Vathyam, R.; Wondimu, E.; Das, S.; Zhang, C.; Hayes, S.; Tao, Z. M.; Asefa, T., *J. Phys. Chem. C* 2011, 115 (27), 13135-13150.
[163] Parambadath, S.; Rana, V. K.; Moorthy, S.; Chu, S. W.; Park, S. K.; Lee, D.; Sung, G.; Ha, C. S., *J. Solid State Chem.* 2011, 184 (5), 1208-1215.
[164] Parambadath, S.; Rana, V. K.; Zhao, D. Y.; Ha, C. S., *Micropor. Mesopor. Mater.* 2011, 141 (1-3), 94-101.
[165] Moorthy, M. S.; Park, S. S.; Fuping, D.; Hong, S. H.; Selvaraj, M.; Ha, C. S., *J. Mater. Chem.* 2012, 22 (18), 9100-9108.
[166] Kao, H. M.; Chung, C. H.; Saikia, D.; Liao, S. H.; Chao, P. Y.; Chen, Y. H.; Wu, K. C. W., *Chem.-Asian J.* 2012, 7 (9), 2111-2117.
[167] Wu, H. Y.; Shieh, F. K.; Kao, H. M.; Chen, Y. W.; Deka, J. R.; Liao, S. H.; Wu, K. C. W., *Chem.-Eur. J.* 2013, 19 (20), 6358-6367.
[168] Moorthy, M. S.; Kim, M. J.; Bae, J. H.; Park, S. S.; Saravanan, N.; Kim, S. H.; Ha, C. S., *Eur. J. Inorg. Chem.* 2013, (17), 3028-3038.

In: Comprehensive Guide for Mesoporous Materials, Volume 4 ISBN: 978-1-63482-092-9
Editor: Mahmood Aliofkhazraei

Chapter 11

THE APPLICATION OF ORDERED MESOPOROUS (ORGANO-) SILICA'S AS A STATIONARY PHASE IN CHROMATOGRAPHY

Matthias Ide and Pascal Van Der Voort
COMOC, Department of Inorganic and Physical Chemistry,
University of Ghent, Krijgslaan, Ghent, Belgium

ABSTRACT

A promising field of application for ordered mesoporous (organo-)silica materials is the separation sciences, *i.e.,* HPLC (High Performance/Pressure Liquid Chromatography). From an economic point of view, chromatographic columns posses small amounts of (organo-)silica materials. This infers that the material's cost, which is higher for ordered mesoporous (organo-)silica materials than for their non-ordered equivalents, is not an issue.

The higher surface area of the materials strongly increases the retentive power of the stationary phase, up to four times when compared to one of the best retaining commercial columns.

Their uniform pore size on the other hand can mean a huge advantage in size exclusion chromatography, where well-defined pore systems can be combined to generate a high performance size exclusion column.

1. INTRODUCTION

To be applied as a packing material in an HPLC column, the ordered mesoporous (organo-)silica particles have to exhibit a monodisperse spherical morphology. Therefore the most common synthesis pathways to generate spherical particles will be discussed.

The large surface area ordered mesoporous particles express leads to a higher contact area between the analytes in the column and the stationary phase. Therefore one of the firstly expected influences of ordered mesoporous materials on the separation of analyte mixtures is a substantial increase in retention.

The use of bridged organosilanes has been well adopted in the synthesis of PMOs (periodic mesoporous organosilicas). However, as a stationary phase this has not been used extensively thus far. Especially in terms of hydrolytical stability the application of organosilanes proves to be a huge advantage.

2. Producing Spherical Mesoporous Particles

The ideal particles for chromatography exhibit a particle size between 1.5 and 5 μm and are very monodisperse. The methods described in literature to produce ordered mesoporous spherical silica particles are the Stöber method, pseudomorphic transformation, and spray drying.

2.1. Stöber's Method to Produce Silica Spheres

Ref.	Type	Size range (nm)	SA (m^2/g)	VpT (cc/g)	Dp (nm)	Shape
[1]	Classic Synthesis	50-2000	-	-	-	Uniform Spheres
[2]	MCM-41	400-1100	1079-1204	0.77-0.93	2.81-3.28	Irr./agglom.
[3]	MCM-41	400-1100	980-1450	0.63-1.20	1.76-5.04	Irr./agglom.
[4]	MCM-41	300-2500	739-1013	0.32-0.41	1.78-2.13	Uniform Spheres
[5]	MCM-41	150-1290	1070-1204	0.59-0.83	2.07-3.69	Uniform Spheres
[6]	MCM-41	10-1520	740-1280	-	1.9-2.1	Uniform Spheres
[7]	MCM-48	-	900-1600	0.5-0.9	2-3	Uniform Spheres
[8]	MCM-41 MCM-48	500-2000	>1000	>1	2-4.5	Uniform Spheres
[9]	MCM-48	24-350	-	-	1.9-2.5	Agglom. Spheres
[10]	MCM-41	0.1-5 mm	484-1100	0.24-0.61	1.4-2	Glass Spheres
[11]	MCM-41	6800	750	0.59	3.3	Polydisp. Spheres
[12]	MCM-41	-	-	-	-	Rods Gyroids
[13]	MCM-41	500-2300	1019	0.72	2.26	Uniform Spheres
[14]	MCM-41	2000-6000	1042	0.7	2.7	Polydisp. Spheres
[15]	SBA-15	1000	696-710	1.27-1.45	8-9.9	Irr. Spheres
[16]	SBA-15	3000-6000	660-853	0.58-1.74	3.5-9.8	Polydisp. Spheres

Ever since the discovery of highly ordered mesoporous materials and their said potential in chromatography, monodisperse spherical particles of these materials are of great interest. [1]

Stöber's method employs a water/alcohol/ammonia/tetraalkoxysilicate system wherein the alcohol has the role of a dispersing agent. By continuous shaking or by use of an ultrasonic bath the formed particles remain in suspension. The dependence of the particle size, shape, and monodispersity on co-solvent concentration, the pH value, the reaction time, the stirring procedure and the interactions between the silica source and the co-solvent was extensively investigated. Particles in the size range of 50 nm to 2 μm were easily achieved but larger particle size usually led to a broadened particle size distribution. [1]

The main issue with this technique is that when it is modified by adding a surfactant to the mixture, the dispersive power of the alcohol tends to fail and the final spheres agglomerate.

In 1997 the group of Klaus Unger has modified the original Stöber method by adding surfactants to the reaction mixture. This led to particles between 400 to 1100 nm in size. They were irregularly agglomerated into shapes of several tens of micrometers. The particles exhibited high specific surface areas (±1000 m^2g^{-1}) and narrow pore size distributions in the 3 nm range.[2,3] Others replaced the co-solvent with methanol and traded in ammonia for sodium hydroxide. This resulted in very monodisperse particles with average particle size diameters just below 1 μm with comparable pore properties. [4, 5]

In 2007 an extended SEM study was conducted on the influence of this co-solvent on the particle size. An inverse relation was observed between the particle size and the solvent concentration. [6] The particles are too small to be used as a packing material in HPLC.

In 1999 the synthesis of cubically ordered mesoporous spheres was published by the Unger group. [7, 8] The particles, with an Ia3d space group, were synthesized through a modified Stöber method. In 2004 cubically and hexagonally ordered silica spheres with radially oriented pores were reported by Rankin et al. [9] A thorough TEM study was conducted and a mechanism for the formation of spherical particles with radial pores was proposed.

While the Stöber method is, theoretically, a process in a basic medium, other groups have managed to produce spheres in an acidic medium by using hydrochloric acid as a catalyst. The properties of these particles were roughly the same except for the particle diameter; which extended to the mm range. [10-12]

In 1998 superficially mesoporous ordered spherical particles were reported by Büchel et al. [13] The particle cores were only 420 nm in diameter and synthesized according to the method prescribed by Stöber.[1] The cores were subsequently stirred in a solution of water, ammonia, a silica source, and a surfactant, generating a porous shell of only 75 nm thick. The surface areas up to 350 m^2g^{-1} were lower than fully mesoporous particles, but nevertheless exhibited a narrow pore size distribution around 3.8 nm.

In the same year Qi et al. introduced a novel pathway to synthesize spherical mesoporous silica materials by using a mixture of CTAB and the non-ionic surfactant Brij 56 as structure directing agent. Particle diameters of 2 to 6 μm were obtained with pore sizes of 2.6 nm and surface areas up to 1000 m^2g^{-1}. [14]

In 2000 Zhao et al. reported the synthesis of spherical SBA-15. It comprised a modification of the standard SBA-15 synthesis pathway by adding CTAB as a co-surfactant and ethanol as a co-solvent. The particle size was fairly monodisperse around 1 μm but the spherical shape was rather distorted by large humps of silica grown on them like barnacles (figure 1f). The particles exhibited a hexagonal pore structure with a surface area of 710 m^2g^{-1}. [15]

Three years later Ma et al. achieved the synthesis of 5 μm sized SBA-15 particles. Their only drawback was that they were too agglomerated to be used in HPLC. [16] Figure 1 shows the original particles by Zhao et al. (f) and the updated particles by Ma et al. (a to d).

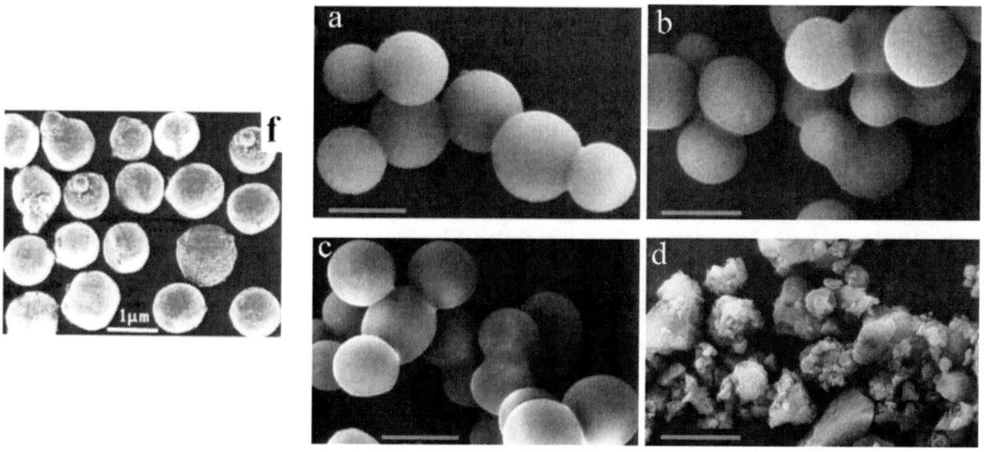

Figure 1. SEM pictures of spherical SBA-15 reported by Zhao et al. (f).[15] SEM images of mesoporous SiO_2 obtained at: (a) 80° C for 5h; (b) 80℃ for 5h and 120℃ for 12h; (c) 80℃ for 5h and 130℃ for 12h; (d) 130℃ for 12h. Scale bar: 5 μm. [16].

2.2. Pseudomorphic Transformation Retains the Spherical Morphology

Ref.	Type	Base particle	Particle size (μm)	SA (m^2/g)	VpT (cc/g)	Dp (nm)	Surfactant
[17]	MCM-41	Nucleosil	5	-	0.93	3.9	CTAB
		Lichrosphere 60	10	-	0.91	3.8	
		Lichrosphere 100	15	-	1.00	3.6	
		Sylopol	50	-	0.80	3.9	
		Silica gel	130	-	0.80	3.8	
[18]	MCM-41	Aerosil 200 Home made	10-41	685-1070	1.20-1.96	5.8-9.6	C_{18}TAB/ TMB
[19]	MCM-41	Nucleosil 120-5	5	916	0.61	3.7	CTAB
		Nucleosil 100-5	5	913	0.79	3.9	
		LiChrospher 100	15-20	911	0.72	3.8	

A second method, described by Martin et al., is a pseudomorphic route.[17] Herein commercial silica spheres (Nucleosil 100-5) were bought and stirred in an alkaline solution comprising sodium hydroxide, water, and CTAB. After a certain reaction time at 110℃ under autogenous pressure the surfactant has templated an ordered pore system inside the spheres without destroying the spherical morphology (figure 2). The change in pore morphology was clearly visible on the X-ray diffractograms and the N_2-adsorption isotherms. Although this seems an interesting approach, it also introduces an additional step in the production of ordered mesoporous silica spheres. It is also limited by the use of standard commercial particles. [18, 19]

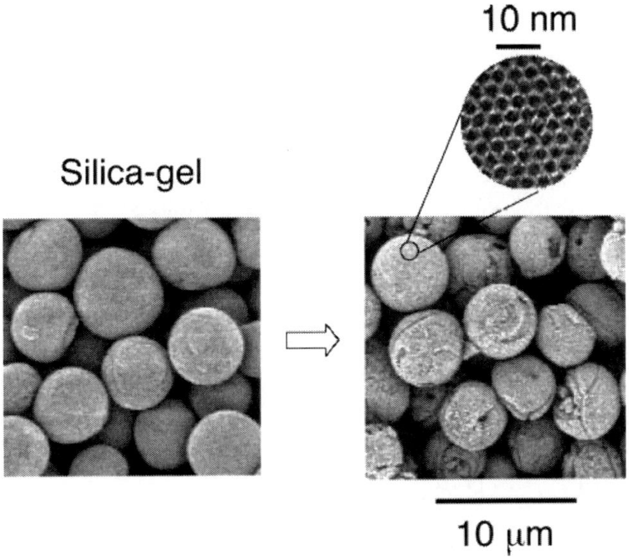

Figure 2. SEM pictures of Nucleosil 100-5 silica gel and its transformation by pseudomorphic synthesis into MCM-41 (Nu100-MCM-41). [19].

2.3. Spray Drying, the Route for Up Scaling to Industrial Proportions

The final method to induce a spherical particle morphology is an aerosol-assisted synthesis, during which the precursor solution is atomized inside a hot air stream.

2.3.1. Aerosol Assisted Evaporation Induced Self Assembly

As the name predicts Aerosol Assisted Evaporation Induced Self Assembly (AAEISA) is based on the pure EISA process. Figure 3 shows this process for the formation of thin films of both ionic and non-ionic templated mesoporous materials. Brinker et al. first described the process in 1999 where it was proven for MCM-41 thin films. [20].

During the process of solvent evaporation the surfactant concentration increases until the critical micelle concentration (CMC) is reached. Above this concentration the surfactant starts forming micelles. CTAB for instance has two CMC values: the first one (0.0009 mol/L in water) for the formation of spherical micelles and a second one for the formation of rod like micelles. The speed of the process depends on the solvent used but it is inherently a slow process.

Spray drying, on the other hand, is a fast process, from spray tip to particle collection takes roughly six seconds on a laboratory scale dryer. It is evident that it is hard to combine a slow EISA process with the speed of an aerosol-assisted synthesis such as spray drying.

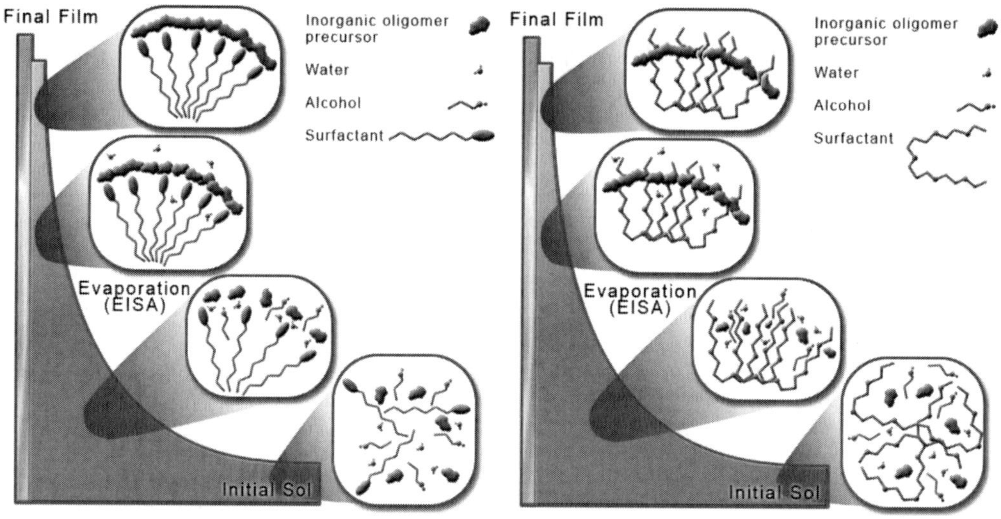

Figure 3. The EISA process for a thin film with an ionic (A) and a non-ionic template (B).

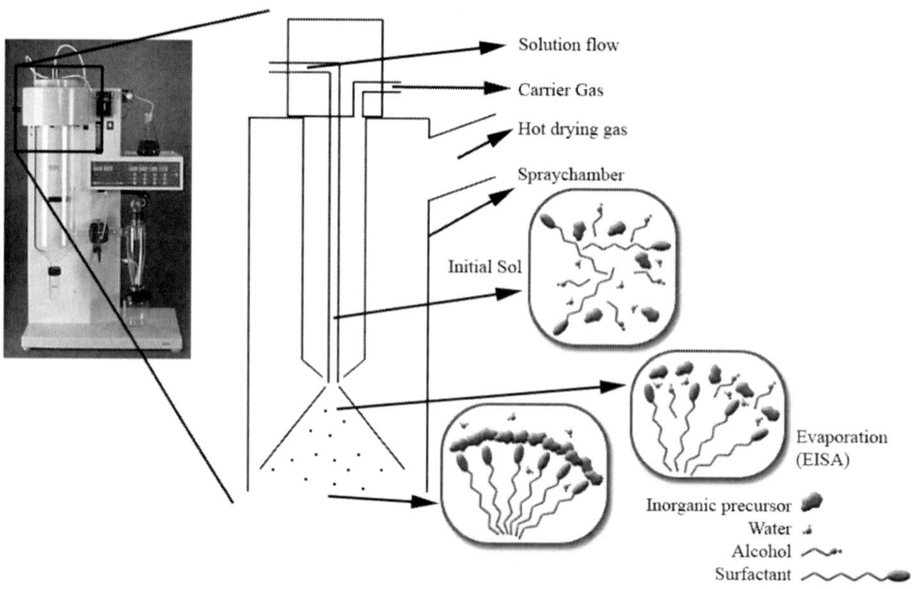

Figure 4. The EISA process applied for the spray drying process.

Therefore literature has reported only materials with the small surfactant CTAB, which were successfully spray dried into mesoporous particles with pore properties equal to those made by a hydrothermal flask synthesis. Figure 4 shows the essence of a spray-drying set-up based upon a commercial Büchi system and an M41S type synthesis procedure.

As a way to somehow predict particle morphology, Vehring et al. applied Peclet numbers to the spray drying process. [21] Peclet numbers in spray drying are defined as the evaporation rate of the solvents in a droplet over the diffusion speed of the components in the droplet solution. Peclet numbers smaller than or equal to one usually means a fast diffusional motion of the solutes with a small radial speed of receding droplet surface. The solutes remain fairly evenly distributed in the droplet during the evaporation, which means small surface enrichment with solid particles, mostly spherical particles as a result.

Peclet numbers larger than one infers the opposite, which leads to shell particles. These can then either remain shells or they can implode with a wrinkled shape as a result.

2.3.2. M41S Type Materials

Ref.	Surfactant	Size range (µm)	SA (m^2/g)	VpT (cc/g)	Dp (nm)	Shape	Spray set up
[22]	CTAB Brij-56 P123	0.1-0.5	1256 514 478	-	1.8 4.1 5.5	Polydisperse Spheres	Homemade: bi-fluid head
[23]	Brij-56 P123 + swellers	0.1-0.5	75-480	0.53-0.98	3.9-14.4	Small spheres	Homemade: bi-fluid head
[24]	CTAB Brij58	5-10	600 516	-	2.23 2.82	Uniform Spheres	VOAG
[25]	CTAB Nanoparticle agglom.	0.2-1 10-25	1162-1588	-	3.2 47-607 2300-15200	Polydisperse Spheres Spherical agglomerates	Commercial Bi-fluid head
[27]	CTAB	1.20-2.96	-	-	2.95-3.65	Polydisperse Spheres	Commercial Bi-fluid head
[28]	CTAB	1-35	866-1204	0.7-0.8	3.2-3.5	Polydisperse Spheres	Commercial Bi-fluid head

In 1999 the group of Brinker reported the first spray dried mesoporous material. [22, 23] The fully porous mesoporous spheres were produced with a homemade aerosol reactor set-up. They made three types of materials based on three different surfactants.

The hexagonally structured silica had a surface area of 1256 m^2g^{-1}, 1.8 nm pores and was made with CTAB.

They also used two larger surfactants, Brij 56 and Pluronic P123 delivering materials with respective surface areas of 514 m^2g^{-1} and 478 m^2g^{-1} and pore sizes of 4.1 nm and 5.5 nm. The pore structure of these last two materials was vesicular. The particle sizes remained very small with a maximum around 500 nm.

In 2002 Rao et al. published a method remotely linked to an aerosol assisted synthesis which produced very uniform spheres in the size range of 5 to 10 µm. [24] They used a vibrating orifice aerosol generator (VOAG) to produce the droplets. The only drawback of this procedure is that the vibration of the VOAG breaks the micelles which leads to inferiorly ordered pore systems.

In 2003 the Unger group reported the first synthesis of spray dried MCM-48 particles by the agglomeration of nanospheres. [25] First the mesoporous nanospheres were produced according to a modified Stöber method, which were then suspended and fed to a spray drier. They were agglomerated to particles between 10 and 25 µm in size with all the characteristics of M41S mesoporous particles. Due to this synthesis pathway both mesopores and macropores are introduced into the spherical particles. Because of this "plural" porosity these are so called hierarchical porous particles.

An in – situ SAXS (Small Angle X-ray Scattering) study of the spray drying process revealed two important influences on the mesostructuring of the particles. [26] First of all a shift to lower pore sizes occurred when higher drying temperatures were applied. Secondly it seemed that ethanol rich solutions promoted a better mesostructure formation than water rich solutions. The explanation for this could be found in the difference in silica condensation rate. Water promotes the kinetically controlled inorganic growth over the self-assembly arrangement hence resulting in poorly resolved mesophases.

It was not before 2004 when Alonso et al. published their paper on morphological and textural control of spray dried mesoporous silica-based spheres, that an extensive study of various factors was performed. [27] The time during which the siloxane sols were aged and their solvent content was varied. Very short aging times resulted either in agglomerated spheres or no spheres at all and very long aging times resulted in larger broken spheres suggesting an initial droplet size of about 20 μm. Longer aging times seemed to have a detrimental effect on the mesophase formation. As for the solvent concentration, lower amounts of solvent resulted in an increase of sphericity of the particles.

In 2011 an experimental design was set up of the spray drying of M41S materials by Ide et al. [28]. Six factors were varied and their influence on the spherical morphology and the pore structure was reported. The main factor that influenced the morphology was the amount of water in the solution, the higher the water content the smaller and more uniform the silica spheres were. The main factor influencing the pore properties was the surfactant concentration. Additionally the surfactant concentration had a negative influence on the morphology. The authors came to the conclusion that the surface tension was the main factor influencing the particle morphology.

2.3.3. SBA-Type Materials

While the first spray dried SBA-15 type material was reported in 1999 by the group of Brinker [22], it took five years until a successful follow up communication could be published. In 2004 the group of Sanchez reported the spray drying synthesis of an ordered hierarchical mesoporous core-shell spherical material based on the non-ionic triblock co-polymer Pluronic F127. [29] The final material exhibited a hexagonal structure with a surface area of 390 m^2g^{-1} and particle sizes far below 1 μm.

Ref.	Surfactant	Size range (μm)	SA (m^2/g)	VpT (cc/g)	Dp (nm)	Shape	Spray set up
[22]	CTAB Brij-56 P123	0.1-0.5	1256 514 478	-	1.8 4.1 5.5	Polydisperse Spheres	Homemade: bi-fluid head
[29]	P123	<<1	390	0.26	6.7	Small spheres Core-shell	Homemade: bi-fluid head
[30]	CTAB P123 P104	1-12	1106 316 382	0.73 0.54 0.52	2.5 8.4 5.9	Polydisperse Spheres	Homemade: bi-fluid head
[31]	Brij-58 P123 F127	0.1-1.2	753 508 567	-	3.5 22.4 10	Polydisperse Spheres	Homemade: bi-fluid head
[32]	F127	1.158-2.236	302-327	-	6-7	Polydisperse Spheres	Commercial Bi-fluid head
[33]	P123		190-292	0.15-0.68	3.1-23	-	Commercial Bi-fluid head

Various research groups attempted to synthesize SBA type materials via spray drying but none of them managed to produce particles with the same large surface areas or uniform pore systems as can be made by a hydrothermal synthesis. The particle size distributions were also very broad, ranging from sub micron particles up to particles of several micrometers in size. Pluronic F127 seemed to be a better surfactant to successfully obtain templated mesoporous spheres through spray drying. [30, 31]

In 2009 the first investigation of the industrial mass-production of spherical mesoporous particles by spray drying was published. [32] The materials were spray dried from a mixture of TEOS, HCl, Pluronic F127, and ethanol. The largest spheres were 2.2 μm in size with a standard deviation of 0.11 μm. The pore system is guessed to be cubical (the XRDs were not conclusive) with a pore size of 6 nm and a maximum surface area of 327 m^2g^{-1}. Around the same time Pitchumani et al. published the spray drying synthesis of mesoporous silica and silica-alumina catalysts using industrial raw materials, such as nano-sized boehmite (Captal B (Al_2O_3) and stabilized colloidal silica. [33] There was no information on the particle morphology. The pore system was said to be equal to SBA-15 with a maximum surface area of 292 m^2g^{-1} and a pore size of around 10 nm.

2.3.4. PMO-Type Materials

Up until now there are no spray dried PMO materials reported. However, in 2005 Alonso et al. reported the first co-condensed organically functionalized mesoporous silica spheres. [34] They described a two-step synthesis, in a first step TEOS, HCl, CTAB and isopropanol were mixed and aged. The formed siloxane oligomers were then mixed with methyltriethoxysilane and subsequently spray dried. Polydisperse discrete spheres with a hexagonal mesophase were formed. In a subsequent article this procedure was extensively studied in order to trace the effects of various factors on the organization of matter and the formation of perfect and uniform spheres.[35] The best sample had well defined micro metric spheres with an average diameter of 1.2 μm and a standard deviation of 0.74 μm. Furthermore the particles showed a clear core shell structure in a way that the shell had a different pore structure (hexagonal, following the form of the particle) than the core (hexagonal but without clear direction) (figure 5).

Ref.	Precursors	Size range (μm)	SA (m^2/g)	VpT (cc/g)	Dp (nm)	Shape	Spray set up
[34]	MTES/TEOS	1-3	-	-	-	Polydisperse Spheres Agglomerates	Commercial Bi-fluid head
[35]	MTES/TEOS	1.2-2.96	-	-	-	Polydisperse Spheres	Commercial Bi-fluid head
[36]	MPTMS/TEOS	0.89-1.76	690-940	-	2.5-3.5	Polydisperse Spheres Agglomerates	Commercial Bi-fluid head
[37]	BTSE/TEOS BTEB/TEOS	0.89-1.17	316-673	-	6.4 10.8	Polydisperse Spheres	Homemade: bi-fluid head

Additionally when spray dried at temperatures around 100℃, the solvent (isopropanol or water) played a huge role in the formation of spheres or collapsed spheres. At this temperature, the evaporation of the solvent in the droplets slowed down due to the water content and deformation of the droplets occurred, forming collapsed spheres and shells. Since slower evaporation should, according to Peclet numbers[21], lead to more solid particles, there must be more than just solvent evaporation playing a role in particle morphology of these materials. Unfortunately due to the complexity of the reaction mixtures no unambiguous research has been conducted to clarify this.

Thiol functionalized mesoporous M41S type materials were spray dried by a co-condensation of TEOS and MPTMS (mercaptopropyltrimethoxysilane) by the group of Alonso.[36] The materials were subjected to a post-synthesis treatment ranging from calcination (removing every organic function including the mercaptopropylgroup), solvent extraction with an ethanol/HCl mixture, and a solvothermal and hydrothermal treatment prior to extraction.

The sulfonated materials were tested with Ag^+ sorption. This proved, together with NMR data, that the thiol functions were homogeneously distributed over the pore surface and were accessible to other chemicals. Surface areas ranged from 1200 m^2g^{-1} for the calcined sample to about 850 m^2g^{-1} for the extracted materials. Post-treatments were performed to prevent the particles from losing their long-range pore ordering by directly extracting the surfactant from the pores.

A post-synthesis curing at 150℃ stabilized the texture of the materials enough to allow an extraction procedure. The first SBA type co-condensed mesoporous spheres produced from mixtures of ethane bridged silanes with TEOS, were spray dried by Yamauchi et al. [37] The particles had a higher surface area with higher ethane content. They were the first who managed to produce SBA-type pore systems with surface areas up to 673 m^2g^{-1} and a pore size of 11 nm. The particles retained a spherical morphology with a mean particle size around 1 μm and a particle size deviation of 0.50 μm.

3. The Influence of Order and Increased Surface Area in Chromatographic Columns

The performance of chromatographic columns is strongly influenced by particle design. Uniform particles allow for a more homogeneous packing of a column and reduce the Eddy diffusion through the column, which leads to a reduced peak dispersion. Since chromatography is in essence a diffusion controlled process, the architecture of the pores in the material plays an important role in both the efficiency and the retention experienced with a particular type of packing material in a column.

Current silica based stationary phases exhibit a broad pore size distribution. This means that no two pathways throughout the particle are identical and that not every functional group is equally accessible. To reduce diffusion times through the particle, superficially porous particles were introduced.

These core-shell particles exhibit an improved chromatographic performance but it cannot be completely attributed to this reduced diffusion time. However, inherent to this particle design a reduced phase ratio leads to a lower ability to load sample and to a reduced retention. [38] To deal with this drawback an ordered pore system with a strongly increased surface area could offer a solution.

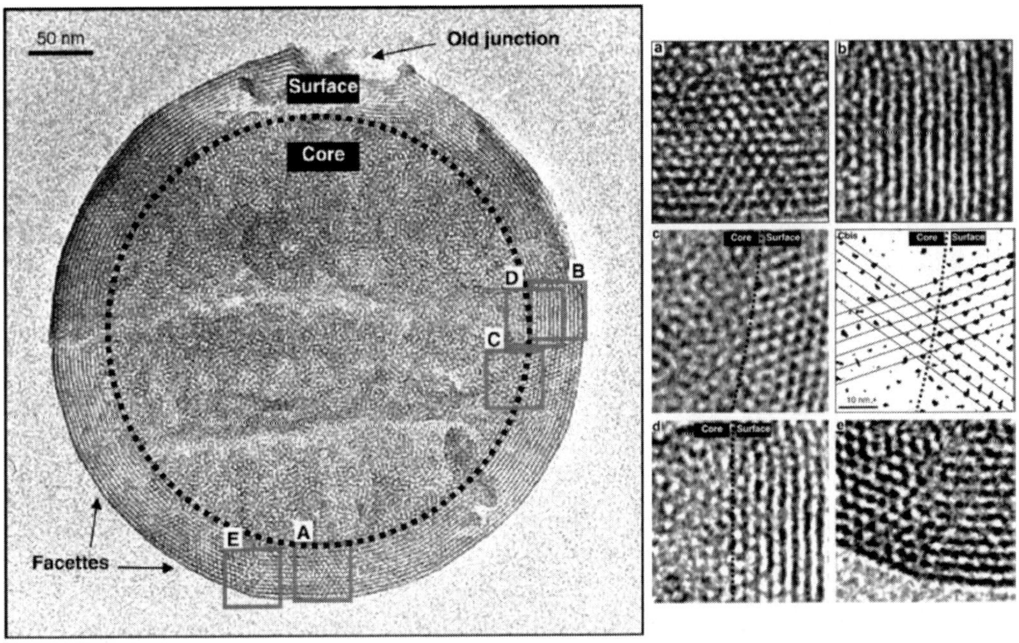

Figure 5. TEM micrograph of an ultra microtomed sphere. The dashed circle represents the border between the ordered domain found at the surface and the non-ordered domain found in the core of the sphere. The marked spots (from A to E) show clearly the P6/mm structure in the outer layer of the material [35].

To improve the separation efficiency the particle size has been ever decreasing over the last four decades from 60 μm to the 1.6 μm particles now available. Two consequences can be attributed to this reduction in particle size: first of all the plate number in liquid chromatography is inversely proportional to two times the particle size in packed columns. Secondly, the backpressure of the column will increase because smaller particles will induce smaller inter particle voids leading to an increased pressure necessary to percolate the same flow through the column.

To somehow counteract this, superficially porous particles can be used to obtain the same efficiency as small particles without the need for high-pressure systems. Therefore an ordered and uniform porosity with a high specific surface area can also increase the retention of packed columns. This increased retention could broaden the scope of reversed phase stationary phases to accommodate the separation of more polar compounds without the need for a different column. Additionally it could give open tubular liquid chromatography and lab-on-a-chip applications the needed push to become a validated technique.

Today HPLC (High Performance/Pressure Liquid Chromatography) is the most used chromatographic technique worldwide. HPLC equipment is the third most sold analytical instrument after the pH meter and the analytical balance. Liquid chromatography was introduced over a hundred years ago by Tswett with the separation of carotene, xanthophylls and chlorophylls. [39-43]

Over this period the technique underwent many changes and additions including thin-layer chromatography, paper chromatography, ion-exchange chromatography, (capillary) gas chromatography, gel chromatography, HPLC, supercritical fluid chromatography, and electrochromatography.

3.1. Silica Hydrolysis: HPLC's Heel of Achilles

Particle and bed stability have always been an issue with silica based columns as water is very often and ever increasingly used in (green) chromatography. The hydrothermal stability is related to two chemical functions of silica: silanol groups and siloxane bridges. During the hydrolysis of silica in an aqueous medium, water molecules easily adsorb to the silica surface by means of hydrogen bonding to the surface silanol groups. In a second step these water molecules break the surface siloxane bridges, resulting into two new silanol groups (figure 8). This process continues until the entire silica structure has been dissolved. [44]

Due to this susceptibility of silica materials to hydrolysis, conventional column performance is easily affected when using harsh conditions such as extreme pH values or higher temperatures in combination with a highly aqueous mobile phase. This instability of silica packing materials is reflected in reduced column efficiency and a reduced retention as a function of the number of chromatographic runs. Column efficiency is reduced by dissolution of the particles in the column which leads to a collapsed bed, affecting the analyte flow path. This leads to deteriorated peak shapes and even peak splitting. Chromatographic retention (**k**) is affected by release and loss of stationary phase. Currently, in reversed phase HPLC a hydrophobic stationary phase is grafted on the silica particles. The octadecyl group is thereby typically anchored to the silica surface by means of ideally 3 siloxane bridges. Therefore this functional group is as susceptible to siloxane cleavage as the entire structure is.

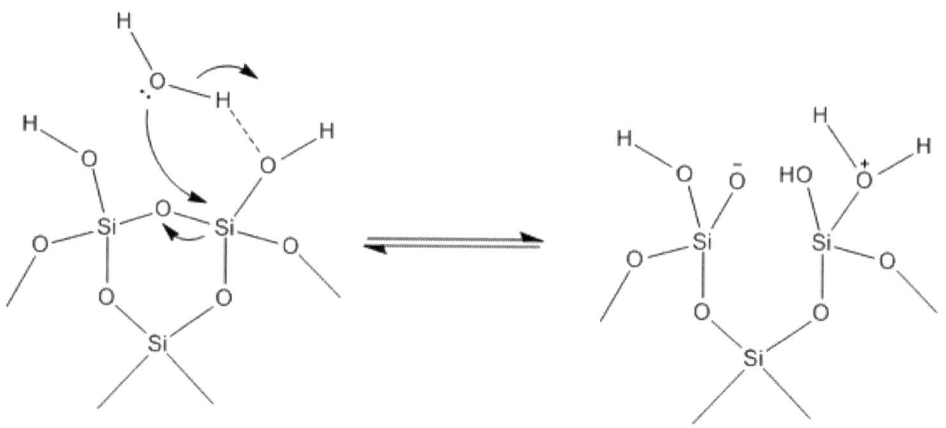

Figure 6. The degradation of silica under the influence of water hydrolysis.

To halt silica hydrolysis, several solutions have been proposed. Endcapping (grafting of trimethylsilyl- groups) was the first solution and more recently researchers started to add more carbon to the stationary phase with semi and full hybrid types of organosilica [45-47], even fully carbonated stationary phases [48] were tried.

3.2. Ordered Porosity in Chromatography

3.2.1. Normal Phase HPLC

The first analyte analysis by mesoporous particles was performed under normal phase conditions. The group of Unger et al. compared MCM-41 particles with commercial alumina, titania, zirconia, and Lichrosphere Si 100 in NPLC. [49] The various stationary phases were tested for the separation of several bases with an n-heptane/2-propanol mixture (99.5/0.5; v/v).

Ref.	Type	Size range (μm)	SA (m^2/g)	VpT (cc/g)	Dp (nm)	Shape
[49]	MCM-41	5-10	615	0.68	4.4	Crushed particles
[50]	MSU-1	0.5-12	891	0.45	2.5	Polydisperse Spheres
[51]	MSU-X MCM-48 MCM-41	10-70 20-50 4-10	495 1124 1186	- - -	6.1 2.2 2.4	Polydisperse Spheres
[52]	Benzene PMO	1-6	1180	1.12	5.4	Polydisperes Spheres
[53]	MCM-41 MCM-48 MTS-TMB MTS-TMB/decane	10 12 10 10	907 1307 1009 657	0.76 1.02 1.72 0.92	3.7 3.4 6.8 5.7	Monodisperse Polymorphic: Lichrosphere 60

The particle sizes were 5 μm for the Lichrospher, alumina, and titania materials and 10 μm for the zirconia material. The MCM-41 material was hydrothermally synthesized and ground to obtain particles in a size range between 5 and 10 μm. This means that these particles are most probably non-spherical and irregularly shaped. A separation of the same analytes was performed on the MCM-41 column and on the Lichrospher and it was concluded that the latter was performing better. However, the columns had a different length, which makes a comparison nearly impossible. Additionally the particle size was different and the MCM material was most likely not spherical. Therefore the different materials cannot be compared and the effect of the ordered porosity or the larger surface area cannot be thoroughly assessed.

MSU-1 particles (made with a Tergitol 15S12 PEO surfactant) were compared to a commercial Nucleosil material by Boissière et al. for normal phase HPLC. [50] The Nucleosil and the MSU-1 particles were packed in 25 cm columns. The chromatographic evaluation was performed under isocratic conditions (dichloromethane/n-hexane; 2/98; v/v) at a flow rate of 1.5 mL/min. It was reported that the retention factors were 4.2 to 6.7 times higher for the MSU material and the plate numbers were 1.6 to 2.1 times higher for the Nucleosil particles. The lower plate numbers were attributed to a larger longitudinal diffusion through the particles but this did not seem to be the only reason. Possibly, the broad particle size distribution of the MSU-1 particles played a major role in this. The higher retention times were mostly attributed to the larger surface area.

The authors compared the MSU results with the results obtained by Gallis et al. [51]. As both the MSU and the MCM-41 material (depicted as APMS in figure 7) exhibited substantial peak broadening while the MCM-48 and the nucleosil packing did not. Therefore they concluded that in order to obtain sharp elution peaks there had to be a strong interconnectivity in the pore system. This meant that, according to the authors, cubically ordered systems are the better option for chromatography. However, as the column dimensions, particle sizes and chromatographic conditions were different for every type of column, no conclusion other than a higher retention for the mesoporous materials can be drawn from this article.

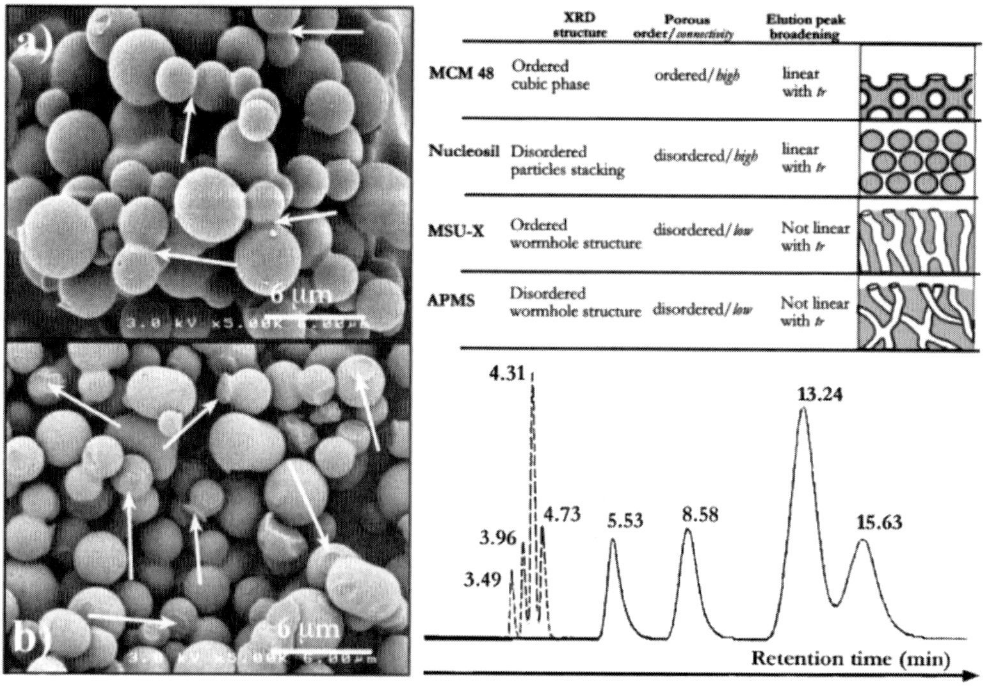

Figure 7. Left: A SEM micrograph depicting the irregular MSU particles. Top right: A table showing the pore ordering and connectivity for four types of tested materials, MCM-48 and APMS were previously reported by Gallis et al.[51] Bottom right: The separation of benzene, naphthalene and biphenyl on a Nucleosil column (dashed line) and an MSU column. [50].

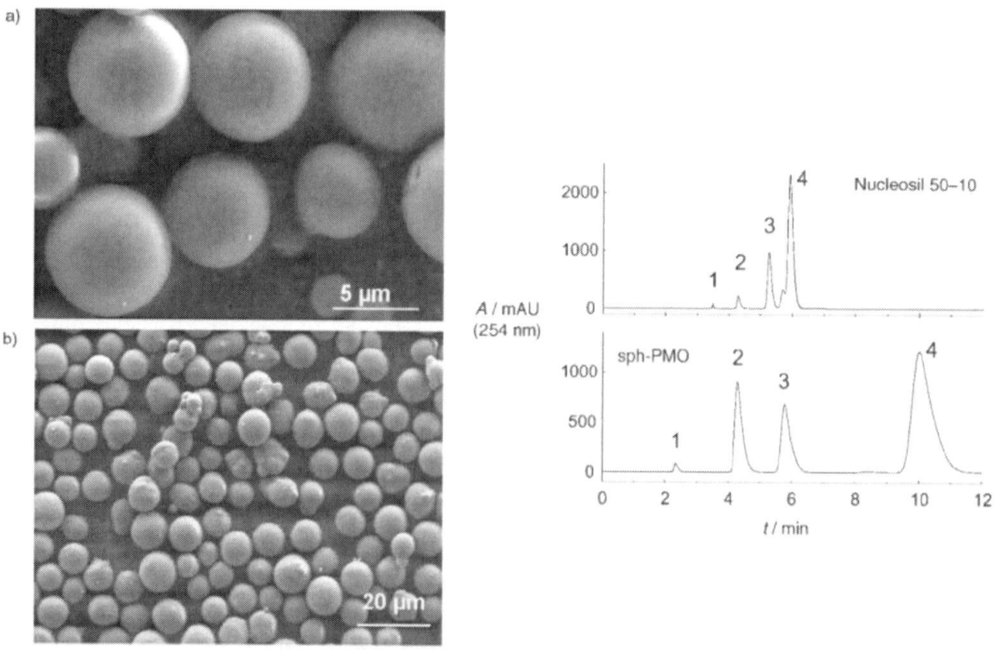

Figure 8. Left: SEM images of spherical particles of a) the sph-PMO and b) Nucleosil 50-10. Right: Chromatograms of mixture 1 (benzene (1), naphthalene (2), biphenyl (3), and phenanthrene (4)) separated with n-hexane at a flow rate of 1 mL/min on the Nucleosil 50-10 column or at a flow rate of 2 mL/min on the sph-PMO column. [52].

Again the effect of the pore structure cannot be thoroughly assessed. The first PMO material to be used as a packing in chromatography was reported in 2006 by the group of Fröba. [52] A benzene PMO material (poly-$O_3Si-C_6H_6-SiO_3$) was used without further functionalization and due to the high organic content of the material, both normal phase and reversed phase separations could be conducted on the column, without subsequent functionalization.

The benzene PMO material was compared with a commercial Nucleosil 50-10 (5 nm pores, 10 µm particles). Both materials were slurry packed under the same conditions in 25 cm, 4 mm ID stainless steel columns. Both columns were chromatographically evaluated under isocratic conditions with n-hexane as eluent. Figure 8 shows the SEM micrographs of both the PMO and the Nucleosil material together with the chromatograms of the separation of polyaromatic hydrocarbons. From the chromatograms it is clear that the PMO material exhibits a higher retention than the nucleosil particles. Additionally the peaks are much broader, so the PMO material does not seem to reach its theoretical efficiency potential.

In the same year Galarneau et al. reported a comparison between normal MCM-41, MCM-48, and pore expanded MCM-41. [53] All the materials were designed by the pseudomorphism syntheses previously reported by the same group. Lichrospher 60 was used as the base material for the pseudomorphic synthesis. It has a particle size of 10 µm, which was retained during the conversion to MCM-41, MCM-48 and two types of large pore MCM-41 (MTS-TMB and MTS-TMB/decane).

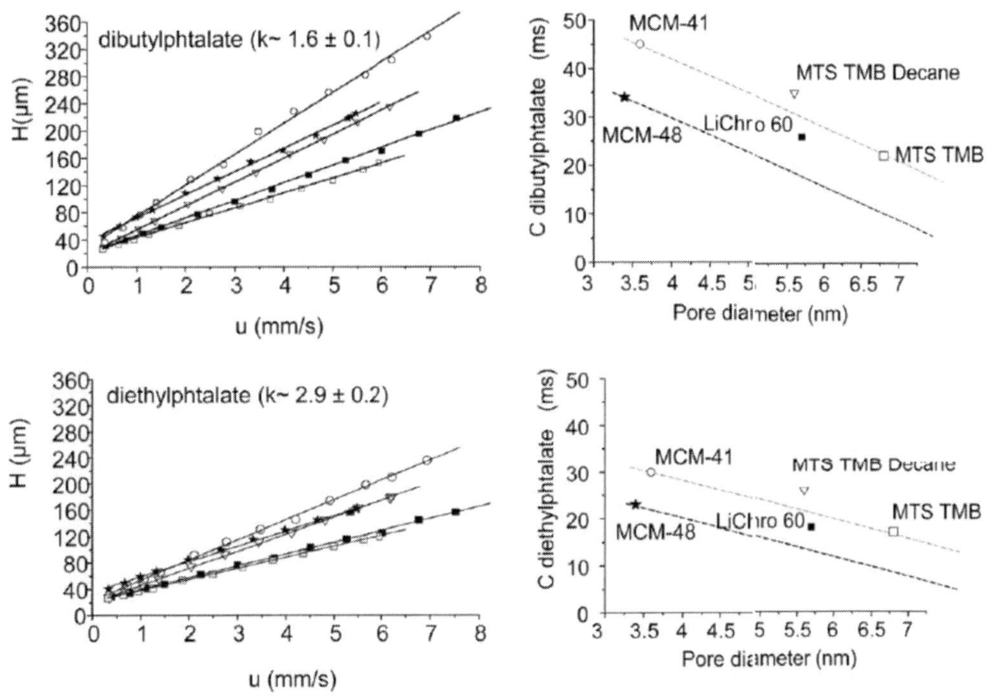

Figure 9. Left: H vs. u curves for diethylphtalate and for dibutylphtalate for an adapted composition of the mobile phase allowing to reach similar retention factors for Lichrospher 60 (filled square) and its pseudomorphs MCM-48 (star), MCM-41 (circle), large pore MCM-41 (MTS-TMB) (empty square) and (MTSTMB/decane) (triangle). Right: Evolution of the C parameter of the H vs. u equation as a function of the pore diameter. [53].

All the materials were slurry packed into 12.5 cm columns with an internal diameter of 4mm. With this comparison the authors claimed to have proven that the difference in mass transfer could be attributed to the pore size and pore structure. Lower C values were observed for MCM-48 in comparison with MCM-41 because of the higher connectivity in the mesopore structure for MCM-48. Additionally, lower C values were also observed for the large pore MCM-41 materials when compared to the small pore MCM-41 particles (figure 9). Since no chromatograms were presented and the van Deemter curves seem to be incomplete it is hard to comply with the authors on these claims.

3.2.2. Reversed Phase HPLC

Ref.	Type	Size range (μm)	SA (m^2/g)	VpT (cc/g)	Dp (nm)	Shape
[16]	SBA-15	3-6	660-853	0.58-1.74	3.5-9.8	Polydisp. Spheres
[19]	MCM-41	5-20	911-916	0.61-0.79	3.7-3.9	Monodisperse Polymorphic: Lichrosphere 60
[28]	MCM-41	1-35	866-1204	0.7-0.8	3.2-3.5	Polydisperse Spheres
[51]	MSU-X MCM-48 MCM-41	10-70 20-50 4-10	495 1124 1186	- - -	6.1 2.2 2.4	Polydisperse Spheres
[54]	SBA-15	2	800	1.05	8.1	No SEM
[55]	SBA-15	<2-16	640-947	0.46-1.21	4.0-9.0	Polydisperse Spheres/polyhedron
[56]	SBA-16	2-100	632-1076	0.55-1.19	7.7-8.3	Polydisperse Spheres
[57]	Benzene-PMO	1.8-2.5	530-749	0.63-0.72	4.3-5.2	Uniform spheres
[58]	Benzene-PMO	1.5-3.5	193-835	0.26-0.7	1.8-10	Uniform spheres
[59]	Amino-PMO	1.5-10	765	0.585	4.09	Polydisperse Spheres

The first report on improved retention in RPLC that could be obtained by the large accessible surface area in these porous materials was published in 1999 by Gallis et al. [51] The ordered mesoporous material was an MCM-41 synthesized hydrothermally under acidic conditions. The material exhibited a surface area of 1186 m^2g^{-1}, a pore diameter of 2.4 nm and a particle size between 4 and 10 μm.

This material was fitted with an octyldimethyl group to obtain a reversed phase stationary phase. The material was compared with a commercial Hypersil-C8 column for the separation of uracil, benzene, naphthalene and biphenyl. The separation was performed under isocratic conditions (H_2O/MeOH; 35/65; v/v) at a flow rate of 1 mL/min and a column temperature of 30℃. The MCM-41 type column exhibits a higher retention, but the efficiency is very poor. This can be attributed to the broad particle size distribution (figure 10).

In 2003 the first report of an SBA-15 type material being applied as packing material was published. [54] The SBA-15 material was functionalized with dimethyloctadecylchlorosilane to obtain a C18 modified SBA-15 with a surface area of 800 m^2g^{-1}, a pore volume of 1.05 $cm^3g^{-1}$and 8.1 nm pores.

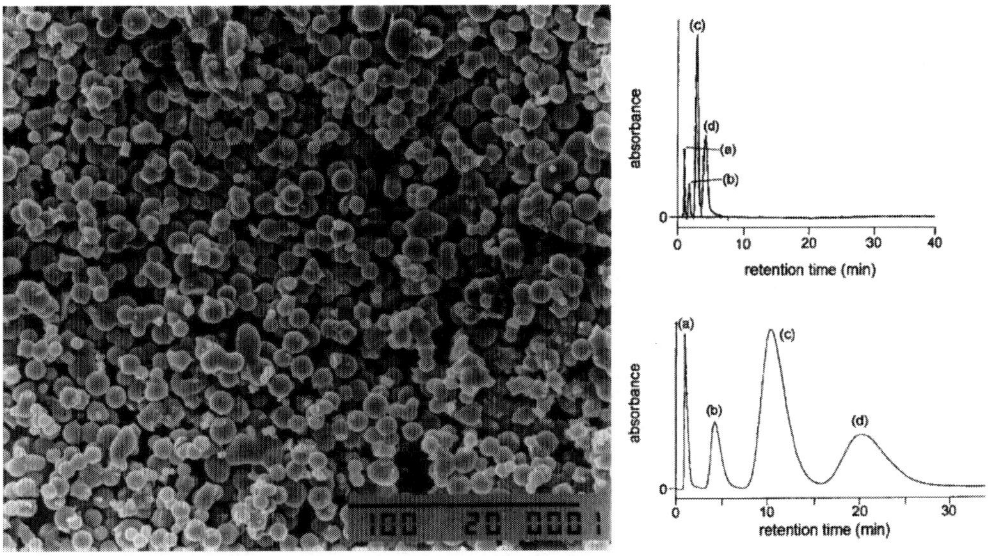

Figure10. Left: A SEM micrograph depicting the irregular particles. Right: The separation of uracil (a), benzene (b), naphthalene (c) and biphenyl (d) on a commercial Hypersil-C8 (top) and a C8 modified MCM-41 column (bottom). [51].

Although no SEM micrographs were shown, the materials were claimed to exhibit a non-spherical but uniform particle size. They were slurry packed in a 10 cm, 250 μm ID capillary column.

On this column four biomolecules (cystein (CYS), glutathione (GSH), 6-thiopurine (6-TP) and dopamine (DA)) were evaluated using a 0.1M phosphorate buffer (pH = 3), no flow rate was given. This was compared to a commercial column. Since no separation conditions were provided for this commercial column, no discernible conclusions can be drawn from the comparison. Nevertheless, the authors claim to observe a clear improvement in the retention time and the separation efficiency. The only conclusion that can be drawn here is that the C18-SBA-15 material successfully separates the four biomolecules.

To perform a triptic digest of myoglobin and a separation of four proteins, the C18- SBA-15 was slurry packed in a 2 mm I.D. steel column (no column length was specified). The chromatogram of the four proteins (lysozyme, bovine serum albumin (BSA), myoglobin, and ovalbumin) shows six peaks of which the first shoulder peak is the heme-group dissociated from myoglobin. The last peak was designated to be an impurity from the ovalbumin sample. What this separation shows is that fairly large molecules (entire proteins) are retained on a column with a pore diameter of 8.1 nm, however without a column death time marker it is impossible to state whether the molecules enter the pores or not.

In the same year two other groups reported roughly the same but added a separation of small aromatic compounds and polyaromatic hydrocarbons.[16,55] Ma et al. reported the synthesis of spherical C18-SBA-15 with a surface area of 522 m^2g^{-1}, a pore volume of 1.06 $cm^3g^{-1}$and 6.1 nm pores. The particles consisted of coagulated spheres. The particles were slurry packed in a 5 cm stainless steel column with a 4.6 mm ID. The chromatograms on the right depict a separation of naphthalene, biphenyl and phenantrene (a) under isocratic conditions at 0.5 mL/min with a water/acetonitrile mixture (30/70; v/v). The second chromatogram (b) shows the separation of three proteins: lysozyme, human serum albumin

(HSA) and bovine serum albumin (BSA). This separation was also performed under isocratic conditions at a flow rate of 0.5 mL/min with water (0.1%TFA)/acetonitrile (0.1%TFA) as the eluent (composition not provided) (figure 11).

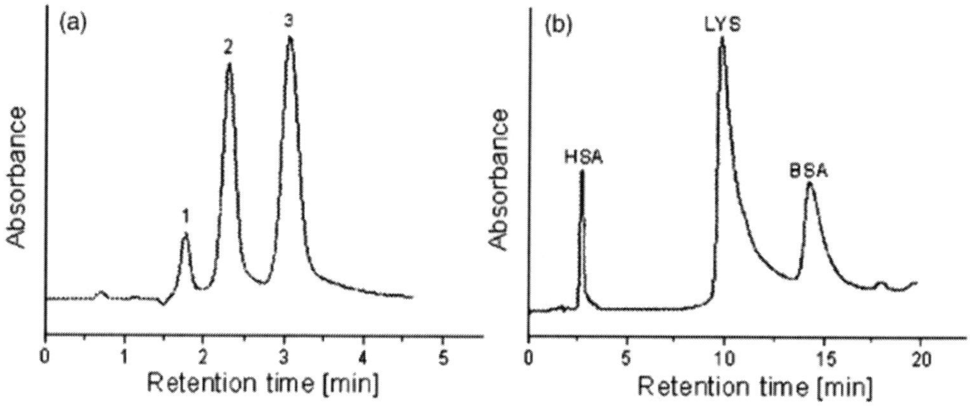

Figure 11. Chromatograms of three aromatic molecules (a) and three proteins (b) with a C18-SBA-15 column. Peaks 1-3 in (a) correspond to naphthalene (1), biphenyl (2), and phenanthrene (3), whereas the three peaks in (b) correspond to human serum albumin (HSA), lysozyme (LYS), and bovine serum albumin (BSA), respectively. [16].

Mesa et al. reported the synthesis of C18 functionalized SBA-15 with a surface area of 811 m^2g^{-1}, a pore volume of 0.91 $cm^3g^{-1}$and 7.6 nm pores. The material was slurry packed in a 15 cm column with a 4.6 mm internal diameter. The column was evaluated for the separation of a mixture of phenol, benzene, toluene, naphthalene, anthracene, and benzopyreen.

The next year an MCM-41 type material produced from commercial Nucleosil100 (Nu100MCM-41, 5 µm) and LiChrospher 100 (LiC-MCM-41, 5 µm) spheres via morphosynthesis was used in reversed phase HPLC after grafting with an octyl-chain.[19] Nu100MCM-41 exhibited a surface area of 913 m^2g^{-1}, a pore volume of 0.79 $cm^3g^{-1}$and a pore diameter of 3.9 nm, while LiC-MCM-41 had a surface area of 911 m^2g^{-1}, a pore volume of 0.78 $cm^3g^{-1}$and a pore diameter of 3.8 nm.

The materials were slurry packed in stainless steel columns of 5 cm with an internal diameter of 4 mm. A conventional silica-based stationary phase was compared to the MCM-41 materials, a ZorbaxSBD-C8 with 5 µm particles in a 15 cm 4.6 mm ID column. For the first time van Deemter curves were provided and showed an optimal plate height of 45 µm for the LiChrosphere MCM-41 column (figure 12A) and a plate height of 12 µm for the Nucleosil 100 MCM-41 material. The optimal linear velocities were 0.04 cm/s and 0.10 cm/s respectively. The MCM-41 stationary phases were reported to have very low mass transfer coefficients, much like monolithic columns, which was attributed to the uniform pore system. Additionally their retention was much higher than the non-ordered commercial materials (see figure 12B).

In 2008 Yang et al. reported the synthesis of spherical SBA-16 particles with a surface area of 632 m^2g^{-1}, a pore volume of 0.70 $cm^3g^{-1}$and a pore diameter of 8.0 nm. They were used as packing material for the reversed phase separation of aromatic molecules.[56] The materials were functionalized with a C8 chain and slurry packed in a 15 cm stainless steel column with a 4.6 mm ID.

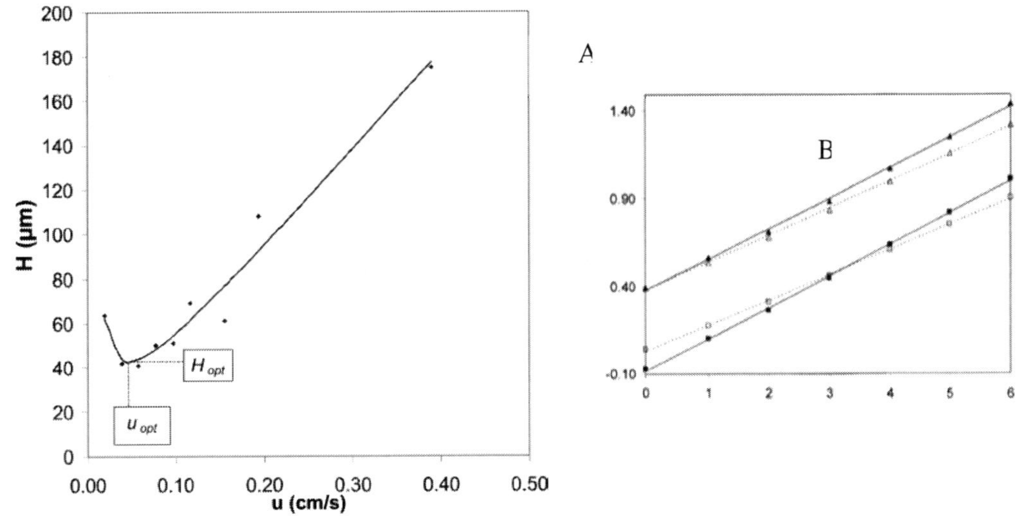

Figure 12. A: Experimental plate height H values versus linear velocity u for the LiC-MCM-41 phase on propylbenzene with water/methanol (30/70; v/v) as mobile phase. B: Log(k) versus the number of carbons in the alkyl chain of alkylbenzene on the Nucleosil Nu100MCM-41 phase in water/methanol (30/70; v/v) (full triangle), in water/acetonitrile (40/60; v/v) (hollow triangle) and on a Zorbax SBD-C8 in water/methanol (30/70; v/v) (full square) and in water/acetonitrile (40/60; v/v) (hollow square) at 30°C. [19].

The column was evaluated isocratically for the separation of a mixture of aromatic hydrocarbons (benzene, naphthalene, phenanthrene, and pyrene). The eluent was methanol/water (65/35; v/v). The maximum obtained plates were 2250 of the theoretical 12500. This discrepancy could probably be attributed to the broad size distribution of the particles.

Zhang et al. reported the synthesis of highly uniform para and meta oriented phenylene-bridged PMO spheres. [57] The materials were slurry packed into 50 x 2.1mm ID stainless steel columns without modification. The phenylene-bridges were the stationary phase for RPLC. The columns were both evaluated for a mixture of 11 compounds under the same gradient conditions. The synthesis procedure was based on a modified Stöber method. The materials itself have a pore size of 5.2 and 4.3 nm, a surface area of 550 and 750 m^2g^{-1} and 0.72 and 0.63 cm^3g^{-1} respectively for the para and meta material. The meta oriented material showed more retention for the same analysis. The authors attributed this to the fact that a para oriented organosilane has its functional phenylene group inside the walls whilst the meta oriented phenylene groups tend to stick out, improving the interaction with the analytes. However since both materials differ substantially in SA, it is more likely that this is the real reason of the retention difference.

In 2009 the same group reported the same phenylene-bridged material but with expanded pores and a C-18 chain grafted to the surface, designated as PHS-C18.[58] The pore size of the material was 8.5 nm with a surface area of 354 m^2g^{-1} and a pore volume of 0.63 cm^3g^{-1}. This resulted in uniform particles with a particle size of 2 µm and subsequently in adequate separations with a plate height of 10 µm, the best one reported so far. Additionally the retention stability of this material in comparison with the pure silica (S-C18) was reported (figure 13).

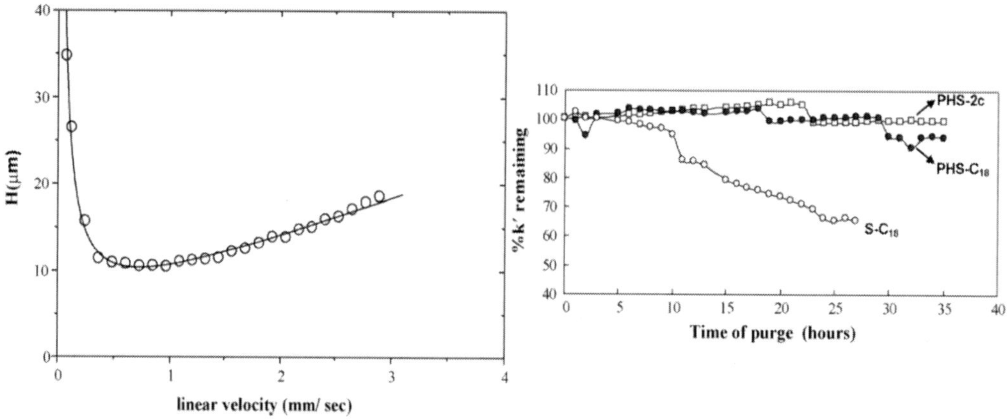

Figure 13. Left: Plot of theoretical plate height (H) vs. linear velocity. Column: PHS-C18 (2.1 mm x 50 mm); mobile phase: acetonitrile/water (55/45; v/v). 30°C. Test probe: phenanthrene. Right: Loss of retention for columns PHS-C18, PHS, and S-C18 as a function of purge time of 50 mM TEA solution: k' measured for phenanthrene. [58].

The columns were exposed to a 50 mM triethylamine (TEA) solution (pH 10) for 60 minutes (0.2 mL/min, 50°C) and then purged with water and acetonitrile (0.2 mL/min, 50℃) for10 minutes each. Phenanthrene was used as a probe molecule to test the efficiency after every run (flow rate: 0.2 mL/min, isocratic; acetonitrile/water, 60/40, v/v; 50℃). The authors concluded that the phenyl group reduced bleeding of the C-18 stationary phase into the eluent whilst the retention of the compared pure silica dropped significantly as a function of time.

In 2011 Li et al. reported the synthesis of a co-condensed organosilane with spherical morphology with both an aminopropyl silane and an ethane bridged silane.[59] The same aging procedure as described by Zhang et al. [58] was used, but now anthracene was used as a probe molecule.

Due to the fact that the aminopropyl function was incorporated into the wall it did not wash out as easily as the grafted aminopropyl material. Here the resistance to leaching of the function was ascribed to the ethane bridge, which makes the silica structure more hydrophobic and thus less susceptible to hydrolysis.

In the same year Ide et al. compared a C-18 functionalized MCM-41 column to 3 commercial columns (Zorbax SB C18, Luna C18 and an X-bridge C18) with comparable column dimensions and particle size dimensions. Their main conclusion was that, compared to the best retaining commercial column, an MCM-41 column with a surface area of 1000 m^2g^{-1} exhibited twice as much retention (figure 14).

The increase in retention corresponded well with the increase in surface area, both for the commercial columns as for the MCM-41 column. It corresponded less with the C-loading of the columns, confirming that the main influence on the retention is the surface area.

The pore size of ordered periodic mesoporous materials is very narrowly defined and their pore size can be tuned within the size range of 2 to 50 nm. Due to this uniform pore size distribution it can be imagined that the family of ordered periodic materials could offer some benefits in size exclusion chromatography. The narrow pore size imposes a narrow size cut-off in an analysis of *i.e.*, a polymer mixture. At the moment such narrow pore size is not available. With a column consisting of known ratios of different pore sizes, mixtures of polymers could be separated in one chromatogram.

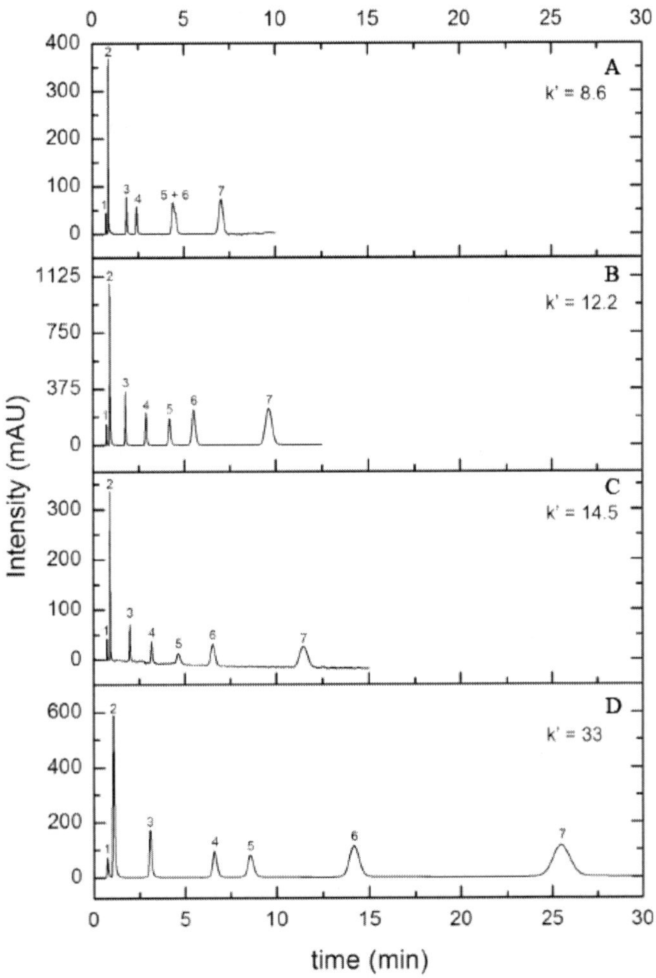

Figure 14. Analysis of a mixture containing uracil (1), caffeine (2), phenol (3), 1-phenyl-1-ethanone (4), propyl-4-hydroxybenzoate (5), benzene (6) and toluene (7) on an MCM-41 column (D), a Luna C18 (2) (C), a Zorbax SB C18 (B) and on an X-Bridge (D) column. The separations were performed isocratically with 40% acetonitrile/60% water as eluent. [28].

3.2.3. Size Exclusion Chromatography

Ref.	**Type**	**Size range (μm)**	**SA (m^2/g)**	**VpT (cc/g)**	**Dp (nm)**	**Shape**
[51]	MSU-X MCM-48 MCM-41	10-70 20-50 4-10	495 1124 1186	- - -	6.1 2.2 2.4	Polydisperse Spheres
[60]	MCM-41	10	756.3	0.818	4.3	Polydisperse Spheres
[61]	Amino-PMO MCM-41	4-10 1-50	1130 1260	0.647 1.02	2.92 3.38	No SEM
[63]	MCF	1-10	-	1.6-2.4	26-49.5	Polydisperse spheres
[64]	MCM-41	10-12	639-1042	1.63-2.17	5.4-6.3	Uniform spheres Polymorph

The group of Unger et al. used MCM41 particles for the first time in a chromatographic application in 1996. [60] The columns (25 cm x 8 mm ID) were packed with either membranes or non-spherical MCM-41 particles and tested for their ability to perform size exclusion chromatography. The MCM-41 materials performed rather poorly because two pore systems were observed, the primary pore system consisting of the uniform pores through the particle (4.3 nm) and the secondary pore system made up of the inter-particle porosity (figure 15A). Therefore the polystyrene standards were roughly separated into two batches that partially overlapped each other. The main reason this did not work well was probably due to the fact the particles were not spherical and secondly not uniform, causing irregularities in the flow pattern through the columns, and leading to peak broadening.

Nassivera et al. reported the first proof of a sharper cut-off in size exclusion chromatography in 2002. [61] The main reason for this can be found in more uniform particle packings. Two types of MCM-41 materials (designated as APMS-30 and MCM-41) were prepared according to previously reported methods. [51, 62]. The APMS-30 (SA: 1130 m^2g^{-1}, Vp: 0.6 cm^3g^{-1}, Dp: 2.9 ±1 nm, 4-10 μm particles) and the MCM-41 (SA: 1260 m^2g^{-1}, Vp: 1.0 cm^3g^{-1}, Dp: 3.4 ± 0.5 nm, 1-50 μm particles) were compared with a commercial Nucleosil-50 (SA: 369 m^2g^{-1}, Vp: 0.8 cm^3g^{-1}, Dp: 11.6 ± 11 nm, 4-6 μm particles) for the size exclusion separation of polystyrene standards of various molecular weight. The materials were packed into 20 cm stainless steel columns with a 4.6 mm ID. The size exclusion experiment was performed at a flow rate of 1 mL/min with tetrahydrofuran (THF) as eluent. Figure 15C shows the SEC curve for the APMS packing with a uniform pore diameter of 2.9 nm and the MCM-41 packing with a uniform pore diameter of 3.3 nm. It shows that there is no total permeation limit for both mesoporous columns for which no apparent explanation is given by the authors.

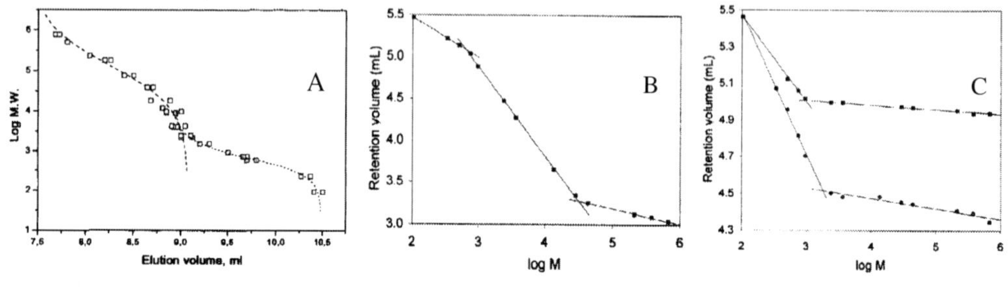

Figure 15. A: The SEC curve of an MCM-41 packing by the group of Unger et al.[60]; B The calibration curve for the separation of polystyrene standards on the Nucleosil column, on the APMS column (C, top) and the MCM -41 column (C, bottom). The total permeation (1), the size exclusion (2) and the total exclusion (3) zones are designated. [61].

To further update the range of ordered mesoporous size exclusion stationary phases Han et al. published the use of ultra large pore ordered mesoporous (MCFs or mesocellular foams) packings with uniform pore diameters varying from 20-50 nm.[63] Three MCF materials were synthesized, designated as MCF-26 (dp = 50 nm), MCF-17 (dp = nm) and MCF-10 (dp = 26 nm). They were slurry packed into 25 cm stainless steel columns with an ID of 4.6 mm.

The packed materials were evaluated for the size exclusion separation of polystyrene standards with molecular weights ranging from 104 up to 770000 Da. The mobile phase was THF and the flow rate was 1.0 mL/min (figure 16).

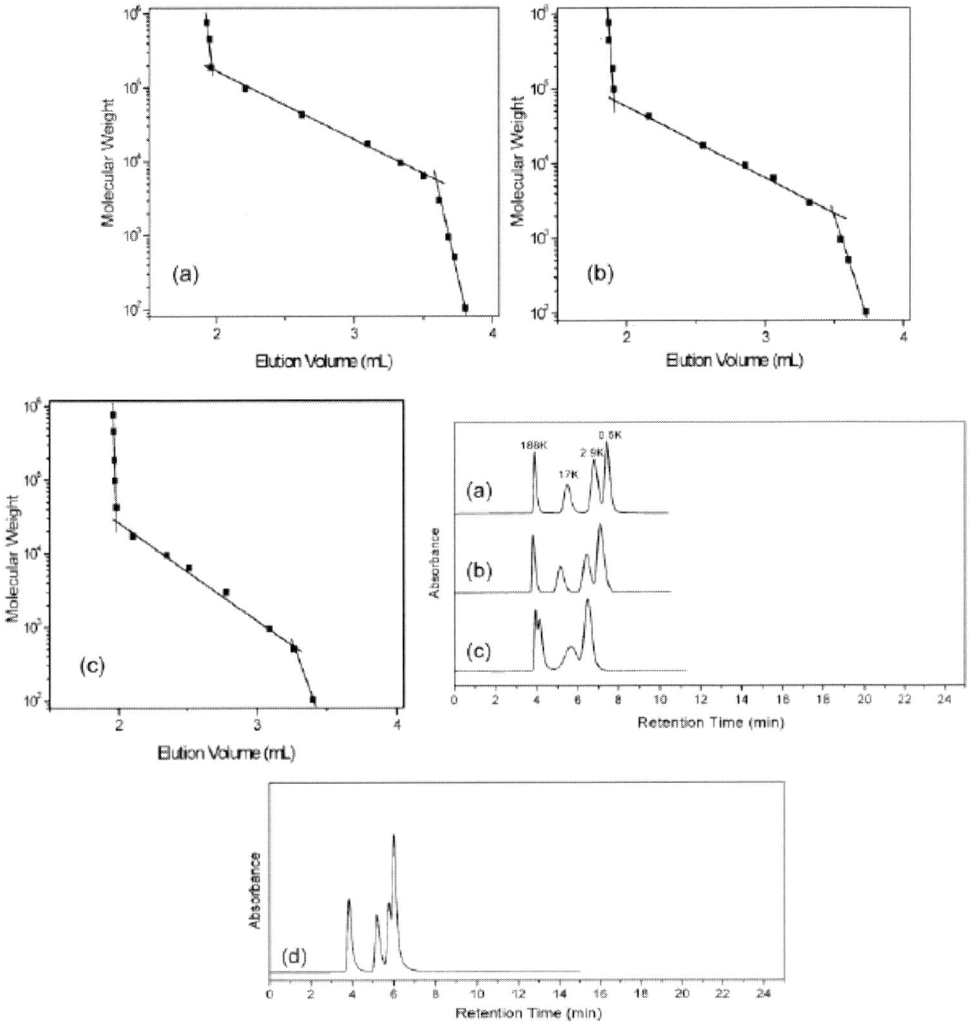

Figure 16. The separation of polystyrene standards by 3 different MCF columns (a = MCF-26, b = MCF-17, c= MCF-10) and the Zorbax PSM 300 column (d). [63].

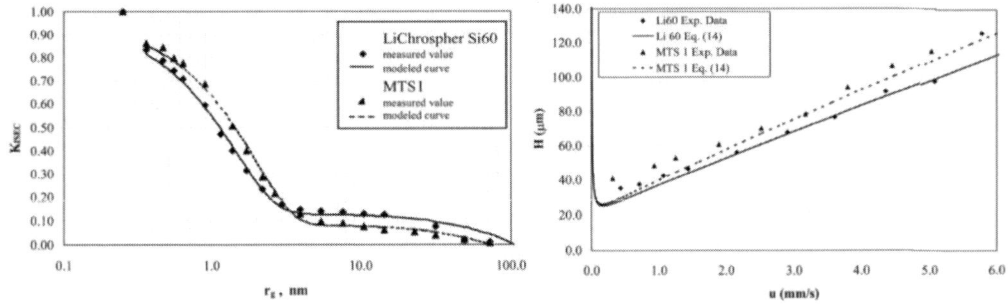

Figure 17. The SEC curves and the H versus u plots for both the original Lichrospher60 and the pseudomorphic column. [64].

The materials were compared to a commercial size exclusion column (Zorbax PSM 300) with a similar molecular weight range as the one of the tested columns (50 nm pores). The MCF column depicted a larger selectivity, which was attributed to the higher pore volume of the material. The column efficiency of the MFC column was lower (9715 plates vs 11210 plates) than the commercial columns, which was, by the authors, attributed to a broader particle size distribution.

Bayram-Hahn reported the pseudomorphical transformation of commercial particles (Lichrospher 60) to ordered mesoporous (MTS) particles.[64] The original Lichropsher 60 base material had a surface area of 769 m^2g^{-1}, a pore volume of 0.9 $cm^3g^{-1}$and a pore size of 4.5 ± 1.3 nm. The pseudomorphic material exhibited a surface area of 1042 m^2g^{-1}, a pore volume of 2.2 $cm^3g^{-1}$and a pore size of 5.4 ±0.6 nm. The MTS material clearly has a larger pore volume, surface area and a more narrow pore size distribution. The materials were slurry packed into 12.5 cm columns with a 4 mm ID prior to their evaluation for size exclusion chromatography. The SE separation was performed on polystyrene standards with a molecular weight ranging from 162 to 23,300,000 Da. The flow rate was 0.2 mL/min with THF as the eluent. Due to the fact that the original particles have a uniform particle size distribution, the column efficiency for both types of columns is the same but the size exclusion efficiency is improved (figure 17). SEC columns seem to exhibit a sharper molecular weight cut-off than the traditional SEC materials. This delivers a cleaner cut in the molecular size distribution of the mixtures but also means that no more than two sizes of molecules can be separated without an array of different pore sizes. This could be a drawback of ordered mesoporous materials as a packing material for SEC.

3.2.4. Chiral HPLC

Ref.	Type	Size range (μm)	SA (m^2/g)	VpT (cc/g)	Dp (nm)	Shape
[65]	MCM-41	0.1-1	955	-	2.4	Polydisperse Spheres
[66]	MCM-41 MCM-48	5-15 0.1-0.5	1000 1000	- -	- -	Polydisperse Spheres
[67]	Ethane PMO	3	829-965	0.63-0.87	2.78-3.02	Polydisperse Spheres

The first report of chiral HPLC performed on MCM-41 modified with a chiral selector was published by the Belgian group of Pierre Jacobs.[65,66] The particles were synthesized through a modified Stöber method, which resulted in a broad particle size distribution. The material was functionalized with γ-aminopropyltriethoxysilane and (R)-α-(1-naphtyl)ethylamine as a chiral selector. A commercial Lichrosorb-NH_2 was also fitted with the chiral selector and both materials were slurry packed in 10 cm columns (no I.D. was provided).

A chiral separation of dinitrobenzoyl-naphtylethylamine (DNB-NEA) was performed with n-hexane/1,2-dichloroethane/ethanol (40/12/3) at a flow rate of 0.6 mL/min. The MCM-41 material exhibited a substantially higher retention. In spite of the good resolution (RS = 2.6 vs 3.6), the chromatographic efficiency was poor, which was attributed to the broad particle size distribution. Unfortunately no chromatograms for the commercial material were provided, which makes a comparative assessment impossible.

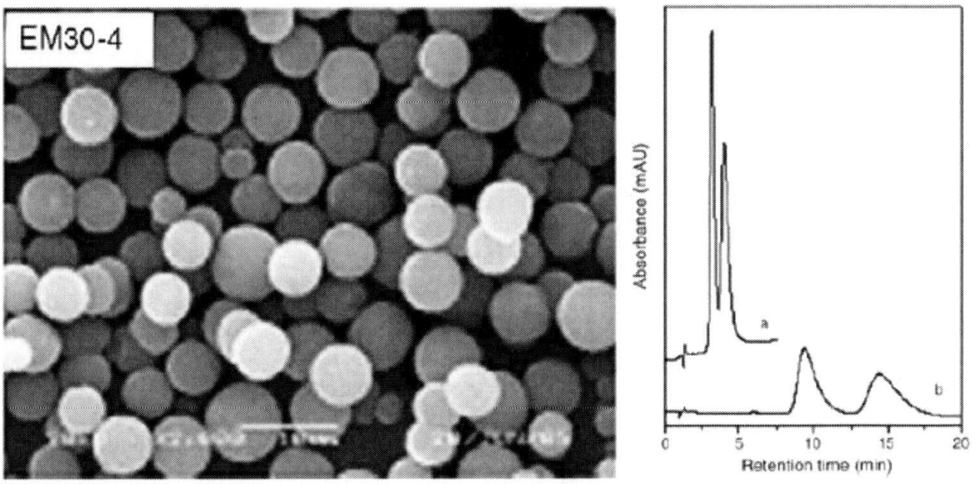

Figure 18. Left: SEM micrograph of the packed chiral material. Right: Chromatograms for the resolution of DL-valine on columns packed with (a) trans-(1R,2R)-diaminocyclohexane-Kromasil and (b) EM30-4 Conditions: column 100 mm x 4.6 mm I.D.; mobile phase: methanol/0.1mM $Cu(OAc)_2$ and 50mM KH_2PO_4 aqueous solution = 10/90 (v/v); flow rate: 0.8 mL/min; detection: 254 nm UV; temperature: 25 ℃. [67].

In 2007 Zhu et al. reported a comparison between a grafted commercial material and a co-condensed organosilica for the separation of chiral groups. [67] The used chiral selector was trans-(1R,2R)-diaminocyclohexane and was grafted on commercial Kromasil material. The co-condensed organosilica was synthesized from a mixture of an ethane bridged silane and N-[3-(triethoxysilyl)propyl]-trans-(1R,2R)-diaminocyclohexane. This way the loading of the chiral group was greatly enhanced resulting in a superior resolution when compared to the grafted material. The co-condensed organosilica had a surface area of 829 m^2g^{-1}, a pore volume of 0.6 $cm^3g^{-1}$and a pore size of 2.8 nm. The mesoporous column exhibits a larger retention for the racemates, but the column efficiency was poor. This poor efficiency was attributed to a broad particle size distribution (figure 18).

Conclusion

Ordered mesoporous materials have been widely researched as potential stationary phases in chromatography. Currently, no systematic research has been conducted to unambiguously ascertain the influence of ordered porosity on the separation of analytes over these materials. To compare the materials with commercially available columns there are more factors than just particle design playing a role. *i.e.*, to compare to a commercial column filled with 1.7 μm fully porous particles, one must conceive a column with 1.7 μm ordered mesoporous particles with the same size distribution and packed in the same way as the commercial column.

To investigate the influence of the type of order (hexagonal, cubical, ...), that should be the only factor that varies. In other words, the surface area should be the same for the different types of order.

To ascertain this, the research done with morphosynthesized particles has the highest potential to answer these fundamental questions.

The only viable conclusion that can be drawn now is that the larger surface area of ordered mesoporous materials induces stronger retentive properties of the column in a more or less one to one relation.

Within this conclusion we want to raise a different question. General chromatographic theory states that the analytes send over a column enter the pores of the packing and are retained there. However this might be correct for large pores with broad pore size distributions, this might not be the case when smaller uniform pore systems are considered. A thought arises that for analytes, and the carrying eluent, to enter the pores this should occur through diffusion because the needed pressure to force a liquid through 2 to 10 nm pores is much larger than what standard pumps these days can provide. Since the speed of diffusion is much lower than the mobile phase velocity, we doubt that analytes actually diffuse through the entire particle when going through a column.

To conclude, ordered mesoporous particles are indeed promising as stationary phases in chromatography, but for them to make it to the market some fundamental questions should be answered unambiguously. Therein lies an interesting field of research for academics.

REFERENCES

[1] Stöber, W.; Fink, A.; Bohn, E. *Journal of Colloid and Interface Science* 1968, *26*, 62–69.

[2] Grun, M.; Lauer, I.; Unger, K. *Advanced Materials* 1997, *9*, 254–257.

[3] Grun, M.; Unger, K. K.; Matsumoto, A.; Tsutsumi, K. *Microporous and Mesoporous Materials* 1999, *27*, 207–216.

[4] Yano, K.; Fukushima, Y. *Journal of Materials Chemistry* 2003, *13*, 2577–2581.

[5] Yano, K.; Fukushima, Y. *Journal of Materials Chemistry* 2004, *14*, 1579–1584.

[6] Shimura, N.; Ogawa, M. *Journal of Materials Science* 2007, *42*, 5299–5306.

[7] Schumacher, K.; Grun, M.; Unger, K. K. *Microporous and Mesoporous Materials* 1999, *27*, 201–206.

[8] Kumar, D.; Schumacher, K.; von Hohenesche, C.; Grun, M.; Unger, K. *Colloids and Surfaces a-Physicochemical and Engineering Aspects* 2001, *187*,109–116, 2nd International TRI/Princeton Workshop on Characterization of Porous Materials, Princeton, New Jersey, Jun 19-21, 2000.

[9] Tan, B.; Rankin, S. E. *Journal of Physical Chemistry B* 2004, *108*, 20122–20129.

[10] Huo, Q. S.; Feng, J. L.; Schuth, F.; Stucky, G. D. *Chemistry of Materials* 1997, *9*, 14–17.

[11] Yang, H.; Vovk, G.; Coombs, N.; Sokolov, I.; Ozin, G. A. *Journal of Materials Chemistry* 1998, *8*, 743–750.

[12] Yang, H.; Coombs, N.; Ozin, G. *Nature* 1997, *386*, 692–695.

[13] Buchel, G.; Grun, M.; Unger, K. K.; Matsumoto, A.; Tsutsumi, K. *Supramolecular Science* 1998, *5*, 253–259.

[14] Qi, L. M.; Ma, J. M.; Cheng, H. M.; Zhao, Z. G. *Chemistry of Materials* 1998, *10*, 1623–1626.

[15] Zhao, D. Y.; Sun, J. Y.; Li, Q. Z.; Stucky, G. D. *Chemistry of Materials* 2000, *12*, 275–279.

[16] Ma, Y. R.; Qi, L. M.; Ma, J. M.; Wu, Y. Q.; Liu, O.; Cheng, H. M. *Colloids and Surfaces a-Physicochemical and Engineering Aspects* 2003, *229*, 1–8.

[17] Martin, T.; Galarneau, A.; Di Renzo, F.; Fajula, F.; Plee, D. *Angewandte Chemie-International Edition* 2002, *41*, 2590–2592.

[18] Lefevre, B.; Galarneau, A.; Iapichella, J.; Petitto, C.; Di Renzo, F.; Fajula, F.; Bayram-Hahn, Z.; Skudas, R.; Unger, K. *Chemistry of Materials* 2005, *17*,601–607.

[19] Martin, T.; Galarneau, A.; Di Renzo, F.; Brunel, D.; Fajula, F.; Heinisch, S.; Cretier, G.; Rocca, J. L. *Chemistry of Materials* 2004, *16*, 1725–1731.

[20] Brinker, C. J.; Lu, Y. F.; Sellinger, A.; Fan, H. Y. *Advanced Materials* 1999, *11*, 579–585.

[21] Vehring, R.; Foss, W.R.; Lechuga-Ballesteros, D. *Journal of Aerosol Science* 2007, *38*, 728-746.

[22] Lu, Y. F.; Fan, H. Y.; Stump, A.; Ward, T. L.; Rieker, T.; Brinker, C. J. *Nature* 1999, *398*, 223–226.

[23] Fan, H. Y.; Van Swol, F.; Lu, Y. F.; Brinker, C. J. *Journal of Non-Crystalline Solids* 2001, *285*, 71–78.

[24] Rao, G. V. R.; Lopez, G. P.; Bravo, J.; Pham, H.; Datye, A. K.; Xu, H. F.; Ward, T. L. *Advanced Materials* 2002, *14*, 1301–1304.

[25] Lind, A.; von Hohenesche, C. D.; Smatt, J. H.; Linden, M.; Unger, K. K. *Microporous and Mesoporous Materials* 2003, 66, 219–227.

[26] Boissière, C.; Grosso, D.; Amenitsch, H.; Gibaud, A.; Coupe, A.; Baccile, N.; Sanchez, C. *Chemical Communications* 2003, 2798–2799.

[27] Alonso, B.; Douy, A.; Veron, E.; Perez, J.; Rager, M. N.; Massiot, D. *Journal of Materials Chemistry* 2004, *14*, 2006–2016.

[28] Ide, M.; Wallaert, E.; Van Driessche, I.; Lynen, F.; Sandra, P.; Van Der Voort, P. *Microporous and Mesoporous Materials* 2011, *142*, 282-291.

[29] Areva, S.; Boissiere, C.; Grosso, D.; Asakawa, T.; Sanchez, C.; Linden, M. *Chemical Communications* 2004, 1630–1631.

[30] Andersson, N.; Alberius, P. C. A.; Pedersen, J. S.; Bergstrom, L. *Microporous and Mesoporous Materials* 2004, *72*, 175–183.

[31] Hampsey, J. E.; Arsenault, S.; Hu, Q. Y.; Lu, Y. F. *Chemistry of Materials* 2005, *17*, 2475–2480.

[32] Yamauchi, Y.; Gupta, P.; Sato, K.; Fukata, N.; Todoroki, S. I.; Inoue, S.; Kishimoto, S. *Journal of the Ceramic Society of Japan* 2009, *117*, 198–202.

[33] Pitchumani, R.; Heiszwolf, J. J.; Schmidt-Ott, A.; Coppens, M. O. *Microporous and Mesoporous Materials* 2009, *120*, 39–46, 3rd International Workshop on New Challenges for Nanoporous Materials Sep 24-26, 2007 Alicante, Spain.

[34] Alonso, B.; Clinard, C.; Durand, D.; Veron, E.; Massiot, D. *Chemical Communications* 2005, 1746–1748.

[35] Alonso, B.; Veron, E.; Durand, D.; Massiot, D.; Clinard, C. *Microporous and Mesoporous Materials* 2007, *106*, 76–94.

[36] Melnyk, I. V.; Zub, Y. L.; Veron, E.; Massiot, D.; Cacciaguerra, T.; Alonso, B. *Journal of Materials Chemistry* 2008, *18*, 1368–1382.

[37] Yamauchi, Y.; Suzuki, N.; Gupta, P.; Sato, K.; Fukata, N.; Murakami, M.; Shimizu, T.; Inoue, S.; Kimura, T. *Science and Technology of Advanced Materials* 2009, *10*.

[38] Fanigliulo, A.; Cabooter, D.; Bellazzi, G.; Tramarin, D.; Allieri, B.; Rottigni, A.; Desmet, G. *Journal of Separation Science* 2010, *33*, 3655–3665.
[39] Nue, U. *HPLC Columns: Theory, Technology and Practice*; John Wiley and Sons, 1997.
[40] Poole, C. F.; Schuette, S. A. *Contemporary practice of chromatography*; Elsevier Science BV, 1984.
[41] Tswett, M. *Proceedings of the Warsaw Society of Naturalists, Biology Section* 1905, 14, 20–39.
[42] Tswett, M. *Berichte der Deutschen botanischen Gesellschaft* 1906, 24, 316–326.
[43] Tswett, M. *Berichte der Deutschen botanischen Gesellschaft* 1906, 24, 384–393.
[44] Vansant, E.; Voort, P. V. D.; Vrancken, K. *Characterization and Chemical Modification of the Silica Surface*; Studies in Surface Science and Catalysis; Elsevier Science BV, 1995; Vol. 93.
[45] Chitty, M.C.; Farkas, T.; Loo, L.; Rustamov, I.M.; Welch, E. pH stable chromatographic media using templated multilayer organic/inorganic grafting, U.S. Patent 7,563,367, October 1, 2004.
[46] Wyndham, K. D.; Iraneta, P. C.; Walsh, D.; Alden, B. A.; Walter, T. H.; Muriithi, B.; Turner, J.; Cook, J.; Boissel, C.; McCall, S.; Neue, U. Research in Hybrid Organic/Inorganic Packing Materials. 19-23 June 2011.
[47] Wyndham, K. D.; O'Gara, J. E.; Walter, T. H.; Glose, K. H.; Lawrence, N. L.; Alden, B. A.; Izzo, G. S.; Hudalla, C. J.; Iraneta, P. C. *Analytical Chemistry* 2003, 75, 6781–6788.
[48] Thermo Scientific, Hypercarb^{TM} HPLC Columns Technical Guide: http://www.interscience.be/promotiesites/hypersil/topics/promotiesites/hypersil/nieuws/hypercarb_technical.pdf
[49] Grun, M.; Kurganov, A. A.; Schacht, S.; Schuth, F.; Unger, K. K. *Journal of Chromatography A* 1996, *740*, 1–9.
[50] Boissiere, C.; Kummel, M.; Persin, M.; Larbot, A.; Prouzet, E. *Advanced Functional Materials* 2001, *11*, 129–135.
[51] Gallis, K. W.; Araujo, J. T.; Duff, K. J.; Moore, J. G.; Landry, C. C. *Advanced Materials* 1999, *11*, 1452–1455.
[52] Rebbin, V.; Schmidt, R.; Froba, M. *Angewandte Chemie-International Edition* 2006, *45*, 5210–5214.
[53] Galarneau, A.; Lapichella, J.; Brunel, D.; Fajula, F.; Bayram-Hahn, Z.; Unger, K.; Puy, G.; Demesmay, C.; Rocca, J. *Journal of Separation Science* 2006, *29*, 844–855.
[54] Zhao, J. W.; Gao, F.; Fu, Y. L.; Jin, W.; Yang, P. Y.; Zhao, D. Y. *Chemical Communications* 2002, 752–753.
[55] Mesa, M.; Sierra, L.; Lopez, B.; Ramirez, A.; Guth, J. L. *Solid State Sciences* 2003, *5*, 1303–1308.
[56] Yang, L.; Wang, Y.; Luo, G.; Dai, Y. *Particuology* 2008, 6, 143–148.
[57] Zhang, Y.; Jin, Y.; Dai, P.; Yu, H.; Yu, D.; Ke, Y.; Liang, X. *Analytical Methods* 2009, *1*, 123–127.
[58] Zhang, Y. P.; Jin, Y.; Yu, H.; Dai, P. C.; Ke, Y. X.; Liang, X. M. *Talanta* 2010, *81*, 824–830.
[59] Li, C.; Di, B.; Hao, W.; Yan, F.; Su, M. *Journal of Chromatography A* 2011, *1218*, 408–415.

[60] Kurganov, A.; Unger, K.; Issaeva, T. *Journal of Chromatography A* 1996, *753*, 177–190.

[61] Nassivera, T.; Eklund, A. G.; Landry, C. C. *Journal of Chromatography A* 2002, *973*, 97–101.

[62] Tolbert, S. H.; Landry, C. C.; Stucky, G. D.; Chmelka, B. F.; Norby, P.; Hanson, J. C.; Monnier, A. *Chemistry of Materials* 2001, *13*, 2247–2256.

[63] Han, Y.; Lee, S. S.; Ying, J. Y. *Chemistry of Materials*2007, *19*, 2292–2298.

[64] Bayram-Hahn, Z.; Grimes, B. A.; Lind, A. M.; Skudas, R.; Unger, K. K.; Galarneau, A.; Lapichella, J.; Fajula, F. *Journal of Separation Science* 2007, *30*, 3089–3103.

[65] Thoelen, C.; Paul, J.; Vankelecom, I. F. J.; Jacobs, P. A. *Tetrahedron Asymmetry* 2000, *11*, 4819–4823.

[66] Thoelen, C.; Van de Walle, K.; Vankelecom, I. F. J.; Jacobs, P. A. *Chemical Communications* 1999, 1841–1842.

[67] Zhu, G. R.; Jiang, D. M.; Yang, Q. H.; Yang, J.; Li, C. *Journal of Chromatography A* 2007, *1149*, 219–227.

In: Comprehensive Guide for Mesoporous Materials, Volume 4 ISBN: 978-1-63482-092-9
Editor: Mahmood Aliofkhazraei

Chapter 12

MAKING PERIODIC MESOPOROUS ORGANOSILICAS FUNCTIONAL MATERIALS

Paula Ferreira[1,*], Cláudia Bispo[1], Mirtha A. O. Lourenço[1], José R. B. Gomes[2], Nicolas Bion[3], Karine De Oliveira Vigier[3] and François Jérôme[3]

[1]CICECO, Department of Materials and Ceramic Engineering, University of Aveiro, Aveiro, Portugal
[2]CICECO, Department of Chemistry, University of Aveiro, Aveiro, Portugal
[3]University of Poitiers, CNRS, UMR7285, Institut de Chimie des Milieux et Matériaux de Poitiers (IC2MP), Poitiers, France

ABSTRACT

Periodic Mesoporous Organosilicas (PMOs) were first described in 1999 and gained immediately a great interest due to their high potential of application in catalysis, adsorption, and separation fields. These materials are formed by condensation of a bissilylated organic bridge precursor in the presence of a structure-directing agent to create porosity. They present unique structural features such as narrow pore distribution, high specific surface areas, alternating hydrophobic and hydrophilic character and adjustable hydrophobicity. PMOs offer the possibility of tuning the material properties by modifying the organic bridge of the precursors or by post-synthesis method. Furthermore, the publication of the phenylene PMO in 2002 presenting, not only meso-scale order but also, molecular-scale periodicity gave to this family an additional stability feature.

This Chapter presents a systematic revision of what has been made in the last fifteen years on the development of acid and basic PMOs for catalytic and adsorption purposes. Considering the complexity of the PMOs and the need of using computation tools to understand and predict experimental findings, we revise for the first time all the work performed concerning the modeling of the PMOs structure and to improve the understanding of their adsorption and catalytic properties.

* Corresponding authors: Paula Ferreira, phone: +351 234401419, e-mail: pcferreira@ua.pt. José R. B. Gomes, +351 234401423, e-mail: jrgomes@ua.pt.

Looking ahead, the Chapter ends with a roadmap of PMO development towards functionality for specific applications.

INTRODUCTION

Periodic mesoporous organosilicas (PMO) were introduced in 1999, when three different research groups independently reported the synthesis of mesoporous materials with a framework formed by alternating organic bridges and silane units [1, 2, 3]. These materials present a great enhancement in terms of structure/properties when compared to the periodic mesoporous silicas of the MS41 family [4, 5]. As the PMOs are prepared from bissilylated precursors containing an organic bridge, a large content of organics can be incorporated into the framework of the material while keeping a very ordered structure and the pore channel network [6, 7]. The properties of the PMOs can easily be tuned by selecting the organic moiety of the precursor (Figure 1), which would change the hydrophilic-hydrophobic character of the material as well as its chemical, mechanical and physical properties.

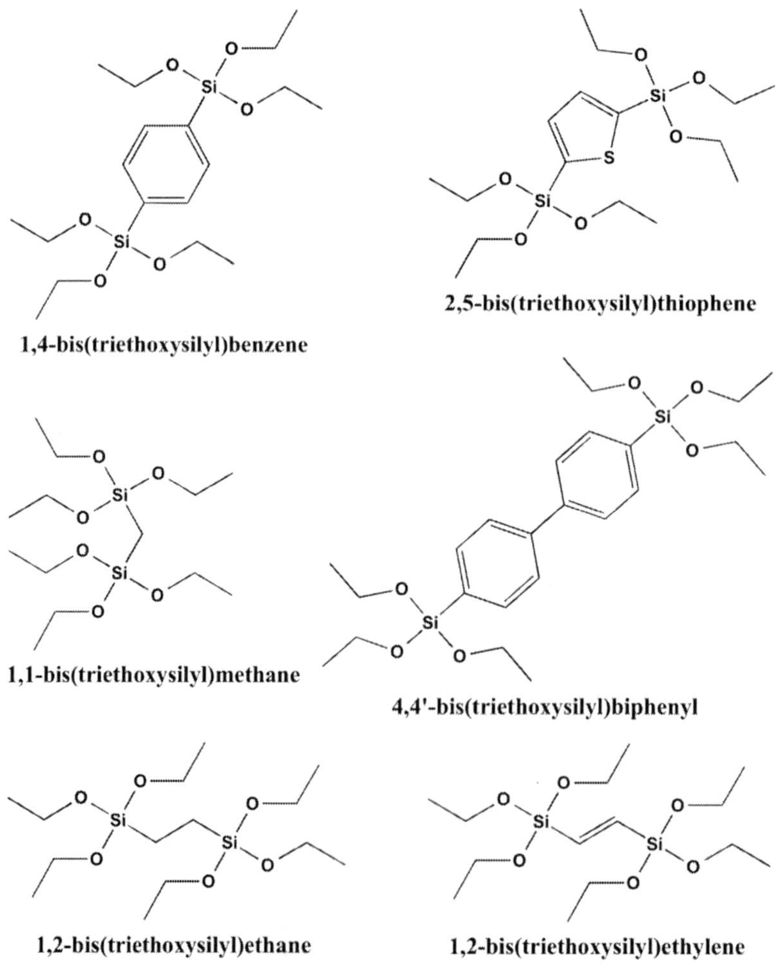

Figure 1. Molecular structures of some common precursors used in the synthesis of PMO materials.

Another important feature is the homogeneous distribution of the organic content and the possibility of achieving molecular-scale periodicity in the walls along the channels direction as reported by Inagaki's group, when a phenyl group was used as bridge [8]. The PMOs are typically prepared using template molecules as structure-directing agents. Cooperative assembly of the silica precursors associated to liquid crystal templating is the invoked mechanism beyond the formation of the mesoporous structure [4, 5, 9, 10, 11]. The use of different cationic alkyltrimethylammonium surfactants with alkyl chain lengths from 14 to 18 was proved to be efficient on preparation of phenylene PMO with different pore sizes, preserving the meso- and molecular-scale periodicities and the thermal stability up to 550 °C [12]. The use of amphiphilic block-copolymers was also reported on the preparation of PMOs; however, for example, in the case of the phenyl PMO, the use of Pluronic® 123 P123 (polyethylene oxide–polypropylene oxide) does not allow the achievement of molecular-scale periodicity [13]. One of the remarkable features of PMOs is the high thermal and hydrothermal stability [14]. A recent study compares the thermal stability of PMOs containing different organic bridges and shows that the aromatic bridges were decomposed at temperatures higher than the aliphatic ones. The contraction of the unit cell with increasing temperature was also minor in the case of the aromatics. With exception of ethylidene, the phenyl, biphenyl and ethylene bridges were stable to the hydrothermal treatment at room and reflux temperature. Apart from changing the bissilylated organic bridge precursor, PMOs have been changed by post-synthesis methodologies or used as ligand walls to bind other organic functions, or even metal complexes. Chemical vapor deposition was used in 2005 to bind tricarbonyl complexes of molybdenum and chromium to the phenyl ring of the PMO framework [15]. The authors reported 15% of conversion of the phenylene rings as a result of their low reactivity within the PMO framework. In 2010, some of us reported the grafting of molybdenum tricarbonyl complex into the phenylene moieties of the PMO using a solution of $Mo(CO)_6$ [16]. The metal loading was proved to be dependent on the reaction time, with a maximum degree of phenyl modification achieved after 24 hours of reaction (comparable to the 15% of Kamegawa et al. [15]).

We also reported the introduction of $Cr(CO)_3$ by liquid phase deposition but, in this case, the metal loading was quite low [17]. In 2007, Inagaki and co-workers [18] published the amination of 28% of the phenyl rings in two steps involving the formation of the nitrated Ph-PMO, followed by the reduction to the aminated Ph-PMO.

Recently, the *N*-alkylation of the amino-functionalized material could be achieved with about 87% conversion by using a potassium iodide catalyzed method and microwave-assisted heating [19]. The amine groups located at the phenyl rings of the PMO can be further used to attach organometallic complexes such as acetylferrocene [20].

Another example is the replacement of the acetonitrile labile ligands of the $[(CH_3CN)_3RuCp]PF_6$ complex, by reaction of ruthenium with the phenyl rings, to give supported organoruthenium complexes of the type $[PhRuCp]PF_6$, where Cp represents cyclopentadienyl groups [21]. A PMO embracing the 2-(4-triethoxysilyl)phenyl)-5-triethoxysilylpyridine precursor was used as a chelating ligand for $[Ru(bpy)_2Cl_2]\cdot 2H_2O$ (where bpy is bipyridine) [22]. Table 1 summarizes the most relevant PMO materials described in the literature since 1999. The introduction of other chemical functionalities into PMOs has also been performed by exploring one-pot synthesis involving co-condensation of bissilylated organic bridged precursors and monossilylated ones containing a functional group or a group that can easily be transformed into the desired species.

Table 1. Summary of some PMOs described in literature

Year	PMO	Application	References[a]
2005	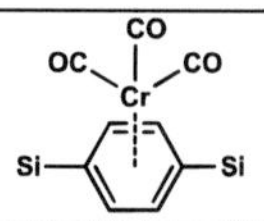	This material is proposed to be applied in catalysis	[15]
2010	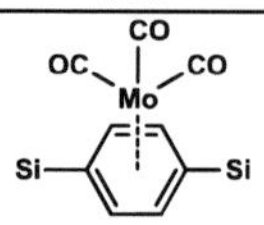	Epoxidation of cis-cyclooctene, 1-octene, trans-2-octene and (R)-(+)-limonene	[16]
2010	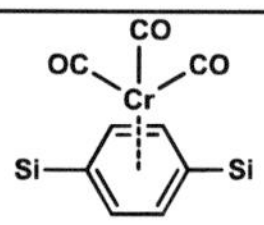	Oxidation of cis-cyclooctene	[17]
2008	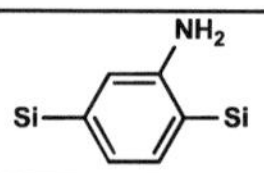	Knöevenagel condensation reaction of malononitrile with benzaldehyde	[18]
2013	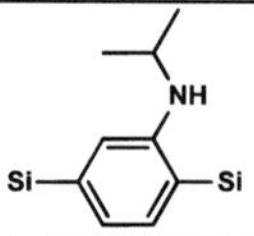	Synthesis method suggested to prepare new PMOs for adsorption and catalytic applications	[19]
2013	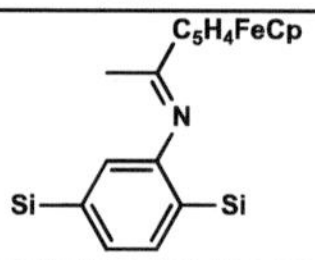	Oxidation of styrene	[20]
2011	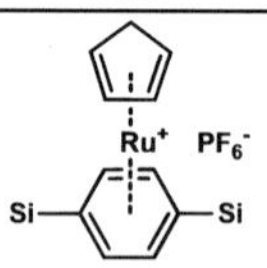	Hydrosilylation of 1-hexyne	[21]
2011	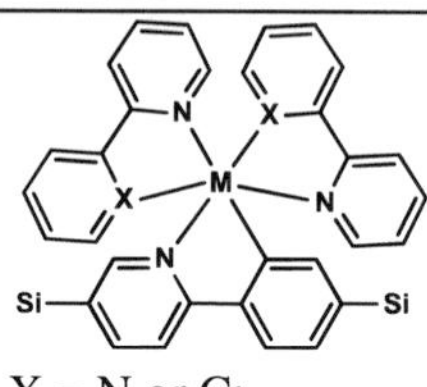 X = N or C; M = Ru or Ir	Photochemical and photophysical systems	[22]

[a]Correspond to the references in the text.

Another possibility can be the post-synthesis grafting of an interesting silylated functional group by reaction with the surface silanols. Figure 2 schematizes the different methods to achieve functional PMO materials.

Several reviews about PMOs have been published in the last fifteen years especially devoted to the description of the synthesis methods [23–27], applications [28–30], making a point in terms of future [31, 32].

In this Chapter, we intend to go into detail in terms of the preparation and applications in catalysis and in adsorption properties of acid and basic functionalized PMO materials, and we will finish by describing the contributions of the computational tools to the understanding and development of PMOs.

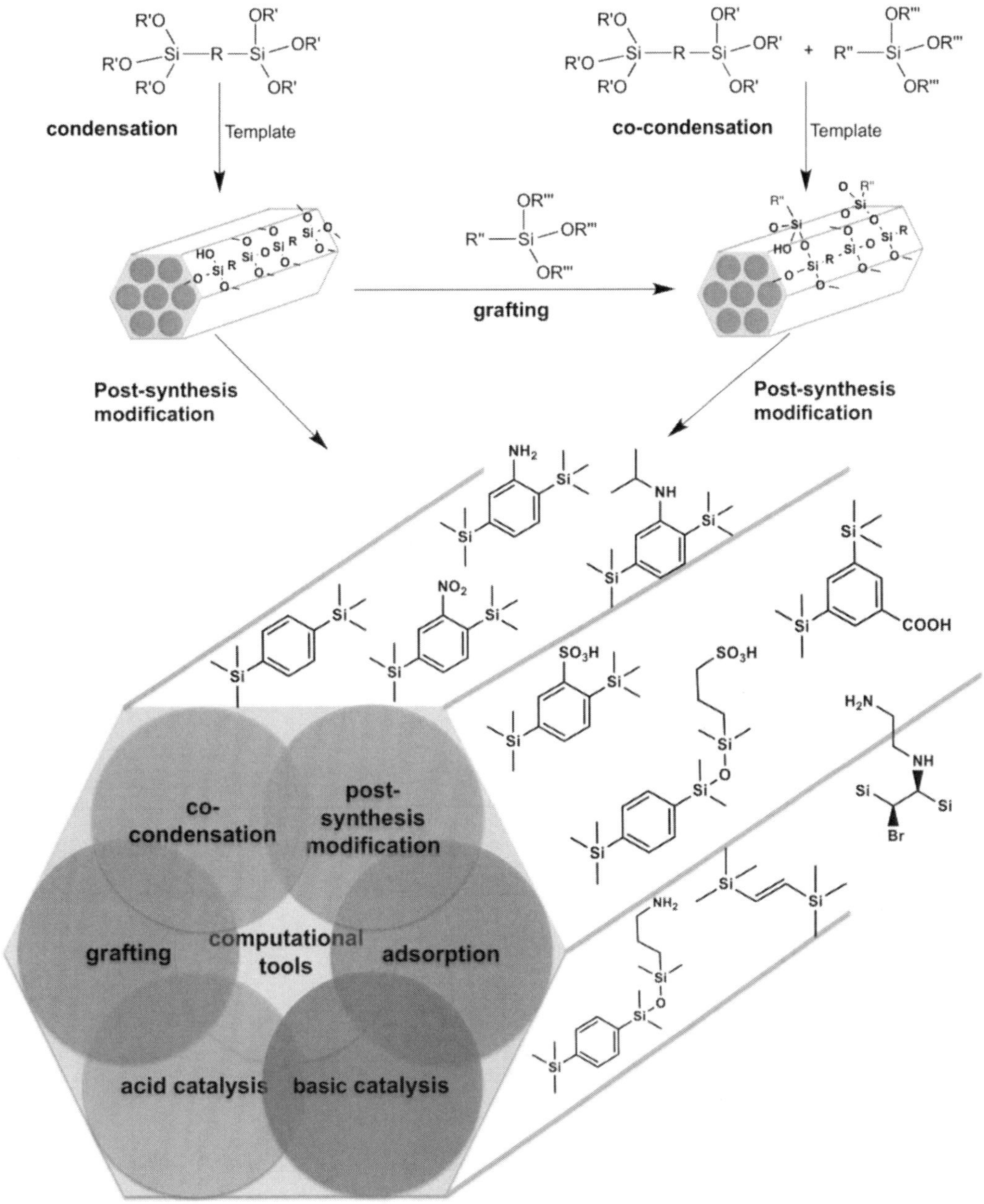

Figure 2. Different pathways for the synthesis of functional PMOs.

Acid PMOs

This section presents a brief review of the most relevant acid-functionalized PMO materials (acid PMOs) and their application in catalysis and adsorption (Table 2).

The introduction of acid groups in the framework of the PMOs was reported for the first time by Inagaki et al. [8] in 2002, when the aromatic ring of the phenyl-PMO (Ph-PMO) was sulfonated using a 25% SO_3/H_2SO_4 solution (SO_3H-Ph-PMO, PMO 1, Table 2).

Table 2. Summary of PMOs described in literature containing acid groups

PMO	Precursor 1	Precursor 2	SDA[a]/pH medium	XRD reflections	References[b]
1[c]	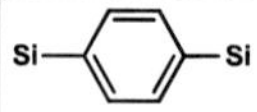 Ph-PMO	SO_3/H_2SO_4	ODTMAC/ basic	(100); (110); (200)	[8]
2	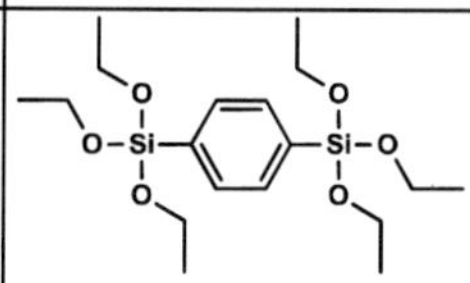 1,4-bis(triethoxysilyl) benzene (BTEB)	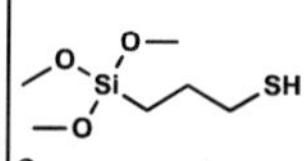 3-mercaptopropyl trimethoxysilane (MPTMS)	ODTMAC/ basic	(100); (110); (200)	[33, 43]
			ODTMAB/ basic		[48]
			P123/ acid		[52]
3	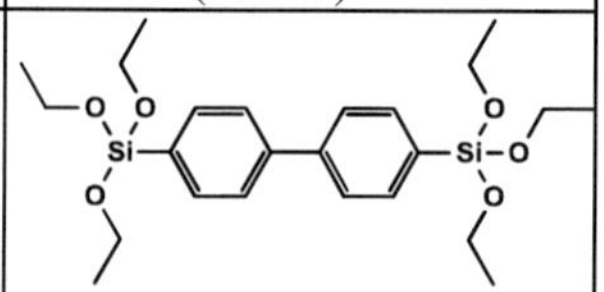 4,4'-bis(triethoxysilyl) biphenyl (BTEBP)		ODTMAC/ basic	(100); (110); (200)	[34]
4	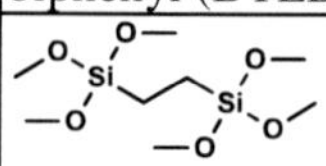 1,2-bis(trimethoxysilyl) ethane (BTME)		ODTMAC/ basic P123/ acid	(100); (110); (200)	[35, 43]
5	or 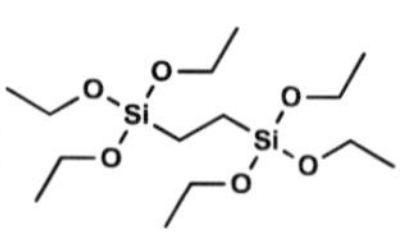 1,2-bis(triethoxysilyl) ethane (BTEE)		ODTMAC/ basic P123/ acid	(100); (110); (200)	[36, 41, 47, 51, 52]
6[c]	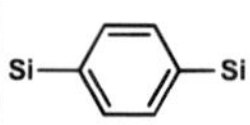 Ph-PMO	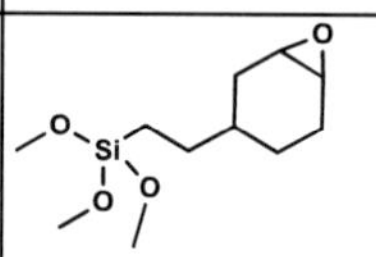 2-(3,4 epoxycyclohexyl) ethyltrimethoxysilane or 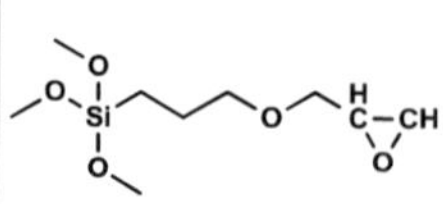 3-glycidoxypropyl trimethoxysilane	ODTMAC/ basic	(100); (110); (200)	[39]
7[c]	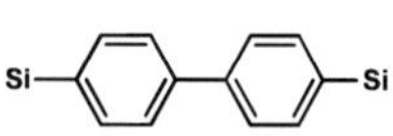 biphenylene-PMO				

PMO	Precursor 1	Precursor 2	SDA[a]/pH medium	XRD reflections	References[b]
8	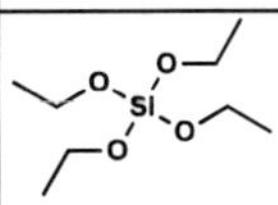 tetraethylorthosilicate (TEOS)	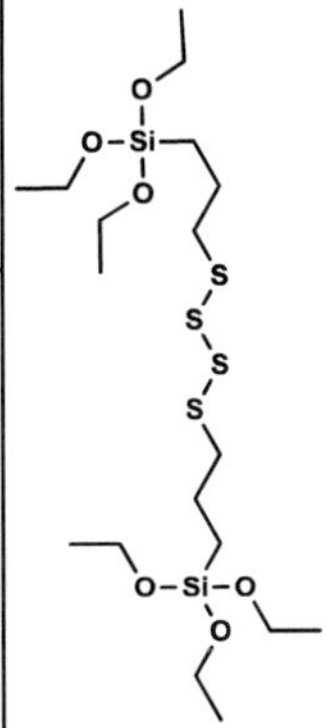 bis[3-triethoxysilyl) propyl]tetrasulfide	Brij-76/ acid	(100)	[40]
9	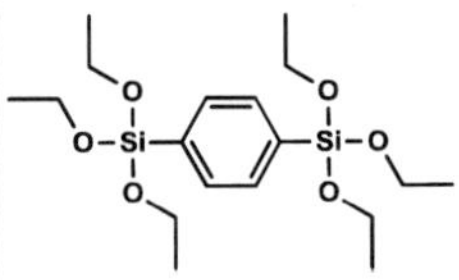 1,4-bis(triethoxysilyl) benzene (BTEB)				
10	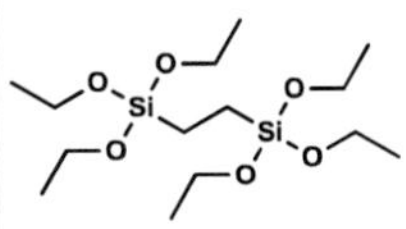 1,2-bis(triethoxysilyl)-ethane (BTEE)	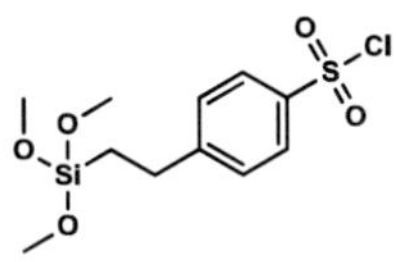 2-(4-chlorosulfonylphenyl) trimethoxysilane (CSPTMS)	P123/ acid	(100); (110); (200)	[41]
11[c]	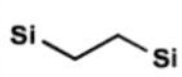 ethane-PMO	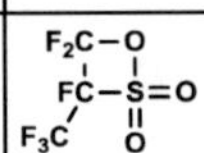 1,2,2-trifluoro-2-hydroxy-1-trifluoromethylethane sulfonic acid β-sultone	P123/ acid	(100); (110); (200)	[42]
12	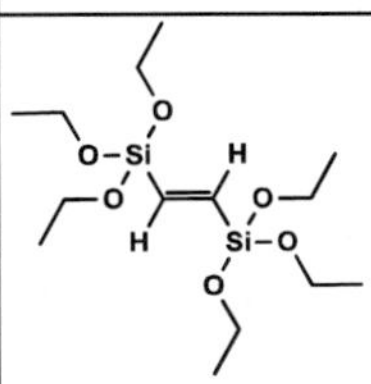 1,2-bis-(triethoxysilyl)-ethene	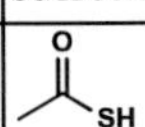 thioacetic acid	P123/ acid	(100); (110); (200)	[49]
13[c]		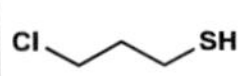 3-chloro-1-propanethiol			[50]
14	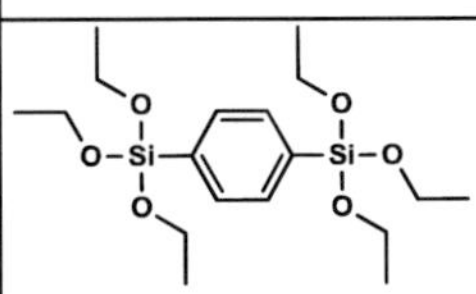 1,4-bis(triethoxysilyl)-benzene (BTEB) + TEOS	 zirconocene dichloride	P123/ acid	(100); (110); (200)	[53]

Table 2. (Continued)

<table>
<tr><th>PMO</th><th>Precursor 1</th><th>Precursor 2</th><th>SDA[a]/pH medium</th><th>XRD reflections</th><th>References[b]</th></tr>
<tr><td>15</td><td>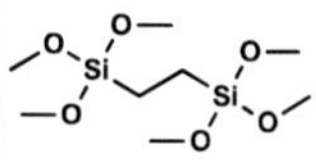
bis(trimethoxysilyl) ethane (BTME)</td><td rowspan="2">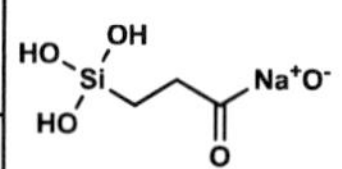
carboxyethylsilanetriol sodium salt (CES)</td><td rowspan="2">P123/ acid</td><td rowspan="2">(100); (110); (200)</td><td>[54]</td></tr>
<tr><td>16</td><td>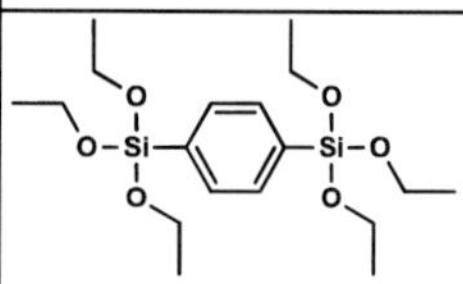
bis(triethoxysilyl) benzene (BTEB)</td><td>[55]</td></tr>
</table>

[a]SDA – structure-directing agent; [b]Ref. corresponds to the references in the text. [c]Materials prepared by post-functionalization. P123 – Pluronic® 123 - triblock copolymer based on poly(ethylene glycol)-poly(propylene glycol)-poly(ethylene glycol) ($EO_{20}PO_{70}EO_{20}$) nonionic surfactant; Brij® 76 – Polyethylene glycol octadecyl ether nonionic surfactant; ODTMAC – octadecyltrimethylammonium chloride; ODTMAB – octadecyltrimethylammonium bromide.

An acid loading of 0.4 mmol/g was achieved. The sulfonic acid group was thermally stable up to 500 °C in air or nitrogen atmosphere. In the same year, Inagaki and co-workers [33] described the synthesis of another acid-PMO (here after designated Ph-PMO-PrSO_3H, PMO 2, Table 1) prepared by co-condensation of 1,4-bis(triethoxysilyl)benzene (BTEB) and 3-mercaptopropyltrimethoxysilane (MPTMS), followed by oxidation of the thiol groups with concentrated nitric acid. In this case, the acid content was up to 0.77 mmol/g while maintaining the structural integrity. Nevertheless, the thermal stability of SO_3H groups in the material was decreased to *ca.* 300 °C. The Ph-PMO-PrSO_3H material represents a breakthrough in the structural viewpoint as it conjugates the molecular-scale periodicity and the presence of flexible sulfonic acid arms on the pores, instead of attached to the phenyl ring. It was verified a decrease of the intensity of d_{100} reflection, typical from the 2D hexagonal arrangement, with the increase of the acid loading. Later, the same group reported the synthesis of other PMO with a biphenylene bridge and Pr-SO_3H acid functionalities (PMO 3, Table 2) [34]. This material exhibits very high hydrophobicity and a maximum acid concentration of 0.99 mmol/g. The application in catalysis of PMOs with sulfonic acid functionalities located either in the walls or in the pores, depending on the organic bridges used, is widely described in the literature especially in water-generating reactions.

In 2005, Inagaki reported the employment of the propyl sulfonic acid ethane bridged PMO (Et-PMO, PMO 4, Table 2), synthesized by the co-condensation of the precursors 1,2-bis(trimethoxysilyl)-ethane (BTME) and MPTMS in the presence of the octadecyltrimethylammonium chloride (C_{18}TMACl) as structure directing agent, in the esterification of the acetic acid with ethanol [35]. Et-PMO with an acid loading of 0.98 mmol/g showed comparable esterification activity to commercial Nafion-H with similar acid loading (0.80 mmol/g). The unique product obtained was the ethyl acetate and the yield depended on the number of acid groups [35]. In another study with an Et-PMO (PMO 5, Table 2) synthesized with P123 as template, it was demonstrated that, the catalytic activity is not only dependent on the acid groups content but also on the hydrophilic/hydrophobic character of the surface [36]. The presence of a small fraction of ethane bridges in the materials contributes to an improvement of the hydrothermal stability and catalytic activity. When the Ph-PMO with Pr-SO_3H (Ph-PMO-PrSO_3H) groups was employed in this reaction, the ethyl acetate yield was significantly enhanced [37]. The authors compared also the performance of materials obtained by grafting and co-condensation, being the grafted PMO the best catalyst. They explained this result on basis of the ion-exchange capacity, the structural integrity and mesoscopic nature of the materials. Although, the SO_3H groups play an important role in the catalytic results, the very good performance is assigned to the unique hydrophilic-hydrophobic surface properties on the crystal-like pore walls [37]. Recently, Ph-PMO sulfonated in the aromatic ring (Table 2, PMO 1) was used in the esterification of acetic acid with ethanol, concluding that the material was more active than Amberlyst-15 when using water as solvent, showing the better tolerance of PMO to water [38]. Kapoor et al. [39] reported, in 2008, the preparation of sulfonic acid functionalized periodic mesoporous silicas (with phenylene or biphenylene organic moieties) via covalent attachment of 2-(3,4 epoxycyclohexyl)ethyltrimethoxysilane or 3-glycidoxypropyltrimethoxysilane followed by reaction with sulfite ions and mild hydrochloric acid (Table 2, PMOs 6 and 7). The PMOs prepared had a significant high SO_3H loading (between 1.18 and 1.36 mmol/g) while preserving the mesoporous order. A reduction of the intensity of the d_{100} indexed to the 2D hexagonal unit cell was observed in comparison to the pristine materials. The acid PMOs prepared using this method showed to be very active in catalytic esterification, acylation, and condensation reactions. For this enhanced catalytic behavior contributes the high number of relatively isolated acid sites as well as the hydrophobicity; the later clearly has an effect on the water local concentration. To evaluate the effect of the proximity between sulfonic acid sites in a Ph-PMO, recently bis[3-triethoxysilyl)propyl]tetrasulfide was co-condensed with tetraethylorthosilicate (TEOS) or BTEB (Table 2, PMOs 8 and 9), leading to the formation of SO_3H sites by the cleavage of the tetrasulfide bridges [40]. These new materials were also applied in the referred esterification reaction and show catalytic activities similar to Amberlyst-15; however, the stability against deactivation and swelling was better in the PMO with phenyl bridges [40].

In 2009, it was described the use of PMOs as acid catalysts in the acetalization of heptanal by 1-butanol [41]. Thus, arene sulfonic acid Et-PMO (Table 2, PMO 10), propylsulfonic acid Et-PMO (Table 2, PMO 4) and arene sulfonic acid SBA-15 were tested as catalysts and their performances were compared with those of HY zeolite and Amberlyst-15. It was verified that the acid mesoporous materials showed the utmost heptanal conversions when compared to HY zeolite and Amberlyst-15. The arene sulfonic acid functionalized materials showed the highest activities probably due to the large pore diameters, and to the high hydrophobicity and strength of the acid sites [41]. This reaction was also catalyzed by perfluoroalkylsulfonic acid Et-PMOs resulting from the grafting of Et-PMO with 1,2,2-trifluoro-2-hydroxy-1-trifluoromethylethane sulfonic acid β-sultone (Table 2, PMO 11) [42].

The use of water as a solvent in organic reactions presents obvious advantages, i.e., it is the cheapest, non-toxic, non-flammable and clean solvent known, increasing the economic practicability of the process. The first example of the application of sulfonated PMOs in catalysis in water was published by Fukuoka et al. [43]. The authors reported the highest tolerance, recyclability, conversion and turn over frequency (TOF) on the hydrolysis of starch and sucrose of acid-functionalized PMOs comparatively to the classic solid catalysts

(Amberlyst-15, Nafion and HZSM-5). Some of us reported, in 2009, the remarkable activity and stability of Ph–PMO–PrSO_3H (Table 2, PMO 2) on the condensation reaction of indole with benzaldehyde in water [44]. Ph–PMO–PrSO_3H in comparison with other solid acid catalysts, showed the best catalytic activity (TOF = 943 h^{-1}) and product yield up to 95%.

Moreover, the activity was preserved even after six catalytic cycles, in contrast with the activity of propylsulfonic acid functionalized SBA-15 and HMS, which dropped significantly after the 1st cycle. With the aim of searching for a deep understanding of the remarkable activity of the Ph-PMO-PrSO_3H materials on the condensation of indole with benzaldehyde, we performed another study using high-resolution solid state NMR techniques where the catalytic performance was related with the different acid loadings of the materials [45]. It was verified that, for proton content above 0.40 mmol/g, the acid loading and the TOF are linearly correlated, which is due to cooperative effects between the SO_3H groups. Concerning the acid strength, it was verified a "deactivation" effect of water that was more prominent for samples with proton loadings higher than 0.40 mmol/g. However, the effect of water on the acid strength was nearly negligible when Ph-PMO-SO_3H exhibited a low proton loading ($\leq$ 0.21 mmol/g) due to the proximity between the SO_3H groups and the hydrophobic benzene ring, as confirmed by the NMR experiments, thus leading to a catalytic activity surprisingly high. Recently, we reported a detailed study relating the role of acidity and hydrophobicity in the mentioned catalytic activity [46]. SBA-15 with propyl sulfonic acid groups and silylated, Ph-PMO-PrSO_3H (Table 2, PMO 2) and silylated on the surface, and a Ph-PMO sulfonated in the aromatic ring (SO_3H-Ph-PMO, Table 2, PMO 1) were compared. According to our observations the interactions between surface silanols and sulfonic acid groups do not affect the catalytic activity in the case of the sulfonated PMOs, while in the case of sulfonic SBA-15 they seem to reduce the activity. Moreover, for the Ph-PMO-PrSO_3H family of materials, the TOF is found to increase with increasing SO_3H loading because of an enhancement of the lateral SO_3H---O_3(H)S interactions that drastically reinforce the acid strength. As such interactions cannot happen in the SO_3H-Ph-PMO catalyst, the activity of this catalyst is lower even for similar acid loading. A low accessibility of the SO_3H active sites to the reactants, as it appeared in the water/toluene dynamic adsorption, could also result in a decrease of the catalytic activity. In 2012, Dumesic et al. [47]. compared the catalytic activity of propylsulfonic acid-functionalized SBA-15, propylsulfonic acid-funcionalized Et-PMOs (Table 2, PMO 7) and non-ordered propylsulfonic acid functionalized silica-gel in the production of 5-hydroxymethylfurfural (HMF) *via* fructose dehydration reaction in aqueous medium. The ordered structure materials were the most active in catalysis. The authors verified that the acid functionalized Et-PMOs were the most stable to the hydrolytic cleavage of the acid site. Recently, some of us reported the use of propylsulfonic acid functionalized Ph-PMOs (Table 2, PMO 2) with different acid content in the fructose dehydration reaction to HMF [48]. It was verified that the Ph-PMO material with relatively low acid content (0.36 mmol/g) was more active than propylsulfonic acid SBA-15 with an acid loading of 0.89 mmol/g, keeping the same activity after two consecutive catalytic cycles [31]. ^{31}P solid-state NMR of adsorbed probe molecules was used to study the acid strength of the different materials, showing that the most active materials displayed a distribution of Brönsted acid sites, being probably the weakest sites the most robust and active in the catalytic reaction in presence of water [48].

Recently, it was reported the synthesis of a 100% thiol containing PMO material through the preparation by aminolysis of the bissilylated 1-thiol-1,2-bis(triethoxysilyl)ethane, followed by condensation of this precursor in the presence of a P123 block co-polymer as template (Table 2, PMO 12) [49]. This PMO is very ordered with an intense low angle (100) peak and a pore diameter of 6.2 nm. So far, no reports on the oxidation of the thiol groups in this material were published.

The same group reported the synthesis of a propylsulfonic acid functionalized PMO by a sequence of modification reactions of trans-ethenylene bridged PMO (Table 2, PMO 13). It is remarkable that the initial mesostructural order of the trans-ethenylene bridged PMO is not affected by the sequential modification of the material. In all steps, the formed materials showed the three typical reflections (100), (110) and (200) of the 2D hexagonal mesoporous structure. The propylsulfonic groups are directly located on the carbon atoms of the organic bridges that become chiral [50].

The material had an activity comparable to that of Amberlyst-15 in the acetylation of glycerol and the activity remained at least up to four catalytic runs. A decrease in the initial reaction rate was verified, however the material reached the equilibrium after 5 h, leading to the same acetylation level as the fresh pristine material.

In 2011, Zhu et al. [51] reported a new bifunctional Au-nanoparticle-acid catalyst resulting from the in situ reduction of Au^{3+} in the propylthiol functionalized Et-PMO (called PMO 5a as it results from the modification of PMO 5). PMO 5a has tiny Au nanoparticles well dispersed in the propyl sulfonic acid Et-PMO. The mesoporous order is retained after the introduction of the Au particles with a positive shift of the (100) peak due to enhancement of the wall thickness. This observation indicates that the Au particles are mainly located onto the pore surface rather than onto the outer surface. This material exhibited high activity and selectivity in water-medium intramolecular hydroamination, styrene oxidation, and three-component coupling reactions. Furthermore, the catalyst could be conveniently recycled and reused at least 10 times.

Karimi et al. [52] reported the synthesis of two novel water tolerant PMOs with phenylene or ethyl as organic bridges, both functionalized with methylpropyl sulfonic acid. These PMOs (Table 2, PMOs 3 and 4) were applied in biodiesel production through the esterification reaction of various fatty acids. As a result, it was possible to produce biodiesel with high yields, confirming that the materials are recyclable and do not lose catalytic activity even after three cycles. Melero et al. [53] reported the synthesis of Lewis acid functionalized Et-PMOs (Table 2, PMO 14) with zirconium metal sites for biodiesel production through the esterification of free fatty acid water containing feedstock. The insertion of zirconium into the PMO framework led to a strong decrease in all the textural properties of the materials (with the exception of microporosity) with a great decrease of the mesopore size. The zirconium loading was dependent on the organic content and directly influenced the catalytic activity. The catalytic activity of the materials was retained up to 20 wt.% of water content (relative to the starting oil), due to the hydrophobic organic bridges of the PMOs.

Apart from sulfonic acid functionalized PMOs, some examples on the preparation of carboxylic acid materials are described in the literature. In 2009, Fiorilli et al. [54] described the one-pot synthesis of carboxylic acid functionalized Et-PMO (Table 2, PMO 15). Later in 2011, PMOs with phenylene organic bridges and acid contents up to 3.08 mmol/g were reported (Table 2, PMO 16) [55]. Carboxylic acid functionalized PMOs seem very promising but their application still needs to be explored [56].

Although the acid-functionalized PMOs are mainly applied in catalysis it is possible to find in literature some applications in adsorption. An Et-PMO functionalized with different loadings of carboxylic acid groups, through co-condensation (Table 2, PMO 15), was applied in the adsorption of an anticancer drug called doxorubicin, presenting very good adsorption efficiency, which was proportional to the acid content [57].

Propylsulfonic acid Ph-PMOs (Table 2, PMO 2) were applied in the adsorption of a pesticide, mesosulfuron, in water, showing a maximum removal efficiency of 95% [58].

Table 3 summarizes the applications of all the acid-PMO materials described.

Table 3. Catalytic applications of acid-functionalized PMOs

PMO	Year	Applications		References [a]
		Catalysis	**Adsorption**	
1	2011	Esterification of acetic acid with ethanol	-	[38]
	2013	Condensation of indole with benzaldehyde in aqueous medium	-	[46]
2	2005	Esterification of acetic acid with ethanol	-	[37]
	2009	Condensation of indole with benzaldehyde	-	[44]
	2012			[45]
	2013			[46]
	2014	Fructose dehydration to HMF	-	[48]
	2014	-	Adsorption of mesosulforon in aqueous medium	[58]
3	2012	Biodiesel production by transesterification of fatty acids	-	[52]
4	2005	Esterification of acetic acid with ethanol	-	[35]
	2012	Biodiesel production by transesterification of fatty acids	-	[52]
5	2005	Esterification of acetic acid with ethanol	-	[36]
5a	2011	Water medium intramolecular hydromination	-	[51]
		Styrene oxidation	-	
		Three component coupling reactions	-	
6	2008	Esterification of acetic acid with ethanol	-	[39]
		Acylation of acetic acid with benzyl alcohol	-	
		Condensation of phenol with acetone	-	
7	2008	Esterification of acetic acid with ethanol	-	[39]
		Acylation of acetic acid with benzyl alcohol	-	
	2012	Fructose dehydration to HMF	-	[47]
8	2013	Esterification of acetic acid with ethanol	-	[40]
9				
10	2008	Acetalization of heptanal by 1-butanol	-	[41]
11	2009	Acetalization of heptanal by 1-butanol	-	[42]
		Liquid phase self-condensation of heptanal	-	
12	2013	Not applied but $[SO_3H]$ = 6.7 mmol/g.		[49]
13	2013	Acetylation of glycerol	-	[50]
14	2013	Biodiesel production from highly acidic water-containing feedstock	-	[53]
15	2012	-	Doxorubicin	[57]
16	2011	Not applied but promising for various fields.		[56]

BASIC PMOS

Basic groups functionalized PMOs, mainly PMOs functionalized with amine groups (hereby designated amine-PMOs), are another class of advanced PMO materials that have been proposed for applications in adsorption and catalysis. The introduction of the amine group in the PMOs framework is made across the co-condensation reaction, where one of the precursors contains the amino group (Table 4), or by post-functionalization of the pristine material (Table 5). Studies dealing with these two different ways of producing amine-PMOs will be separately reviewed below.

Co-Condensation

The first attempt to obtain PMOs with amine groups was made in 2001 by Burleigh and co-workers [59] aiming at achieving a material useful in ion metal adsorption. A co-condensation reaction of 1,2-bis(triethoxysilyl)ethane (BTEE) and [3-(2-aminoethylamino) propyl] trimethoxysilane (EDATMS) in the presence of CTAC structure agent under basic conditions gave rise to PMO 17 that exhibits single (100) reflection (Table 4). The authors also added metal salts of nickel, copper or zinc during the synthesis which complexed with the available amines. The process used is known as molecular imprinting. The surfactant and metal ions removal leads to the formation of a material with cavities displaying high selectivity and strong affinity for the metal ion template, which is revealed during the adsorption of the Cu(II), Ni(II) and Zn(II) ions.

In 2006, Wahab et al. [60] reported the formation of well-defined 2D hexagonal order PMOs fibers when BTEE was mixed with 1-[3-(trimethoxysilyl)propyl] urea (UREDO) or (3-triethoxysilylpropyl) isocyanate (ICP) in the presence of CTAB/Brij30 surfactant mixture under basic conditions (Table 4, PMOs 18 and 19). They showed the influence of the amount of the nitrogen containing precursor in the formation of the PMOs meso-scale order. Disordered materials were formed with molar percentage in nitrogen precursors higher than 25 mol%. This was noticed by the disappearance of the typical (110) and (200) reflections of the 2D hexagonal structure on the XRD patterns, being only observed a broad d_{100} reflection for both PMOs 18 and 19 (Table 4). The basic nature of UREDO or ICP precursors led to an increase of the pH during the synthesis, which hindered the organization of the micelles in a 2D-hexagonal arrangement. The influence of the reaction temperature on the formation and the internal pore architecture was also observed. The change of the reaction temperature to 75 °C originated pore channels aligned parallel to fiber axis with hexagonal arrangement.

The PMOs formed with TEOS and 1,1'-(butane-1,4-diyl)bis(3-(3-(trimethoxysilyl) propyl)urea) or with bis[3-(trimethoxysilyl)propyl]amine (BTMS-PA) in the presence of CTAB surfactant under basic conditions possess a well-defined hexagonal order (Table 4, PMOs 20 and 21, respectively) [61, 62].

Table 4. Summary of PMOs prepared by co-condensation using silylated precursors containing nitrogen

PMO	Precursor 1	Precursor 2	SDA[a] / pH medium	XRD reflections	References [b]
17	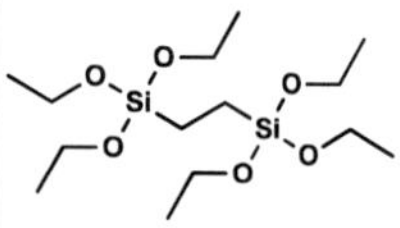 1,2-bis(triethoxysilyl) ethane (BTEE)	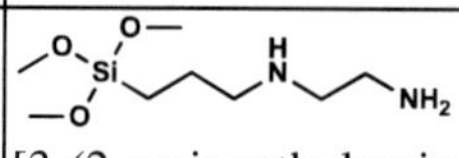 [3-(2-aminoethylamino)propyl] trimethoxysilane (EDATMS)	CTAC/ basic	(100)	[59]
18		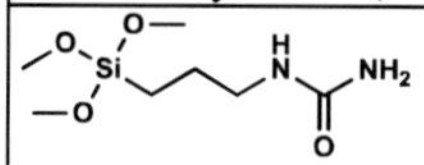 1-[3-(trimethoxysilyl)propyl] urea (UREDO)	CTAB/ Brij30/ basic	(100)	[60]
19		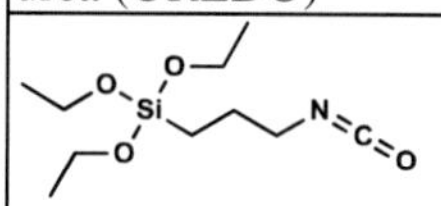 (3-triethoxysilylpropyl) isocyanate (ICP)			
20	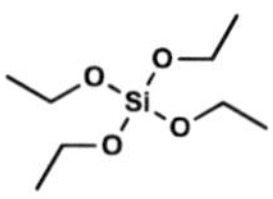 tetraethoxysilane (TEOS)	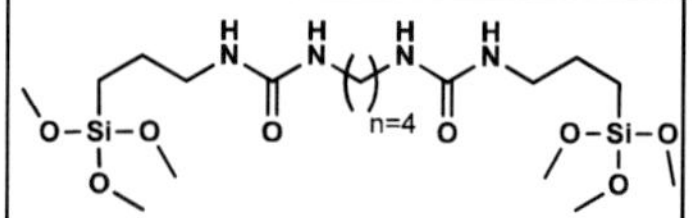 1,1'-(butane-1,4-diyl)bis(3-(3-(trimethoxysilyl)propyl)urea) (Urea-derived organosilane)	CTAB/ basic	(100); (110); (200); (210)	[61]
21		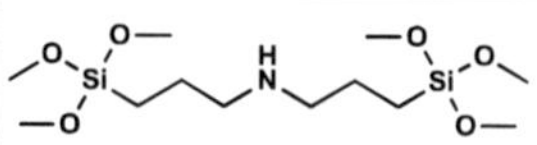 bis[3-trimethoxysilyl)propyl] amine (BTMS PA)		(100); (110); (200)	[62]
22			P123 or F127 / acid	disordered	[63]
23	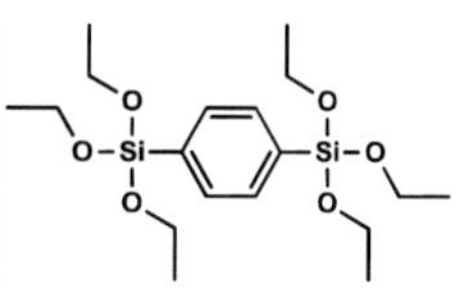 1,4-bis(triethoxysilyl) benzene (BTEB)	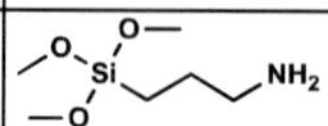 (3-aminopropyl) trimethoxysilane (APTMS)	ODTMAC/ basic	(100); (110); (200)	[64]
24		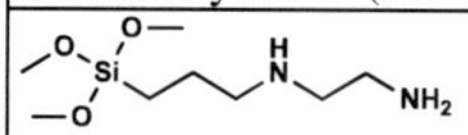 (EDATMS)			
25		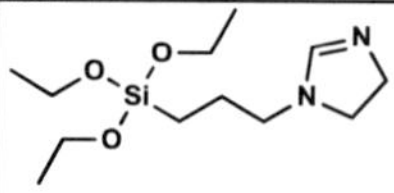 N-(3-triethoxysilylpropyl) 4,5-dihydroimidazole (ImTES)		(100) / fibers	
26	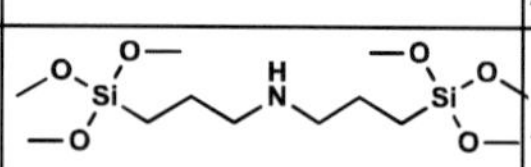 (BTMS-PA)	-	P123/ acid	disordered	[65] [66] [67]

[a]SDA – structure-directing agent; [b]Reference corresponds to the references in the text. CTAC - cetyltrimethylammonium chloride; CTAB - cetyltrimethylammonium bromide; P123 and F127 – Pluronic® block co-polymers; ODTMAC – octadecyltrimethylammonium chloride.

Benitez et al. [61] demonstrated the high metal binding capacity of PMO 20 (Table 4) for the Fe(III) adsorption when comparing with other silica materials, with no blockage of the pores.

PMO 21 (Table 4) displayed uniform nanospheres morphology as a result of the integration of amine groups in the framework walls, and it was successfully applied as heterogeneous organocatalyst in water-medium Knoevenagel condensation reactions and solvent-free Henry reactions [62]. The authors showed the easy recyclability of this material that could be used several times without significant decrease in activity. PMO materials prepared with TEOS and BTMS-PA could also be synthetized using different surfactants such as P123 and F127 under acid conditions (Table 4, PMO 22) [63]. The enhancement of the amount of BTMS-PA led to disordered materials with different pore sizes. The best PMO PA material was used to immobilize chloroperoxidasc (CPO) enzyme, which could be reused 20 times with little loss in activity.

Coutinho et al. [64] reported the synthesis of Ph-PMOs containing propylamine, propylethylenediamine and propyldihydroimidazole groups (Table 4, PMOs 23, 24 and 25, respectively). The synthesis of these three amine-PMOs materials was remarkable since they display both meso- and molecular-scale up to 10 mol% of functional groups. When the amine functional precursor loading increases above that value, the meso-order is lost while the molecular periodicity is maintained.

Ha and co-workers [65] synthetized a PMO material by condensation of BTMS PA using P123 structure agent under acid conditions (Table 4, PMO 26). This material has no pore order, but possesses hollow spherical morphology. The PMO 26 displayed hydrophilic silicate layers and it was tested, together with Ph-PMO and biphenyl-PMO materials, in the adsorption of lysozyme. Ha and co-workers observed that the amine moieties of the former material do not interact with the lysozyme. They suggested that this enzyme prefers to interact with materials having high hydrophobic character such as Ph-PMO and biphenyl-PMO. Another reason is related with the silanols loading observed in each material. The PMO 26 only possesses 30% free silanols groups against about 70% of silanols observed in the other two PMOs. In this way the authors found that it was necessary to take into account the balance between the hydrophilic and hydrophobic character of the materials, which leads to differences in the adsorption. The same authors [66] also reported the use of these three materials in the adsorption of glycine, L-lysine and isoleucine. The importance of surface properties of the materials was observed in this report. L-lysine, a strong hydrophilic amino acid, was mainly adsorbed by PMO 26 while isoleucine, a strong hydrophobic amino acid, was strongly adsorbed by biphenyl-PMO. Although, glycine has epicene properties (i.e., it works in hydrophilic and hydrophobic environments), it was preferentially adsorbed by hydrophobic Ph- and biphenyl-PMOs than by PMO 26. The pore size diameter of the materials also had significant contribution in the adsorption of the amino acids presented above. The authors showed the importance of electrostatic interactions between PMOs and the adsorbed amino acids. Furthermore, they observed that the maximum amount of amino acid adsorbed occurs in the isoelectric point of the amino acid. More recently, Ha and co-workers [67] explored the capability of these three PMOs to adsorb nicotine from aqueous solution. The nicotine exhibits hydrophobic character being strongly adsorbed by Ph-PMO. The authors showed again the strong dependence of isoelectric point and hydrophobicity of the PMOs materials in the adsorption of nicotine.

Post-Functionalization

As discussed earlier in this chapter, the use of amine precursors during the PMO synthesis by co-condensation leads to changes in reaction pH, which hinders the formation of well-organized pore structures. The introduction of amine functional groups into the pores of PMOs materials by post-functionalization reaction is extremely interesting due to the opportunity of introducing new functionalities in a very organized material. The introduction of amine functional groups in the PMO can be done at the free silanols groups of the material using an amine silylated precursor or can be made by direct modification of the organic moieties with amine sources. Different amine-functionalized PMOs prepared by direct modification of the organic moieties in the parent PMOs are summarized in Table 5. PMO 27 was synthetized by bromine addition and successive ethylenediamine (EDA) substitution of the ethylene bridge in the ethylene-PMO, while PMO 28 (Table 5) was prepared by direct substitution of EDA in the chloro-propyl-ethane-PMO [68]. Both PMOs were prepared by Nakai et al. [68] and exhibit disordered pores. The authors showed that the pristine ethene-PMO exhibited 2D hexagonal pore order. The bromination reaction had very small influence on the 2D hexagonal mesostructure. In contrast, the diffraction peaks were diffuse after the reaction with EDA, implying an extensive modification of the mesostructure, even though the conversion due to substitution is much smaller than that due to bromination. In the case of PMO 28 (Table 5), only a broad low angle diffraction peak is visible in the XRD patterns and the substitution of chloride by EDA did not affect the structure. These two PMO materials were tested in the adsorption of arsenate. The authors observed that the distance between the amine functional group and the surfaces of the pore walls had influence in the adsorption of the arsenate. PMO 28 displayed higher capacity in the adsorption of arsenate when compared with PMO 27 due to the sufficient distance between the propyl and the silanol groups, which prevented that these groups interacted with each other. In this way, the authors suggest that the chemical nature of the propyl group is independent of the pore wall surface, which contrasts with the behavior of PMO 27.

The high density of amine groups in the PMO 28 (2.8 $mmol\cdot g^{-1}$), can be also responsible for the high capacity in the adsorption of the arsenate when compared to the amine density of only 0.87 $mmol\cdot g^{-1}$ in the PMO 27.

As described before in this Chapter, the benzene bridge of a PMO can be aminated by electrophilic aromatic substitution (Table 5, PMO 29) [18] The 2D hexagonal symmetry of the pores corresponding to the meso-order of the material in the benzene PMO suffers some destruction during the amination reaction, being only observed the single d_{100} reflection in the XRD pattern, but the molecular periodicity is unchanged during the functionalization procedure. The NH_2-Ph-PMO was successfully applied in the Knoevenagel condensation reaction of malononitrile with benzaldehyde. The possibility to reutilize this material for further use was also revealed.

The modification of the amine group in the NH_2-PMO into secondary amines was recently explored by our group [19]. The *N*-alkylation of the amine-phenylene moieties with 2-bromopropane was accomplished using microwave heating in the presence of potassium iodide (Table 5, PMO 30). The (100) reflection in the XRD patterns still almost unchanged and the molecular-scale periodicity is maintained in the prepared material. The microwave heating revealed to be a rapid way to functionalize aromatic amines in a PMO without destroying its meso- and molecular-scale periodicities.

Table 5. Amine post-functionalization of PMOs

PMO	Base PMO	Intermediate PMO	Functional PMO	XRD reflections	References[a]
27	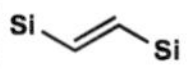 ethene-PMO	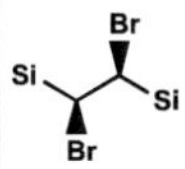 trans-bromination	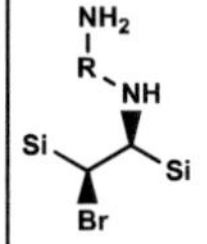 R = $(CH_2)_2$	disordered	[68]
28	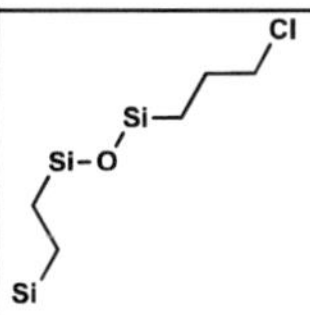 Cl-Pr-Ethane-PMO	-	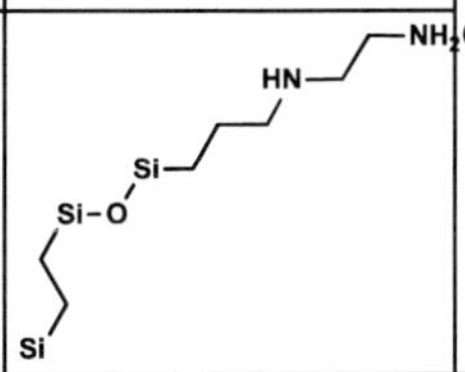		
29	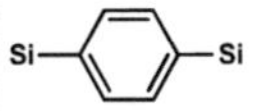 Ph-PMO	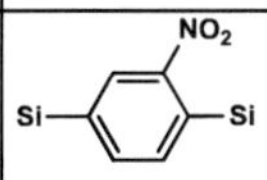 nitration	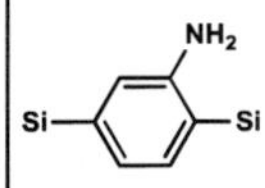	(100)	[18]
30	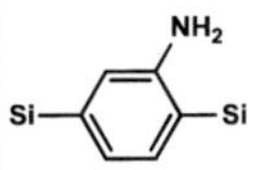 NH_2-Ph-PMO	-	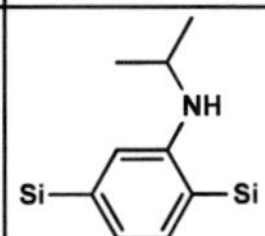	(100)	[19]
31		-	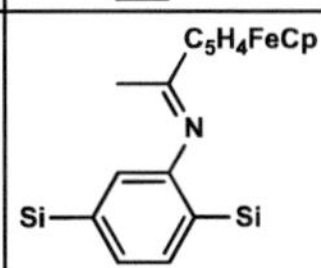	(100)	[20]
32	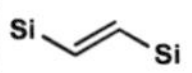 ethene-PMO	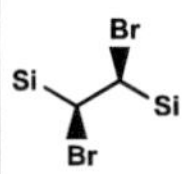 trans-bromination	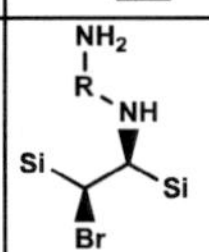 R = $(CH_2)_4$	(100); (110); (200)	[69]
33			R = $(CH_2)_6$		
34			R = $(CH_2)_{12}$		
35			R = $(CH_2)_2NH(CH_2)_2$	(100)	
36			R = $(CH_2)_2NH(CH_2)_2NH(CH_2)_2NH(CH_2)_2$		

[a]Reference corresponds to the references in the text.

This functionalization could be extended to the preparation of new materials tuned with different amino-alkyl groups for a desired application.

Recently, the amine groups attached to the phenylene rings of PMO were used to anchor acetylferrocene (Table 5, PMO 31) [20]. About 15% of the amine groups of the pristine material were used to achieve the incorporation of 0.21 $mmol \cdot g^{-1}$ of iron in the material. The final material was applied in the oxidation of styrene in the presence of hydroperoxides (oxidant species), leading to 25-27% yield in benzaldehyde. A remarkable leaching was verified due to water sensitivity of the azomethine linkage.

De Canck et al. [69] reported the bromination of ethylene bridges followed by nucleophilic substitution of the bromide by diaminobutane (DAB), diaminohexane (DAH), diaminododecane (DADD), diethyltriamine (DETA) and tetraethylenepentamine (TEPA), aiming the preparation of materials (Table 5, PMOs 32 - 36, respectively) to be used as CO_2 adsorbents. By varying the reaction conditions, e.g. temperature and reaction time, different nitrogen content was achieved for each functionalized PMO. The optimal conditions to attain maximum nitrogen content were room temperature and three hours of reaction time. DAB substitution reaction resulted in higher nitrogen content than with DADD. The difficulties obtained in the substitution of bromide by DADD were suggested to be due to the poor diffusion of the dodecane chain within the pores of the PMO. Moreover, the nitrogen content of the material is lower for nucleophiles with long alkyl chains than for the other cases considered. All amine materials presented a good 2D hexagonal mesoporous order, with exception of the PMO prepared with the TEPA ligand, which exhibited just a single reflection in the XRD patterns assigned to the (100) of the 2D hexagonal arrangement. The materials also revealed a Type IV isotherm curve typically observed in mesoporous materials.

The authors observed a decrease on the S_{BET}, V_p and d_p values upon functionalization, which indicates the anchorage of the functional groups inside the pores. The PMOs 32 - 36 (Table 5) were tested for CO_2 adsorption in order to understand which of the following properties, nitrogen loading, variation in chain length and the presence of primary and secondary amines, are the most important to obtain a good adsorbent. When compared with the aminopropyl functionalized SBA-15, De Canck et al. [69] found that PMO 34 (Table 3) presented slightly higher adsorption efficiency with a CO_2/N ratio of 0.42. PMO 34, with the lowest nitrogen loading (0.33 mmol/g), the longest alkyl chain, and the smallest amount of secondary amines, seems to be the most interesting PMO for CO_2 adsorption. It was suggested that the reason for the better performance of this material is related with the presence of amines, sufficiently close to each other so that they could interact simultaneously with the same CO_2 molecule. Moreover, it was proposed that the hydrophobic interactions between the dodecane chains could also play an important role. If large amounts of amines exist in the material, hydrogen bonding between the amine groups is more probable to occur, which deactivates the amine moieties for CO_2 adsorption.

Additionally, hydrophobic interactions between long alkyl chains were suggested to avoid chain folding, preventing the formation of hydrogen bonding between amines and silanols groups; hence, with long alkyl chains, amines are free to interact with CO_2.

In summary, the amine-PMOs can be successfully used and reused in adsorption and catalysis applications. Table 6 summarizes the potential and tested applications of the nitrogen based PMOs discussed in this Chapter.

The development of amine-PMO materials with optimal features for different applications needs to take into account several different factors, such as, nitrogen loading, surface area, pore volume and pore size, distances between amines and between amines and silanols, and the hydrophilic/hydrophobic character of the framework.

The introduction of the amine groups into the PMO frameworks can be made using different methods. The post-functionalization procedures have drawn attention owing to the functionalization of the PMO organic moieties.

The possibility to extend typical homogeneous phase reactions to functionalize hybrid materials has been shown and opens an avenue to tune the materials properties for new applications.

Table 6. Potential applications for nitrogen containing PMOs

PMO	Year	Applications		References[a]
		Adsorption	Catalysis	
17	2001	Cu(II), Ni(II); Zn(II)	-	[59]
18 19	2006	Suggested for transport and deliver of chemical or biological species; catalysis; chemosensing and chromogenic discrimination.		[60]
20	2006	Fe (III)	-	[61]
21	2012	-	Water-medium Knoevenagel condensation reactions; solvent-free Henry reactions	[62]
22	2007	Immobilization chloroperoxidase (CPO) enzyme	-	[63]
23 24 25	2008	Interesting for both catalytic and adsorption applications.		[64]
26	2009	Lysozyme	-	[65]
	2010	Glycine; L-lysine; Isoleucine		[66]
	2011	Nicotine from aqueous solution		[67]
27 28	2007	Arsenate	-	[68]
29[b]	2008	-	Knoevenagel condensation reaction of malononitrile with benzaldehyde	[18]
30	2013	Synthetic method that can be extended to the preparation of new materials tuned with different amino-alkyl groups for adsorption and/or catalysis		[19]
31	2013	-	Oxidation of styrene	[20]
32 33 34 35 36	2013	CO_2	-	[69]

[a]Reference corresponds to the references in the text. [b]It is also suggested as a precursor material for the design of highly functionalized PMOs with strong shape-selectivity for catalytic and adsorption processes.

ACID-BASIC PMOS

Some recent publications describe the preparation of PMO containing simultaneously acid and basic functions for catalytic applications. The seminal work reporting the synthesis of mesoporous organosilica structures with opposite functional properties was carried out by Alauzun et al. [70] in 2006 and it was highlighted by Jaroniec [71]. Two precursors containing a disulfide function and a protected aminopropyl group were added to the tetraethoxysilane for the self-assembly reaction. The disulfide bridge was transformed into thiol groups and oxidized to sulfonic acid functions. The final step was the de-protection of the aminopropyl group. However only the accessibility of the basic site was checked in the reaction with acrylamide. It was verified that in aprotic solvents both basic and acid groups could coexist, while in protic solvents, neutralization of the amino groups rapidly occurred. The first work describing sequential acid and basic reactions was made by the group of Thiel [72], in 2010. These authors introduced the sulfonic acid group in the hydrophobic part (by direct sulfonation of the aromatic ring) and the amine group NH_2 protruding the channels, through co-condensation reaction of the BTEB with 3-aminopropyltrimethoxysilane. X-ray diffraction patterns showed the characteristic reflections at low diffraction angles assigned to the mesoporous structure of the material and the medium range peaks related to the molecular-scale periodicity, proving the structural integrity of the materials after functionalization. The accessibility of the amino and of the acid groups was evaluated on the catalytic conversion of benzaldehyde dimethyl acetal into 2-nitrovinyl benzene. This reaction involves a first acid-catalyzed step, which is followed by a base-catalyzed step. A quantitative yield of 2-nitrovinyl benzene was achieved after 20 h, which confirmed the presence of well-isolated acid and base functionalities in the PMO, i.e., catalytic sites were not destroyed through neutralization reaction [72]. Sasidharan et al. [73] willing to investigate which would be the best locations for both acid and base functionalities, prepared two bifunctional catalysts by the combination of co-condensation and post-synthesis modification (Figure 3). In one of the catalysts (Figure 3a), 1,2-bis(triethoxysilyl)ethene precursor was co-condensed with 3-aminopropyl triethoxysilane using Pluronic P123 as template, followed by the addition of the sulfonic group in the organic bridge. In the second catalyst (Figure 3b), 1,2-bis(triethoxysilyl)ethene was co-condensed with MPTMS, the $–NH_2$ group was grafted across –C=C– through epoxidation followed by amination reaction. The two materials were tested in a sequential acid-base one-pot reaction comprising acid-catalyzed deacetalization of dimethylacetal to yield the respective aldehyde followed by base catalyzed nitroaldol (Henry) reaction. The catalytic results showed that the amine groups located in the hydrophobic environment exhibited lower activity due to decreased basicity when compared with that of the hydrophilic SiO_2 network. Both catalysts were recycled four times successively without loss of activity.

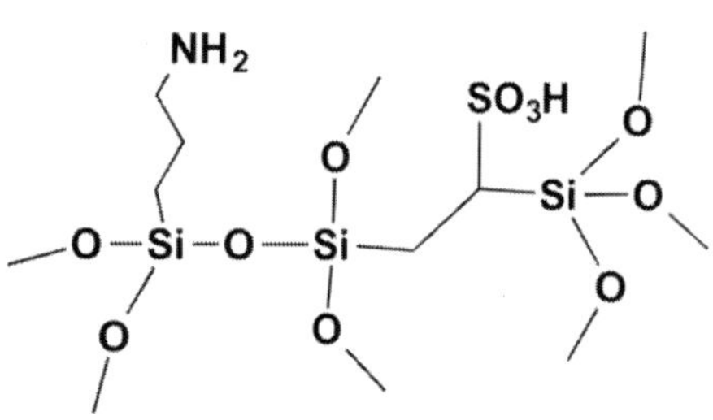

a

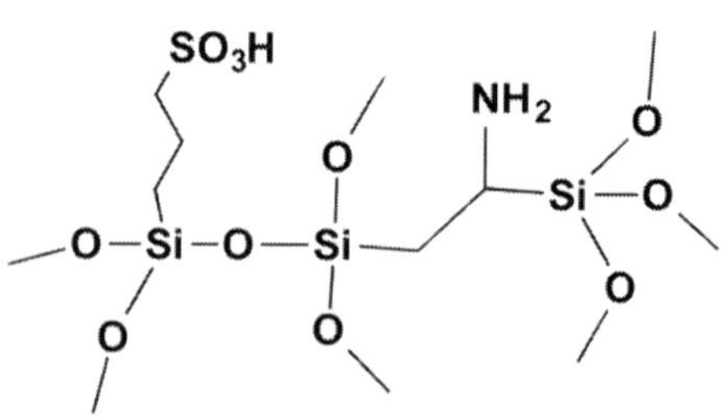

b

Figure 3. Bifunctional catalysts prepared by Sasidharan et al. [73].

THEORETICAL STUDIES OF PMOS

Modern computational approaches were used in the past few years to aid the interpretation of experimental data concerning the structure of PMOs and to improve the understanding of their adsorption and catalytic properties. Most studies rely on approaches based on the density functional theory (DFT) together with small, finite size, molecular models [74–79] but, more recently, studies based on classical molecular dynamics (MD) [80] or on models repeated periodically in three-dimensions [80–82] appeared in the literature. These computational studies are reviewed below.

Models and Methods

The extended structure of the PMOs was modeled either by consideration of a small portion of the material in vacuum, the so-called cluster model approach, or by repetition of a unit cell in the three spatial dimensions. In practice, both approaches have advantages and limitations. Obviously, the former can only be used to study properties that have a quite localized nature, unless other more sophisticated strategies are used to embed the cluster for mimicking the effects of the extended structure of the material in the effective cluster. The use of one or the other approach usually involves a step of validation where calculated properties are compared with benchmark data.

For more extended details, please refer to recent reviews on the application of computational approaches to study other classes of materials [83, 84].

Computational methods based on the density functional theory have emerged as the tools of choice for studying materials, which includes the PMO materials. This is due to the fact that DFT approaches are very competitive when compared with wave function methods in terms of the ratio between the quality of the calculated data and the computational resources required to perform the calculations. Note that the most recent DFT methods are able to tackle systems dominated by dispersion interactions affecting weakly bonded systems, which was very probably the most important drawback associated with the computational approaches based on DFT [85]. However, DFT or wave function methods cannot be used to model systems composed by many particles since the computational cost is prohibitively high. Thus, the way out is usually the reduction of the size of the system (cluster model or small unit cell repeated periodically) or the reduction of the complexity of the calculations by using additional approximations as in semi-empirical or classical approaches [86].

Spectroscopy

The surface properties of a PMO material having both meso- and molecular-scale periodicity based on the 1,4-bis(triethoxysilyl)benzene precursor (Figure 1) were characterized by Onida et al. [74] from a combined Fourier transform infrared (FTIR) spectroscopy and DFT study performed for the bare material and also for the material interacting with several probe molecules (e.g. CO, H_2CO, NH_3, etc.). Comparison of vibrational shifts was made against values calculated for systems based on amorphous silica.

The calculations were used to aid the assignment of the experimentally observed IR modes and were performed at the B3LYP/6-31+G* level of theory, without BSSE or dispersion corrections, using the molecular models represented in Figure 4 for PMO and for amorphous silica. Onida et al. found that O-H and Si-O stretching modes in Ph-PMO appeared at lower frequencies than those corresponding to the SiOH groups in amorphous silica, and concluded that hydroxyl species in the PMO material consist of non-interacting isolated silanols [74]. Additionally, since the shifts in the O-H frequency of SiOH groups interacting with probe molecules were smaller in Ph-PMO than in silica, they concluded that the silanols in the former material are less acidic than those in the latter. The lower acidity in Ph-PMO was found to have a small effect (less than 2 kJ/mol) in the calculated energies of interaction for the probe molecules directly interacting with the SiOH groups of the models in Figure 4. Nevertheless, from adsorption isotherms obtained experimentally, they found for acetone that the interaction with Ph-PMO was stronger than with silica, which was interpreted in terms of secondary interaction between the slightly acidic methyl groups of acetone and the electronic cloud of the aromatic benzene-like rings. These observations lead the authors to suggest that both the organic and inorganic components played a role in dictating the surface behavior [74]. Unfortunately, the calculations done for CO, H_2CO and NH_3 probe molecules were not extended to acetone to test these assumptions, probably due to the limited size of the models considered and to the lack of dispersion in the B3LYP method.

DFT calculations, also with the B3LYP hybrid approach, were used by Hoffman et al. [75] to support the assignment of the most prominent bands in the Raman and IR spectra of four different PMOs based on the 1,4-bis(triethoxysilyl)benzene precursor as in reference [74] but also on the 1,2-bis(triethoxysilyl)ethane, 2,5-bis(triethoxysilyl)thiophene, and 4,4'-bis (triethoxysilyl)biphenyl precursors (Figure 1). Their aim was the unraveling of the vibrational spectroscopic features of PMO materials and their respective precursors with emphasis on the Si-C stretching modes. The calculations performed by Hoffman et al. [75] considered molecular models based on the condensation of 2 (benzene, biphenyl, and thiophene), 3 (benzene and thiophene) or 4 (benzene and ethane) monomers of the precursors shown in Figure 1. The optimized configurations of the condensed models show several hydrogen bonds between the terminal OH groups.

In the case of the precursor with a thiophene-like ring, the models were found to be accurate for calculation of Raman peaks but not for IR peaks, since some vibrational modes expected to be active were not found in the calculated spectra, probably due to a low degree of condensation of the model. Based on the calculations, many bands between 500 cm^{-1} and 1200 cm^{-1} assigned by some authors as Si-C modes were suggested by Hoffman et al. to be more likely due to various Si-O-H in-plane and out-of-plane bending modes and C-H in-plane deformation modes. In fact, the latter authors went a step further and concluded that a typical Si-C stretching mode does not exist in these materials due to the coupling with other modes and, therefore, Si-C stretching is not a characteristic group frequency [75].

In fact, the calculations performed for the benzene tetramer model, the model providing best agreement with experimental data, showed some modes resembling pseudo Si-C stretching vibrations, which involve the movement of the terminal $Si(OCH_2CH_3)_3$ motifs, but the latter isolated groups are unlikely in PMO materials. In summary, they found that from precursors to corresponding PMO materials, wavenumber shifts and intensity changes occur and, therefore, modes are almost unpredictable and each material is a single case since no analogy with different organic bridges was found.

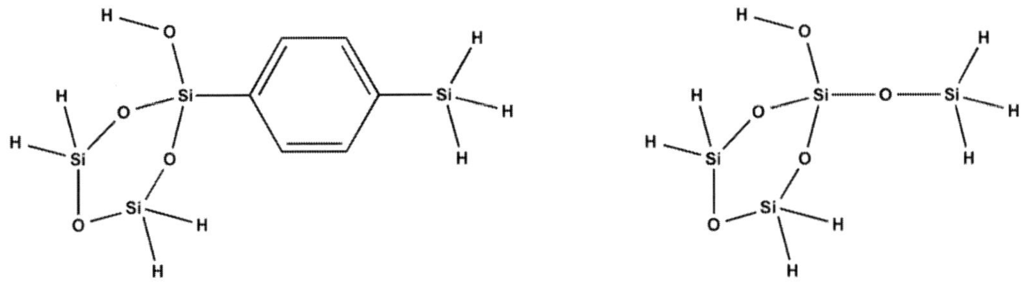

a b

Figure 4. Molecular structures of cluster models for PMO (a) and for amorphous silica (b).

Thus, vibrational spectroscopy was found to be an untrustworthy approach for monitoring the integrity of the Si-C bonds in PMO or related organosilica materials and the authors suggest solid-state NMR for such purposes [75].

Adsorption

As already introduced in the experimental sections above, the organic linkers in the PMOs were thought as targets for the functionalization of these materials for gas sorption and/or gas separation. Several interesting atmospheric molecules, e.g. CO_2 and CH_4, interact quite weakly with most materials and, therefore, the correct calculation of the energetics of adsorption requires the consideration of the dispersion effects. Camarota et al. [76] combined variable temperature infrared (VTIR) spectroscopy and B3LYP-D calculations (B3LYP + dispersion corrections based on the ideas of Grimme et al. [87]) to study the interaction of CO, a probe molecule widely used for IR surface studies, with phenylene PMO (benzenesilica). The PMO was modeled with the same cluster model employed by Onida et al. [74] (Figure 4). Therefore, despite the calculations have been done with a different code, i.e., CRYSTAL06 code, and slightly different basis sets, the most important difference between these two works is the inclusion of the dispersion effects in the study by Camarota et al. [76]. The latter authors found three types of adsorbed CO species at low coverage, namely, i) physisorbed CO interacting with benzene rings, which act as Lewis bases, and H-bonded CO interacting with a silanol group, which act as Lewis acidic centers, via the ii) carbon or via the iii) oxygen atoms. In the case of the H-bonded configurations, they suggest that some residual lateral interaction with the benzene rings also takes place. Based on the VTIR data, the standard enthalpies of interaction were determined to be 13.2 $kJ\cdot mol^{-1}$, ~12 $kJ\cdot mol^{-1}$ and 10.6 $kJ\cdot mol^{-1}$ for configurations ii, iii) and i), respectively. The corresponding interaction energies obtained from the cluster model B3LYP-D calculations were 15.5 $kJ\cdot mol^{-1}$, 11.4 $kJ\cdot mol^{-1}$ and 8.1 $kJ\cdot mol^{-1}$, respectively. Despite the absence of BSSE and thermal (at T=298.15 K) corrections, and the consideration of a small molecular fragment to model the PMO, the agreement between experimental and computational data is extraordinary. Also to note is the difference between the B3LYP-D and B3LYP interaction energies calculated for CO interacting by its carbon atom with the silanol group. The value calculated by Camarota et al. [76] is of about 10 $kJ\cdot mol^{-1}$ larger than that reported by Onida et al. [74]. Additionally, the comparison of the calculated and experimental frequencies showed significant differences, which confirms previous results [75] where it was found that quite large models of the PMO structures were needed to compute these spectroscopic features.

More recently, Martinez and Pacchioni studied the interaction of CO, CO_2 and CH_4 with two periodic models for the walls of benzenesilica [81], i.e., the flat areas of the nanopores of the three-dimensional periodic lattice, which were built by taking into account the experimental structural information available in the literature [8, 88]. The two models have different number of silanols on the surface and are characterized by two different sequences of the silica rings. In the first model, all the rings are six-member cycles with a ratio between the fully condensed silica species (T^3) and silica species with one silanol group (T^2) equal to 1:1 (Figure 5a), while in the second model, composed by six- and four-member cycles, the T^3/T^2 ratio is 2:1 (Figure 5b). The former structure presents one silanol every five Si-O bonds while, in the latter, one silanol exists per eight Si-O bonds. The aim of the computational work of Martinez and Pacchioni [81] was to understand if any of the CO, CO_2 and CH_4 gases could be stored or separated by PMO materials. In fact, calculations can be useful to identify the sites in the PMO nanochannels where the interaction is stronger, and to suggest ways to increase the adsorption capability by modifying in a desired way the structure of the material (e.g. functionalization of the material). Additionally, the authors refer that the choice of these three molecules was also based on their intrinsic features, i.e., small permanent dipole moment manifested by CO, absence of dipole but presence of a quadrupole moment in CO_2 and apolar nature of CH_4.

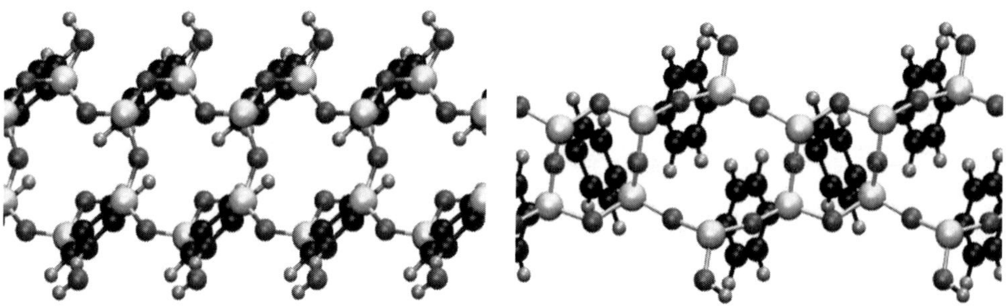

a b

Figure 5. Models of the walls of benzenesilica formed by six-member rings (a) or formed by a sequence of six- and four-member rings (b). Adapted from reference [81] with optimized atomic positions kindly supplied by the authors.

The periodic DFT calculations were performed with two different approaches, namely, PW91 and PBE exchange-correlation functionals, and with the consideration of dispersion corrections [87]. The VASP code was used together with plane-wave basis sets for the valence electrons, while the core electrons were mimicked with the projector-augmented plane-wave (PAW) approach. From the calculated data it was found that the three molecules interacted preferentially in a physisorbed state with the silanol groups. For CO adsorbing on the more realistic model composed by six- and four-member rings, the interaction energies calculated with the PBE-D and PW91-D functionals were 14.3 $kJ \cdot mol^{-1}$ and 15.7 $kJ \cdot mol^{-1}$, respectively. Interestingly, these energies are very close to the value, c.f. 15.5 $kJ \cdot mol^{-1}$, calculated by Camarota et al. [76] using a cluster model and a different exchange-correlation functional (see above). The PW91-D interaction energy calculated by Martinez and Pacchioni for CO in the less realistic model was 20.7 $kJ \cdot mol^{-1}$. The larger energy calculated for CO interacting with the model based only on six-member rings is due to the fact that in this model CO simultaneously interacts with two silanols, while in the more realistic model, CO interacts only with a silanol group. The PW91-D values calculated for CO_2 were 26.1 $kJ \cdot mol^{-1}$ and 15.4 $kJ \cdot mol^{-1}$ on the less and more realistic models, respectively, while smaller energies were calculated for methane, 15.8 $kJ \cdot mol^{-1}$ and 10.6 $kJ \cdot mol^{-1}$, respectively. The dispersion contribution to the calculated interaction energies was ~5-8 $kJ \cdot mol^{-1}$ in the case of the more realistic model, and between ~7-11 $kJ \cdot mol^{-1}$ in the case of the model based just on six-member rings.

Functionalization

The DFT calculations by Martinez and Pacchioni suggested that CO and CO_2 have virtually identical interaction with the Ph-PMO walls, which seems to be slightly stronger than CH_4. These same authors checked the influence of the functionalization of the material in the adsorbate-substrate interaction by introducing chemical modifications either in the inorganic and the organic parts of the PMO with fully condensed silica species, i.e., model of the wall having only six-member rings [81]. The modifications introduced were i) substitution of one Si by a Ti atom, which changed slightly the Brönsted acidity of the silanol groups, ii) removal of the proton from one of the silanols, which has been replaced by a sodium ion in the opposite side of the wall (creation of a negatively charged silanol group), iii) substitution of Si by Al and introduction of a sodium atom that formally donates one electron to the Al impurity, iv) substitution of –CH– in the benzene ring by –N– corresponding to the substitution of a benzene linker (phenylene) by pyridine and v) substitution of phenylene (benzene linker) by biphenylene. From the energies calculated for the interaction between the three molecules and the parent or modified PMO walls, Martinez and Pacchioni concluded that the modifications introduced in the inorganic part, including charged defects, have little impact on the adsorption properties and generally results in reduced binding [81]. However, modifications in the organic part resulted in higher interaction energies for CO and CO_2 and in lower interaction energy for CH_4 when phenylene was replaced by pyridine. It was also found a reduction in the calculated interaction energies when the molecules interacted exclusively with the organic parts on going from the material with phenylene rings to the material with biphenylene linkers, but the energy of interaction with the silanols was unchanged and these remained as the preferred adsorption sites [81].

Importantly, the identification of the local nature of the adsorbate-substrate interaction suggests that a rapid screening of the adsorption properties of inorganic and organic parts of the network of PMOs can be efficiently done using molecular or cluster models, which is supported by the similar interaction energies calculated for CO by periodic and cluster model approaches as described above [76, 81].

More recently, Sharifi et al. [82] studied the proton conductivity of Ph-PMO functionalized with sulfonic acid groups by several experimental and theoretical techniques. Experimentally, these authors synthesized the Ph-PMO and then grafted the SO_3H groups at different locations, i.e., i) just on silica, ii) just on phenylene rings and iii) both at silica and phenylene moieties. They characterized the materials by X-ray diffraction, neutron scattering and argon adsorption isotherms. The proton conductivities were obtained from impedance spectroscopy measurements. Theoretically, they considered either dry and wet (100 or 150 water molecules) PMO models having 6 benzene rings and 16 $Si(OR)_3OH$ groups, R=H or C or Si, and performed several quantum mechanical (semi-empirical DFTB approach) molecular dynamics simulations at two temperatures (350 and 450 K).

The proton conductivity was found to be higher for Ph-PMO with sulfonic groups grafted at the phenylene rings or both at phenylene and silica groups than for pristine Ph-PMO or Ph-PMO with sulfonic groups grafted only at silica positions. The calculations show that the deprotonation of the sulfonic acid group and dissolution of the proton into bulk water, even under ideal conditions, is severely inhibited in the Ph-PMO material. In the case of proton transport between neighboring sulfonic acid groups, the calculations predict potential energy barriers between 10 $kJ \cdot mol^{-1}$ and 34 $kJ \cdot mol^{-1}$, with these limits corresponding to optimal wet and dry conditions [82]. So, isolated sulfonic acid groups act as proton traps, but if a sufficiently high concentration of these groups exists on the surface of Ph-PMO, proton conduction along the pore wall will compete with proton conduction in bulk water [82].

Catalysis

The initial steps responsible for CO_2 activation in the reaction mechanism of propylene coupling with CO_2 catalyzed by a PMO material based on the TBTS organic linker, Figure 6, were studied by Prasetyanto et al. making use of the DFT-RPBE approach [77]. Basically, these authors studied the interaction of CO_2 with the TBTS linker and found direct binding of the carbon atom of the CO_2 adsorbate with one of the nitrogen atoms of the triazine ring, together with the formation of a $R_2NH \cdots OCO$ contact between the adsorbate and a neighboring amine group, which was the most favorable situation. The resulting pseudo-cyclic carbamate structure (Figure 7a) was compatible with the mechanistic pathway supported by their experimental studies [77]. They found two other structures involving also the simultaneous interaction with two nitrogen atoms from the linker (Figures 7b and 7c) that were also compatible with the experimental findings.

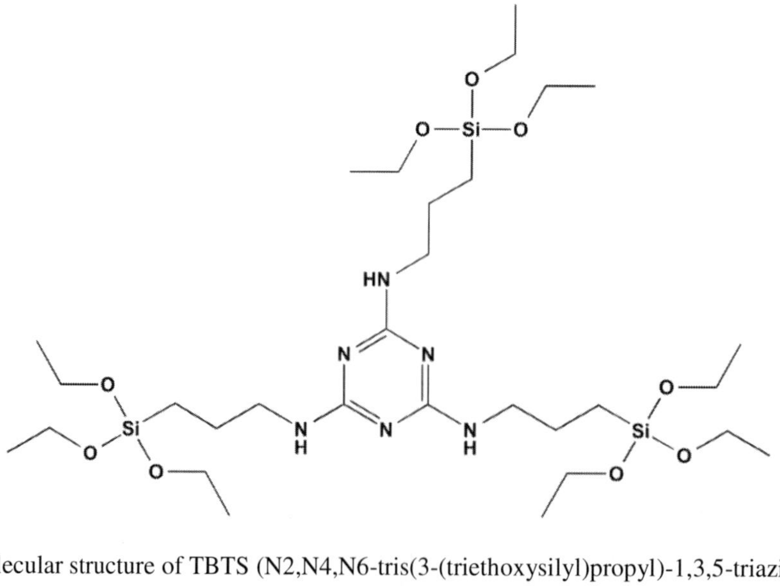

Figure 6. Molecular structure of TBTS (N2,N4,N6-tris(3-(triethoxysilyl)propyl)-1,3,5-triazine-2,4,6-triamine).

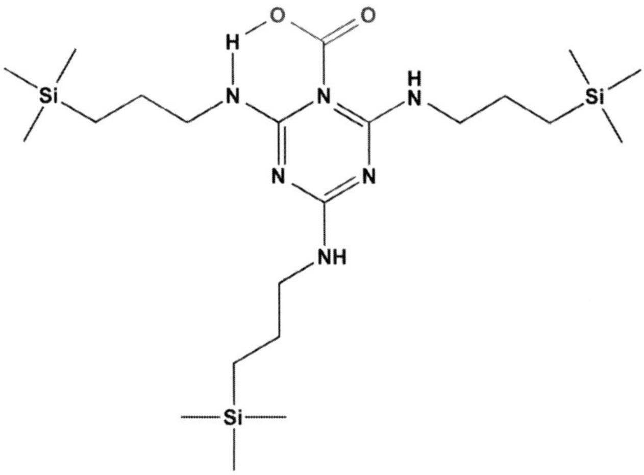

a

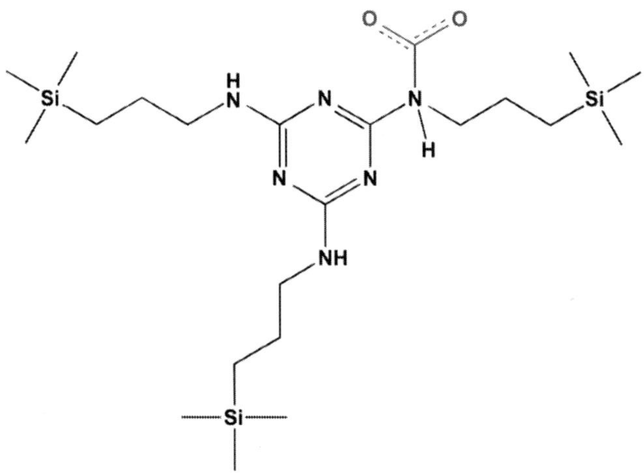

b

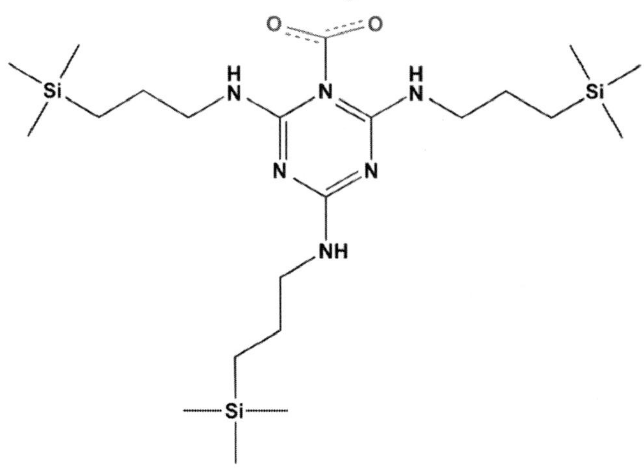

c

Figure 7. (Continued).

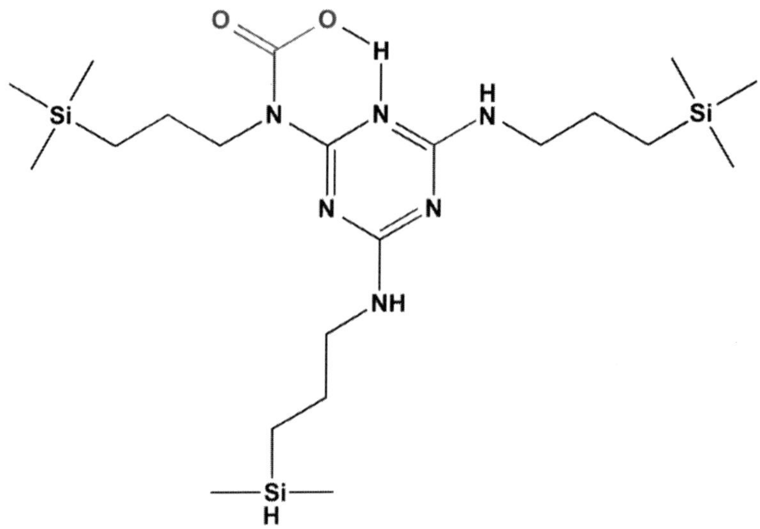

d

Figure 7. Possible routes for CO_2 activation by TBTS-PMO. Adapted from reference [77].

Interestingly, the calculations suggested that the species shown in Figure 7d, which was reported by other authors, is less probable since the reaction yielding this intermediate from the separated fragments is endothermic at the contrary of the reactions yielding the intermediates shown in Figures 7a-7c [77].

Synthesis

The initial stages of the synthesis of PMO materials, i.e., the interaction between silica monomers or small oligomers and condensation of these species to yield organosilica, were also studied by means of computational approaches.

Shirai et al. performed DFT studies to study the occurrence of Si-C bond cleavage in various organosilane precursors, with general structure $RSi(OH)_3$, which are depicted in Figure 8 [78]. These authors considered the B3LYP/6-311+G(d,p) approach to calculate the proton affinities of the carbon atom at the ipso-position and of the carbanion generated after Si-C bond cleavage in the precursors. The proton affinities at these two sites were chosen as indices for Si-C bond stability under acidic and basic conditions, respectively.

The predicted stabilities were found to be in qualitative agreement with the experimental results. In fact, the calculations suggested the terphenyl and N-methylcarbaxole R groups to be the least stable under acidic conditions, and the terphenyl and anthracene R groups to be the least stable under basic conditions. Interestingly, Si-C bond cleavage was seen experimentally for these cases, while the Si-C bond in the precursors with the other R groups was more robust. Thus, under acidic conditions, the calculations confirm that the reaction leading to Si-C cleavage proceeds via a σ-complex formed by H^+ introduction at the carbon atom at the ipso-position in the organosilane precursors (Figure 9a) while, under basic conditions, the reaction proceeds via the formation of an intermediate complex where the OH^- species binds directly the Si atom of the organosilane precursors (Figure 9b).

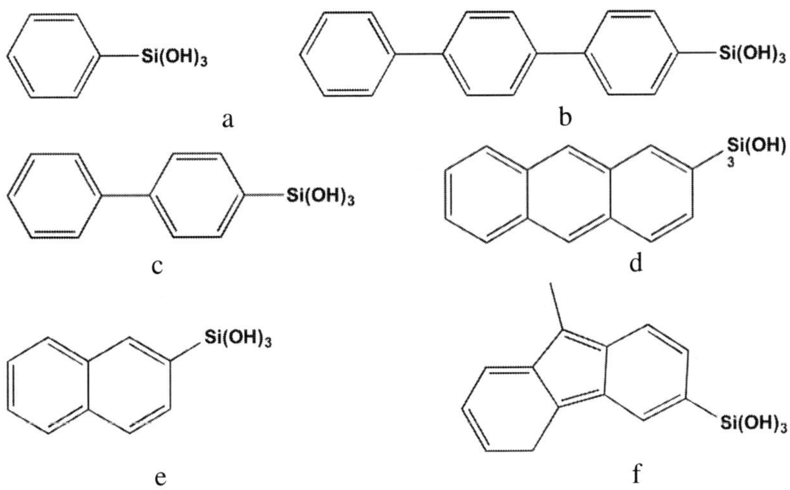

Figure 8. Models of organosilane precursors with general structure $RSi(OH)_3$, with R derived from: a) phenyl; b) terphenyl, c) biphenyl, d) anthracene, e) naphthalene and f) N-methylcarbazole. Adapted from reference [78].

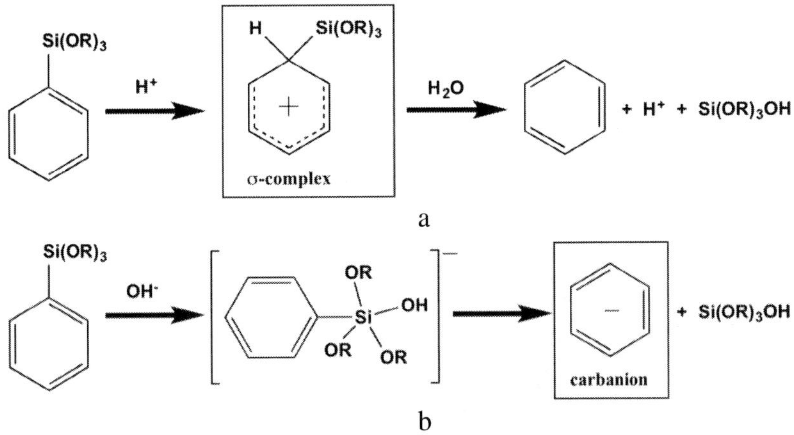

Figure 9. General mechanism for Si-C bond cleavage in the reaction of organosilane polycondensation to organosilica under a) acidic and b) basic conditions proposed in reference [78].

Very recently, some of us have been involved in the study on the role of the organic linker in the early stages of the templated synthesis of PMO materials [80]. In that work, classical molecular dynamics (MD) simulations were performed for surfactant-bromide-water solutions containing several organosilicate precursors based on 1,2-bis(triethoxysilyl)ethane, 1,2-bis(triethoxysilyl)ethylene or 1,4-bis(triethoxysilyl)benzene (Figure 1).

These solutions contain the basic building blocks used in the experimental synthesis of PMOs and the choice of these three precursors is justified by the different degree of molecular-scale periodicity found in the resulting materials. In fact, benzene, ethylene and ethane organic linkers were found to lead to high degree, to a lower degree, and to no molecular-scale periodicity, respectively [8].

Molecular-scale periodicity is also absent in PMOs synthesized from 1,1-bis(triethoxysilyl)methane. The tentative explanation for the differences observed between the structures of the phenylene- or ethylenesilicas and ethane or methanesilicas was the different size of the organic linker and the strength of the interaction between the precursor molecules [8]. Additionally, it was suggested that the molecular-scale periodicity was promoted by the mesostructure resulting from the self-assembly of the surfactants. However, a few years later, phenylenesilica and biphenylsilica hybrid materials presenting ordered molecular scale periodicity were synthesized without the use of surfactants [89].

The syntheses of these PMOs are usually carried out in alkaline media, and therefore, the organosilicates are deprotonated to a significant extent. Force field parameters for the negatively charged silicate groups were not available in the literature and had to be developed a priori [79, 90–92].

The analyses carried out for the trajectories resulting from the MD simulations by Futamura et al. [80] showed that the organic and inorganic parts of the benzenesilicates and of the ethylenesilicates were clearly inserted between the cationic surfactant head-groups in the micelles, while in the case of ethanesilicates, only the inorganic moiety interacted with the micelle surface. Interestingly, this behavior resulted in the promotion of micelles having more surfactants (i.e., larger mass average micelle size along the simulation) in the former than in the latter cases. These observations were suggested to be due to a more efficient screening of the surfactant head-group repulsion, which promotes the decrease of the surface curvature and hence the formation of larger micelles [80].

Futamura et al. found also that in the case of the solutions containing benzenesilicates or ethylenesilicates, the ordering of the organosilicates was much higher than in the case of the solutions with ethanesilicates. In fact, they found that in the former cases the organosilicates lie with their silicate-organic-silicate axis parallel to the micelle surface while, in the latter situation, the ethanesilicates prefer to orient their silicate-organic-silicate axis perpendicularly to the surface with much less order.

This suggests that the presence or absence of molecular-scale periodicity in the pore walls of PMOs is dictated at the very early stages of the templated synthesis process. Crucial for such periodicity is the strength of the surfactant micelle-organosilica interaction and the relative flexibility and orientation of the organic linker, i.e., more rigid organic fragments that interact strongly with the surfactant micelle surface will have higher degrees of molecular-scale periodicity than more flexible organic fragments interacting weakly with the surfactant micelles.

CONCLUSION

The previous sections of this Chapter described in detail what was done in the last 15 years in terms of development of acid, basic and acid-basic PMOs. In the case of the acid PMOs, the propyl sulfonic acid functionalization was the most explored and only recently different sulfonic acid groups were developed. Only two examples of the introduction of carboxylic acid groups are reported in literature. In the case of the basic PMOs the most explored functionalization was based on introduction of aminopropyl groups but other basic functions are also described.

In both acid and basic PMOs, it was seen that the inclusion of high content of functional groups had a negative effect on the organization of the final material. It was verified that in the case of the acid PMOs, they have been mainly applied in catalysis, while the basic PMOs have been essentially used in adsorption of gases, metals and organic compounds such as nicotine. The computational tools have proved to be very useful not only on the understanding of the PMO synthesis and functionalization, but also on predicting properties and mechanisms involved in the application of these materials.

In the field of catalysis, PMO-based materials offer noticeable advantages over purely mesoporous siliceous materials. First of all, their higher hydrophobicity together with their crystalline structure in the case of phenylene and biphenylene-bridged structures clearly contribute to improve their stability in water making possible their direct use in water or in water-generating reactions.

In this context, PMOs are predicted to be promising solid catalysts for biomass processing for which catalyst deactivation by water is an important scientific issue. Although hydrophobicity is often pointed out by authors as a key parameter to explain the superior stability and activity of PMOs in the presence of water, it is clear that deeper characterizations have revealed that other parameters such as catalytic sites dispersion and location play also a pivotal role by governing the strength of catalytic sites and reactant adsorption rates. In addition, the possibility to (1) functionalize PMO materials with either acid, basic or redox sites and (2) to prepare multifunctional surfaces thanks to different chemical modifications of the organic and siliceous parts of PMO, as well as, introduction of metal nanoparticles or metal complexes, offer great means to design solid catalysts capable of creating complexity from simple structures in a minimum of reaction steps.

Considering the adsorption applications, PMOs are still very unexplored. Much needs to be investigated in the field of adsorption/capture and separation of gases. The possibility of making PMOs on demand, designing the organic moieties in order to specifically possess affinity to a certain gas and not to the other, seems to be an open avenue. The reviewed literature showed the great potential of PMOs to adsorb metals and organic molecules from water media taking advantage of the hydrophilic-hydrophobic properties of PMOs. New PMO materials, novel adsorbates and adsorption mechanisms need to be studied. PMOs can be also interesting as drug carriers for controlled released of pharmaceuticals, as they can be designed for distinct adsorption-desorption behavior.

Looking ahead, after 15 years of PMOs, this family of materials keeps very alive, and seems to be very promising in a variety of fields, namely, in catalysis and in adsorption, especially in water media or in which water is formed.

To the best of our knowledge PMOs are not yet at the market despite all expectations related to their application as low dielectric constant materials. It is our opinion that commercialization of mesoporous materials still requires the overpassing of a few obstacles in the future such as (1) reduction of the costs related to their synthesis, (2) improvement of thermal and mechanical stability, especially in the presence of polar media and (3) enhancement of the morphological control.

Clearly, at this stage of research, commercialization of mesoporous materials will require investment in fundamental research and development since it is generally admitted in the industry that creation of novel market for materials or chemicals takes at least 20 years.

ACKNOWLEDGMENTS

This work is supported in the framework of projects PTDC/EQU-EQU/099423/2008, (FCOMP-01-0124-FEDER-010345), PTDC/QUI-QUI/113678/2009 (FCOMP-01-0124-FEDER-015644) and PEst-C/CTM/LA0011/2013 (FCOMP-01-0124-FEDER-037271) co-financed by the Fundação para a Ciência e a Tecnologia (FCT), European Regional Development Fund (FEDER) and QREN-COMPETE. JRBG and PF hold *Investigador FCT* positions. MAOL acknowledges FCT for the PhD grant ref. SFRH/BD/80883/2011.

REFERENCES

[1] Inagaki, S.; Guan, S.; Fukushima, Y.; Ohsuna, T. *J. Am. Chem. Soc.* 1999, 121, 9611–9614.

[2] Melde, B.; Holland, B. *Chem. Mater.* 1999, 3302–3308.

[3] Asefa, T.; MacLachlan, M. J.; Coombs, N.; Ozin, G. A. *Nature* 1999, 402, 867–871.

[4] Kresge, C.; Leonowicz, M.; Roth, W. *Nature* 1992, 359, 710–712.

[5] Beck, J. S.; Vartuli, J. C.; Roth, W. J.; Leonowicz, M. E.; Kresge, C. T.; Schmitt, K. D.; Chu, C. T. W.; Olson, D. H.; Sheppard, E. W. *J. Am. Chem. Soc.* 1992, 114, 10834–10843.

[6] MacLachlan, M.; Asefa, T.; Ozin, G. *Chem. Eur. J.* 2000, 6, 2507–2511.

[7] Asefa, T.; Yoshina-Ishii, C.; MacLachlan, M. J.; Ozin, G. A. *J. Mater. Chem.* 2000, 10, 1751–1755.

[8] Inagaki, S.; Guan, S.; Ohsuna, T.; Terasaki, O. *Nature* 2002, 416, 304–307.

[9] Kresge, C. T.; Roth, W. J. *Chem. Soc. Rev.* 2013, 42, 3663–3670.

[10] Petkovich, N. D.; Stein, A. *Chem. Soc. Rev.* 2013, 42, 3721–3739.

[11] Firouzi, A.; Kumar, D.; Bull, L.; Besier, T.; Sieger, P.; Huo, Q.; Walker, S.; Zasadzinski, J.; Glinka, C.; Nicol, J.; Et, A. *Science* 1995, 267, 1138–1143.

[12] Bion, N.; Ferreira, P.; Valente, A.; Gonçalves, I. S.; Rocha, J. *J. Mater. Chem.* 2003, 13, 1910–1913.

[13] Goto, Y.; Inagaki, S. *Chem. Commun.* 2002, 2410–2411.

[14] Esquivel, D.; Jiménez-Sanchidrián, C.; Romero-Salguero, F. J. *Mater. Lett.* 2011, 65, 1460–1462.

[15] Kamegawa, T.; Sakai, T.; Matsuoka, M.; Anpo, M. *J. Am. Chem. Soc.* 2005, 127, 16784–16785.

[16] Coelho, A. C.; Balula, S. S.; Bruno, S. M.; Alonso, J. C.; Bion, N.; Ferreira, P.; Pillinger, M.; Valente, A. A.; Rocha, J.; Gonçalves, I. S. *Adv. Synth. Catal.* 2010, 352, 1759–1769.

[17] Coelho, A. C.; Balula, S. S.; Antunes, M. M.; Gerganova, T. I.; Bion, N.; Ferreira, P.; Pillinger, M.; Valente, A. A.; Rocha, J.; Gonçalves, I. S. *J. Mol. Catal. A Chem.* 2010, 332, 13–18.

[18] Ohashi, M.; Kapoor, M. P.; Inagaki, S. *Chem. Commun.* 2008, 841–843.

[19] Lourenço, M. A. O.; Siegel, R.; Mafra, L.; Ferreira, P. *Dalt. Trans.* 2013, 42, 5631–5634.

[20] Gomes, A. C.; Lourenço, M. A. O.; Bruno, S. M.; Ferreira, P.; Valente, A. A.; Pillinger, M.; Gonçalves, I. S. *J. Organomet. Chem.* 2014, 751, 501-507.

[21] Kamegawa, T.; Saito, M.; Watanabe, T.; Uchihara, K.; Kondo, M.; Matsuoka, M.; Anpo, M. *J. Mater. Chem.* 2011, 21, 12228–12231.

[22] Waki, M.; Mizoshita, N.; Tani, T.; Inagaki, S. *Angew. Chemie* 2011, 50, 11667–11671.

[23] Kapoor, M. P.; Inagaki, S. *Bull. Chem. Soc. Jpn.* 2006, 79, 1463–1475.

[24] Fujita, S.; Inagaki, S. *Chem. Mater.* 2008, 20, 891–908.

[25] Whitnall, W.; Asefa, T.; Ozin, G. A. *Adv. Funct. Mater.* 2005, 15, 1696–1702.

[26] Hoffmann, F.; Cornelius, M.; Morell, J.; Fröba, M. *Angew. Chem. Int. Ed.* 2006, 45, 3216–3251.

[27] Mizoshita, N.; Tani, T.; Inagaki, S. *Chem. Soc. Rev.* 2011, 40, 789–800.

[28] Hoffmann, F.; Fröba, M. *Chem. Soc. Rev.* 2011, 40, 608–620.

[29] Van der Voort, P.; Esquivel, D.; De Canck, E.; Goethals, F.; Van Driessche, I.; Romero-Salguero, F. J. *Chem. Soc. Rev.* 2013, 42, 3913–3955.

[30] Hatton, B. D.; Landskron, K.; Hunks, W. J.; Bennett, M. R.; Shukaris, D.; Perovic, D. D.; Ozin, G. A. *Mater. Today* 2006, 9, 22–31.

[31] Wang, W.; Lofgreen, J. E.; Ozin, G. A. *Small* 2010, 6, 2634–2642.

[32] Hatton, B.; Landskron, K.; Whitnall, W.; Perovic, D.; Ozin, G. A. *Acc. Chem. Res.* 2005, 38, 305–312.

[33] Yang, Q.; Kapoor, M. P.; Inagaki, S. *J. Am. Chem. Soc.* 2002, 124, 9694–9695.

[34] Kapoor, M. P.; Yang, Q.; Goto, Y.; Inagaki, S. *Chem. Lett.* 2003, 32, 914–915.

[35] Yang, Q.; Kapoor, M. P.; Shirokura, N.; Ohashi, M.; Inagaki, S.; Kondo, J. N.; Domen, K. *J. Mater. Chem.* 2005, 15, 666–673.

[36] Liu, J.; Yang, Q.; Kapoor, M. P.; Setoyama, N.; Inagaki, S.; Yang, J.; Zhang, L. *J. Phys. Chem. B* 2005, 109, 12250–12256.

[37] Yang, Q.; Kapoor, M. P.; Inagaki, S.; Shirokura, N.; Kondo, J. N.; Domen, K. *J. Mol. Catal. A Chem.* 2005, 230, 85–89.

[38] Esquivel, D.; Jiménez-Sanchidrián, C.; Romero-Salguero, F. J. *J. Mater. Chem.* 2011, 21, 724–733.

[39] Kapoor, M. P.; Fujii, W.; Kasama, Y.; Yanagi, M.; Nanbu, H.; Juneja, L. R. *J. Mater. Chem.* 2008, 18, 4683–4691.

[40] López, M. I.; Esquivel, D.; Jiménez-Sanchidrián, C.; Romero-Salguero, F. J. *ChemCatChem* 2013, 5, 1002–1010.

[41] Rat, M.; Zahedi-Niaki, M. H.; Kaliaguine, S.; Do, T. O. *Micropor. Mesopor. Mater.* 2008, 112, 26–31.

[42] Dubé, D.; Rat, M.; Shen, W.; Béland, F.; Kaliaguine, S. *J. Mater. Sci.* 2009, 44, 6683–6692.

[43] Dhepe, P. L.; Ohashi, M.; Inagaki, S.; Ichikawa, M.; Fukuoka, A. *Catal. Lett.* 2005, 102, 163–169.

[44] Karam, A.; Alonso, J. C.; Gerganova, T. I.; Ferreira, P.; Bion, N.; Barrault, J.; Jérôme, F. *Chem. Commun.* 2009, 7000–7002.

[45] Siegel, R.; Domingues, E.; De Sousa, R.; Jérôme, F.; Morais, C. M.; Bion, N.; Ferreira, P.; Mafra, L. *J. Mater. Chem.* 2012, 22, 7412–7419.

[46] Bispo, C.; Ferreira, P.; Trouvé, A.; Batonneau-Gener, I.; Liu, F.; Jérôme, F.; Bion, N. *Catal. Today* 2013, 4–11.

[47] Tucker, M. H.; Crisci, A. J.; Wigington, B. N.; Phadke, N.; Alamillo, R.; Zhang, J.; Scott, S. L.; Dumesic, J. A. *ACS Catal.* 2012, 2, 1865–1876.

[48] Bispo, C.; Vigier, K. D. O.; Sardo, M.; Bion, N.; Mafra, L.; Ferreira, P.; Jérôme, F., *Catal. Sci. Technol.* 2014, 4, 2235-2240.

[49] Esquivel, D.; van den Berg, O.; Romero-Salguero, F. J.; Du Prez, F.; Van Der Voort, P. *Chem. Commun.* 2013, 49, 2344–2346.

[50] De Canck, E.; Dosuna-Rodríguez, I.; Gaigneaux, E.; Van Der Voort, P. *Materials* 2013, 6, 3556–3570.

[51] Zhu, F.-X.; Wang, W.; Li, H.-X. *J. Am. Chem. Soc.* 2011, 133, 11632–11640.

[52] Karimi, B.; Mirzaei, H.; Mobaraki, A. *Catal. Sci. Technol.* 2012, 2, 828–834.

[53] Sánchez-Vázquez, R.; Pirez, C.; Iglesias, J.; Wilson, K.; Lee, A. F.; Melero, J. A. *ChemCatChem* 2013, 5, 994–1001.

[54] Fiorilli, S.; Camarota, B.; Perrachon, D.; Bruzzoniti, M. C.; Garrone, E.; Onida, B. *Chem. Commun.* 2009, 4402–4404.

[55] Ting, C.; Chung, C.-H.; Kao, H. *Chem. Commun.* 2011, 47, 5897–5899.

[56] Han, L.; Terasaki, O.; Che, S. *J. Mater. Chem.* 2011, 21, 11033–11039.

[57] Kao, H.; Chung, C.-H.; Saikia, D. *Chem. Asian J.* 2012, 7, 2111–2117.

[58] Oladejo, G. S.; Bispo, C.; Bion, N.; Ferreira, P.; Batonneau-Gener, I. *Micropor. Mesopor. Mater.* 2014, 200, 117–123

[59] Burleigh, M.; Dai, S.; Hagaman, E.; Lin, J. *Chem. Mater.* 2001, 13, 2537–2546.

[60] Wahab, M. A.; Imae, I.; Kawakami, Y.; Kim, I.; Ha, C.-S. *Micropor. Mesopor. Mater.* 2006, 92, 201–211.

[61] Benitez, M.; Das, D.; Ferreira, R. *Chem. Mater.* 2006, 18, 5597–5603.

[62] Zhu, F.; Yang, D.; Zhang, F.; Li, H. *J. Mol. Catal. A Chem.* 2012, 363-364, 387–397.

[63] Hudson, S.; Cooney, J.; Hodnett, B. K.; Magner, E. *Chem. Mater.* 2007, 19, 2049–2055.

[64] Coutinho, D.; Xiong, C.; Balkus, K. J. *Micropor. Mesopor. Mater.* 2008, 108, 86–94.

[65] Park, M.; Park, S. S.; Selvaraj, M.; Zhao, D.; Ha, C.-S. *Micropor. Mesopor. Mater.* 2009, 124, 76–83.

[66] Shin, J.; Park, S.; Selvaraj, M.; Ha, C.-S. *J. Porous Mater.* 2012, 19, 29–35.

[67] Shin, J. H.; Park, S. S.; Ha, C.-S. *Colloids Surf. B. Biointerfaces* 2011, 84, 579–84.

[68] Nakai, K.; Oumi, Y.; Horie, H.; Sano, T.; Yoshitake, H. *Micropor. Mesopor. Mater.* 2007, 100, 328–339.

[69] De Canck, E.; Ascoop, I.; Sayari, A.; Van Der Voort, P. *Phys. Chem. Chem. Phys.* 2013, 15, 9792–9.

[70] Alauzun, J.; Mehdi, A.; Reyé, C.; Corriu, R. J. P. *J. Am. Chem. Soc.* 2006, 128, 8718–8719.

[71] Jaroniec, M. *Nature* 2006, 442, 638–640.

[72] Shylesh, S.; Wagener, A.; Seifert, A.; Ernst, S.; Thiel, W. R. *Angew. Chem. Int. Ed.* 2010, 49, 184–187.

[73] Sasidharan, M.; Fujita, S.; Ohashi, M.; Goto, Y.; Nakashima, K.; Inagaki, S. *Chem. Commun.* 2011, 47, 10422–10424.

[74] Onida, B.; Borello, L.; Busco, C.; Ugliengo, P.; Goto, Y.; Inagaki, S.; Garrone, E. *J. Phys. Chem. B* 2005, 109, 11961–11966.

[75] Hoffmann, F.; Gungerich, M.; Klar, P. J.; Fröba, M. *J. Phys. Chem. C* 2007, 111, 5648–5660.

[76] Camarota, B.; Ugliengo, P.; Garrone, E.; Arean, C. O.; Delgado, M. R.; Inagaki, S.; Onida, B. *J. Phys. Chem. C* 2008, 112, 19560–19567.
[77] Prasetyanto, E. A.; Ansari, M. B.; Min, B.-H.; Park, S.-E. *Catal. Today* 2010, 158, 252–257.
[78] Shirai, S.; Goto, Y.; Mizoshita, N.; Ohashi, M.; Tani, T.; Shimada, T.; Hyodo, S.; Inagaki, S. *J. Phys. Chem. A* 2010, 114, 6047–6054.
[79] Futamura, R.; Jorge, M.; Gomes, J. R. B. *Theor. Chem. Acc.* 2013, 132, 1–10.
[80] Futamura, R.; Jorge, M.; Gomes, J. R. B. *Phys. Chem. Chem. Phys.* 2013, 15, 6166–6169.
[81] Martinez, U.; Pacchioni, G. *Micropor. Mesopor. Mater.* 2010, 129, 62–67.
[82] Sharifi, M.; Köhler, C.; Tölle, P.; Frauenheim, T.; Wark, M. *Small* 2011, 7, 1086–1097.
[83] Gomes, J. R. B.; Fajín, J. L. C.; Cordeiro, M. N. D. S.; Teixeira, C.; Gomes, P.; Pillai, R. S.; Novell-Leruth, G.; Toda, J.; Jorge, M. In: *Density Functional Theory: Principles, Applications and Analysis*; Maurin, J.; Pelletier, J. M., Eds.; Nova Science Publishers, Inc.: New York, 2013; pp. 1–58.
[84] Fischer, M.; Gomes, J. R. B.; Jorge, M. *Mol. Simul.* 2014, 40, 537-556.
[85] Ramalho, J. P. P.; Gomes, J. R. B.; Illas, F. *RSC Adv.* 2013, 3, 13085–13100.
[86] Jensen, F. *Introduction to Computational Chemistry*; First edit.; John Wiley and Sons: Chichester, England, 2004.
[87] Grimme, S. *J. Comput. Chem.* 2006, 27, 1787–1799.
[88] Comotti, A.; Bracco, S.; Valsesia, P.; Ferretti, L.; Sozzani, P. *J. Am. Chem. Soc.* 2007, 129, 8566–8576.
[89] Okamoto, K.; Goto, Y.; Inagaki, S. *J. Mater. Chem.* 2005, 15, 4136–4140.
[90] Gomes, J. R. B.; Cordeiro, M. N. D. S.; Jorge, M. *Geochim. Cosmochim. Acta* 2008, 72, 4421–4439.
[91] Jorge, M.; Gomes, J. R. B.; Cordeiro, M. N. D. S.; Seaton, N. A. *J. Am. Chem. Soc.* 2007, 129, 15414–15415.
[92] Jorge, M.; Gomes, J. R. B.; Cordeiro, M. N. D. S.; Seaton, N. A. *J. Phys. Chem. B* 2009, 113, 708–718.

In: Comprehensive Guide for Mesoporous Materials, Volume 4 ISBN: 978-1-63482-092-9
Editor: Mahmood Aliofkhazraei

Chapter 13

CARBON XEROGELS AS MESOPOROUS SUPPORTS IN THE PREPARATION OF CATALYTIC SYSTEMS FOR ELECTROCHEMICAL APPLICATIONS

C. Alegre, M.E. Gálvez, R. Moliner and M.J. Lázaro
Instituto de Carboquímica, CSIC. Miguel Luesma Castán,
Zaragoza, Spain

ABSTRACT

Resorcinol–formaldehyde gels can be prepared under a great variety of synthesis conditions leading to the production of organic and carbon gels such as aerogels, cryogels or xerogels – differing in the gel drying process: super-critically, freeze drying or sub-critically. If the synthesis conditions are appropriately chosen and the gel is mechanically strong enough to withstand the capillary pressures, subcritical-conventional drying can be carried out without involving major changes in the structure of the organic material. Moreover, it is possible to fine-tune the textural properties of these materials during its sol–gel process based synthesis.

Therefore, synthesis-tailored carbon xerogels of high porosity (>80%), high surface area (400-800 m^2/g) and controlled pore size distribution (1 to 50 nm), can be easily prepared. These properties are fundamental requirements in the application of carbon materials as supports in the preparation of catalytic systems, especially within the field of fuel cell technologies. Pore structure in fuel cell electrodes affects mass transportation of fuel and exhausts, highly affecting the fuel cell performance. In fact, several studies point out the key importance of a mesopore-enriched porous structure in carbon materials for fuel cells mass transportation processes.

The present chapter aims to summarize the influence of the different synthesis parameters, such as reactants ratio (precursor/formaldehyde), the catalyst concentration (precursor/catalyst ratio) and type (basic and acid), and the type of solvent, on the final carbon xerogel properties, in view of their utilization as supports in the preparation of Pt catalysts for their application as electrodes in direct methanol fuel cells (DMFCs).

1. Carbon Xerogels and Their Synthesis

Carbon gels are porous materials with a reticular structure formed by interconnected spherical carbon nanoparticles. These materials are obtained by the carbonization of an organic gel obtained from the polycondensation of hydroxylated benzenes (resorcinol, phenol, catechol, etc) and aldehydes (formaldehyde, furfural, etc). The porosity of these materials is located both inside the particles (composed of micropores) and between the particles, giving rise to mesopores and macropores [Yoshizawa, 2003], as shown in the scheme of Figure 1.

This type of nanostructure is responsible for their unique properties, that allow the application of these materials in several fields, such as supercapacitors [Saliger, 1998; Pekala, 1998; Li, 2003; Li, 2006, Fernández, 2012; Moreno-Castilla, 2012] adsorbents [Meena, 2005; Maldonado-Hódar, 2007], thermal and acoustic insulators [Mrowiec-Biało, 1998; Smith, 1998], and as catalysts supports for different catalytic applications such as fuel cells [Glora, 2001; Smirnova, 2005; Arbizzani, 2007; Job, 2009; Mahata, 2007; Machado, 2012; Maia, 2012; Rodrigues, 2012]. Moreover, due to the possibility of fine-tuning their textural properties during the sol–gel process employed for their preparation, these materials have attracted enormous attention, given that one may change the ratio of micro-, meso- and macropores and, accordingly, their specific surface areas over a wide range, in order to tailor any particular application [Gomes, 2004; Job, 2005]. Besides, carbon gels can be synthesized in different forms, monoliths, powder, films, microspheres, etc [Yamamoto, 2002; Horikawa, 2004; Moreno-Castilla, 2005; Zhang, 2012], favoring its application in very different fields.

Resorcinol–formaldehyde gels have been extensively studied, since their introduction by Pekala in 1989 [Pekala, 1989]. Numerous works were published in recent years describing the great variety of synthesis and processing conditions which lead to the production of organic and carbon gels such as aerogels, cryogels or xerogels—differing in the gel drying process: supercritically, freeze drying or subcritically—as well as how these conditions affect the final texture and structure of these materials [Al-Muhtaseb, 2003].

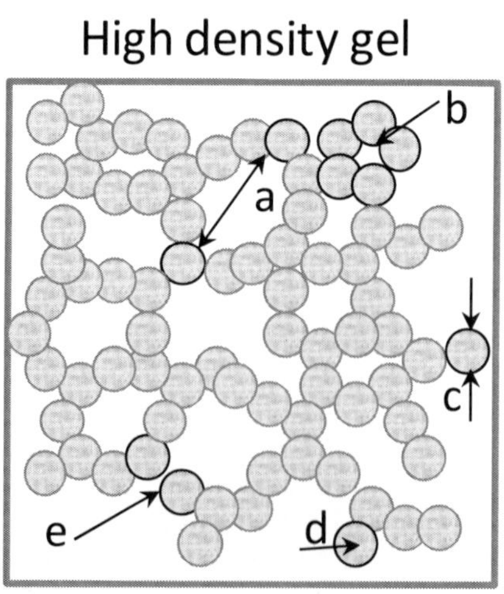

Figure 1. Scheme of carbon gel microstructure: (a) mesopores between chains of interconnected particles (b) micropore formed between several particles (c) individual particle (d) micropores inside particles (e) neck between adjacent particles.

It is generally observed that subcritical drying, i.e., conventional drying, may result in a substantial collapse of the porous structure of the initial organic gel, however, if the synthesis conditions are appropriately chosen and the gel is mechanically strong enough to withstand capillary pressures, conventional drying can be carried out without involving major changes in the structure of the organic material [Al-Muhtaseb, 2003; Leonard, 2005].

In particular, the uniqueness of the textural properties of these synthesis-tailored carbon materials lies in their high porosity (>80%), high surface area (400–1200 m^2 g^{-1}) and controlled pore size distributions (1 to 50 nm). These properties are fundamental requires in the application of carbon materials as supports in the preparation of catalytic systems, and especially within the field of fuel cell technologies [Zeng, 2012[a], Zeng, 2012[b]].

In fact, depending on the application, pore size distribution can play a definitive role in catalytic activity and stability. Pore structure in fuel cell electrodes affects mass transportation of fuel and exhausts, highly affecting the fuel cell performance [Du, 2007]. Several studies point out the key importance of a macro-mesopore enriched porous structure in carbon materials for fuel cells mass transportation processes [Du, 2007; Antolini, 2009; Liang, 2008].

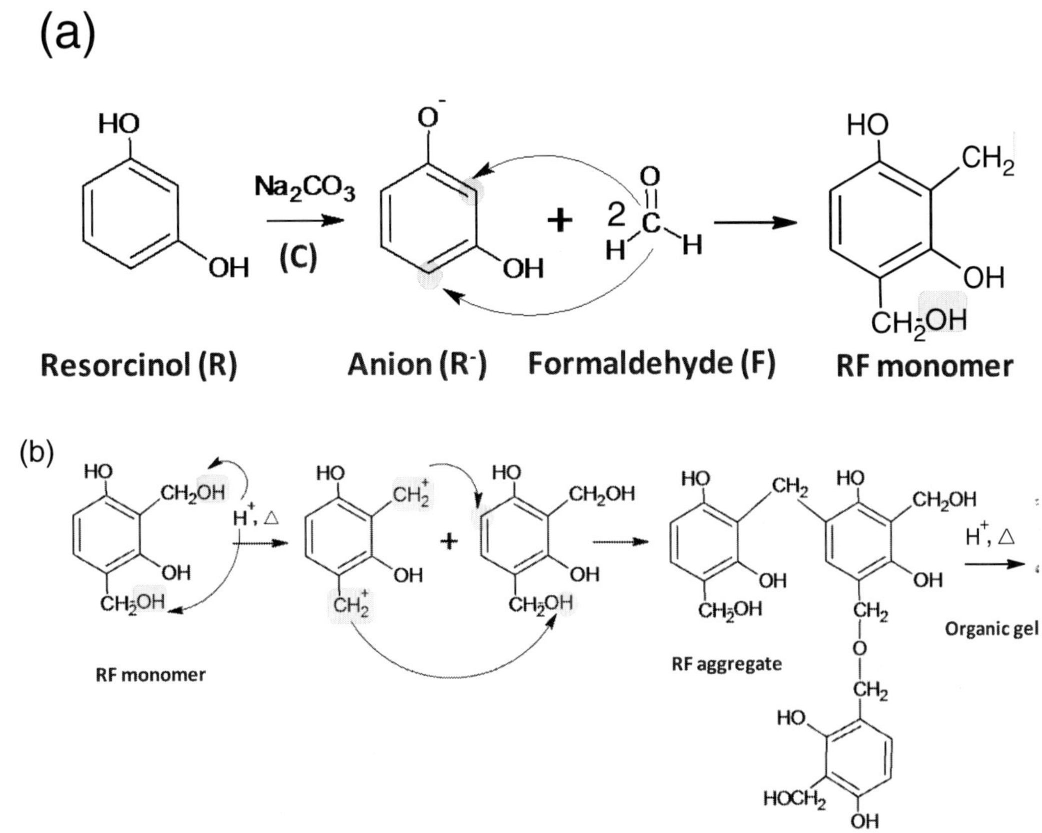

Figure 2. Sol-gel synthesis mechanism, leading to the production of the organic gels: (a) Resorcinol-Formaldehyde addition mechanism (b) Condensation of RF monomers and aggregates, yielding the organic gel.

This fact stands out as a unique advantage of using carbon gels as electrocatalyst supports, moreover in comparison to the mostly microporous carbon blacks normally used in the preparation of the commercially available catalytic systems.

The synthesis of carbon xerogels is based on the polycondensation of resorcinol (R) and formaldehyde (F), normally in stoichiometric molar ratio, R/F = 0.5, and in the presence of a basifying agent, responsible for the deprotonation of the resorcinol molecule, commonly named "catalyst" in the existing literature.

There are nevertheless some studies in which an acidic catalyst was used in the sol-gel synthesis [Berthon, 2001; Brandt, 2004; Fairén-Jiménez, 2006]. Reactions taking place involve two steps: (i) an addition reaction between the deprotonated resorcinol (resorcinol anion) and formaldehyde, as shown in Figure 2a; and (ii) the condensation of the resorcinol-formaldehyde monomers, as shown in Figure 2b, what leads to the formation of the organic gel.

Generally, this organic gel is submitted to a solvent exchange, in order to substitute water (generally employed as solvent) with an organic solvent, with a lower surface tension constant ((for example, $\gamma_{Water(20\ °C)} = 72{,}75\cdot 10^{-3}\ N\cdot m^{-1}$; $\gamma_{Acetone(20\ °C)} = 23{,}7\cdot 10^{-3}\ N\cdot m^{-1}$)) [Job, 2005]. This treatment allows preserving the porous structure of the original gel.

Subsequently, the organic gel is submitted to a drying process that determines the type of material obtained:

- Aerogels, dried by super-critical drying (after solvent exchanges: water to acetone or ethanol, and then to CO_2), the gels are dried under supercritical conditions);
- Cryogels, obtained by freeze-drying (after freezing, the solvent is removed by sublimation under low pressure).
- Xerogels, dried under sub-critical conditions by means of evaporative drying, or convective drying [Leonard, 2005; Zubizarreta, 2008].

Finally, the gel is submitted to a pyrolysis treatment in an inert atmosphere, generally between 600 and 1200 ° C [Al-Muhtaseb, 2003].

Among all the synthesis parameters, those having a greater influence on the properties of the gel are: i) the pH of the starting mixture, ii) the concentration and type of reactant-precursors (the use of different substances such as cresol, furfural, etc., has been moreover considered [Li, 2001]) (iii) the drying conditions selected. Other parameters, such as the temperature and time of gelation and curing [Job, 2006[a]; Job, 2006[b]], the carbonization temperature [Kuhn, 1998] and the type of solvent used [Qin, 2001; Fairén-Jiménez, 2008], have as well a relatively important effect on the final gel properties.

Pekala et al. [Pekala, 1989; Pekala, 1990; Pekala, 1992[a]; Pekala, 1992[b] Gross, 1997; Pekala, 1998] carefully studied the relationship between the modification of each synthesis parameter and its particular influence on the properties of the final material obtained. These studies focused mainly on describing in detail the different reactions involved in the formation of the organic aerogels and the subsequent carbon aerogel obtained upon drying and calcination.

The variation of density, textural properties such as surface area and pore volume, mechanical, acoustical, optical, thermal and electrical properties, was evaluated as a consequence of the modification of synthesis parameters such as the molar ratio resorcinol-to-basifying agent (sodium carbonate), R/C, and the dilution ratio (solvent-to-reactant ratio). They concluded that the cross-linking extent and the surface area of the gel have a determinant effect on the rest of the above mentioned macroscopic properties of the final product. Concretely, these studies pointed out that using high amounts of basifying agent (low R/C around 50) yields materials with high surface areas but low pore volumes, due to pore collapse occurring to a greater extent.

On the other hand, low amounts of basifying agent (R/C around 300) result in materials possessing somehow lower surface area but increased pore volume. With respect to dilution ratio, lower values lead to shorter gelation times which results in a more dense final material. Subsequent studies considered the use of other precursors such as melamine [Pekala, 1992] or phenol [Pekala, 1995], as well as the influence of the carbonization temperature, which mainly influences the final surface area of the carbon material [Reinnacher, 1998]. Saliger and co-workers [Saliger, 1997] further studied the effect of the different synthesis parameters on the properties of the final product obtained and confirmed that it is possible to prepare carbon materials, both aerogels and xerogels, with high surface areas, minimizing the collapse of the pore structure, optimizing the gelation time and using even lower R/C ratios among 500 and 1500, even at higher solid concentrations than those considered by Pekala and co-workers.

It is as well remarkable the work performed by Job, Pirard and co-workers. They carefully optimized the synthesis procedure for obtaining high surface area carbon aerogels and xerogels, by means of selecting the appropriate pH, gelation and curing time and temperature, dilution ratio, etc., as well as the drying conditions (sub-critical drying), with the aim of avoiding the collapse of the gel porous structure [Job, 2004; Job, 2005; Leonard, 2005; Job, 2006[a]; Job, 2006[b] b; Job, 2007].

They further claimed that it is possible to tailor the material properties, i.e., in terms of pore size distribution, through a proper selection of the synthesis parameters. Menéndez and Arenillas recently presented a new microwave-based drying procedure which shortens considerably the overall synthesis process of carbon xerogels, maintaining their textural properties, similar to those of a conventional sub-critically dried materials [Zubizarreta, 2008; Menéndez, 2012].

Figueiredo and co-workers have studied in the last years the modification of the surface chemistry of carbon xerogels having different textural properties [Gomes, 2004; Samant, 2005; Figueiredo, 2006; Mahata, 2008]. Moreno-Castilla and Maldonado-Hódar have synthesized carbon aerogels and xerogels doped with different metals, such as Pt, Ag or Pt, as sorbents for the elimination of volatile organic compounds (VOCs) [Maldonado-Hódar, 1999; Moreno-Castilla, 2005].

2. Influence of the Synthesis Parameters in the Carbon Xerogel Properties

In the following sections, the influence of each separate synthesis parameter on the different physico-chemical properties of the produced carbon xerogel, will be analyzed in detail. A summary of these observations is presented in Table 1.

Table 1. Summary of observations on the influence of synthesis parameters on carbon xerogel properties

Parameter	*Influence*			*References*
R/C molar ratio	Influences initial pH of precursor mixture	Low R/C = high pH	Difficult aggregation and growth, small polymeric particles, less compact organic gel, higher tendency to collapse	Pekala, 1990; Tamon, 1998; Al-Muhtaseb, 2003; Job, 2004
		Medium-high R/C = low pH	Big polymeric particles, more compact organic gel, yielding a carbon material with higher pore volume	Job, 2005; Job, 2006; Zubizarreta, 2009
		High R/C = low pH	Upon atmospheric drying, gel structure starts to collapse	Alegre, 2012[a]
R/F molar ratio	Influences dilution and gelation rates	Low R/F = excess formaldehyde	Higher dilution of the reactants, near gelation limit, higher tendency to collapse	Al-Muhtaseb, 2003, Alegre, 2012[a]
		High R/F = excess resorcinol	Problems in gelation, microporous materials	Tamon, 1997, Maldonado-Hódar, 1999, Alegre, 2012[a]
Type of solvent	Acetone vs. water	More controlled evaporation and slower gelation, bigger primary particles, less compact gel		Fairén-Jiménez, 2008, Alegre, 2012[a]
Other	H_2SO_4 vs. Na_2CO_3	Enhanced crosslinking, more compact gel		Alegre, 2012[a]

2.1. Influence of the Molar Ratio Resorcinol-to-Sodium Carbonate (Initial pH)

The properties of the carbon xerogels depend to a very important extent on the relative amount of sodium carbonate (basifying agent, catalyst) in the initial mixture. This parameter determines at the same time the initial pH of the mixture of the precursors, since the main role of sodium carbonate is to provide a basic environment that promotes the deprotonation of resorcinol. Therefore one can expect that the relative amount of sodium carbonate will strongly determine the polymerization extent in the organic gel. Several studies [Pekala, 1990; Tamon, 1998; Al-Muhtaseb, 2003; Job, 2004] claim that a high amount or sodium carbonate, i.e., low resorcinol-to-sodium carbonate molar ratio, R/C around 50, or, correspondingly, high initial solution pH, results in a higher number of monomers resorcinol-formaldehyde (RF), see Figure 2a, due to favored and fast deprotonation of the resorcinol ring. Deprotonation is followed by the condensation and growth of these monomers, Figure 2b, yielding resorcinol-formaldehyde aggregates (RF). This aggregation reaction is however favored at low pH, that is, just the contrary of what one can expect at low R/C molar ratios. Therefore, high pH will difficult this aggregation and growth mechanism resulting in small polymeric particles, having sizes around 3-5 nm [Job, 2004]. The presence in the sol of these small primary particles generates an organic gel which is less compact [Sharma, 2009, Job, 2004]. Pekala et al. [Pekala, 1990] pointed out as well that these small size primary particles lead to organic gels having acceptable surface area and adequately developed porous structure that, however, tends to collapse more easily during the pyrolysis process resulting in a carbon material with a pore size distribution centered in smaller pore sizes. Figure 3 depicts this polymerization scenario at low R/C molar ratios, i.e., high pH

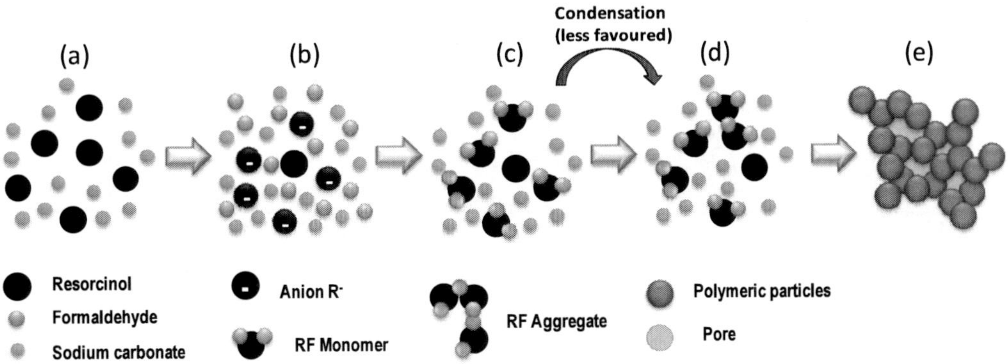

Figure 3. Representative scheme of the formation of an organic gel synthesized with low R/C ratio (≈50), where (a) activation of the resorcinol ring by the Na_2CO_3; (b) mixture between resorcinol anion and formaldehyde; (c) aggregate formation (by means of addition of formaldehyde to resorcinol); (d) aggregate condensation; (e) organic gel.

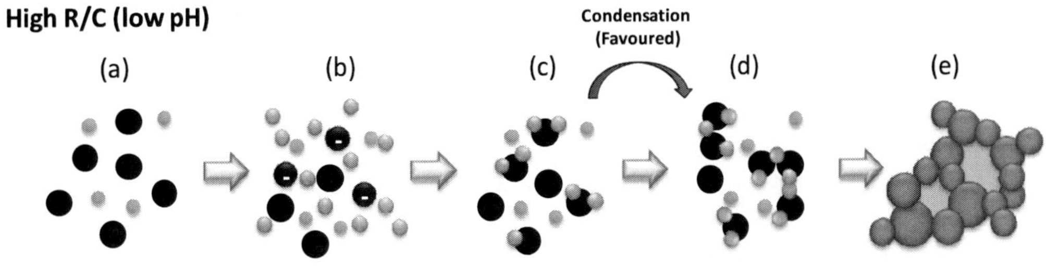

Figure 4. Representative scheme of the formation of an organic gel synthesized with high R/C ratio (≈800), where (a) activation of the resorcinol ring by the Na_2CO_3; (b) mixture between resorcinol anion and formaldehyde; (c) aggregate formation (by means of addition of formaldehyde to resorcinol); (d) aggregate condensation; (e) organic gel.

On the other hand, high resorcinol-to-sodium carbonate, R/C, molar ratios, i.e., low pH, result in big polymeric particles, among 16-200 nm diameter, yielding a carbon material with higher pore volume. More dense and therefore resistant organic gel structure results in a lower tendency to pore collapse during the pyrolysis process as shown in Figure 4.

Alegre et al. synthesized carbon xerogels using different resorcinol to sodium carbonate (R/C) molar ratios, namely 50, 800 and 1500 [Alegre, 2012[a]]. They reported that the carbon material synthesized using R/C = 800, possessed higher mesoporous volume and showed increased mean pore size in comparison to the one prepared at R/C = 50. This fact can be explained in terms of the mechanism depicted in Figures 3 and 4. However, they observed that a further increase of R/C molar ratio from 800 to 1500 resulted in a carbon xerogel of substantially lower pore volume and much lower surface area in comparison to the materials synthesized at R/C values of 50 and 800, contrarily to what was reported in the literature [Job, 2005; Job, 2006; Zubizarreta, 2009]. It is important to remark that, Alegre et al. employed an atmospheric evaporation procedure for the drying of the organic gels synthesized in their work. This more aggressive drying, vis-à-vis i.e., supercritical drying or a controlled-pressure sub-critical drying, may result in a more extensive collapse of the pore structure, especially at extremely high R/C ratios yielding wide mesopores and macropores that do not withstand the evaporation forces generated upon drying. Pore collapse will therefore result in a material with a less developed porous structure. In any case, Alegre and co-workers were able to synthesize carbon xerogels with surface areas around 400-600 m^2/g and considerably high mesopore volume, i.e., 0.96 cm^3/g, accounting almost 90% of the pore volume, average pore size 23.3 nm, in the case of the carbon xerogel prepared using R/C=800 (resorcinol-to-formaldehyde molar ratio, R/F = 0.5).

The relative amount of sodium carbonate, i.e., R/C molar ratio, has as well a very strong influence in the structural properties of the synthesized carbon materials. Alegre et al. [Alegre, 2012[a]] analyzed by means of X-ray photoelectron spectroscopy (XPS) and Raman spectroscopy the structural differences of a series of carbon xerogels prepared using R/C = 50, 800 and 1500. They observed in general higher ordering degree for the carbon materials synthesized using higher R/C, i.e., the carbon xerogels prepared using R/C = 800 and 1500 presented a C1s band in XPS spectra with a marked C-C contribution, whereas for the carbon xerogel prepared using R/C = 50, contribution of functionalized C (-C-O and –C=O) was much more significant. Raman confirmed XPS results, showing that the relative intensity of the bands D and G (I_D/I_G decreased in the case of the xerogels synthesized at higher R/C ratios. The authors explain this fact as a consequence of the slower gelation process taking place at low sodium carbonate concentrations, resulting in a denser and more cross-linked organic gel, with a high degree of polymerization that upon pyrolysis yielded a more organized material, presenting a lower number of structural defects.

Moreover, Alegre and co-workers [Alegre, 2012[a]] presented as well in their work, the results of the measurement of the electrical conductivity of their synthesized carbon xerogels. Such measurements provided further evidence on the higher ordering degree in some of the synthesized materials. In agreement with the results obtained in the characterization of these materials by means of XPS and Raman spectroscopy, the carbon xerogels prepared at R/C = 800 and 1500 presented higher conductivity within the measured pressure interval, pointing in fact to a less defected structure in these cases.

Oxygen content, i.e., presence of O heteroatoms, either coming from leftovers of unreacted material or in the form of oxygen surface functionalities, was found to be lower for a R/C ratio of 800 [Alegre, 2012[a]], as a consequence of higher degree of polymerization and in complete agreement with the more organized and C-C enriched structure revealed by XPS and Raman analyses.

2.2. Influence of the Molar Ratio Resorcinol-to-Formaldehyde

Textural and structural properties can be also modified by changing the molar ratio resorcinol-to-formaldehyde in the sol-gel synthesis. Al-Muthaseb and co-workers [Al-Muhtaseb, 2003] described how a resorcinol-to-formaldehyde molar ratio, R/F, under the stoichiometric value of 0.5, results in an excess of formaldehyde, and therefore, in a higher dilution of the reactants. This higher dilution effect results in bigger primary particle and aggregate sizes, near the gelation limit of sol. Therefore, upon drying, the organic gel structure suffers from a more dramatic pore collapse resulting in a material with smaller average pore size, and lower total and mesopore volumes. Alegre et al. confirmed this observation [Alegre, 2012[a]]. They synthesized carbon xerogels using R/F = 0.25, 0.5 and 1. The carbon material produced at R/F = 0.25 possessed smaller mean pore size and lower pore volume when compared to the xerogel synthesized using the stoichiometric resorcinol-to-formaldehyde molar ratio, R/F = 0.5. Tamon et al. [Tamon, 1997] previously reported similar collapse of the mesopore structure, due to the presence of remaining formaldehyde after sol-gel polycondensation.

On the other hand, a further increasing R/F molar ratio up to a value of 1, i.e., excess resorcinol, resulted in a predominantly microporous carbon material [Alegre, 2012[a]]. Tamon et al. [Tamon, 1997] described the difficulties in drying the gel prepared using R/F = 1, due to incomplete gelation of the sol. Maldonado-Hódar et al. [Maldonado-Hódar, 1999] reported lower surface areas and pore volume corresponding to pore diameter higher than 50 nm, for carbon aerogels synthesized using increasing R/F molar ratios.

Regarding the influence of the resorcinol-to-formaldehyde molar ratio, R/F, on the structural properties of carbon xerogels, Alegre et al. [Alegre, 2012[a]] observed a somehow more ordered structure in the case of the carbon material synthesized using the stoichiometric ratio, R/F = 0.5, than both the carbon xerogels prepared using the under-stoichiometric R/F =0.25 or over-stoichiometric R/F = 1 ratios. Note that, however, these three materials were synthesized using a R/C ratio = 50, far from the optimal value of 800 found to lead to a more organized xerogel structure. In fact, Raman evidenced almost no difference among this three materials. Almost no differences were moreover observed in terms of electric conductivity. This fact may help us to conclude that, therefore, resorcinol-to-sodium carbonate molar ratio, R/C, plays a more decisive role in the sol-gel cross-linking mechanism and determines more strongly the final organic gel structure than the molar ratio resorcinol-to-formaldehyde, R/F, at least under relative sodium carbonate excess and fast resorcinol deprotonation conditions. Nevertheless, considerably higher oxygen content was measured for the xerogel prepared using R/F = 1 in comparison to the one synthesized at the stoichiometric ratio, i.e., 4 wt% vs. 1.9 wt% respectively. This higher oxygen content points to a far from complete polymerization in the case of this material.

2.3. Influence of the Type of Solvent

Alegre et al. [Alegre, 2012[a]] prepared a carbon xerogel partially substituting water by acetone during the sol-gel synthesis procedure. These authors observed that the presence of acetone as solvent in the synthesis mixture resulted in a more controlled evaporation, due to the lower surface tension of acetone in comparison to water. Moreover, most probably the presence of acetone in the media slows down the gelation rate, yielding higher primary particle size and a less compact gel, to a much lesser extent, but having a similar effect than the increase in resorcinol-to-sodium carbonate molar ratio, R/C. Fairén-Jiménez and co-workers described a similar influence of acetone as solvent in the preparation of carbon aerogels [Fairén-Jiménez, 2008]. Previous studies [Qin, 2001] reported that the addition of a certain amount of acetone to the main solvent, i.e., water, results in a slower gelification. Moreover, it is well know that resorcinol-formaldehyde aggregates are less soluble in acetone (aptrotic solvent) than in water. Due to this fact, nucleation regime is shorter, yielding smaller size primary particles and therefore a gel with narrower porosity. Berthon-Fabry and co-workers [Berthon-Fabry, 2004] claimed that due to the low solubility of sodium carbonate in acetone, one can expect a slower formation of hydroxymethyl groups, which will result in an incomplete gelification of the mixture. In fact, as a consequence of this, it is normal to add an acid and adjust the pH during the gelification step. In this way, Alegre et al. [Alegre, 2012[a]] added HNO_3 2N until a stable pH = 6 was attained. Acid addition provides protons that react with the hydroxyl group in the hydroxymethyl molecules, forming methylene bridges leading to a higher crosslinking extent [Pekala, 1989]. As a result of slower gelification and lower degree of crosslinking, the carbon xerogel synthesized partially substituting water by acetone as solvent shows a less ordered structure, as evidenced by the XPS and Raman analyses perfomed by Alegre and co-workers [Alegre, 2012[a]].

2.4. Other Parameters: Catalyst and Type of Precursors

Alegre et al. [Alegre, 2012[a]] reported as well some results on the influence of employing sulfuric acid instead of sodium carbonate as the basifying agent. Sometimes referred as “catalyst”, this component of the sol-gel mixture will be in charge of further accelerating the gelation process [Al-Muhtaseb, 2003]. Alegre et al. observed in fact faster gelation when adding H_2SO_4, leading to a highly crosslinked and thus more dense organic gel structure which yields a completely microporous carbon material [Alegre, 2012[a]]. XPS and Raman evidenced however a low ordering degree in the structure of this carbon xerogel.

3. Influence of the Carbon Xerogel Textural, Structural and Chemical Properties in the Features and Electrochemical Behavior of Pt Catalysts

As already pointed out during the abstract and introduction of this chapter, the particular textural, structural and chemical properties of each of the carbon xerogels prepared through the several possible modifications of their synthesis procedure, result in different physicochemical features and electrochemical behavior of supported Pt catalysts. Table 2 summarizes the observations described herein.

Table 2. Influence of textural and structural properties on the features of carbon xergogel-supported Pt-catalysts

Parameter	*Influence*			*References*
Textural properties	Pt distribution and pore blockage	Less developed porous structure	Higher tendency to pore blockage, lower degree of dispersion, bigger Pt crystal size	Alegre, 2012[a]; Alegre, 2012[b]
Structural properties	Ordering degree and polymerization degree	High degree of ordering (graphitic)	Higher interaction between active phase and support, higher degree of dispersion, lower Pt particle size and more reduced Pt active phase	

ITextural properties, i.e., pore size distribution, of the carbon xerogel strongly determines Pt distribution and pore blockage occurring as a consequence of the loading of the active metal. This becomes especially true for electrocatalysts, which are normally prepared using high metal loads between 20-80%. Therefore, adequate pore size distribution, with an important part of it belonging to mesopores, preferably wide mesopores, is always desirable for the preparation of such catalytic systems [Arbizzani, 2007; Job, 2006; Alegre, 2012[b]].

Alegre et al. [Alegre, 2012[a]; Alegre, 2012[b]] observed pore blockage occurring to a higher extent when the carbon xerogel employed as support possessed a less developed porous structure, consequence of using a lower resorcinol-to-sodium carbonate molar ratio, R/C = 50. When using the carbon material synthesized using a higher R/C ratio, i.e., R/C = 800, higher pore volume, and above all wider pores, i.e., average pore diameter 23.3 nm in comparison to 11.7 nm determined for the material prepared at R/C = 50, active phase was more adequately distributed within the porous structure of this material, leading to a lower deterioration of the textural properties of the material upon Pt loading pointing to lower extent of pore blockage.

Pt crystal size was found as well to be strongly dependent on the carbon xerogel properties. In this case, most probably a combination of textural properties favoring metal distribution and structural properties leading to a better interaction between active phase and support. Figure 5 shows an schematic representation of carbon xerogels properties, affecting crystal size. Different carbon xerogels are shown in X axis, synthesis under different R/C and R/F ratios. Different synthesis conditions lead to carbon materials with different properties (see I_D/I_G ratio calculated from Raman spectra and atomic percentage of C-C bonds obtained from XPS (on left Y axis vs crystal size of the catalyst supported on these carbon materials shown. on the right Y axis, is The lower crystal sizes are obtained for catalysts supported on carbon xerogels with high ordering degree and high atomic percentage of C-C bonds , *i.e.*, those synthesized under stoichiometric R/F and high R/C ratio. Data are an adaptation of the results recently presented by Alegre and co-workers [Alegre, 2012[a]; Alegre, 2012[b]]. For example, enhanced textural properties in combination with a more ordered structure in the case of the carbon xerogel prepared using a R/C ratio of 800 resulted in a lower Pt crystal size, i.e., 3.6 nm vs. 5.3 nm measured for the catalyst supported on the carbon xerogel prepared at R/C = 50.

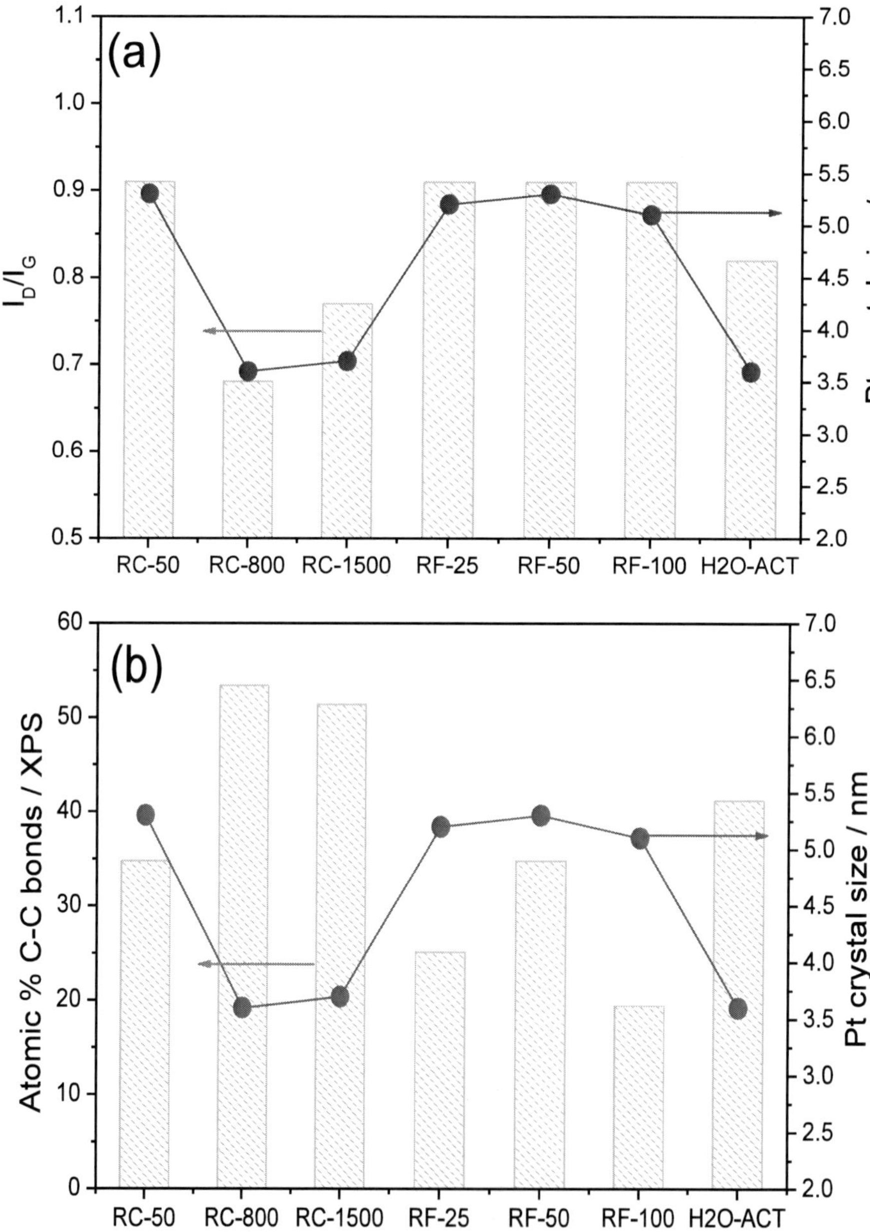

Figure 5. Carbon xerogels properties related to crystal size obtained for carbon-xerogel supported catalysts (a) I_D/I_G ratio obtained from Raman spectra (left y axis) vs. Pt crystal size (right y axis); (b) Atomic percentage of C-C bonds obtained from XPS (left y axis) vs. Pt crystal size (right y axis).

TEM images acquired for this series of catalysts, shown in Figure 6, confirmed these Pt crystal sizes calculated from the XRD patterns and evidenced a more uniform distribution of the metallic particles when using the carbon xerogel possessing the more developed textural properties and more organized structure.

Moreover, XPS analysis pointed to a more favorable Pt reduction when using highly ordered carbon xerogel (i.e., synthesized at high ratios, R/C = 800 and R/C=1500 when compared to the one prepared using R/C = 50), as shown in Table 1.

Alegre et al. [Alegre, 2012[a]; Alegre, 2012[b]] studied both the methanol oxidation reaction (MOR) and the oxygen reduction reaction (ORR) in a half-cell for previously mentioned Pt-catalysts.

Figure 7 shows chronoamperometric tests at 600 mV for the MOR (Figure 7a) and polarization curves for the ORR (Figure 7b). As shown, catalyst supported on the carbon support with the highest pore volume and the highest order degree, *i.e*, CXG-RC-800 presented the highest activity towards both reactions. Its best performance was ascribed to a combination of appropriate crystal size, high amount of reduced Pt and high pore volume, favoring reactants diffusion.

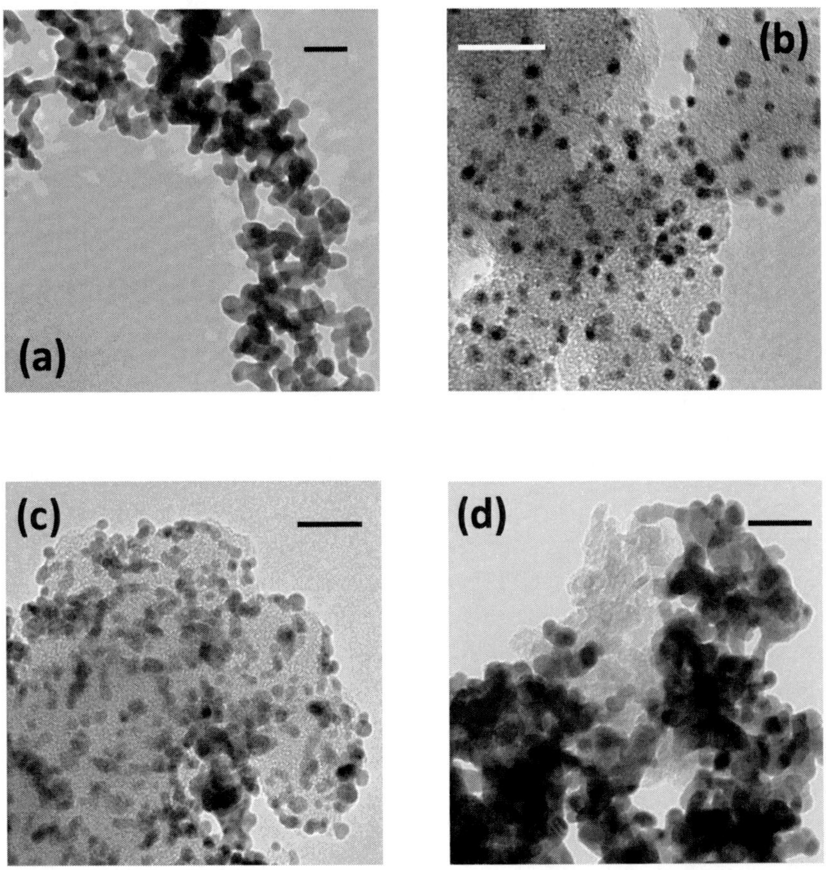

Figure 6. TEM micrographs for catalysts supported on: (a) Pt/CXG-RC-50 (b) Pt/CXG-RC-800 and (c) Pt/CXG-RC-1500 (d) Pt/CXG-RF-100.

Table 1. Pt crystal size determined by means of XRD, and relative intensity of Pt species, determined from XPS for carbon xerogel-supported catalysts

Catalyst	Pt crystal size (nm)	Pt^0	PtO	PtO_2
Pt/CXG-RC-50	5,3	40,1	45,3	14,6
Pt/CXG-RC-800	3,6	65,6	26,8	7,6
Pt/CXG-RC-1050	3,7	66,8	16,3	16,9
Pt/CXG-RF-25	5,1	48,0	41,2	10,8
Pt/CXG-RF-50	5,3	40,1	45,3	14,6
Pt/CXG-RF-100	5,2	43,5	33,2	23,3

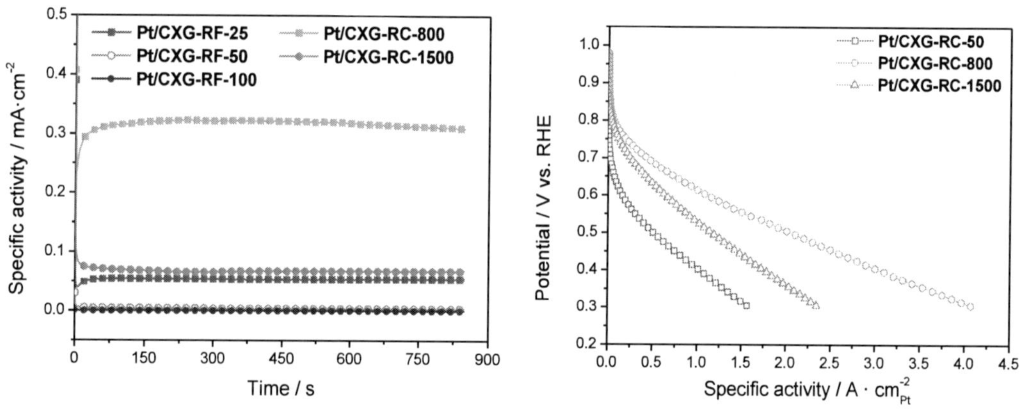

Figure 7. (a) Chronoamperometries at 600 mV in a 2M CH_3OH + 0.5M H_2SO_4 solution (b) polarization curves in a 0.5M H_2SO_4 solution feeding pure O_2, for carbon-xerogel-supported catalysts. Room temperature.

4. Potential Commercialitaion

Due to the high novelty of these materials, to date and to the best of our knowledge, there is not any published study in the existing literature reporting the cost and potential commercialization of carbon aerogels or xerogels. However, a rough estimation performed recently, taking into account the synthesis route we followed and the reactants employed, yields an approximate value around 0.50 $ g^{-1} for our carbon xerogels. This estimated relatively low production cost forecasts a highly perspective potential commercialization for these materials, if they were synthesized at an industrial level, for the different applications considered so far, i.e., absorbents and catalyst supports.

Conclusion

Carbon xerogels can be synthesized under a great variety of synthesis conditions. Initial pH is the key factor determining the final properties of these carbon materials. Synthesis conditions of the gels strongly influence the textural and structural properties of the final carbon material. Non-stoichiometric mixtures of resorcinol and formaldehyde yield partially polymerized gels of weak and poorly developed structure. Faster gelation in the presence of an acidic catalyst results in a too compact gel, with a low extent of development of its porous structure. On the other hand, an increase in resorcinol/catalyst molar ratio results in slower gelation, higher polymerization extent resulting in highly mesoporous materials (R/C = 800) or even macroporous (R/C = 1500), which further presented less defected—a more ordered—structure. Synthesis conditions also determine their surface chemistry.

Properties of carbon xerogels determine final properties of Pt catalysts. For example high Pt crystal sizes are obtained for those catalysts prepared using the low polymerized and less ordered carbon xerogels, whereas Pt crystal sizes around 3.5 nm are obtained for the carbon materials possessing lower concentration of defects in their structure.

With respect to the catalytic activity of these carbon xerogel-based Pt-catalysts in both the electrochemical methanol oxidation (MOR) and oxygen reduction reactions (ORR), more ordered supports resulted in higher values of specific activity, due to enhanced metal-support interaction when using carbon materials with a lower concentration of defects, which was most probably due to an increased presence of effectively reduced Pt on their surface.

Acknowledgments

Authors acknowledge the financial support for the research work developed and compiled in this chapter from the Ministry of Economy and Competitiveness under the project CTQ2011-28913-C02-01.

References

Alegre C., Sebastián D., Baquedano E., Gálvez M.E., Moliner R., Lázaro M.J. Tailoring Synthesis Conditions of Carbon Xerogels towards Their Utilization as Pt-Catalyst Supports for Oxygen Reduction Reaction (ORR). *Catalysts,* 2 (2012[a]) 466-489.

Alegre C., Gálvez M.E., Baquedano E., Pastor E., Moliner R., Lázaro M.J. Influence of support's oxygen functionalization on the activity of Pt/carbon xerogels catalysts for methanol electro-oxidation. *International Journal of Hydrogen Energy,* 37 (2012[b]) 7180-7191.

Al-Muhtaseb S.A., Ritter J.A. Preparation and Properties of Resorcinol-Formaldehyde Organic and Carbon Gels. *Advanced Materials,* 15 (2003) 101-114.

Antolini E. Carbon supports for low-temperature fuel cell catalysts. *Applied Catalysis B,* 88 (2009) 1-24.

Arbizzani C., Beninati S., Manferrari E., Soavi F., Mastragostino M. Cryo- and xerogel carbon supported PtRu for DMFC anodes. *Journal of Power Sources,* 172 (2007) 578-586.

Berthon S., Barbieri O., Ehrburger-Dolle F., Geissler E., Achard P., Bley F., Hecht A.M., Livet F., Pajonk G.M., Pinto N., Rigacci A., Rochas C. DLS and SAXS investigations of organic gels and aerogel. *Journal of Non-Crystalline Solids,* 285 (2001) 151-161.

Berthon-Fabry S., Langohr D., Achard P., Charrier D., Djurado D., Ehrburger-Dolle F Anisotropic high-surface-area carbon aerogels. *Journal of Non-Crystalline Solids,* 350 (2004) 136-144.

Brandt R., Fricke J. Acetic-acid-catalyzed and subcritically dried carbon aerogels with a nanometer-sized structure and a wide density range. *Journal of Non-Crystalline Solids,* 350 (2004) 131-135.

Du, H.; Li, B.; Kang, F.; Fu, R.; Zeng, Y. Carbon aerogel supported Pt–Ru catalysts for using as the anode of direct methanol fuel cells. *Carbon,* 45, (2007) 429–435.

Fairén-Jiménez D., Carrasco-Marín F., Moreno-Castilla C. Porosity and surface area of monolithic carbon aerogels prepared using alkaline carbonates and organic acids as polymerization catalysts. *Carbon,* 44 (2006) 2301-2307.

Fairén-Jiménez D., Carrasco-Marín F., Moreno-Castilla C. Inter- and Intra-Primary-Particle Structure of Monolithic Carbon Aerogels Obtained with Varying Solvents. *Langmuir,* 24 (2008) 2820-2825.

Figueiredo J.L., Pereira M.F.R, Serp P., Kalck P., Samant P.V., Fernandes J.B. Development of carbon nanotube and carbon xerogel supported catalysts for the electro-oxidation of methanol in fuel cells. *Carbon,* 44 (2006) 2516-2522.

Gomes H.T., Samant P.V., Serp Ph., Kalck Ph., Figueiredo J.L., Faria J.L. Carbon nanotubes and xerogels as supports of well-dispersed Pt catalysts for environmental applications. Applied Catalysis B: *Environmental,* 54 (2004) 175-182.

Gross A.F., Vajo J.G., Van Atta S.K., Olson G.L. Enhanced Hydrogen Storage Kinetics of LiBH4 in Nanoporous Carbon Scaffolds. *Journal of Physical Chemistry B,* 112 (2008) 5651–5657.

Job N., Pirard R., Marien J., Pirard J.P. Porous carbon xerogels with texture tailored by pH control during sol-gel process. *Carbon,* 42 (2004) 619-628.

Job N., Théry A., Pirard R., Marien J., Kocon L., Rouzaud J.N., Béguin F., Pirard J.P. Carbon aerogels, cryogels and xerogels: Influence of the drying method on the textural properties of porous carbon materials. *Carbon,* 43 (2005) 2481-2494.

Job N., Sabatier F., Pirard J.P., Crine M., Lèonard A. Towards the production of carbon xerogel monoliths by optimizing convective drying conditions. *Carbon,* 44 (2006[a]) 2534-2542.

Job N., Panariello F., Marien J., Crine M., Pirard J.P., Lèonard A. Synthesis optimization of organic xerogels produced from convective air-drying of resorcinol-formaldehyde gels.. *Journal of Non-Crystalline Solids,* 352 (2006[b]) 24-34.

Job N., Panariello F., Crine M., Pirard J.P., Lèonard A. Rheological determination of the sol–gel transition during the aqueous synthesis of resorcinol-formaldehyde resins. Colloids and Surfaces A: *Physicochemical and Engineering Aspects,* 293 (2007) 224-228.

Kuhn J., Brandt R., Mehling H., Petricevic R., Fricke J. In situ infrared observation of the pyrolysis process of carbon aerogels. *Journal of Non-Crystalline Solids,* 225 (1998) 58-63.

Léonard A., Job N., Blacher S., Pirard J.P., Crine M., Jomaa W. Suitability of convective air drying for the production of porous resorcinol–formaldehyde and carbon xerogels. *Carbon,* 43 (2005) 1778–1814.

Li, W.C., Lu, AH., and Guo, SC.; Characterization of the microstructures of organic and carbon aerogels based upon mixed cresol–formaldehyde. Carbon 39 (2001) 1989-1994.

Liang C., Sha G., Guo S. Resorcinol-formaldehyde aerogels prepared by supercritical acetone drying. *Journal of Non-Crystalline Solids,* 271 (2000) 167-170.

Liang, C.; Li, Z.; Dai, S. Mesoporous Carbon Materials: Synthesis and Modification.. *Angew. Chem. Int. Ed.,* 47 (2008) 3696–3717.

Lu X., Arduini-Schuster M.C., Kuhn J., Nilsson O., Fricke J., Pekala R.W. Thermal Conductivity of Monolithic Organic Aerogels. *Science,* 255 (1992[b]) 971-972.

Machado B.F., Gomes H.T., Serp Ph., Kalck Ph., Figueiredo J.L., Faria J.L. Carbon xerogel supported noble metal catalysts for fine chemical applications. *Catalysis Today,* 149 (2009) 358-364.

Machado B.F., Morales-Torres S., Pérez-Cadenas A.F., Maldonado-Hódar F.J., Carrasco-Marín F., Silva A.M.T., Figueiredo J.L., Faria J.L. Preparation of carbon aerogel supported platinum catalysts for thc selective hydrogenation of cinnamaldehyde. Applied Catalysis A: *General,* 425-426 (2012) 161-169.

Mahata N., Silva A.R., Pereira M.F.R., Freire C., de Castro B., Figueiredo J.L Anchoring of a [Mn(salen)Cl] complex onto mesoporous carbon xerogels. *Journal of Colloid and Interface Science,* 311 (2007) 152-158.

Mahata N., Pereira M.F.R., Suárez-García F., Martínez-Alonso A., Tascón J.M.D., Figueiredo J.L. Tuning of texture and surface chemistry of carbon xerogels. *Journal of Colloid and Interface Science,* 324 (2008) 150-155.

Maldonado-Hódar F.J., Ferro-García M.A., Rivera-Utrilla J., Moreno-Castilla C. Synthesis and textural characteristics of organic aerogels, transition-metal-containing organic aerogels and their carbonized derivatives. *Carbon,* 37 (1999) 1199-1205.

Menéndez J.A., Juárez-Pérez E.J., Ruisánchez E., Calvo E.G., Arenillas A. A microwave-based method for the synthesis of carbon xerogel sphere. *Carbon,* 50 (2012) 3555-3560.

Moreno-Castilla C., Maldonado-Hódar F.J. Carbon aerogels for catalysis applications: an overview. *Carbon,* 43 (2005) 455-463.

Mrowiec-Białło J., Jarzbski A. B., Lachowski A. I., Malinowski J. J. Two-component aerogel adsorbents of water vapour. Journal of Non-Crystalline Solids, 225 (1998) 184-187.

Pekala R.W. Organic aerogel from polycondensation of resorcinol with formaldehyde. *Journal of Material Science,* 24 (1989) 3221-3227.

Pekala R.W., Alviso C.T., LeMay J.D Organic aerogels: microstructural dependence of mechanical properties in compression. *Journal of Non-Crystalline Solids,* 125 (1990) 67-75.

Pekala R.W., Alviso C.T., Kong F.M., Hulsey S.S. Aerogels derived from multifunctional organic monomers. *Journal of Non-Crystalline Solids,* 145 (1992[a]) 90-98.

Pekala R.W., Alviso C.T., Lu X., Gross J., Fricke J. New organic aerogels based upon a phenolic-furfural reactio. *Journal of Non-Crystalline Solids,* 188 (1995) 34-40.

Pekala R.W., Farmer J.C., Alviso C.T., Tan T.D., Mayer S.T., Miller J.M., Dunn. B Carbon aerogels for electrochemical applications. *Journal of Non-Crystalline Solids,* 225 (1998) 74-80.

Qin G., Guo S. Preparation of RF organic aerogels and carbon aerogels by alcoholic sol–gel process. *Carbon,* 39 (2009) 1935-1937.

Reichenauer G., Emmerling A., Fricke J., Pekala R.W. Microporosity in carbon aerogels. *Journal of Non-Crystalline Solids,* 225 (1998) 210-214.

Saliger R., Bock V., Petricevic R., Tillotson T., Geis S., Fricke J. Carbon aerogels from dilute catalysis of resorcinol with formaldehyde. *Journal of Non-Crystalline Solids,* 221 (1997) 144-150.

Samant P. V., Rangel C. M., Romero M. H., Fernandes J. B., Figueiredo J. L. Carbon supports for methanol oxidation catalyst. *Journal of Power Sources,* 151 (2005) 79-84.

Sharma C.S.; Kulkarni M.M.; Sharma A.; Madou A. Synthesis of carbon xerogel particles and fractal-like structures. *Chemical Engineering Science,* 64 (2009) 1536-1543.

Tamon H.; Ishizaka H.; Mikami M.; Okazaki M. Porous structure of organic and carbon aerogels synthesized by sol-gel polycondensation of resorcinol with formaldehyde. *Carbon,* 35 (1997) 791-796.

Tamon H., Ishizaka H. Porous characterization of carbon aerogels. Carbon, 36 (1998) 1397-1409.

Zeng J., Francia C., Dumitrescu M.A. , Monteverde Videla A.H.A., Vijaykumar S.I., Specchia S. Spinelli P. Electrochemical Performance of Pt-Based Catalysts Supported on Different Ordered Mesoporous Carbons (Pt/OMCs) for Oxygen Reduction Reaction. *Ind. Eng. Chem. Res,,* 51 (2012[a]) 7500-7509.

J. Zeng, C. Francia, C. Gerbaldi, M. A. Dumitrescu, S. Specchia, P. Spinelli Smart synthesis of hollow core mesoporous shell carbons (HCMSC) as effective catalyst supports for methanol oxidation and oxygen reduction reactions. *Journal of Solid State Electrochemistry,* 16 (2012[b]) 3087-3096.

Zhang L., Liu H., Wang M., Chen L. Structure and electrochemical properties of resorcinol–formaldehyde polymer-based carbon for electric double-layer capacitors. *Carbon,* 45 (2007) 1439-1445.

Zubizarreta L., Arenillas A., Menéndez J.A., Pis J.J., Pirard J.P., Job N. Microwave drying as an effective method to obtain porous carbon xerogels. *Journal of Non-Crystalline Solids,* 354 (2008) 4024-4026.

Zubizarreta L., Arenillas A., Menéndez J.A., Pis J.J. Carbon materials for H_2 storage. *International Journal of Hydrogen Energy,* 34 (2009) 4575-4581.

In: Comprehensive Guide for Mesoporous Materials, Volume 4 ISBN: 978-1-63482-092-9
Editor: Mahmood Aliofkhazraei

Chapter 14

A NEW ROUTE TO ACHIEVE TARGETING MATERIALS: FROM INORGANIC-ORGANIC HYBRID PRECURSORS TO INORGANIC POROUS MATERIALS

***Rui Wu*[1,2] *and Bin Zhang*[1,3*]**

[1]Department of Chemistry, School of Science, Tianjin University
[2]School of Chemical Engineering and Technology, Tianjin University, China
[3]Collaborative Innovation Center of Chemical Science and Engineering, Tianjin, China

ABSTRACT

Transforming from inorganic-organic hybrid precursors into inorganic compounds is a newly emerged but versatile and efficient route to achieve inorganic porous materials, as the extraction of organic components in hybrid precursors is helpful for generating porous structure. This novel route has triggered increasing attention in recent years, as one can easily prepare porous materials with good control of morphology, structure and composition, as well as enhanced properties. This chapter reviews progress in this field, first introducing several methods to synthesize inorganic-organic hybrids, especially those that can be used as the starting materials for this transformation reaction, then dividing all transformation strategies into several categories: solid state decomposition, organic component stripping, couple of chemical transformation and organic component stripping, sonication, acidolysis and light irradiation, followed by highlight of superior properties according to different unique (mainly porous) structures of as-prepared products. Also, some prospects will be put forward briefly at the end of this chapter.

* E-mail: bzhang@tju.edu.cn.

1. Introduction

Inorganic materials, especially those composed of porous transition metal compounds, are of great interest due to their unique physical and chemical properties, together with their potential applications in fields of photocatalysis, solar cells, energy storage, biological assays and separations, detections and sensors, TFTs, supercapacitors and LEDs [1, 2]. While many efforts have been focused on the controlled synthesis of 1D, 2D and 3D porous morphologies using traditional methods including chemical vapor decomposition (CVD), electrodeposition (ED), template-directed synthesis, laser-assisted catalytic growth (LCG) and so on [3, 4], there do exist some shortcomings, such as harsh conditions, low efficiency, lattice defects and low electron-transfer rate. So it is emergent to put forward a general, facile and more effective way to synthesize inorganic porous materials with good control of shape, structure and composition as well as enhanced properties. In order to better control these parameters, diverse effective methods have arisen. Among them, thermal decomposition is a most traditional one, while chemical transformation is a most novel one. Thermal decomposition can be utilized in both solid and liquid phase and result in certain objective materials, namely solid state decomposition and organic component stripping called in this chapter, respectively. J. Yu and coworkers synthesized titania thin films from $(NH_4)_2TiF_6$ aqueous solution addition of boric acid by the new liquid-phase deposition (LPD) method. The effect of following use of calcination at different temperatures on their photocatalytic activity was investigated in detail [5]. Chemical transformation can be divided into four categories as reported by Xia group: alloy formation, galvanic replacement, cation exchange and anion exchange [6]. D. H. Son et al. have showed a versatile cation exchange route to expand the range of materials with diverse compositions, structures and morphologies [7]. However, despite the rapid development in fields of chemical synthesis, the fabrication of porous structure is still mainly based on traditional template method, inevitably introducing redundant and even poisonous elements, which enlarges the complexation of the whole reaction system and does not meet the conception of green chemistry.

Recently, inorganic-organic hybrid materials have been undergoing a rapid development since the pioneer work by researchers at Mobil Research and Development Corp. in 1992 [36, 37]. As one of the most promising classes of materials, they are not only the simple mixture of inorganic building blocks and organic components, they combine the stiffness, thermal stability and excellent properties such as electronic, magnetic and optical of inorganic blocks with the structural diversity, flexibility and processability of organic components to generate a synergetic effect, leading to several enhanced known and even some unknown properties [21], which make them good candidates in supercapacitor [8], gas sensors [9, 10], therapeutic and diagnostic imaging [11], light-controlled p-n junction [12] and direct methanol fuel cells [13]. Except the application of themselves directly, they can also act as precursors to transform into inorganic counterparts. It's well known that organic components play a key role in the formation of inorganic-organic hybrids, representing a general route to prepare 3D networks from inorganic 1D chain or 2D slabs [18-20]. So using different methods, such as thermal decomposition and chemical transformation mentioned above, to extract organic components from inorganic-organic hybrids may possibly generate inorganic porous compounds with singular shapes, architectures and reinforced properties that are much more difficult to achieve via traditional direct inorganic routes, where the extraction procedure of organic components plays an important role in the preparation of porous structure.

Based on this speculation, this chapter reviews a novel transformation route to prepare inorganic porous materials from inorganic-organic hybrid precursors, and introduces several strategies, including solid state decomposition, organic stripping strategy, coupling of chemical transformation and organic component stripping, sonication, acidolysis and light irradiation, all of which will be highlight focusing on the crystalline, growth mechanisms and excellent properties with special concentration on porous products. Besides, methods to prepare inorganic-organic hybrids and some prospects will also be discussed concisely.

2. Methods to Synthesize Inorganic-organic Hybrids

As a newly emerged class of compounds, inorganic-organic hybrids must be synthesized before using instead of purchasing directly, so this chapter will first introduce some strategies for the preparation of these hybrids, mainly two parts, solid state reactions and wet chemical reactions.

Evidently, solid mixing of two individual parts will certainly generate a hybrid, namely a solid-state reaction [14, 15]. As a counterpart, wet chemical reactions, mainly hydrothermal and solvothermal ones, are able to generate hybrids with unique morphologies, structures and properties, which can be divided into sol-gel technology and polymeric complexation reactions depending on different sublattices [16]. For those hybrids which can be transformed into inorganic functional materials, organic molecules are always intercalated into inorganic building blocks, clearly resulting from sol-gel procedures, which will be shown in detail as followings, especially the intercalation method, the products of which will be fully highlight later in this chapter.

2. 1. Sol-gel Technology

In case of sol-gel method, the sublattices are always three-dimensional inorganic compounds, while organic groups will be covalently bonded to inorganic building blocks, and the products are always called “inorganic-organic hybrids”.

Notably, aiming at the goal to fabricate original nanomaterials with distinct structures and properties compared with their solid and bulk counterparts, several novel synthetic methods have been fully explored to prepare inorganic-organic hybrids. Recently, the intercalation method is of great interest because it can intercalate organic molecules, especially some small organic amine ones, to 1D and 2D inorganic building blocks, forming a general strategy from 1D chains and 2D slabs into homogeneous 3D stereoscopic architectures composed of not only layered structures but also non-layered ones [17]. What's more, products generated by this method seem no limitation as inorganic building blocks and organic components can be exchanged by many analogic species. Finally, properties are different slightly but inevitably from each other due to the synergetic effect of both inorganic and organic parts, and also, some special ones emerge, such as the quantum confinement effect (QCE) of hybrids consisting of II-VI slabs or chains with small organic amine molecules, which is pretty important because before these examples, QCE only existed for very small, chemically grown and colloids dots, and they would be virtually impossible. For example, Li group have successfully obtained [(MQ) ($L_{0.5}$)] [18] (where M = Mn, Zn, Q = Se, Te and L = en, pda) (Figure 1a and 1c) which are composed of single-layered chains (n = 1), [(MSe) ($L_{0.5}$)] [19] (where M = Zn, Mn and L = en, pda) (Figure1b and 1d) which are composed of single-layered slabs (n = 1), and [(M_2Q_2)(L)] [20] (where M = Cd, Zn, Q = S, Se and L = en, pa, ba, ha) which are composed of double-layered slabs (n = 2) (en = ethylenediamine, pda = propanediamine, pa = propylamine, ba = butylamine, ha = haxylamine). But there is still lack of further investigations of hybrids composed of multi-layered structures, such as the value of n larger than 3 (n = number of atoms in a single layer), and the controlled synthesis of 2D slabs and 3D networks by using of the same 1D inorganic chains. Following finding of QCE seems much more fascinating than the extended structure. In a typical report, the band gap of as-prepared [α-ZnTe(en)$_{0.5}$] is measured to 3.5 eV, indicating a very large blue shift from that of bulk ZnTe (2.1 eV). By using density functional theory (DFT) within local density approximation (LDA), the shift problem is same as the well-known"band gap problem", namely calculated band gap of bulk ZnTe is 1.2 eV while the experimental value is 2.1 eV, and this shift problem can be rationalized that the onset of effective transition edge indeed shifts from 1.2 eV of bulk ZnTe to 2.5 eV of this hybrid [21].

Besides intercalation, there are still some other novel strategies, such as chemical vapor decomposition [22] and the Langmiur-Blodgett technique [23]. The combination of these ones forms a complete sol-gel system that is accessible to inorganic-organic hybrids.

2.2. Polymeric Complexation Reactions

The sublattices used here are always monodimensional organic polymers, with inorganic cations bounded through co-ordination bonds. Products obtained from this category, including self-assembly [24-30], an advanced strategy, are always called "organic-inorganic hybrids", and not widely used in the transformation into inorganic compounds.

It's worth noting that to be convenient, hybrids used in this chapter are all called inorganic-organic ones no matter which categories they come from.

3. Methods to Transform Inorganic-organic Hybrid Precursors into Inorganic Porous Materials

As mentioned above, this new route can be dated back to 1992 since a new family of mesoporous molecular sieves, M41S, including MCM-41 and MCM-48, was discovered by researchers at Mobil. Relevent research has suffered a breakthrough from mesoporous silica to mesoporous metal oxides, and later to transition metal oxides and chalcogenides. While enormous methods have been utilized to fabricate inorganic porous materials, using inorganic-organic hybrids as precursors is a newly emerged route mainly containing following strategies: solid thermal decomposition, organic component stripping, coupling of chemical transformation and component stripping, sonication, acidolysis and light irradiation, all of which will be discussed in detail in this chapter.

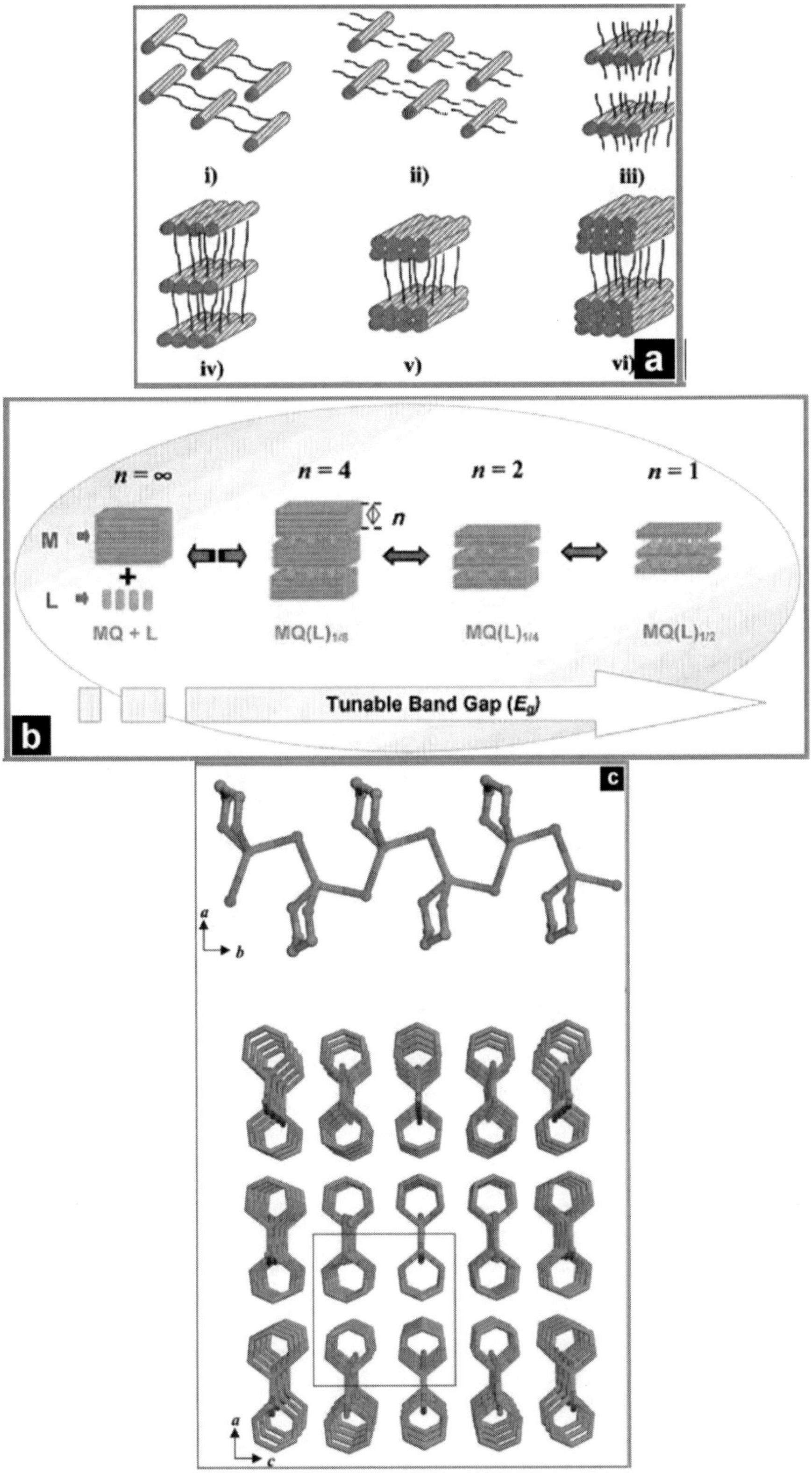

Figure 1. (Continued)

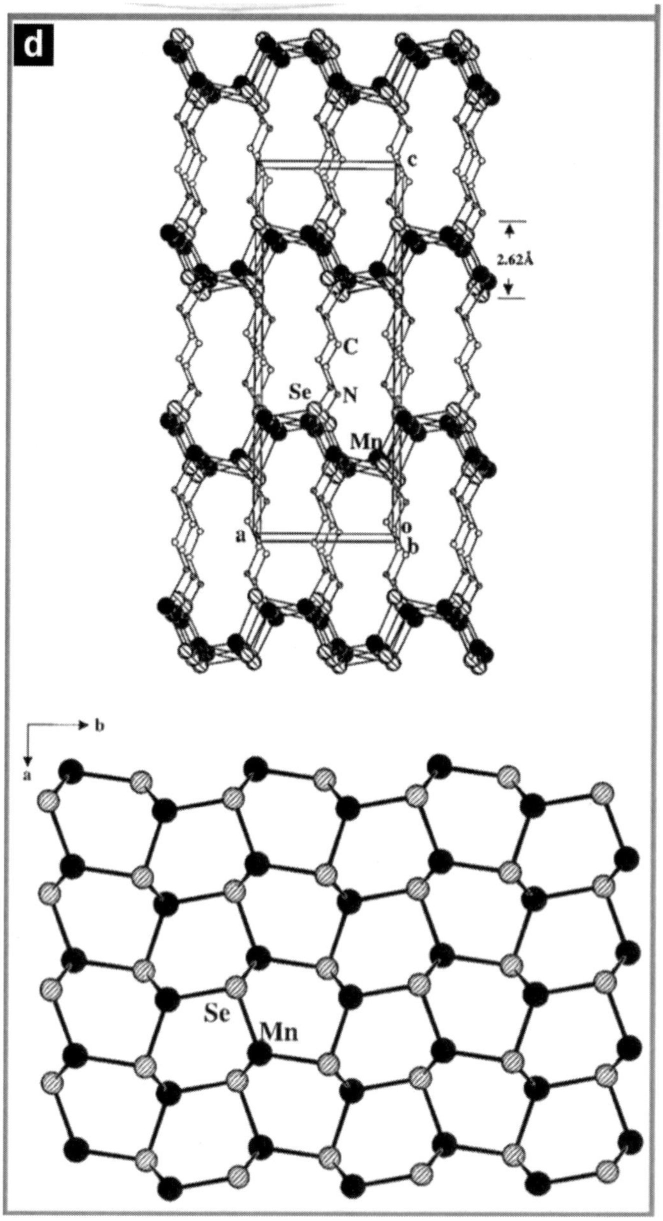

Figure 1. (a) View of 3D structure from 1D chains: i) 1D chains interconnected by organic spacers, ii) 1D chains coordinated, but not interconnected by organic spacers, iii) isolated monatomic layers formed by 1D chains, iv) interconnected monatomic layers formed by 1D chains; v) interconnected bi-layers formed by 1D chains; vi) interconnected tri-layers formed by 1D chains. (b) Variation of a single structure parameter leads to different thickness of II−VI slabs. (c) top: [ZnTe(pda)$_{0.5}$] chains along *b* axis. Light-blue cylinders are Zn, red cylinders Te, blue cylinders N and gray cylinders C, respectively; bottom: View of the packing of 1D chains down *b* axis. (d) [MnSe(en)$_{0.5}$] top: View along *b* axis with unit cells outlined; down: View of 2D slabs along *c* axis. (a) and (c): Reprinted with permission from Reference [18]. Copyright 2003 American Chemical Society. (b): Reprinted with permission from Reference [20]. Copyright 2007 American Chemical Society. (d): Reprinted with permission from Reference [19]. Copyright 2001 American Chemical Society.

3.1. Solid State Decomposition

Solid state decomposition, namely calcination, is a traditional method that always requires high temperature and high pressure but can effectively extract some elements from raw materials due to their instability under these conditions. Its advantages are as outstanding as its disadvantages, and how to better control the morphology and crystalline of products under such harsh conditions remains a great challenge. There have been many reports about the results of TGA of inorganic-organic hybrids, leading to pure inorganic phase, or even a phase transfer during the rising of temperature, indicating calcination of inorganic-organic hybrids is a simple but effective approach to fabricate pure inorganic materials with well phase-selectivity at different times and temperatures [31, 32, 33]. Using varied organic components, researchers have successfully prepared enormous inorganic functional materials.

	Inorganic-organic hybrid precursors	**Method to prepare precursors**	**Inorganic porous materials**	**Method to prepare porous product**	**References**
1	Si source, surfactants	Self-assembly	Mesoporous silica	Solid state decomposition	[36-39]
2	Block copolymer	Polymeric complexation reaction	Mesoporous metal oxide	Solid state decomposition	[46-58]
3	$[Fe_{18}S_{25}](TETAH)_{14}$	Intercalation reaction	Fe_7S_8/ Fe_2O_3	Solid state decomposition	[76]
4	$Mo_3O_{10}(C_6H_8N)_2\cdot 2H_2O$	Intercalation reaction	Mo_2C	Solid state decomposition	[77]
5	S-containing Ce-based hybrid	Intercalation reaction	CeO_2	Solid state decomposition	[81]
6	ZnS-DETA	Intercalation reaction	$Cd_xZn_{1-x}S$	Couple of chemical transformation and component stripping	[98]
7	MoO_3-EDA	Intercalation reaction	MoS_2	Couple of chemical transformation and component stripping	[103]

Figure 2. Schematic illustration of typical previous work from inorganic-organic hybrid precursors into inorganic porous materials, including methods to prepare all of them.

First, this chapter will highlight a special class and some related materials as they are the first examples of transforming inorganic-organic hybrids into inorganic porous materials, namely mesoporous silica and mesoporous metal oxides. By using of tetraethylorthosilicate (TEOS) or tetramethylammonium silicate solution as Si source and cetyltrimethylammonium chloride (CTAC) as surfactant as well as different surfactant to silica molar ratio, MCM-41, MCM-48, thermally unstable lamellar materials and cubic octamer, of which the pores could be tuned between 1.5 to 10 nm, were first successfully synthesized via a liquid crystal template mechanism (LCT) [34, 35] by C. T. Kresge et al. and J. S. Beck et al. [36, 37] (Figure 3a). Meanwhile, Inagaki group demonstrated the reaction between a crystalline single-layered poly-silicate kanemite and aqueous solutions of alkyltrimethylammonium

(ATMA) ions with further calcination to produce mesoporous FSM-16 via intercalation of layered silicates [38, 39] (Figure 3b and c). Also, many reported have been focused on the elaborated synthetic mechanisms of them as well as some detailed differences between them [40-43]. Thus, the fabrication of MCM-41, MCM-48 and FSM-16 together opened the door to the formation of various mesoporous structures. Inspired by this method [44, 45], some metal oxides with different size distribution from ligand-assisted template mechanism (LAT) [46-56] and mesoporous metal oxide using amphiphilic poly(alkylene oxide) block copolymers as structure-directing agents [57, 58] have been synthesized. It's obvious that hybrids used in this period come from self-assembly or even polymeric complexation reactions, but those discussed later are all come from the intercalation reactions.

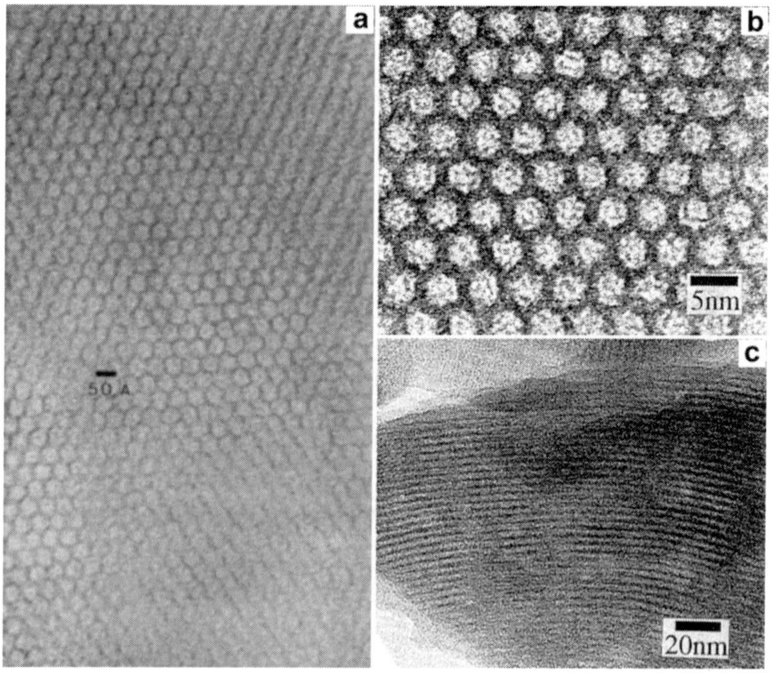

Figure 3. Transmission electron micrographs of (a) MCM-41, and (b), (c) FSM-16. (a): Reprinted with permission from Reference [36]. Copyright 1992 Nature Publishing Group. (b) and (c): Reprinted with permission from Reference [39]. Copyright 1996 The Chemical Society of Japan.

Next, some later achievements will be shown according to different organic molecules. The first case is the use of hydrazine, a small molecule containing the two N atoms in one molecule, which makes it a good participant to form inorganic-organic hybrids. Meanwhile, they can not only be incorporated to chalcogens, but also form $N_2H_5^+$ ions, making inorganic-organic hybrids vary from each other even resulting from a slight difference of conditions [59]. For example, H. Yao and coworkers used plate-like hydrazine-CdTe hybrids with formula of $Cd(N_2H_4)TeO_3$ to fabricate porous CdTe and porous Te with retaintion of macro-structure by calcination at 450 °C and 300 °C, respectively, following by HCl leaching. BET analysis confirmed its mesoporous structure while UV-vis spectra suggested an obvious red-shift of as-prepared porous architectures [64]. It was found that the different activity of N_2H_4 and thermal stability of as-formed inorganic compounds are essential for the two distinct products. These as-prepared porous microcrystals have potential applications in sensor devices or serving as templates to generate other porous materials. Besides, a series of metal chalcogenides transformed from hybrids containing hydrazine molecules have been demonstrated by Mitzi group. For example, γ-InSe can be obtained by decomposition of a precursor with approximate composition of $(N_2H_4)_2(N_2H_5)_2In_2Se_4$ in an inert atmosphere [60]. The as-prepared indium selenides can be used as TFTs via a spin-coated thin-film deposition onto thermally oxidized silicon substrates which are cleaned by a "piranha" process. One key advantage of hydrazine-precursors method is the ability to create high-performance films using a simple one-step and low-cost approach. Besides, it has the extended metal chalcogenide framework broken up into smaller inorganic anions and a counter kind of cations ($N_2H_5^+$) that is volatile and can be removed from the film with small disruption. After realizing the general mechanism of hydrazine-precursors route, D. B. Mitzi has prepared $GeSe_2$, $SnSe_2$ [59], In_2Te_3 [61] and some ternary systems such as $CuInTe_2$ [61] $CuInSe_2$ [62] and $SnS_{1.5}Se_{0.5}$ [63], all of which can be used in TFTs after a similar "piranha" treatment.

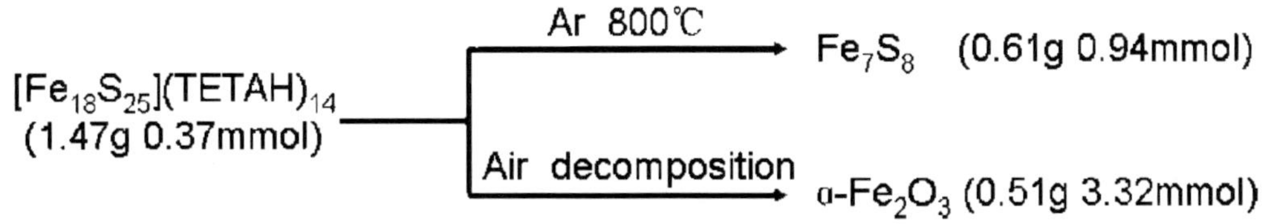

Figure 4. Illustration of the synthetic methods of porous Fe_7S_8 and α-Fe_2O_3 by thermal decomposition of inorganic-organic $[Fe_{18}S_{25}]$ $(TETAH)_{14}$ hybrid precursors. Reprinted with permission from Reference [76]. Copyright 2008 American Chemical Society.

Also, a series organic molecules with analogic structure, namely en, DETA and TETA (EDA = ethylenediamine, DETA = diethylenetriamine, TETA = triethylenetetramine), can be intercalated with inorganic blocks to generate inorganic-organic hybrids, and once calcinated, much research about transition metal oxides and chalcogenides has been fully explored. For example, porous Fe_7S_8 nanowires and α-Fe_2O_3 nanorods have been achieved by calcination of nanoribbon-like $[Fe_{18}S_{25}]$ $(TETAH)_{14}$ (TETAH = protonated TETA) precursor under vacuum and in the air, respectively [76], which represents a great adding toolbox of the combination of small amine molecules and magnetic transition metal semiconductors (Figure 4). Besides, S. Yu and coworkers demonstrated a shape and phase controllable strategy to synthesize pure wurtzide ZnS nanosheets by removing en from lamellar precursors $ZnS(en)_{0.5}$ at elevated temperatures under vacuum, while the further oxidation of as-prepared ZnS approached wurtzide ZnO flake-like dendrites [65, 66]. It is found that the passivation of ZnS surface by en molecules during calcination is a key factor to retain the lamellar morphology instead of breaking into nanorods. Except for zinc oxides and chalcogenides [67-73], $Fe_{1-x}S$ [74] and MoS_x [75] were also achieved by calcination of corresponding en hybrid precursors.

Many other organic molecules are also accessible to this method. Nanoporous Mo_2C with retention of nanowire-like morphology was represented by Tang group by calcinating $Mo_3O_{10}(C_6H_8N)\cdot 2H_2O$ [77] (Figure 5). BET analysis confirmed its mesoporous structure. Further hydrogen evolution reaction demonstrated its highly active and stable activity as eletrocatalyst in energy conversion and storage. It is important to notice that the amine molecules intercalated between inorganic blocks act as both reducing agents and carbon sources, and it's a tunable and general strategy to synthesize various molybdenum carbides and related structures, such as Mo_2C, MoO_2-doped MoC nanowires and Mo-doped Mo_2C

[77, 78]. More recently, Y. Yu et al. designed a facile strategy to synthesize CeO_2 microspheres with hierarchical mircopores, mesopores and macropores simultaneously in a single system through the combination of the calcination of inorganic-organic hybrid precursors with Kirkendall effect during the oxidation processs of sulfide hybrids into oxides, while dopant procedure can happen at the same time to maximize the synergistic effect between dopant and CeO_2 to get excellent catalytic activity [81] (Figure 6). Besides, Y. Li and coworkers utilized a novel vacuum-pyrolysis-carbothermal (VPC) process to fabricate tabular WS_2 [79] and WO_2 [80]. The driving forces for the curling of lamellar sheets can be ascribed to the reduction of interlayer interaction at the edges of sheets and the thermal stress at high temperature which causes the scrolling of the layered sheets with reduced interlayer forces at the edges. Through the intentionally designed preparation of inorganic-organic hybrid sulfides, selenides or tellurides followed by calcination at suitable temperature in air, this conversion method is believed to be broadened into obtain other hierarchical porous metal oxides. Also, by using other kinds of organic components, ZnS [82], GeM_2(M = S, Se) [83], Ga [84] and GaO [85] have been obtained, respectively.

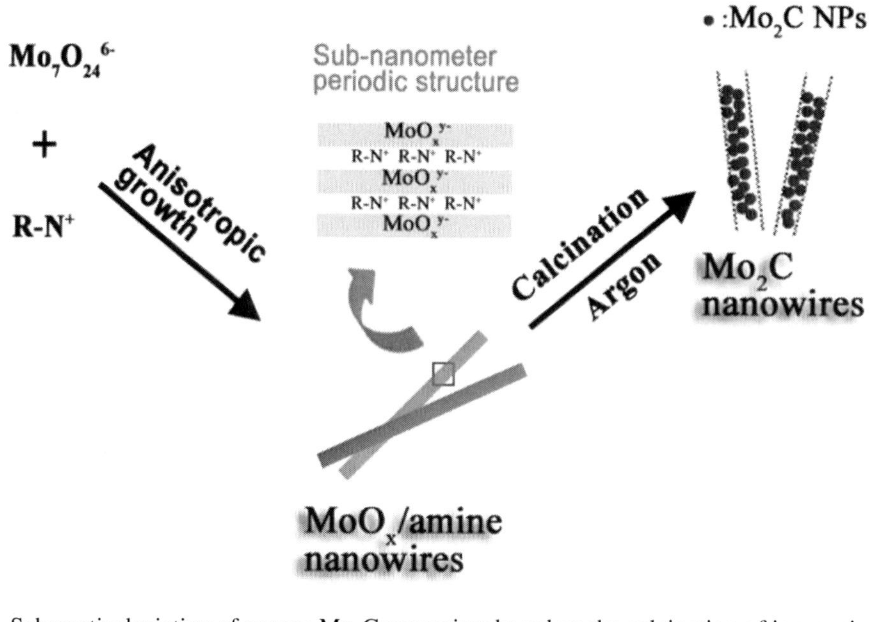

Figure 5. Schematic depiction of porous Mo_2C nanowires based on the calcination of inorganic-organic $Mo_3O_{10}(C_6H_8N)\cdot 2H_2O$ hybrid precursors with retention of 2D macro structure. Reprinted with permission from Reference [78]. Copyright 2009 American Chemical Society.

3.2. Organic Component Stripping

As a similar method to solid thermal decomposition, organic component stripping strategy also requires a relatively high temperature but is special because of the introduction of proper solvents. Different solvents such as distilled water, alcohol and 1-octadecene (ODE) are used according to the different properties of organic components to provide varied temperatures due to their distinct boiling points. This strategy is efficient for some unique structure such as 1D nanorods and hollow nanotubes, however, there are few reports about the fabrication of porous structure. So this chapter will only list some typical examples briefly according to different solvents without description in detail.

For distilled water, a most typical solvent in which many organic components containing amino groups have good solubility, many related works have been reported about generating 1D structures like 1D CdE (E = S, Se and Te) nanorods from the layered-structured $CdE(en)_{0.5}$ precursors [86], both nanorod-like wurtzide and sphalerite ZnS calcination of $ZnS(en)_{0.5}$ precursors [87], VO_x nanotubes [88-91]. For instance, a facile hydrothermal way to obtain 1D CdE (E = S, Se and Te) nanorods has been demonstrated by Z. Deng and coworkers, and the unique 1D nanorod-like structure is determined by the layered-structured $CdE(en)_{0.5}$ precursors [86]. To better rationalize this phenomenon, a refined version of solvent coordination molecular template (SCMT) mechanism was proposed (Figure 7). Organic components are first removed upon the collapse of layered precursors for two possible reasons, strong vibration during en-mediated solvothermal treatment and molecule replacement in a foreign solvent system, which begins from some places near crystal defects or the edges of precursors. While the direct solvothermal synthesis in en will first produce small rods at proper temperature with further recrystallization into individual one, this hydrothermal treatment can directly yield interlinked nanorods.

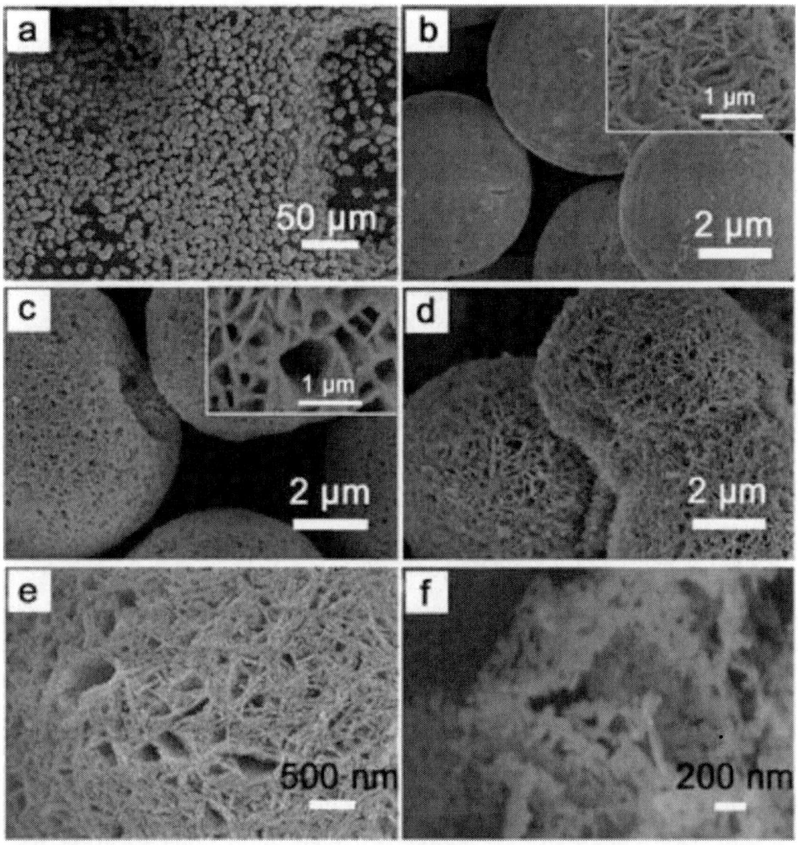

Figure 6. SEM images of inorganic-organic hybrid precursors (a, b) and as-prepared CeO_2 microspheres (c-f) by thermal decomposition at different temperatures: (c) 300°C and (d-f) 800°C. Reprinted with permission from Reference [81]. Copyright 2011 Royal Society of Chemistry.

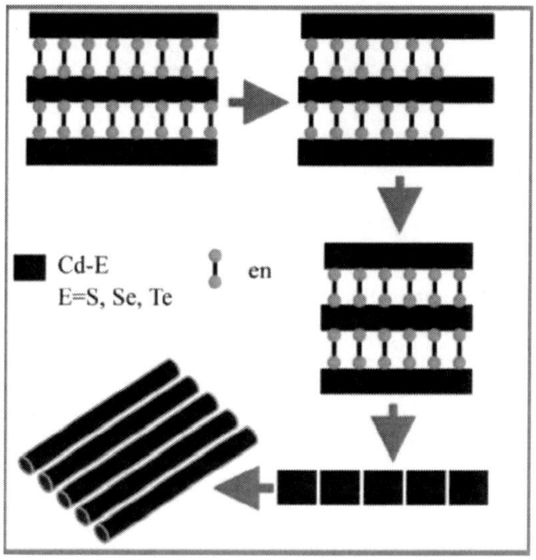

Figure 7. Schematic representation of the revised SCMT mechanism: from layered intermediate to CdE nanorods. Reprinted with permission from Reference [86]. Copyright 2003 American Chemical Society.

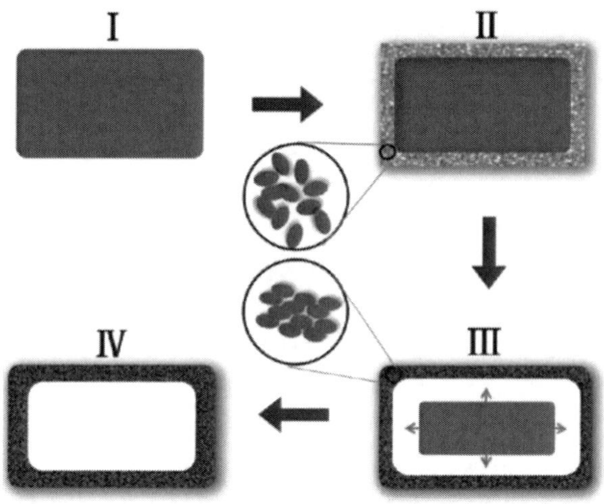

Figure 8. Illustration of the formation of ZnSe nanotubes from $ZnSe(DETA)_{0.5}$ nanobelts. Reprinted with permission from Reference [97]. Copyright 2010 American Chemical Society.

Except for distilled water, many other organic solvents like alcohol, ODE and ethylene glycol (EG) are also accessible to organic component stripping. For example, ZnM(M = S, Se) [92, 93] and EuS [94-96] can be generated using different precursors at proper conditions. The difference by using different solvents was also reported. For example, 1-octadecene (ODE) and ethylene glycol (EG) were used to prepare $Zn_{1-x}Mn_xSe$ nanobelts and nanotubes, respectively, from $[Zn_{1-x}Mn_xSe]$ $(DETA)_{0.5}$ by M. Zhang and coworkers [97]. The formation mechanism of ZnSe (x = 0) nanobelts has already been discussed above, while that of ZnSe (x = 0) nanotubes was carried out by adopting time-dependent reactions. A modified Kirkendall effect coupling with the oriented attachment mechanism is put forward to rationalize this phenomenon (Figure 8). A hollow homogeneous core-shell structure is formed at the early stage of reaction with a gradational inward processs, and finally, the cores are completely evacuated. The outside of nanobelt-like hybrid precursors first reacts with EG through a dehydration reaction, leaving only single individual inorganic ZnSe nanoparticles via a following dissolution-recrystallization process. These ZnSe orient and aggregate to form a compact shell. This procedure continues with the outward diffusion of hybrid precursors as well as the adsorption of newly recrystallized ZnSe nanoparticles on the shell through an oriented-attachment process. The iso-oriented ZnSe nanotubes are due to many processes such as the Kirkendall effect, the dissolution of precursors, recrystallization, oriented attachment of newly decomposed ZnSe nanoparticle and an Ostawald ripening process.

3.3. Coupling of Chemical Transformation and Component Stripping

Traditional methods to get porous materials always include defects in products, as well as harsh conditions and surplus elements. In order to overcome such disadvantages, modified and novel methods have emerged recently, and some of them indeed bring breakthroughs, especially chemical transformation, which is a straightforward, versatile and efficient approach as it can easily control the composition, morphology and structure of materials. Compared with solid state decomposition and organic component stripping methods, coupling of chemical transformation and component stripping strategy can bring new elements into the product system, which also provide a possibility to generalize this method, as well as extract organic parts at the same time. Although chemical transformation can be divided into four categories [6], this chapter will only focus on cation exchange and anion exchange. Despite that chemical transformation has been applied in variety of systems, only few examples can be found forthe usage of inorganic-organic hybrids as precursors to get inorganic porous materials. Herein, four specific examples will be presented in detail according to the structures of products, systematically describing the synthetic routes and proposed mechanisms.

Porous nanostructures, resulting in larger specific area and much more active sites, have attacked considerable attention due to their superb morphology-dependent properties, as well as their intriguing applications. The simultaneously control of their composition and pore size together with a well-crystal-like morphology is undergoing examining. Y. Yu and coworkers fabricated nanoporous single-crystal nanosheet-like $Cd_xZn_{1-x}S$ via cation exchange reaction of inorganic-organic hybrid ZnS-DETA precursors with Cd^{2+} ions [98]. The composition and pore size of as-prepared $Cd_xZn_{1-x}S$ could be adjusted by simple changing solvents and the reaction concentration of Cd^{2+}, namely the ratio of hybrid precursors to cation, and even pure nanoporous CdS nanosheets could be achieved using excessive Cd^{2+}. What's more important, this strategy has been approved to be a general one to prepare single-crystal nanoporous metal oxides and selenides. The growth mechanism was also investigated as following: time-dependent reactions reveal that Zn^{2+} has been exchanged quickly and irreversibly, indicating that the cation exchange is the enthalpy-driven reaction; there are two reasons for the formation of pores: the rapid kinectic and stain release due to the lattice mismatch between ZnS and CdS, and the dissolution of DETA molecules in solution, namely deionized water (Figure 9a). Except for cation exchange to produce nanoporous structure, anion exchange is also accessible. An excellent example is the fabrication of nanoporous FeP nanosheets. Inorganic-organic hybrid $Fe_{18}S_{25}$-TETA was used as precursors with anion exchange with trioctylphosphine (TOP) as P source [99]. Related research by R. E. Schaak and S. L. Brocks has made to successfully convert metal or metal oxide into metal phosphide, which provides a general guideline for the synthesis of metal phosphides [100, 101]. Y. Xu et al. have extended this guideline to obtain porous transition metal phosphide with controlled morphology via a simple solvothermal route. Reasons for the appearance of voids are the same as those in nanoporous $Cd_xZn_{1-x}S$ mentioned before (Figure 9b). The similarity between the procedure of producing nanoporous $Cd_xZn_{1-x}S$ and FeP proves that using inorganic-organic hybrids as precursors via ion exchange reaction, nanoporous structures can be successfully achieved in a variety of systems because of their unique intrinsic structure, thus providing an efficient but handy route to achieve nanoporous functional materials, which maybe an inevitable addition to traditional methods.

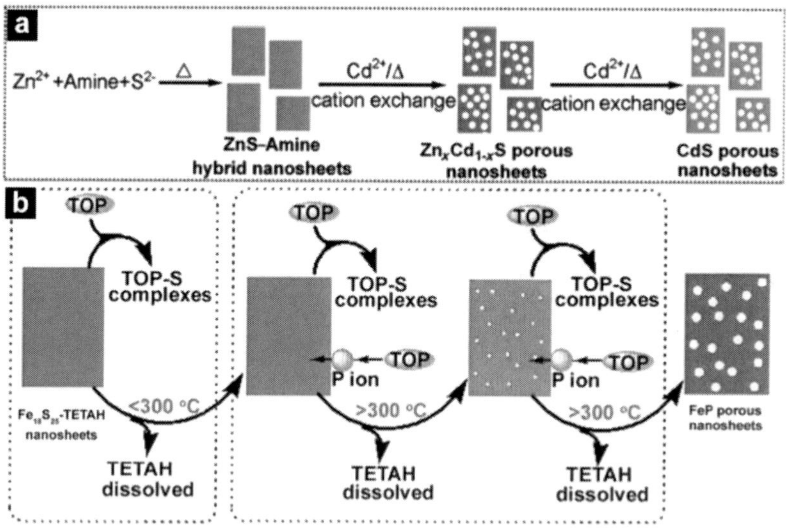

Figure 9. (a) Nanoporous ZnS sheets with adjustable compositeon prepared through cation-exchange strategy of the inorganic-organic ZnS-DETA hybrid sheets. (b) Scheme illustrating the synthesis of the porous FeP nanosheets through anion-exchange strategy of the inorganic-organic $Fe_{18}S_{25}$-TETAH nanosheets with P ions. (a): Reprinted with permission from Reference [98]. Copyright 2012 Wiley-VCH. (b): Reprinted with permission from Reference [99]. Copyright 2013 The Royal Society of Chemistry.

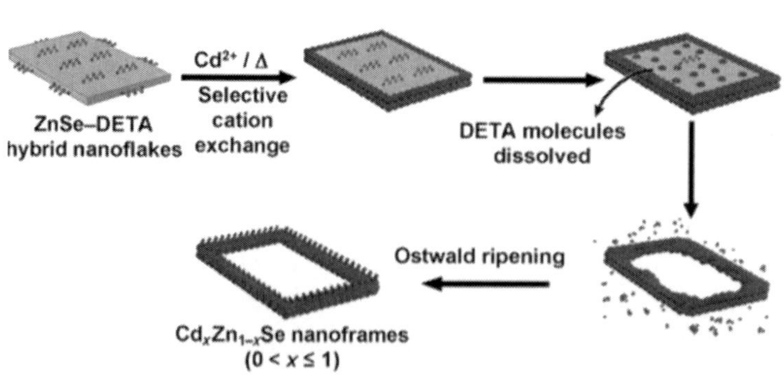

Figure 10. Schematic illustration the synthesis of composition-tunable $Cd_xZn_{1-x}Se$ nanoframes through the selective cation-exchange reaction of unpurified inorganic-organic ZnSe-DETA hybrid sheets with Cd ions. Reprinted with permission from Reference [102]. Copyright 2012 Wiley-VCH.

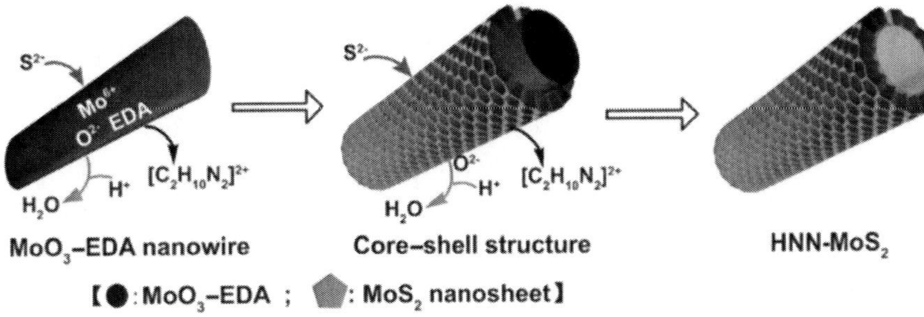

Figure 11. Schematic representation of the synthesis of hierarchical nanosheet-based MoS_2 nanotubes through anion-exchange reaction of the inorganic-organic MoO_3-EDA hybrid nanowires with S ions. Reprinted with permission from Reference [103]. Copyright 2013 Wiley-VCH.

Besides simple cation exchange reaction, a selective cation exchange to produce hollow $Cd_xZn_{1-x}Se$ nanoframes has also been reported [102]. Unpurified ZnSe-DETA as precursors and Cd^{2+} were used in a typical synthesis, and different compositions of products could also be achieved by controlling the molar ratio of the hybrid precursors and cadmium cations. Moreover, this strategy can also be broadened to fabrication of other systems, suggesting it maybe general to produce nanomaterials with skeleton-like structures. Further investigation reveals the unpurified precursor is a key factor, leaving enough DETA molecules absorbed onto hybrids, so that the selectivity is attributed to the selective chemical etching of inorganic nanocrystal rather than the differences in crystal planes of hybrids, and the cation exchange reaction is followed by Ostward ripening mechanism as shown in Figure 10. Such a selective route brings a huge shock to traditional synthetic methods. It opens a new door for designing reactions for a certain morphology and structure.

More recently, hierarchical porous MoS_2 nanotubes stacked by layered nanosheets via anion exchange of MoO_3-EDA nanowire-like hybrids with S ions were prepared by S. Zhuo and coworkers [103]. Traditional methods to prepare MoS_2 always contain toxic H_2S as S source, high temperature and crystalline defects. Such disadvantages have been overcome by using L-cysteine as S source and a simple hydrothermal reaction at only 200 ℃. A time-dependent reaction was utilized to give insight into this anion exchange mechanism. The reaction first happens in the surfaces, leaving internal parts solid hybrids. The formation of inorganic porous tubular MoS_2 is attributed to the outward diffusion of Mo with reaction of sulfur anion and the dissolution of EDA molecules, resulting in hollow nanowires, namely nanotubes (Figure 11). Also, the pore size distribution was determined to be about 30 nm by mesopore BET analysis. This new resultant is interesting because the combination of the facile synthetic route, the low cost of raw materials and the high photoelectrocatalytic property make as-prepared porous MoS_2 the most promising candidate for hydrogen generation, and this can be use as models for many other hierarchical inorganic functional materials.

Some other interesting results have also been achieved. Yu group have synthesized unique Cu_2S-PbS heteronanostructures via partially cation exchange reaction [104], while lead chalcogenide [105], CdSe [106] and $Cu(OH)_2$ [107] were obtained by Y. J. Zhu, T. Hyeon, Y. D. Huh, respectively, via anion exchange reaction. In short, using inorganic-organic hybrids as precursors to achieve inorganic porous materials via ion exchange reactions is a newly emerging but efficient route. Organic parts play an essential role because they are intercalated between inorganic blocks, making the dissolution of organic parts and ion exchange happen simultaneously, which show a promising route to intentionally design the morphology, structure and composition. What's more, due to their unique dependent properties, these materials have potential applications in various fields. Although great prospects have been revealed in this field, as an infant route, there are still many improvements needed to be done. Meanwhile, compared with direct hydrothermal or solvothermal synthesis, this is at least a two-step procedure, so how to express the superiority of this two-step procedure is a challenging focus.

3.4. Sonication

It is well known that each organic part must have a good solubility in several specific solvents. Similar to organic component stripping strategy, sonication maybe a good route for the dissolution of organic components in hybrids at a relatively low one without any changes of inorganic blocks.

Briefly, metastable h-WO_3 [108], β-CuS_2 [109], ZnS [110] and ZnTe [111] have been successfully synthesized using different solvents, while this chapter will highlight some examples about the fabrication of ultrathin nanosheets. Inspired by some natural phenomena like the separation of a batch of papers into many pieces under the effect of external forces, Xie group have generated large-area freestanding single layers of non-layered ZnSe with four-atomic thickness by the sonication of lamellar (Zn_2Se_2)(pa) [112]. The whole route can be illustrated as Figure 12a, after the fabrication of lamellar hybrids consisting of non-layered inorganic compounds and organic species, namely ZnSe and pa here, sonication is essential for the exfoliation procedure, leading to single layer ZnSe surrounded by pa molecules, and the following solvothermal treatment is inevitable for the removal of pa molecules, resulting in final products, ZnSe. The cleavage of ZnSe slabs along (110) plane confirmed by XRD (as shown in Figure 12c) is found to be important for their stability because (110) facets are cleavage and nonpolar planes of II−VI semiconductor compounds. Meanwhile, ultrathin CdS nanosheets with a thickness of about 4 nm via an sonication-induced aqueous exfoliation strategy are successfully prepared by Y. Xu et al. [113]. The whole facile protocol includes the use of inorganic-organic lamellar CdS-DETA hybrids as starting materials, water as dispersion medium and L-cysteine as stabilizing agents as revealed by Figure 12b. It's interesting and also thought-provoking that XRD patterns show the cleavage along (002) planes instead of their nonpolar (110) facets (as shown in Figure 12d). Together with results in case of absence of L-cysteine, it's found that the addition of L-cysteine is a key factor to improve the stability of as-obtained colloid solutions and facilitate the exfoliation of hybrid lamellar precursors.

3.5. Acidolysis

Acidolysis has a similar principle as sonication, but it demands a good solubility of organic parts in acids, meaning a limitation depending on different kinds of organic components. Most issues contain atom N, which will transform into NH_4^+ and are easily dissolve in water. For example, M. Chen and coworkers used diluted HCl solution (pH 1-2) to prepare wurtzide ZnSe from $ZnSe(N_2H_4)_{0.5}$, of which the atom N has an effective influence on the final phase of products [114], while W. Yao et al. synthesized wurtzite ZnS by removal of the amine template from mesostructured wurtzite ZnS/amine nanocomposites [115] using aqueous acetic acid solution (1:10 (v/v) acetic acid/water), in which protic solvents with high polarities are essential to break forces between nanowires. What's more, Y. Sugahara reported the acidolysis of precursor $(C_8H_{17}NH_3)_2WO_4$ with following calcination procedure to generate monoclinic WO_3 through an intermedia of orthorhombic H_2WO_4 (Figure 13) [116]. The as-prepared WO_3 possess a large surface specific area and good photocatalytic activity for water oxidation compared with commercial one.

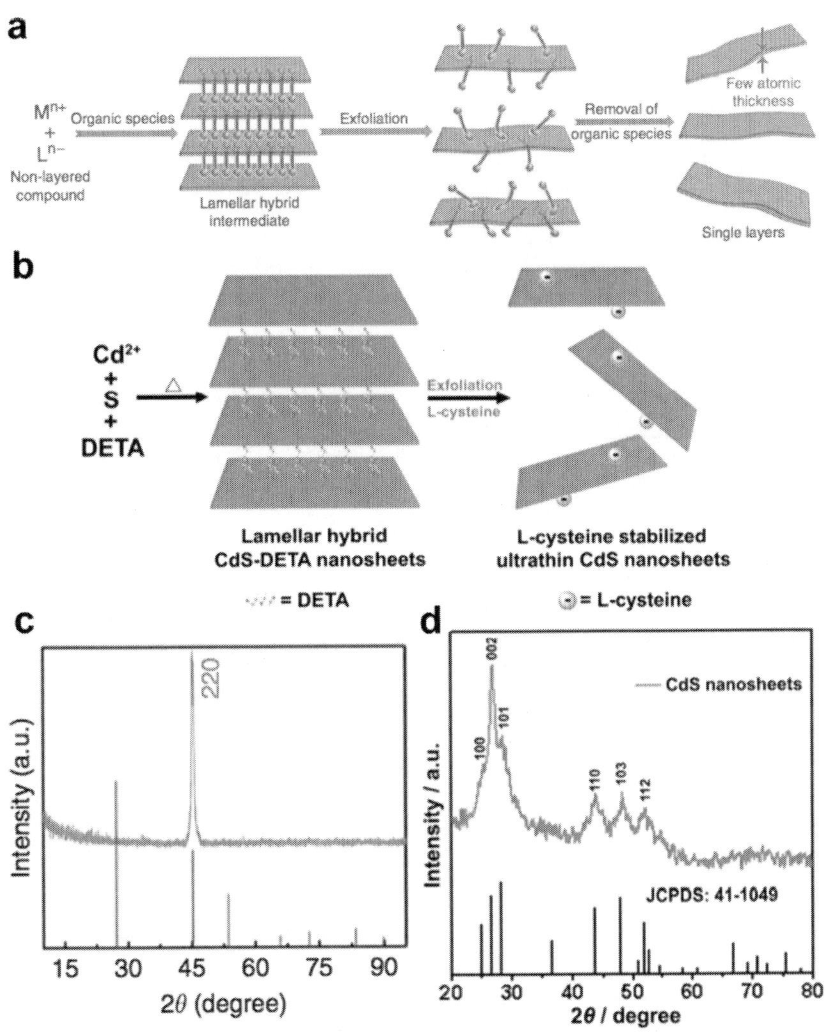

Figure 12. (a) Synthetic route for freestanding single layers of non-layered ZnSe with four-atomic thickness; (b) Scheme illustrating the fabrication of ultrathin CdS nanosheets with a thickness of about 4 nm; XRD patterns for (c) ZnSe, and (d) CdS. (a) and (c): Reprinted with permission from Reference [112]. Copyright 2012 Nature Publishing Group. (b) and (d): Reprinted with permission from Reference [113]. Copyright 2013 the Royal Society of Chemistry.

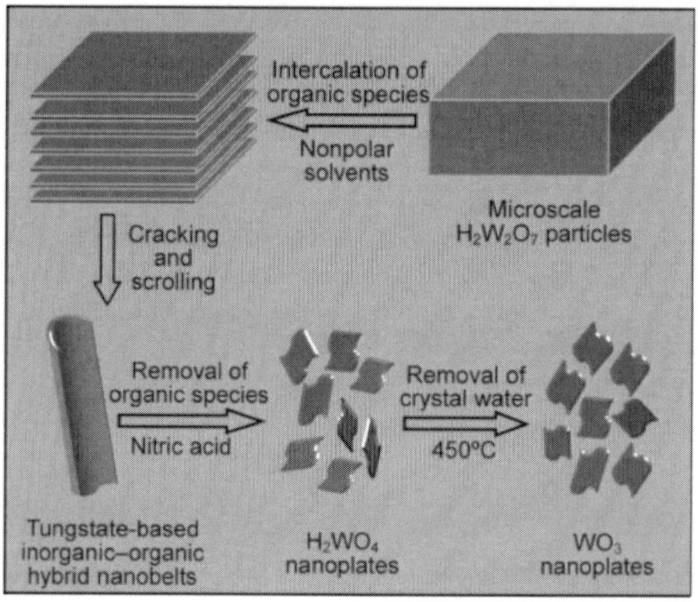

Figure 13. Schematic representation of the conversion process from microscale $H_2W_2O_7$ and tungstate-based inorganic-organic hybrid belts to WO_3 nanoplates via intermediate product of H_2WO_4 nanoplates. Reprinted with permission from Reference [116]. Copyright 2008 Wiley-VCH.

3.6. Light Irradiation

Recently, light irradiation was found to be acceptable for transforming inorganic-organic hybrids into inorganic functional materials, which is called a light-driven solid-state conversion route (LSCR). The inherent driven force is the photogenerated charges in the inorganic blocks, initiating redox reactions to remove the organic parts and rearranging the inorganic parts to form a new compound. Two typical examples will be listed in this chapter. The first is the preparation of porous TiO_2 by Chen group, which have a good property in electron storage [117]. Titanium glycolate (TG) was used as raw materials, and under UV-irradiation, TG could be converted into porous TiO_2 while organic species such as ethanol and aldehyde were detected in the solution due to the decomposition of organic components. EPR was used to prove that photogenerated eletrons are trapped by Ti^{4+} to form Ti^{3+}, consistent with the emergence of intense blue color. After exposure of as-prepared TiO_2 to air, the stored electrons are scavenged by O_2, and these electrons can be recovered with treatment of suitable hole scavengers under UV light irradiation. The second one is the fabrication of Zn_4 $(OH)_6(SO_4){\cdot}nH_2O$ from ZnS-cyclohexylamine (CHA) by light-driven transformation [118]. A complex mechanism was shown by Chen as follows: once under irradiation, electrons generate from the valence band of ZnS to form eletron-hole pairs. The electrons move to the surface of ZnS and are captured by O_2 in water to form oxidative groups, which function as effective oxidizing agents to decompose CHA molecules in ZnS-CHA. Meanwhile, holes directly oxidize CHA molecules. The resulting amines lead to the formation of OH^- by extracting H atom in H_2O. ZnS undergo photocorrosion induced by self-oxidation, producing Zn^{2+} and SO_4^{2-} ions. These three kind of ions together reform $Zn_4(OH)_6(SO_4){\cdot}nH_2O$, reconstructing the structure of inorganic blocks. These as-prepared

$Zn_4(OH)_6(SO_4)\cdot nH_2O$ can be easily converted into ZnO by simple calcination at elevated temperatures.

All these above-mentioned methods form a systematic route from inorganic-organic hybrid precursors into inorganic functional materials. Compared with traditional methods, it is a more effective route while always with facile conditions, complex mechanism and resulting in excellent properties of as-prepared products.

4. Enhanced Properties of As-prepared Inorganic Materials with Unique Structures

One notable advantage of using inorganic-organic hybrids as precursors to prepare inorganic functional materials is that the structures of products vary from each other. Hollow, nanoporous, core-shell, ultrathin and some other intriguing architectures can be easily prepared, while under terms of direct inorganic synthesis they are hard to achieve and the whole synthetic procedure can be more complicated. What's more, these structures are benefit for enhanced properties, as they can provide a larger specific surface area, more active sites and less crystal defects. Materials fabricated by this route have been applied in fields of photocatalysis, electrocatalysis, photoelectrocatalysis, TFTs, magnetism and so on. This chapter will list some of them mainly focus on porous structures.

Nanoporous structure, which has attracted considerable attention because of its improved performance compared with its solid and bulk counterparts, can be achieved because organic components and inorganic building blocks are homogenously located in precursors, and after extracting organic parts, cavity will possibly appear. For chemical transformation, there is another reason because of the rapid kinetic reaction and stain release due to the lattice mismatch before and after transformation. Yu group have prepared porous α-Fe_2O_3 by the calcination of $[Fe_{18}S_{25}](TETAH)_{14}$ in the air, showing a good promising application but with little research in detail, while a series of porous nanomaterials have been synthesized by Zhang group with thorough detection of photocatalytic, electrocatalytic and photoelectroncatalytic properties. For example, composition-tunable $Cd_xZn_{1-x}S$ can be used for photocatalytic hydrogen generation from water splitting (PHWS), and as an example, nanoporous single-crystal $Zn_{0.5}Cd_{0.5}S$ nanosheets ($x = 0.5$) (C-1) is discussed in detail with comparison of $Zn_{0.5}Cd_{0.5}S$ nanorods (C-2), indicating that C-1 shows a stable H_2 generation rate which is about 1.5 times higher than that of C-2 (as seen in Figure 14b), as well as a good stability even after 15 h. The reasons can be cncluded as the narrower band gap and stronger absorption ability of photons of C-1 which can be proved by UV-vis spectra (Figure 14a), the single-crystal structure resulting in less defects where photogenerated electrons and holes recombine and nanoporous structure providing larger specific surface area and more active sites. The first case of using transition metal phosphide for electrochemical hydrogen evolution reaction (HER) is also reported that as-obtained porous FeP nanosheets possess a good activity as revealed by polarization curves and Tafel plots (Figure 14c and 14d), as well as a lower resistance as revealed by electrochemical impedence spectroscopy. Besides, the photoelectrocatalytic performance of nanosheet-based MoS_2 nanotubes containing macropores and mesopores simultaneously is tested, which shows a much higher activity (Figure 14f) and lower resistance compared with commercial ones, due to the enhanced absorption ability of photons (Figure 14e), much more active sites and unique hierarchical porous structures of nanosheets. Finally, they also use hierarchical porous CeO_2, in which macropore, mesopore and mocropore could coexist, as catalytic supports for Cu for the preferential oxidation of carbon monoxide in H_2-reach stream, representing a good activity as well as selectivity (Figure 14g and 14h).

Ultrathin 2D nanomaterials, especially composed of non-layered compounds have attracted numerous attentions due to their unique physical and chemical properties since the discovery of graphene. While much effort has been done and improvement achieved, there are few examples of using inorganic-organic hybrids as precursors. Recently, Y. Sun et al. reported the synthesis of non-layered ultrathin ZnSe exfoliated from (Zn_2Se_2) (pa), the photoelectrochemical catalytic performance of which is much more excellent (Figure 15a), especially for incident photon-to-current conversion efficiency (IPCE), 42.5%, which is much better than that of most existing reports (Figure 15b). Another case of ultrathin CdS nanosheets from CdS-DETA for PHWS was proposed by Y. Xu and coworkers, showing a 6 times higher H_2 generation rate than common CdS nanosheets (Figure 15c and 15d), and the superior properties is ascribed to high density of photocatalytic active sites and quick transfer of photogenerated charge carriers from photogenerated center to the surface of these catalysts because of large surface area provided by ultrathin structure, and the blue shift of the band gap of as-prepared CdS compared with bulk ones due to quantum size effect.

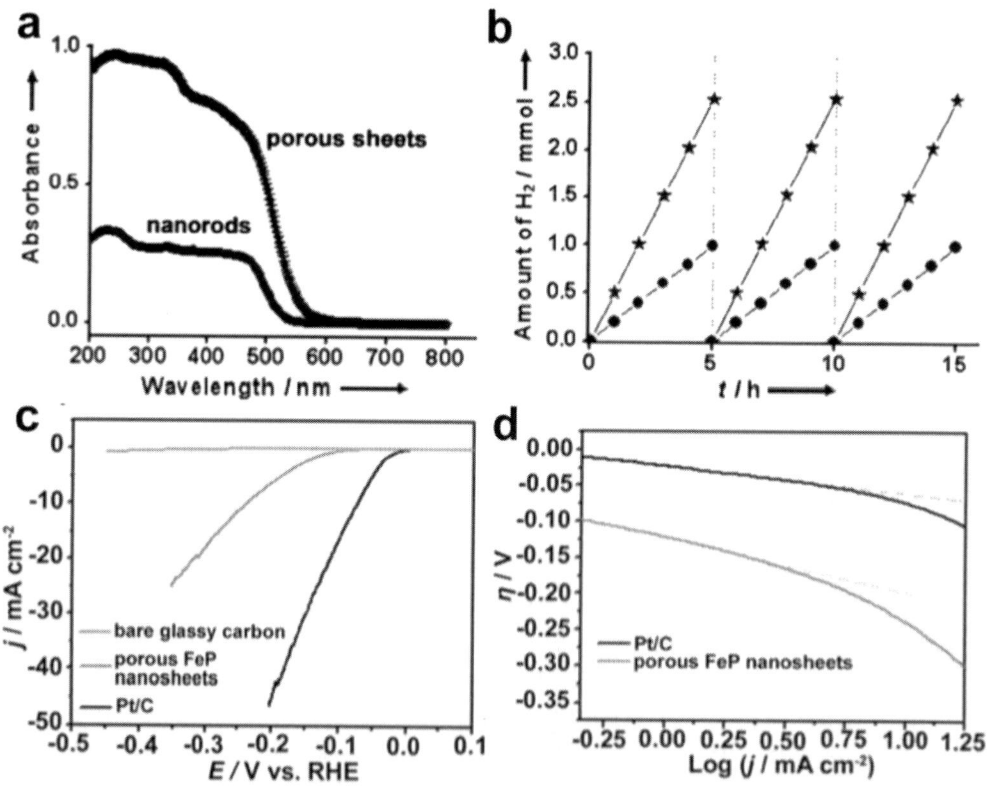

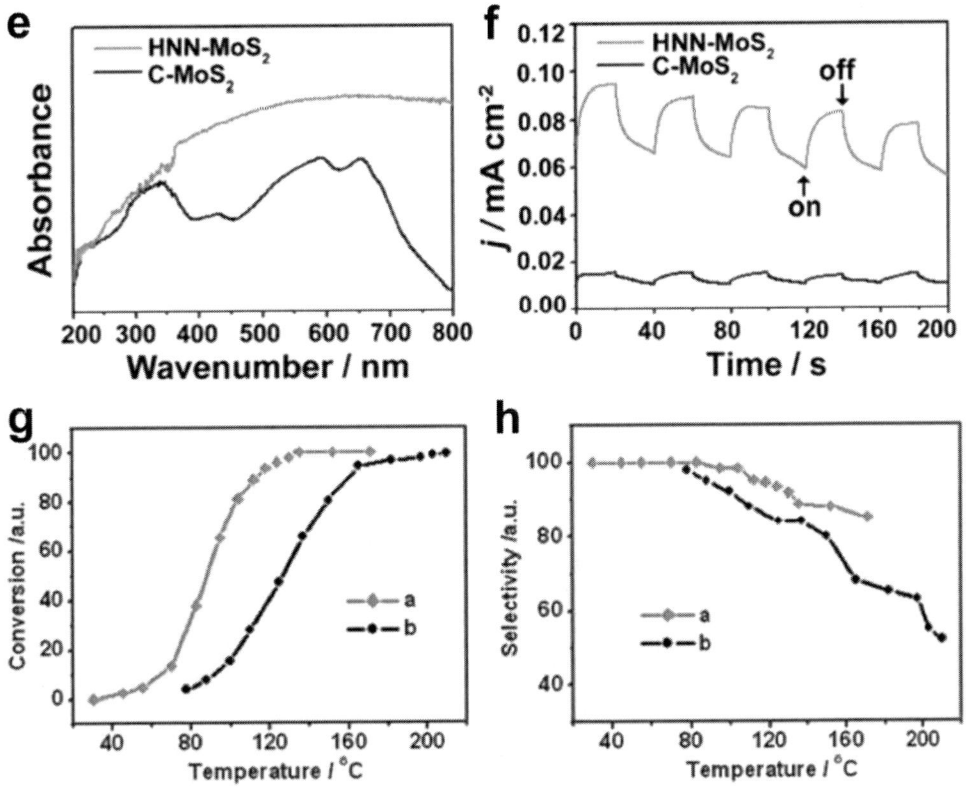

Figure 14. (a) UV/vis diffuse reflectance spectra and (b) time course of PHWS rate under irradiation of visible light of as-prepared $Zn_{0.5}Cd_{0.5}S$ nanorods (●) and $Zn_{0.5}Cd_{0.5}S$ porous nanosheets (the rest); (c) Polarization curves for HER on bare glassy carbon electrode and modified glassy carbon electrode comprising FeP catalysts and Pt/C catalysts; (d) Tafel plots derived from (c); (e) UV/vis diffuse reflectance spectra and (f) photocurrent density response at an applied bias of 0.6 V versus SCE under visible irradiation of as-prepared hierarchical MoS_2 nanotubes (HNN- MoS_2) and commercial ones (C-MoS_2); Catalytic (g) activity and (h) selectivity of Cu loading on hierarchical porous CeO_2 (a in Figure 14g and Figure 14h) and CeO_2 nanoparticles (b in Figure 14g and Figure 14h). (a) and (b): Reprinted with permission from Reference [98]. Copyright 2012 Wiley-VCH. (c) and (d): Reprinted with permission from Reference [99]. Copyright 2013 The Royal Society of Chemistry. (e) and (f): Reprinted with permission from Reference [103]. Copyright 2013 Wiley-VCH. (g) and (h): Reprinted with permission from Reference [81]. Copyright 2011 The Royal Society of Chemistry.

Porous hollow structure is another valid and highly efficient one, and the formation of it is always a re-crystalline procedure coupled with Kirkendall effect, oriented attachment process and Ostwald ripening mechanism. In a typical report, Yu group have fabricated hollow ZnSe nanotubes through the couple of modified Kirkendall effect and oriented attachment, the band gap of which changes from 2.7 eV in bulk to 3.0 eV because of the quantum size effect. More recently, hollow composition-tunable $Cd_xZn_{1-x}Se$ nanoframes composed of single-crystalline nanorods have been prepared by X. Wu and coworkers, and further tests for PHWS reveal a composition-dependent properties and much higher activity compared with $Cd_{0.33}Zn_{0.67}Se$ via a direct inorganic route (Figure 16).

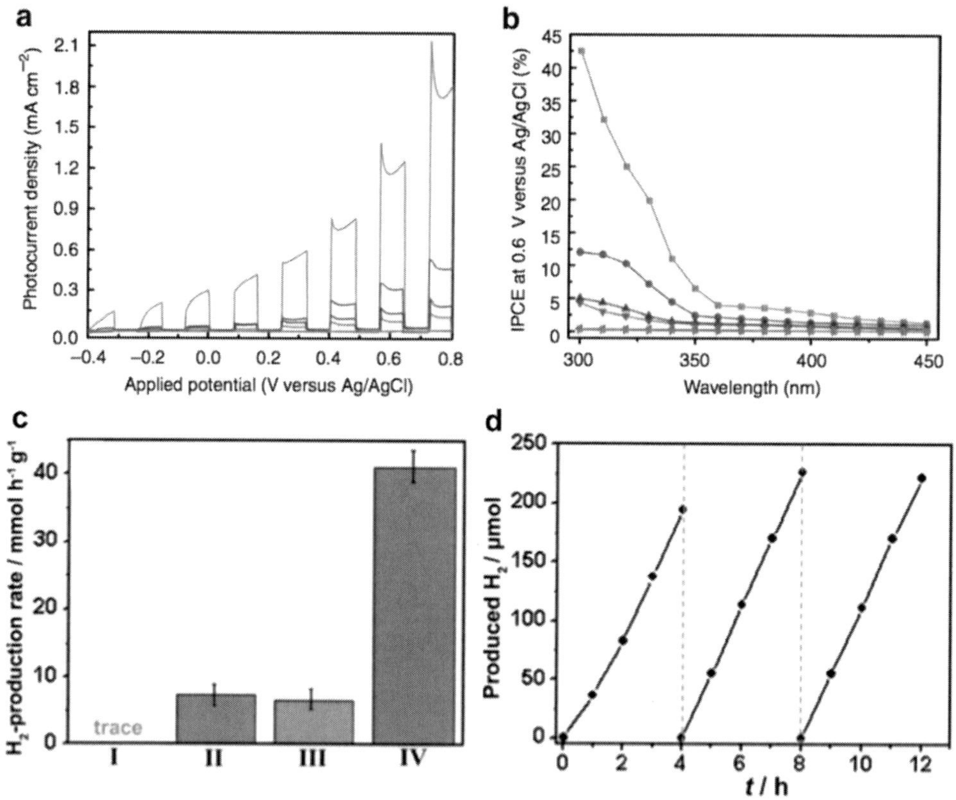

Figure 15. (a) Photocurrent density versus applied potential curves; (b) IPCE (the rectangle, circle, triangle, reverse triangle and darker rectangle plots correspond to ZnSe single layers, ZnSe-pa double layers, quadruple layers and octuple layers, $(Zn_2Se_2)(pa)$ and bulk ZnSe, respectively); (c) Comparison of the PHWS activity of different photocatalysts: CdS nanoparticles (I), CdS-DETA nanosheets (II), CdS nanosheet-based aggregates (III) and ultrathin CdS nanosheets (IV); (d) Time course of PHWS rate of ultrathin CdS nanosheets. (a) and (c): Reprinted with permission from Reference [112]. Copyright 2012 Nature Publishing Group. (b) and (d): Reprinted with permission from Reference [113]. Copyright 2013 The Royal Society of Chemistry.

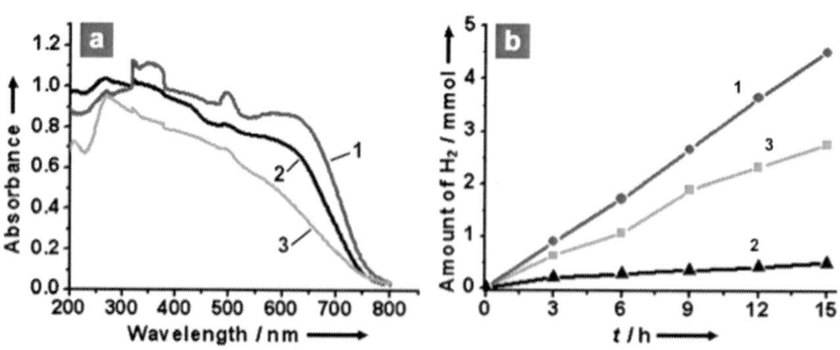

Figure 16. (a) UV/vis diffuse reflectance spectra and (b) time course of PHWS rate under visible irradiation of as-prepared hollow $Cd_{0.33}Zn_{0.67}Se$ nanoframes (1), hollow $Cd_{0.25}Zn_{0.75}Se$ nanoframes (3) and $Cd_{0.33}Zn_{0.67}Se$ nanoparticles (2). Reprinted with permission from Reference [102]. Copyright 2012 Wiley-VCH.

CONCLUSION

In summary, present methods to transform inorganic-organic hybrid precursors into inorganic porous materials require several conditions in common: first, a proper precursor must be fabricated depending on different resultants, for example, if a sheet-like product is demanded, a precursor with sheet-like or cubic structure is always adopted; second, a suitable method is needed to extract organic components from precursors, resulting in some specific characteristics, especially porous structure, and leaving behind inorganic building blocks, while many factors such as temperature, time and solvents should be taken into account; third, the whole transforming procedure is very complicated, and with other valid mechanisms, some effective and general methods to achieve inorganic porous materials have been proposed, such as LCT, LAT, VPC, SCMT and LSCR.

With these standards in mind, the course to extract organic components represents a key challenge. For most cases, an organic molecule is intercalated to one or two inorganic atoms, forming multidimensional inorganic-organic precursors. Different courses can result in different morphologies, phases, and even compositions. This chapter has divided the whole route into six main categories, solid state decomposition, organic component stripping, coupling of chemical transformation and organic component stripping, sonication, acidolysis and light irradiation. Most categories, except the coupling of chemical transformation and organic component stripping, focus on initial inorganic atoms in hybrid precursors, always obtaining porous inorganic blocks under vacuum or metal oxides in the air with retention of morphology, such as nanoporous Mo_2C nanowires from $Mo_3O_{10}(C_6H_8N)\cdot 2H_2O$ under vacuum and porous Fe_2O_3 nanorods from nanoribbon-like $[Fe_{18}S_{25}]$ $(TETAH)_{14}$ in the air. On the other hand, chemical transformation is a newly emerged but handy and efficient route as it can not only preserve morphologies of precursors, but also introduce new elements via ion exchange reactions. Coupling of it with usage of inorganic-organic hybrids as precursors can selectively control the composition of final products by using different amount of raw materials, reaching ternary or even multiple systems and easy to examine the composition-dependent properties. What's more, the extraction procedure can possibly generate porous structure due to the extraction of organic molecules, the rapid strain release, and lattice mismatch before and after the whole transformation. Also, some other interesting structures, such as hollow, incorporated with some other mechanisms such as Kirkendall effect, Ostwald ripening mechanism and oriented attachment, or ultrathin, from an exfoliation procedure, ones can be achieved. All these unique structures lead to much enhanced properties, especially for photocatalytic, electrocatalytic and photoelectrocatalytic hydrogen evolution reactions, and this chapter gives highlight to porous one. Combining all these categories together there forms a systematic transforming route from inorganic-organic hybrids into inorganic porous materials. Compared with some traditional methods, it can easily fabricate numerous structures that are favor for superior properties intentionally, while requires less harsh conditions and less toxic origin materials. Although tremendous developments have been made, further efforts still need to be done to extend this route to be a more major and universal one, focusing on following aspects: a) diverse novel and efficient strategies to extract organic components in inorganic-organic hybrids, such as electron beam irradiation; b) extension of products to some other classes of inorganic porous materials; c) much more generality to fabricate some unique morphologies, especially porous one, with excellent morphology-dependent properties; and d) better understanding of the whole transformation procedure with theoretical calculation and elaborate experiments.

ACKNOWLEDGMENTS

This work was financially supported by the National Natural Science Foundation of China (No. 21373149) and the Innovation Foundation of Tianjin University.

REFERENCES

[1] Hwang, Y. J., Hahn, C., Liu, B., Yang, P. *ACS Nano* 2012, 6, 5060.
[2] Maeda, K., Takata, T., Hara, M., Saito, N., Inoue, Y., Kobayashi, H., Domen, K. *J. Am. Chem. Soc.* 2005, 127, 8286.
[3] Tian, N., Zhou, Z., Yu, N., Wang, L., Sun, S. *J. Am. Chem. Soc.* 2010, 132, 7580.
[4] Yamauchi, Y., Tonegawa, A., Komatsu, M., Wang, H., Wang, L., Nemoto, Y., Suzuki, N., Kuroda, K. *J. Am. Chem. Soc.* 2012, 134, 5100.
[5] Yu, J., Yu, H., Cheng, B., Zhao, X., Yu, J. C., Ho, W. K. *J. Phys. Chem. B* 2003, 107, 13871.
[6] Moon, G. D., Ko, S., Min, Y., Zeng, J., Xia, Y., Jeong, U. *Nano Today* 2011, 6, 186.
[7] Son, D. H., Hughes, S. M., Yin, Y., Alivisatos, A. P. *Science* 2004, 306, 1009.
[8] Murugan, A. V., Viswanath, A. K., Gopinath, C. S., Vijayamohanan, K. *J. Appl. Phys.* 2006, 100, 074319.
[9] Itoh, T., Matsubara, I., Shin, W., Izu, N., Nishibori, M. *Sens. Actuators B: Chem.* 2008, 128, 512.
[10] Polleux, J., Gurlo, A., Barsan, N., Weimar, U., Antonietti, M., Niederberger, M. *Angew. Chem.* Int. Ed. 2006, 45, 261.
[11] Vivero-Escoto, J. L., Huang, Y. T., *Int. J. Mol. Sci.*, 2011, 12, 3888.
[12] Guo, Y., Tang, Q., Liu, H., Zhang, Y., Li, Y., Hu, W., Wang, S., Zhu, D. *J. Am. Chem. Soc.* 2008, 130, 9198.
[13] Gao, M., Gao, Q., Jiang, J., Cui, C., Yao, W., Yu, S. *Angew. Chem.* Int. Ed. 2011, 50, 4905.
[14] Park, D. H., Kim, J. E., Oh, J. M., Shul, Y.G., Choy, J. H. *J. Am. Chem. Soc.* 2010, 132, 16735.
[15] Choy, J. H., Kwon, S. J., Park, G. S. *Science* 1998, 280, 1589.
[16] Portier, J., Choy, J. H., Subramanian, M. A. *Int. J. Inorg. Mater* 2001, 3, 581.
[17] Paek, S. M., Oh, J. M., Choy, J. H. *Chem. Asian J.* 2011, 6, 324.
[18] Huang, X., Li, J., Zhang, Y., Mascarenhas, A. *J. Am. Chem. Soc.* 2003, 125, 7049.
[19] Huang, X., Heulings, H. R., Le, V., Li, J. *Chem. Mater.* 2001, 13, 3754.
[20] Huang, X., Li, J. *J. Am. Chem. Soc.* 2007, 129, 3157.
[21] Huang, X., Li, J., Fu, H. *J. Am. Chem. Soc.* 2000, 122, 8789.
[22] Santra, T. S., Liu, C., Bhattacharyya, T. K., Patel, P., Barik, T. K. *J. Appl. Phys.* 2010, 107, 124320.
[23] Lee. D. K., Kang, Y. S., Lee, C. S., Stroeve, P. *J. Phys. Chem. B* 2002, 106, 7267.
[24] Cong, H., Yu, S. *Current Opinion in Colloid Interface Science* 2009, 14, 71.
[25] Sheng, W., Kim, S., Lee, J., Kim, S. W., Jensen, K., Bawendi, M. G. *Langmuir* 2006, 22, 3782.

[26] Ma, Y., Ali, S. R., Wang, L., Chiu, P. L., Mendelsohn, R., He, H. *J. Am. Chem. Soc.* 2006, 128, 12064.
[27] Huang, J., Ichinose, I., Kunitake, T. *Angew. Chem.* Int. Ed. 2006, 45, 2883.
[28] Herrikhuyzen, J. V., George, S. J., Vos, M. R. J., Sommerdijk, N. A. J. M., Ajayaghosh, A., Meskers, S. C. J., Schenning, A. P. H. J. *Angew. Chem.* Int. Ed. 2007, 46, 1825.
[29] Shim, B. S., Tang, Z., Morabito, M. P., Agarwal, A., Hong, H., Kotov, N. A. *Chem. Mater* 2007, 19, 5467.
[30] Gheith, M. K., Pappas, T. C., Liopo, A. V., Sinani, V. A., Shim, B. S., Motamedi, M., Wicksted, J. P., Kotov, N. A. *Adv. Mater.* 2006, 18, 2975.
[31] Gao, M., Yao, W., Yao, H., Yu, S. *J. Am. Chem. Soc.* 2009, 131, 7486.
[32] Cong, H., Yu, S. *Adv. Funct. Mater.* 2007, 17, 1814.
[33] Mitzi, D. B. *Inorg. Chem.* 2007, 46, 926.
[34] Monnier, A., Schüth, F., Huo, Q., Kumar, D., Margolese, D., Maxwell, R. S., Stucky, G. D., Krishnamurty, M., Petroff, P., Firouzi, A., Janicke, M., Chmelka, B. F. *Science* 1993, 261, 1299.
[35] Beck, J. S., Vartuli, J. C., Kennedy, G. J., Kresge, C. T., Roth, W. J., Schramm, S. E. *Chem, Mater.* 1994, 6, 1816.
[36] Kresge, C. T., Leonowicz, M. E., Roth, W. J., Vartuli, J. C., Beck, J. S. *Nature* 1992, 359, 711.
[37] Beck, J. S., Vartuli, J. C., Roth, W. J., Leonowicz, M. E., Kresge, C. T., Schmitt, K. D., Chu, C. T. W., Olson, D. H., Sheppard, E. W., McCullen, S. B., Higgins, J. B., Schlenker, J. L. *J. Am. Chem. Soc.* 1992, 114, 10834.
[38] Inagaki, S., Fukushima, Y., Kuroda, K. *J. Chem. Soc., Chem, Commum.* 1993, 680.
[39] Inagaki, S., Koiwai, A., Suzuki, N., Fukushima, Y., Kuroda, K. *Bull. Chem. Soc. Jpn.* 1996, 69, 1449.
[40] Vartuli, J. C., Schmitt, K. D., Kresge, C. T., Roth, W. J., Leonowicz, M. E., McCullen, S. B., Hellring, S. D., Beck, J. S., Schlenker, J. L., Olson, D. H., Sheppard, E. W. *Chem. Mater.* 1994, 6, 2317.
[41] Vartuli, J. C., Kresge, C. T., Leonowicz, M. E., Chu, A. S., McCullen, S. B., Johnson, I. D., Sheppard, E. W. *Chem. Mater.* 1994, 6, 2070.
[42] O'Brian, S., Francis, R. J., Price, S. J., O'Hare, D., Clark, S. M., Okazaki, N., Kuroda, K. *J. Chem. Soc., Chem, Commum.* 1995, 2423.
[43] Steel, A., Carr, S. W., Anderson, M. W. *J. Chem. Soc., Chem, Commum.* 1994, 1571.
[44] Huo, Q., Margolese, D. I., Ciesla, U., Demuth, D. G., Feng, P., Gier, T. E., Sieger, P., Firouzi, A., Chmelka, B. F., Schuth, F., Stucky, G. D. *Chem. Mater.* 1994, 6, 1176.
[45] Huo, Q., Margolese, D. I., Ciesla, U., Feng, P., Gier, T. E., Sieger, P., Leon, R., Petroff, P. M., Schuth, F., Stucky, G. D. *Nature* 1994, 368, 317.
[46] Antonelli, D. M., Ying, J. Y. *Angew. Chem.* Int. Ed. 1996, 35, 426.
[47] Antonelli, D. M., Ying, J. Y. *Chem. Mater.* 1996, 8, 874.
[48] Tian, Z., Tong, W., Wang, J., Duan, N., Krishnan, V. V., Suib, S. L. *Science* 1997, 276, 926.
[49] Bagshaw, S. A., Pinnavaia, T. J. *Angew. Chem.* Int. Ed. 1996, 35, 1102.
[50] Wong, M. S., Ying, J. Y. *Chem. Mater.* 1998, 10, 2067.
[51] Kim, A., Bruinsma, P., Chen, Y., Wang, L., Liu, J. *Chem. Comm.* 1997, 33, 161.
[52] Pacheco, G., Zhao, E., Garcia, A., Sklyarov, A., Fripiat, J. J. *Chem. Comm.* 1997, 33, 491.
[53] Ciesla, U., Schacht, S., Stucky, G. D., Unger, K. K., Schüth, F. *Angew. Chem.* Int. Ed. 1996, 35, 541.
[54] Severin, K. G., Abdel-Fattah, T. M., Pinnavaia, T. J. *Chem. Comm.* 1998, 34, 1471.
[55] Putnam, R. L., Nakagawa, N., McGrath, K. M., Yao, N., Aksay, I. A., Gruner, S. M.,

Navrosky, A. *Chem. Mater.* 1997, 9, 2690.
[56] Vaudry, F., Khodabandeh, S., Davis, M. E. *Chem. Mater.* 1996, 8, 1451.
[57] Yang, P., Zhao, D., Margolese, D. I., Chmelka, B. F., Stucky, G. D. *Nature* 1998, 396, 152.
[58] Yang, P., Zhao, D., Margolese, D. I., Chmelka, B. F., Stucky, G. D. *Chem. Mater.* 1999, 11, 2813.
[59] Mitzi, D. B. *Inorg. Mater.* 2005, 44, 3755.
[60] Mitzi, D. B., Copel, M., Chey, S. J. *Adv. Mater.* 2005, 17, 1285.
[61] Mitzi, D. B., Copel, M., Murray, C. E. *Adv. Mater.* 2006, 18, 2448.
[62] Milliron, D. J., Mitzi, D. B., Copel, M., Murray, C. E. *Chem. Mater.* 2006, 18, 588.
[63] Mitzi, D. B., Kosbar, L. L., Murray, C. E., Copel, M., Afzali, A. *Nature* 2004, 428, 299.
[64] Yao, H., Li, X., Ai, S., Yu, S. *Nano Res.* 2010, 3, 81.
[65] Yu, S., Yang, J., Qian, Y., Yoshimura, M. *Chemical Physics Letter* 2002, 361, 362.
[66] Yu, S., Yoshimura, M. *Adv. Mater.* 2002, 14, 296.
[67] Deng, Z., Wang, C., Sun, X., Li, Y. *Inorg. Chem.* 2002, 41, 869.
[68] Ni, Y., Cao, X., Hu, G., Yang, Z., Wei, X., Chen, Y., Xu, J. *Crystal Growth & Design* 2007, 7, 280.
[69] Nasi, L., Caletani, D., Besagni, T., Ferro, P., Fabbri, F., Licci, F., Mosca, R. *J. Phys. Chem. C* 2012, 116, 6960.
[70] Xiong, S., Shen, J., Xie, Q., Gao, Y., Tang, Q., Qian, Y. *Adv. Funct. Mater.* 2005, 15, 1787.
[71] Gao, X., Li, X., Yu, W. *J. Phys. Chem. B* 2005, 109, 1155.
[72] Jang, J. S., Yu, C. J., Choi, S. H., Ji, S. M., Kim, E. S., Lee, J. S. *Journal of Catalysis* 2008, 254, 144.
[73] Liu, J., Guo, Z., Meng, F., Luo, T., Li, M., Liu, J. *Nanotech.* 2009, 20, 125501.
[74] Nath, M., Choudhury, A., Kundu, A., Rao, C. N. R. *Adv. Mater.* 2003, 15, 2098.
[75] Pütz, J., Aegerter, M. A. *Thin Solid Films* 1999, 351, 119.
[76] Zang, Z., Yao, H., Zhou, Y., Yao, W., Yu, S. *Chem. Mater.* 2008, 20, 4749.
[77] Liao, L, Wang, S., Xiao, J., Bian, X., Zhang, Y., Scanlon, M. D., Hu, X., Tang, Y., Liu, B., Girault, H. H. *J. Mater. Chem.* 2010, 20, 2807.
[78] Gao, Q., Zhang, C., Xie, S., Hua, W., Zhang, Y., Ren, N., Xu, H., Tang, Y. *Chem. Mater.* 2009, 21, 5560.
[79] Li, Y., Li, X., He, R., Zhu, J., Deng, Z. *J. Am. Chem. Soc.* 2002, 124, 1411.
[80] Li, Y., Li, X., Deng, Z., Zhou, B., Fan, S., Wang, J., Sun, X. *Angew. Chem.* Int. Ed. 2002, 41, 333.
[81] Yu, Y., Hou, S., Meng, M., Tao, X., Liu, W., Lai, Y., Zhang, B. *J. Mater. Chem.* 2011, 21, 10525.
[82] Jang, J. W., Choi, S. H., Jang, J. S., Lee, J. S., Cho, S., Lee, K. H. *J. Phys. Chem. C* 2009, 113, 20445.
[83] Nath, M., Choudhury, A., Rao, C. N. R. *Chem. Comm.* 2004, 40, 2698.
[84] Song, L., Chen, J., Zhu, L., Xia, J., Yang, J. *Inorg. Mater.* 2011, 50, 7988.
[85] Song, L., Teng, Y., Chen, J. *J. Phys. Chem. C* 2012, 116, 22859.
[86] Deng, Z., Li, L., Li, Y. *Inorg. Chem.* 2003, 42, 2331.
[87] Chen, X., Xu, H., Xu, N., Zhao, F., Lin, W., Lin, G., Fu, Y., Huang, Z., Wang, H., Wu, M. *Inorg. Chem.* 2003, 42, 3100.
[88] Krumeich, F., Muhr, H. J., Niederberger, M., Bieri, F., Schnyder, B., Nesper, R. *J. Am. Chem. Soc.* 1999, 121, 8324.
[89] Muhr, H. J., Krumeich, F., Schönholzer, U. P., Bieri, F., Niederberger, M., Gauckler, L. J., Nesper, R. *Adv. Mater.* 2000, 12, 231.
[90] Niederberger, M., Muhr, H. J., Krumeich, F., Bieri, F., Günther, D., Nesper, R. *Chem.*

Mater. 2000, 12, 1995.
[91] Spahr, M. E., Bitterli, P., Nesper, R., Müller, M., Krumeich, F., Nissen, H. U. *Angew. Chem*. Int. Ed. 1998, 37, 1263.
[92] Dong, Y., Peng, Q., Li, Y. *Inorg. Chem. Comm.* 2004, 7, 370.
[93] Ni, Y., Zhang, L., Zhang, L., Wei, X. *Cryst. Res. Technol.* 2008, 43, 1030.
[94] An, C., Tang, K., Yang, Q., Qian, Y. *Inorg. Chem.* 2003, 42, 8081.
[95] Mirkovic, T., Hines, M. A., Nair, P. S., Scholes, G. D. *Chem. Mater.* 2005, 17, 3451.
[96] Han, W., Gao, M. *Crystal Growth & Design* 2008, 8, 1023.
[97] Zhang, M., Lu, Y., Chen, J., Zhang, T., Liu, Y., Yang, Y., Yao, W., Yu, S. *Langmuir* 2010, 26, 12882.
[98] Yu, Y., Zhang, J., Wu, X., Zhao, W., Zhang, B. *Angew. Chem*. Int. Ed. 2012, 51, 897.
[99] Xu, Y., Wu, R., Zhang, J., Shi, Y., Zhang, B. *Chem. Comm.* 2013, 49, 6656.
[100] Henkes, A. E., Schaak,R. E. *Chem. Mater.* 2007, 19, 4234.
[101] Muthuswamy, E., Brock, S. L. *J. Am. Chem. Soc.* 2010, 132, 15849.
[102] Wu, X., Yu, Y., Liu, Y., Xu, Y., Liu, C., Zhang, B. *Angew. Chem*. Int. Ed. 2012, 51, 3211.
[103] Zhuo, S., Xu, Y., Zhao, W., Zhang, J., Zhang, B. *Angew. Chem*. Int. Ed. 2013, 52, 8602.
[104] Zhuang, T., Fan, F., Gong, M., Yu, S. *Chem. Comm.* 2012, 48, 9762.
[105] Tong, H., Zhu, Y., Yang, L., Li, L., Zhang, L. *Angew. Chem*. Int. Ed. 2006, 45, 7739.
[106] Son, J. S., Wen, X., Joo, J., Chae, J., Baek, S., Park, K., Kim, J. H., An, K., Yu, J. H., Kwon, S. G., Choi, S. H., Kim, Y. W., Kuk, Y., Hoffman, R., Hyeon, T. *Angew. Chem*. Int. Ed. 2009, 48, 6861.
[107] Park, S. H., Lee, Y. J., Huh, Y. D. *Chem. Comm.* 2011, 47, 11763.
[108] Oaki, Y., Imai, H. *Adv. Mater.* 2006, 18, 1807.
[109] Yao, W., Yu, S., Pan, L., Li, J., Wu, Q., Zhang, L., Jiang, *J. Small* 2005, 1, 320.
[110] Chen, C., Zhu, C., Hao, L., Hu, Y., Chen, Z. *Chem. y Lett.* 2004, 33, 898.
[111] Mitzi, D. B. *Inorg. Chem.* 2005, 44, 7078.
[112] Sun, Y., Sun, Z., Gao, S., Cheng, H., Liu, Q., Piao, J., Yao, T., Wu, C., Hu, S., Wei, S., Xie, Y. *Nature Comm.* 2012, 3, 1057.
[113] Xu, Y., Zhao, W., Xu, R., Shi, Y., Zhang, B. *Chem. Comm.* 2013, 49, 9803.
[114] Chen, M., Gao, L. *Mater. Chem. Phys.* 2005, 91, 437.
[115] Yao, W., Yu, S., Wu, Q. *Adv. Funct. Mater.* 2007, 17, 623.
[116] Chen, D., Gao, L., Yasumori, A., Kuroda, K., Sugahara, Y. *Small* 2008, 4, 1813.
[117] Zou, X., Li, G., Wang, K., Li, L., Su, J., Chen, J. *Chem. Comm.* 2010, 46, 2112.
[118] Zou, X., Li, G., Zhao, J., Wang, P., Wang, Y., Zhou, L., Su, J., Li, L., Chen, J. *Inorg. Chem.* 2011, 50, 9106.

In: Comprehensive Guide for Mesoporous Materials, Volume 4 ISBN: 978-1-63482-092-9
Editor: Mahmood Aliofkhazraei

Chapter 15

SBA-16 MESOPOROUS SILICA AS CATALYTIC SUPPORT FOR HYDRODESULFURIZATION CATALYSTS

***E. M. Rivera-Muñoz*[1], *B. Pawelec*[2], *R. Nava*[3], *J. A. Cortés*[4] *and R. Huirache-Acuña*[4,†]**

[1]Centro de Física Aplicada y Tecnología Avanzada, Universidad Nacional Autónoma de México, Querétaro, México
[2]Instituto de Catálisis y Petroleoquímica, CSIC, Madrid, Spain
[3]Facultad de Ingeniería, Universidad Autónoma de Querétaro, Centro Universitario, Cerro de las Campanas, Querétaro, México
[4]Facultad de Ingeniería Química, Universidad Michoacana de San Nicolás de Hidalgo, Ciudad Universitaria, Morelia, Michoacán, México

ABSTRACT

SBA-16 is an interesting mesoporous silica material having nanopores and a large surface area, which is widely employed as catalysts supports, absorbents, drug delivery materials, etc. This material has attracted attention of catalytic researchers because of its 3D structure consisting of ordered and interconnected spherical mesopores which facilitates the transport of reactants and products without pore blockage. Since it has lack of functionality, heteroatoms have been incorporated by direct or post-synthesis methods in order to modify their functionality. The aim of this chapter is to show the state-of-the-art related to the SBA-16-based mesoporous systems as catalytic supports for hydrodesulfurization (HDS), to take stock of knowledge about current formulations and applications in this field.

Keywords: SBA-16, mesoporous, silica, catalytic support, hydrodesulfurization

† E-mail: rafael_huirache@yahoo.it. Tel.: +52 443 3273584.

1. INTRODUCTION

The more stringent environmental regulations enacted throughout the world have increased the need of more active hydrotreating (HDT) catalysts, in the petroleum refining industry [1]. Usually, catalysts used for diesel oil hydrotreatment are γ-Al_2O_3 supported molybdenum or tungsten sulfides promoted with cobalt or nickel (Figure 1).

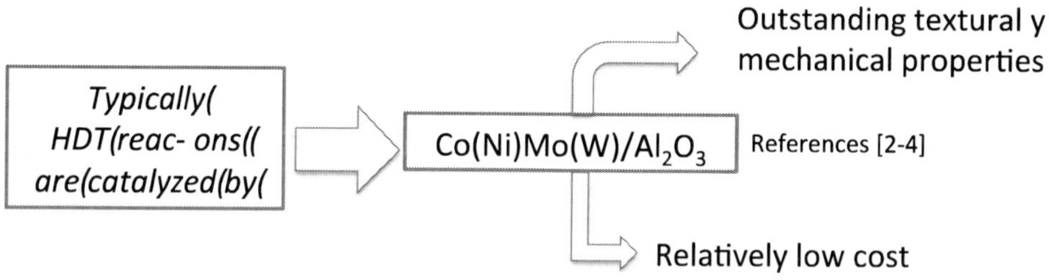

Figure 1. Typically used catalysts for hydrodesulfurization reactions.

Current strategies for the design of novel HDS catalysts often include variations in the support formulation, catalyst preparation method and active phase formulation. Regarding research into new supports, a family of highly ordered mesoporous molecular sieves (MMS) has recently attracted the interest of the research community because of their large pores (2-50 nm) and high surface area which allows higher activity per unit weight [5-29]. By using MMS, different kinds of HDS catalysts have been synthesized (Table 1).

Table 1. Summary of the studies dedicated to mesoporous molecular sieves used as supports of HDS catalysts

Catalyst	Synthesis method	Refe-rence	Findings
CoMo/Al-HMS	Supercritical deposition	[5]	Experimental data indicated higher conversion at higher temperature for the new catalyst with respect to the commercial catalyst.
NiMo/Al-HMS	Supercritical impregnation	[6]	Analysis of the experimental data using Arrhenius' law resulted in the calculation of frequency factor and activation energy of the HDS for all catalysts.
CoMoW/P-SBA-15(16)	Impregnation	[7]	Both CoMoW/SBA-15 and CoMoW/SBA-16 catalysts were more active than a $CoMo/Al_2O_3$ commercial catalyst containing a small amount of phosphorous in its formulation.

Catalyst	**Synthesis method**	**Refe-rence**	**Findings**
HPMo(W)/HMS	Wet impregnation	[11]	It is assumed that supported polyphosphates of Mo and W on mesoporous carriers are good oxide precursors for the active sulfided phase of the catalysts for the HDS of DBT.
Noble metals/ Mesoporous Zeolite	Pore volume impregnation	[19]	Mesoporous zeolites are very good supports for noble metals and make them much more active HDS catalysts than noble metals on normal zeolites, alumina or amorphous silica-alumina.
Pd/Al-MCM-41	Impregnation	[20]	The modification of the siliceous MCM-41 by post-synthesis alumination had a beneficial effect on the hydrodesulfurization of the supported palladium catalysts.
Ni(Co)Mo/Al(Ti)-HMS	Successive impregnation using the pore filling method	[21]	The NiMo and CoMo phases deposited on Al-HMS and Ti-HMS substrates led to higher catalytic activity than with Al(Ti)-free catalysts.
WS_2/Al-HMS	Incipient wetness impregnation method.	[23]	Catalysts indicated that HMS and Al-HMS materials induce outstanding activities in WS_2 and its promoted analogues.

Among them, SBA-16 system (SBA = Santa Barbara Amorphous) has attracted attention of catalytic researchers because of its interesting 3D structure consisting of ordered and interconnected spherical mesopores easy accessible for guest molecules, which facilitates the transport of reactants and products without pore blockage. Recently, some authors have mentioned their use as hydrodesulfurization supports, In addition, SBA-16 substrate shows super large cage-like mesoporous structure with a cubic symmetry (*Im*3*m*), high surface area and high thermal stability, which provides more favorable mass transfer kinetics than the unidirectional pore system of materials such as HMS and SBA-15 [30, 31]. In this chapter, we are going to show the possibilities of the use of SBA-16 silica as support for hydrodesulfurization catalysts.

2. Influence of Support

2.1. Synthesis of Bare SBA-16

Zhao et al. [32] reported the synthesis of mesoporous SBA-16 silica materials, using non-ionic triblock copolymers as template [33]. This type of surfactant is very interesting, because it is easily separated, is nontoxic, biodegradable, and inexpensive. The synthesis conducted with these surfactants usually occurs in low-pH solutions (pH ≈ 2) where the interaction occurs through an $S^0H^+X^-I^+$ mechanism (S^0H^+ being the surfactant hydrogen-bonded to a hydronium ion, X^- the chloride ion, and I^+ the protonated silica) [34].

The SBA-16 material has a 3D cubic arrangement of mesopores corresponding to *Im3m* space group symmetry [32]. This material was initially synthesized in the presence of triblock copolymer Pluronic F127. High-resolution electron microscopy of the SBA-16 demonstrated that each mesopore in the body-centered cubic array is connected with its eight neighbors [35]. Not only because of its pores, but also thermal, mechanical and chemical resistance properties that makes it a preferable choice to other types of supports.

The main way to obtain a well-defined and structured SBA-16 material is to use a surfactant templated polymerization instead of an uncontrolled reaction. The organic-inorganic self-assembly is driven by noncovalent weak bonds such as hydrogen bonds, van der Waals forces, and electrovalent bonds between the triblock copolymer surfactant and inorganic species. Instead of a simple superposition of the weak interaction, an integrated and complex synergistic reaction facilitates the process [30]. Cooperative assembly between poly(ethylene oxide)-poly(propylene oxide)-poly(ethyleneoxide), ($EO_{106}PO_{70}EO_{106}$, Pluronic F127), organic surfactant and inorganic precursors is generally involved, forming inorganic/organic mesostructured composites. In general, the amphiphilic surfactant molecules form a liquid crystal by aggregation in aqueous solution [36, 37]. The formation of the liquid crystal matrix strongly depends on the conditions in the solution. The structure of the liquid crystal is the so called mesostructure. Important parameters for the mesophase formation are for instance the temperature, concentration or the pH-value of the solution.

2.2. Redox and Acid-based Properties

In general, the studies in this field present the relationship between different physical and chemical properties of the support with the activity of different hydrotreating reactions.

Santa Barbara Amorphous No. 16 (SBA-16) is significant mesoporous silica with exclusive and important properties of highly ordered mesopores, hydrothermally stable and thick wall and large surface area which render it as promising catalyst for wide applications. Moreover, functionalization and modification of SBA-16 could enhance and optimize its catalytic activity. The purely siliceous SBA's materials have only silanol groups on their surface, which are of low acid strength and catalytically non-active. Therefore, in order to make these materials useful for catalysis, stronger acidic sites have to be introduced into their framework. Generally, the substitution of heteroatoms with valence lower than silicon creates negative charges in the framework, which can be compensated by protons, thus generating acidity in these materials [41]. The pure siliceous SBA-16 material has no Brønsted acidity. Thus, considerable efforts have been already applied to create acid sites in the silica-based materials by the isomorphous substitution of the Si^{4+} atoms by Al^{3+}ions. Moreover, the substitution of Si^{4+} ions of SBA-16 by foreign ion is an efficient method to enhance the catalyst stability and leads to catalysts with redox properties.

Table 2. Summary of SBA-16 synthesis methods

Synthesis method conditions	Template	Reference	Findings
Acidic conditions over a narrow range of dilute surfactant concentrations (3-5%) at room temperature (RT). The first report by Zhao et al. was described as follows: in a typical preparation, 4.0 g of Pluronic F127 was dissolved in 30 g of water and 120 g of 2M HCl solution with stirring at RT. Then 8.50 g of tetraethyl orthosilicate (TEOS) was added into that solution with stirring for 20 h. The mixture is then aged at 80°C overnight without stirring. The solid product is recovered, washed, and air-dried at RT. Yields are ~98%. Calcination is carried out by slowly increasing temperature from room temperature to 500°C in 8 h and heating at 500°C for 6 h.	Poly(alkylene oxide) triblock copolymer $EO_{106}PO_{70}EO_{106}$ (Pluronic F127)	[32, 33]	Higher copolymer concentrations result in the formation of silica gel, while lower concentrations lead to the formation of amorphous silica. Calcined SBA-16 had a pore size of 54 Å, a pore volume of 0.45 cm^3/g, and a BET surface area of 740 m^2/g. The N_2adsorption-desorption isotherm is type IV with a type-H_1 hysteresis loop.
SBA-16 mesoporous silicas with cubic *Im*3*m* structure were synthesized using Pluronic F127 and its blends with Pluronic P123 triblock copolymer as supramolecular templates.	Pluronic F127 and Pluronic P123	[38]	It was shown that the pore cage diameter in SBA-16 can be enlarged in a wide range not only by increasing the synthesis temperature and time, as previously known, but also by increasing the content of P123 copolymer in the polymer mixture. These three ways allowed to synthesize SBA-16 with nominal mesopore diameters from ~4.5 to 9 nm.
SBA-16 was hydrothermally synthesized by using Brij700 as a structure-directing agent and tetraethylorthosilane (TEOS) as a silica source. In a typical experiment, 0.50 g of Brij700 was dissolved in 50 g of 2M HCl solution at 10-40°C. A chosen amount of 1,3,5-trimethylbenzene (TMB) (between 0.20 and 1.0 g) and 3.5 g of TEOS were added into this solution. The mixture was maintained at 4-40°C for 24 h with vigorous stirring, followed by hydrothermal treatment at 100°C for different times from 24 to 72 h.	Brij700 (polyoxyethylene (100) stearyl ether, $C_{18}H_{37}EO_{100}$)	[39]	A small pore SBA-16 type mesoporous silica with large microporosity and ultra-thick wall.

Table 2. (Continued)

Synthesis method conditions	Template	Reference	Findings
A ternary surfactant mixture with the chemical components of 0.26 g C_8TMAB: 0.3 g SDS: 0.7 g, Pluronic F127 ($EO_{106}PO_{70}EO_{106}$) were stirred in 50 g H_2O at 30°C, where the molar ratio of C_8TMAB/SDS was 1.0. The solution was directly poured into a 200 ml dilute sodium silicate solution with [SiO_2] = 70.0 mM, at pH~3.0. The reaction was continued under static condition at 30°C. After several days of induction time, the solid products were collected by filtration, washed with water, and dried at 100°C	C_8TMAB Sodium dodecyl sulfate (SDS) Pluronic F127	[40]	Faceted single crystals of cubic form were obtained under well-controlled synthetic conditions. Synthetic factors in controlling the crystal morphology include stirring, water content, the ionic surfactants and acidity. Changing these conditions for a slower reaction rate always favors the formation of faceted single crystals.

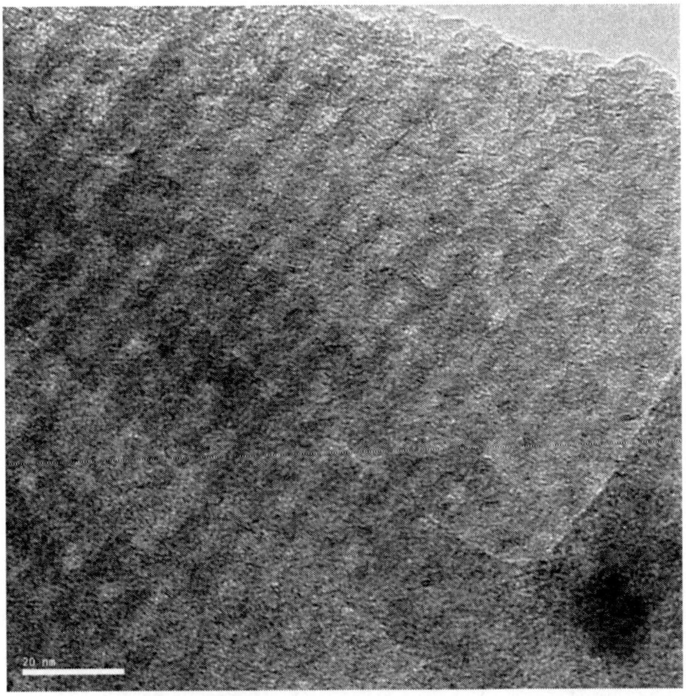

Figure 2. Typical High Resolution Transmission Electron Microscopy (HRTEM) image of SBA-16 mesoporous silica. It can be observed the size and morphology of the 3D structure of ordered and interconnected spherical mesopores.

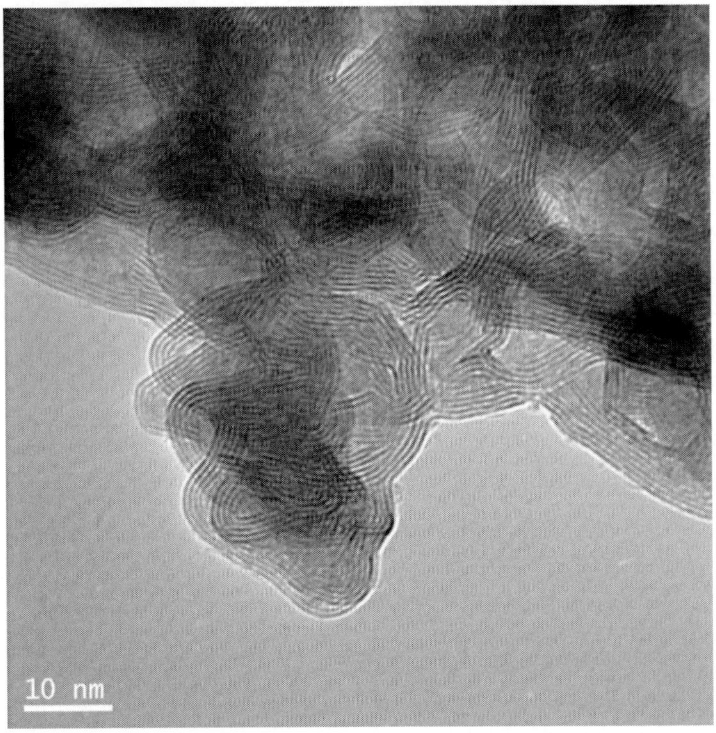

Figure 3. High Resolution Transmission Electron Microscopy Image of fresh NiMoW sulfide catalyst supported on SBA-16.

2.2.1. Al^{3+} Ion Loading

T. Klimova et al. [28] prepared a series of Al-containing mesoporous molecular sieves SBA-16 by direct and post-synthetic methods using different Al sources. NiMo sulfides were employed as active phases to prepare catalysts supported on Al-SBA-16 which were tested in the 4,6-dimethyldibenzothiophene hydrodesulfurization (HDS) reaction (one of the most refractory sulfur compounds in gas oil). They found that the characteristics of the Al-SBA-16-supports strongly depend on the used Al source and the method selected for Al incorporation into the SBA structure. Post-synthetic alumination of SBA-16 by reacting with $AlCl_3$ or Al(i-$PrO)_3$ has a number of advantages in comparison with the incorporation of Al by grafting with $NaAlO_2$ or direct synthesis, among them: a good control of Al content in the sample; little influence on the mesoporous structure of the parent SBA-16; creation of Brønsted acid sites; and stronger interaction with deposited Ni and Mo species providing them better dispersion.

R. Huirache-Acuña and coworkers [42] studied the effect of direct Al-incorporation into SBA-16 and HMS supports for Co(Ni)-Mo-W ternary catalysts. Al-HMS and Al-SBA-16 mesoporous silica substrates were prepared by the one-pot sol–gel synthesis method via the neutral templating pathway using dodecylamine and Pluronic F127 as surfactants, respectively. The oxide precursors were characterized by a variety of techniques (N_2 adsorption–desorption isotherms, XRD, DRS UV-vis spectroscopy and ^{27}Al NMR), whereas sulphided catalysts were characterized by XPS and HRTEM, and tested in the HDS of DBT conducted in a high-pressure batch reactor at T = 350°C and P = 3.1 MPa. ^{27}Al NMR spectra indicate that a larger amount of Al was incorporated into the framework of the HMS than in extraframework positions. Contrary to both Al-HMS-based catalysts, the CoMoW/Al-SBA16 sample shows peaks at 107.6, 66.7 and 58.1 ppm. The latter two peaks indicate that this sample possesses a large proportion of the Al inserted into tetrahedral positions within the framework of SBA-16. The activity results revealed that NiMoW/Al-HMS catalyst was more active than all Co-promoted catalysts (including a commercial CoMo/Al_2O_3 catalyst). The HDS activity was found to be markedly influenced by the textural properties of support and the dispersion of the active phases on the catalyst surface. It was hypothesised that the presence of extra framework $AlNO_3$ phases on the surface of CoMoW/Al-SBA16 compromised its HDS activity because of the formation of a less active "onion-type" Mo(W)S_2 structure. The lowest activity of the CoMoW/Al-SBA16 catalyst among the catalysts studied was likely due to the fact that it records partial blocking of the pores by the guest molecules, as deduced by the normalized BET specific surface area of the oxide catalyst precursors, which helps to decrease the accessibility of the DBT molecules to the active sites. Moreover, a comparison of the average pore diameter of the two CoMoW/Al-HMS and CoMoW/Al-SBA16 catalysts suggested that the higher activity of the former should be linked to its larger pore size (5.0 vs. 1.7 nm), favouring the diffusion of a large DBT molecule into the inner porous structure of this catalyst.

2.2.2. Ti^{4+} Ion Loading

Similarly to the support modification with Al, the HDS activity of the SBA-15-based catalysts could be enhanced by the incorporation of Ti^{4+} ion into SBA-16. The effect of Ti incorporation was investigated by J. C. Amezcua and coworkers [31]. A series of Ti-containing mesoporous SBA-16 supports and the respective NiMo catalysts were prepared to study the effect of the method for support preparation and TiO_2 loading on the dispersion of titania, characteristics of Ni and Mo surface species and their catalytic activities in the 4,6-dimethyldibenzothiophene hydrodesulfurization (HDS). Titania was incorporated to purely siliceous SBA-16 support by two post-synthetic methods: chemical grafting and incipient wetness impregnation. These methods were chosen because Ti incorporation into SBA structure by direct synthesis appears unlikely since these materials are synthesized in strongly acidic media (2 M HCl). Additionally, postsynthetic methods have serious advantages in comparison with Ti incorporation during hydrothermal synthesis of SBA materials: higher Ti loading without substantial loss of the parent SBA structure and texture, location of Ti species on the support surface that makes them accessible for the interaction with NiMo active species and possibility of fine tuning of the characteristics of the deposited TiO_2 species by modifying the preparation procedure used.

The results showed that Ti-SBA-16 supports with different TiO_2 content and dispersion can be obtained without substantial loss of SBA-16 characteristics. The best dispersion of TiO_2 in the SBA-16 pore channels was obtained by the grafting method. However, titania loading in this case was limited to 15.1 wt% TiO_2. Samples prepared by impregnation showed the presence of different Ti species: isolated Ti species and small anatase clusters (inside the mesopore channels of SBA-16) and anatase crystallites (on the external surface), even for low titania loadings. Crystal size of anatase phase increased with titania loading. Authors found that TiO_2 incorporation in the SBA-16 support leads to stronger interaction of Mo and Ni oxidic species with the support providing better dispersion to the sulfided metal species as observed by High Transmision Electron Microscopy (HRTEM). In addition, NiMo catalysts supported on Ti-modified SBA-16 showed high activity in 4,6-dimethyldibenzothiophene HDS.

Asma Tufail Shah et al. [43], presented a study about synthesis of Ti-containing SBA-16 using and evaporation-induced self-assembly method which offered a good alternative for obtaining Ti-containing SBA-16-type mesoporous material with high thermal stability, thick pore walls, and high surface area. Ti-containing SBA-16-type mesoporous silicate material was prepared by the evaporation-induced self assembly method as follows: mix 2 g of F127 with 15 ml of C_2H_5OH, adjusted pH at 1.5 using 2 M HCl solution, stir for 30 min followed by an addition of calculated amount of tetrabutyl titanate ($C_{16}H_{36}O_4Ti$) (to get 5, 10, and 15 wt% of Ti relative to SiO_2), and stir for another 2 h. Adding 4.16 g of TEOS and stirring for 7 h. The resulting gel was transferred into a petri dish to evaporate solvent at room temperature for 2 days, to get a rigid solid. Then gel was heated at 80°C for 12 h to remove ethanol, and calcined at 550°C for 6 h (at heating rate of 1°C/min). XPS and XRD techniques showed that most of the Ti was incorporated into the silica framework. Catalytic test results suggested that the catalyst was highly active in oxidation of refractory bulky organosulfur molecule DBT. And its activity did not decrease even after three recycling runs. This method was low cost, simple, one-step, and nonhydrothermal. The resulting material was highly stable, possessed a large pore diameter, thick walls, and high surface area and showed high activity for oxidative desulfurization of bulky organosulfur compound dibenzothiophene (DBT), and removed sulfur contents up to 99.78% under ambient conditions.

2.2.3. Surface Modification with Different Groups

Transition-metal phosphides are a novel class of compounds which are promising next-generation hydrotreating catalysts. R. Huirache-Acuña et al. [7], studied phosphate-loaded mesoporous SBA-15 and SBA-16 materials as supports of novel ternary Co-Mo-W hydrodesulfurization catalysts. In this study the effects of support morphology (SBA-15 against SBA-16) and its modification with variable amounts of phosphorus were studied. The sulfided catalysts were tested in a reaction of hydrodesulfurization (HDS) of dibenzothiophene (DBT) at 350°C and 3.1 MPa of total H_2 pressure. Regardless of the support, it was found that the presence of phosphorous inhibited the HDS activity, and the catalytic response of both SBA-15- and SBA-16-based systems did not depend on the support morphology. In terms of the reaction rate constants (k), both CoMoW/SBA-15 and CoMoW/SBA-16 catalysts were more active than a $CoMo/Al_2O_3$ commercial catalyst containing a small amount of phosphorous in its formulation. In general, the spent CoMoW/(x)P-SBA-16 catalysts showed a lower coke formation than their SBA-15-based counterparts because the phosphate incorporation to the SBA-16 material led to decrease in coke formation in the final catalysts, as opposite to the behavior of SBA-15-based counterparts. The most significant results obtained by these studies can be summarized as follows: (i) the SBA-16 proved to be a better substrate to developing a large density of $Mo(W)S_2$ clusters on the surface than its SBA-15 counterpart; (ii) the P-free SBA-16 material was most optimized substrate for supporting Co, Mo and W phases; (iii) the irregular oxide Mo^{6+} and W^{6+} particles were precursors of the active phase formed upon sulfidation; (iv) regardless of the support morphology, the addition of phosphate was found to be detrimental for catalytic activity, (v) for the SBA-16-based catalysts the presence of a large amount of phosphate (1.5 and 2.0 wt%) on the support surface inhibited coke formation during the HDS of DBT.

P-containing SBA-16 mesoporous systems were used as supports of ternary Ni–Mo–W hydrodesulfurization (HDS) catalysts by M. A. Guzmán et al. [44]. Characterization by X-Ray Diffraction (XRD) evidenced the formation of crystalline Mo_1-xWxO_3 and $NiMoO_4$ phases on the surface of oxide catalyst precursors with high P-loadings (1.0 and 1.6 wt.%). The UV–vis DRS measurements showed a substantial decrease of energy band gap after P-loading onto the SBA-16 substrate with respect to P-free sample. The sulfided catalysts were tested in separate HDS of dibenzothiophene (DBT) and 4,6-dimethyldibenzothiophene (4,6-DMDBT) reactions performed in a batch reactor at 320°C and H_2 pressure of 5.0 MPa. For both HDS reactions, the initial catalyst activity displayed a volcano-type curve indicating that catalyst behavior depends strongly on the phosphorous loading being all catalysts more active in the HDS of DBT then in the HDS of 4,6-DMDBT. The NiMoW/SBA-16 catalyst loaded with optimized amount of phosphorous (1.0 wt. %) showed superior initial activity than the P-free counterpart. This effect was attributed to the enhancement of active phase dispersion on the support surface, as revealed by High Resolution Transmision Microscopy (HRTEM) measurements. On the other hand, it was observed that the formation of ''onion-type'' $Mo(W)S_2$ phases on the catalyst surface was detrimental for the HDS activity (Figure 3), because a low edge-to-basal sites ratio was obtained.

At reaction time of 5 h, the most active NiMoW/SBA-16 catalyst loaded with 1.0 wt. % of P showed similar activity in both HDS of DBT and 4,6-DMDBT reactions. It was found that HYD route of both reactions was promoted by the support modification with P and due to inhibition of coordinatively unsaturated sites (CUSs) by competitive adsorption of products (mainly H_2S).

The removal of recalcitrant organic sulfur compounds, mainly dibenzothiophene (DBT) and its alkylated derivatives, from petroleum and its fractions is necessary because of industrial and environmental concerns. Oxidative desulfurization is a possible complementary alternative to the conventional hydrodesulfurization processes for the pre-combustion desulfurization of petroleum derived fuels. Oxidative desulfurization has been successfully applied to remove organic sulfur compounds from diesel and light gas oil. The catalytic potential of chloroperoxidase (CPO) immobilized on mesoporous materials was evaluated for the oxidation of 4,6-dimethyldibenzothiophene in water/acetonitrile mixtures by Eduardo Terrés et al. [45]. Two different types of materials were used for the immobilization: a metal containing Al-MCM-41 material with a pore size of 26 Å and SBA-16 materials with three different pore sizes: 40, 90 and 117Å. The SBA-16 (40 Å) did not retain any CPO. The nature and the pore size of the material affected the catalytic activity of the enzyme as well as its stability. Compared to the free enzyme, the thermal stability of CPO at 45°C was two and three times higher than when immobilized on Al-MCM-41and SBA-16 (90 Å), respectively.

3. Influence of Active Phase

A bifunctional catalyst is required for hydrodesulfurization reaction. Since bare SBA-16 has no HDS activity, it is necessary to create active sites. The classical active phases working in the HDS reaction are transition metal sulfides. Transition metal sulfides can be obtained by sulfidation of oxide catalyst precursors or by direct decomposition of different thiosalts. Recently, a few studies on the SBA-16-suported metal phosphides and carbides are reported also, because of their high activity in the HDS reaction.

3.1. Oxide Precursors

The most commonly used method for the preparation of oxide catalyst precursors is support impregnation with the Mo(W) and/or Ni(Co) salt precursors according to the incipient wetness method and employing aqueous solutions of nickel and/or cobalt nitrates, ammonium heptamolybdate and/or ammonium metatungstate. It is known that the catalyst preparation method (co-impregnation or successive impregnation) and type of active phase precursor used have great importance for the final catalyst behavior. Concerning SBA-16-based HDS catalysts, the studies on the influence of catalyst preparation method or type of active phase precursor on the final HDS activity are scarce (Table 3).

The new generations of catalysts, such as NEBULA®, are based on a totally different concept of bulk-like materials. In this sense, research on unsupported NiMoW ternary systems has shown that those catalysts record an HDS activity fourfold greater than that of the supported NiMo and NiW catalysts commercially available today. Because the surface properties of bulk catalysts can be easily tuned by supporting them on different substrates, which allows the modification of the dispersion and coordination of the deposited active ingredient, the challenge is to develop supported ternary systems, which could be more active than bulk-like systems.

Table 3. Summary of HDS catalysts supported on SBA-16

Catalyst	Synthesis method	Reference	Findings
NiMo/Al-SBA-16	NiMo/SBA-16 catalysts were prepared by a standard incipient wetness technique. The calcined supports were impregnated successively using aqueous solutions of ammonium heptamolybdate and nickel nitrate. Mo was impregnated first. After each impregnation, the catalysts were dried (100°C, 24 h) and calcined (500°C, 2 h).	[28]	The work by Klimova et al. [28] has shown that NiMo catalysts supported on Al-SBA-16, prepared by the post-synthesis alumination of SBA-16, record high activity in the HDS of 4,6-DMDBT. The authors attribute this to the good dispersion of Ni and Mo phases and to the bifunctional character of the catalysts. The interaction of Ni and Mo species with the support becomes stronger with Al loading into the SBA-16 by postsynthetic aluminum grafting. According to this, the dispersion of sulfided Mo species increases. However, when SBA-16 is aluminated using $NaAlO_2$, NiMo catalysts with low surface area and activity are obtained.
NiMo/Ti-SBA-16	J. C. Amezcua and coworkers used similar precursors and nominal composition for catalysts: 12 wt. % MoO_3 and 3 wt.% NiO (as reported by Klimova et al. [28]).	[31]	The interaction of Ni and Mo species with the support becomes stronger with Ti loading into theSBA-16. According to this, the dispersion of sulfided Mo species increases.
CoMoW/P-SBA-16	Two series of CoMoW catalysts were prepared using the P-free and P-containing SBA-15 and SBA-16 substrates. Each support was loaded with fixed amounts of molybdenum (8.53 wt% as MoO_3), tungsten (13.75 wt% as WO_3) and (3.81 wt% as CoO), by simultaneous impregnation via incipient wetness method. For this, an aqueous solution containing ammonium heptamolybdate tetrahydrate ($(NH_4)_6Mo_7O_{24}.4H_2O$, Aldrich), ammonium metatungstate ($(NH_4)_6H_2W_{12}O_{40}.xH_2O$, Aldrich) and cobalt nitrate hexahydrate ($Co(NO_3)_2.6H_2O$, Aldrich, 98%) was prepared.	[7]	Although the SBA-15 and SBA-16-supported catalysts could not be appropriate for ultra-low sulfur Diesel applications owning their lower pore size than it is needed for the entrance of any methyl-substituted DBT molecules; useful information was obtained from the catalyst screening in the HDS of DBT. Although the trend in activity cannot be correlated with the textural properties, the acid properties, or more importantly with the support morphology, it was confirmed that the density of the active phases on the SBA-16 support surface (from combined HRTEM and XPS), together with the inhibitor effect of the phosphate on the coke formation, are two important factors which govern the HDS activity of CoMoW catalysts supported on mesoporous silica substrates.

Catalyst	Synthesis method	Reference	Findings
Ni(Co)MoW/Al-SBA-16	Al-SBA16 supports were loaded with fixed amounts of molybdenum, tungsten and cobalt/nickel. The concentrations were calculated to achieve a Mo/W atomic ratio of 0.5 and a Ni(Co)/Mo atomic ratio of 0.3. The catalysts were prepared by simultaneous impregnation via the incipient wetness method. The solids obtained were dried at 85°C for 16 h and then calcined at 500°C for 4h.	[42]	From the HDS of DBT reaction results, the main conclusions of this study are as follows: (i) Ni is more effective as a promoter than Co; (ii) the presence of extraframework $AlNO_3$ phases on the surface of CoMoW/Al-SBA16 might compromise its HDS activity because of the formation of a less active "onion-type" $Mo(W)S_2$ structure.
NiMoW/P-SBA-16	Each support was first phosphorous modified and loaded with fixed amounts of molybdenum (8.53 wt% as MoO_3), tungsten (13.75 wt% as WO_3) and nickel (3.81 wt% as NiO). For this, an aqueous solution containing ammonium heptamolybdate tetrahydrate ($(NH_4)_6Mo_7O_{24}.4H_2O$, Aldrich), ammonium metatungstate ($(NH_4)_6H_2W_{12}O_{40}.xH_2O$, Aldrich) and nickel nitrate hexahydrate ($Ni(NO_3)_2.6H_2O$, Aldrich, 98%) was prepared. The concentrations were calculated to achieve a Mo/W atomic ratio of 0.5 and a Ni/Mo atomic ratio of 0.3.	[44]	The information regarding the structure of Ni-Mo-W catalysts, in their oxidized and sulfided forms, was gained by using several physico-chemical techniques. By using UV-vis, it was observed that the P-containing catalysts showed a band at about 735 nm indicating the presence of octahedrally coordinated Ni^{2+} ions (NiO and/or $NiW(Mo)O_4$ species. A lower intensity of this band with respect to P-free sample suggested some inhibition of the formation of NiO and/or $NiW(Mo)O_4$ species on the P-modified SBA-16 surface.

3.2. Other SBA-16-Loaded Phases

It has been shown that transition metal sulfides of the 2nd and 3rd rows such as Ru, Rh, Os and Ir were especially active during HDS reactions [46]. Similarly, sulfides of Ir, Os and Re were most active in HDN of quinoline [47] and sulfides of Ir and Pt in HDN of pyridine [48]. However, these catalysts have a low tolerance to sulfur and nitrogen poisoning, therefore a previous deep hydrodesulfurization and hydrodenitrogenation charge pre-treatment is required. It is of interest that Ir catalysts display also very high HDN/HDS selectivity. This has already been observed during the study of simultaneous reaction of thiophene and pyridine over a carbon supported Ir catalyst at 280ºC and confirmed more recently for alumina supported Ir catalysts at 320 and 360ºC [49, 50].

Gerardo Simon Balangero Bottazzi et al. [41], studied the catalytic properties of 5 wt. % Ir-containing SBA-16 catalysts (with and without aluminum as heteroatom), in the hydrogenation of tetralin to decalin, in the presence of 100 ppm of N as quinoline, indole and carbazole, and 100 ppm of S as dibenzothiophene and 4,6-dimethyldibenzothiophene at 250ºC and 15 atm of pressure of hydrogen, using a Parr reactor. Ir/SBA-16 and Ir/Al-SBA-16 were prepared by wetness impregnation using Iridium Acetylacetonate as source of Ir. The Ir/SBA-16 catalyst demonstrated high activity measured in tetralin hydrogenation at mild conditions. The catalyst showed a good resistance to sulfur and nitrogen compounds. The inhibiting effect increased in the order: DBT < quinoline < 4,6-dimethyl-DBT < indole < carbazole. The inhibiting effect of the nitrogen/sulfur compounds was strong, but the activity was still higher than with commercial NiMo/alumina catalysts. In addition, Ir/SBA-16 catalysts, maintained a useful catalytic activity, even in the presence of relatively high amounts of sulfur compounds (up to 100 ppm, sulfur basis).

Aguillón-Martínez E. C. et al. [51], studied carbide and nitride phases supported on mesoporous materials like SBA type in order to obtain a better material than commercial catalysts ($NiMoS/Al_2O_3$, Sg = 269 $m^2.gr^{-1}$) specifically in the HDS of dibenzothiophene. The XRD patterns exhibited the presence of carbide and nitride phases in each series, respectively. In the nitride materials, the presence of oxide phases was more evident than in the carbide catalysts. The principal product was biphenyl (BF) for all the analyzed series. This behavior suggested that the DBT desulphurization pathway for the carbide materials was similar to that of sulfide catalysts. Bicyclohexyl (BCH) was analyzed as a product, and cyclohexylbenzene traces (CHB) were determined in a single catalyst (NiMoC-2%P/SBA-16). This behavior was attributed to the hydrogenation character of carbide catalyst reported previously.

A series of WO_3/SBA-16 samples with different W contents were synthesized by MA NA Ji Sheng et al. [52], under hydrothermal conditions using tetraethyl orthosilicate (TEOS) as silica source, ammonium paratungstate as tungsten source, and the mixture of F127 ($EO_{106}PO_{70}EP_{106}$) and P123 ($EO_{20}PO_{70}EP_{20}$) as template in strong acidic medium. The corresponding WxC/SBA-16 (x = 1, 2) catalysts were prepared by the temperature programmed carburization (TPC) method using CH_4/H_2 mixture gases from their oxide precursors. The structures of the samples were characterized by XRD, N_2 adsorption/desorption isotherms, TEM, and FTIR. The catalytic performance of the WxC/SBA-16 (x = 1, 2) catalysts for the hydrodesulfurization (HDS) of thiophene was evaluated in a continuos flow fixed bed quartz reactor. The results indicated that all samples maintained well cage like structure of SBA-16 mesoporous molecular sieve with a cubic symmetry. In the tungsten carbide catalysts, the tungsten carbide was the W_2C phase when n(Si)/n(W) was from 30 to 10. However, the tungsten carbides were a mixture of W_2C and WC phases when n(Si)/n(W) was 7.5. The tungsten carbide catalysts exhibited good catalytic activity for HDS of thiophene.

CONCLUSION

Compared with other mesoporous silicas, SBA-16 material shows important advantages like thicker framework walls, small crystallite size of primary particles, super-large cages and higher thermal stability. One of the most important properties of this mesoporous material is that due to its three-dimensional channel interconnectivity exists more favorable mass transfer, minimizing the transfer problem of bulky reactants and products which may be encountered in other mesoporous materials. On other hand, the ease of the method of preparation, the orderly structure, size and shape of their pores, all these achievable through simple changes in the method of synthesis, makes SBA-16 mesoporous silica a very versatile material, potentially applicable in many areas of science and engineering of materials such as in this case, as support of catalysts used in hydrodesulfurization reactions in petroleum refining processes. This has become in a growing area of research worldwide, same that still being developed and underlines the importance of a revision of the state of the art in this issue, to take stock of knowledge about the methods of synthesis and applications of this porous material to develop new materials and novel applications with interesting physical and chemical properties.

ACKNOWLEDGMENTS

Authors acknowledge the financial support of DGAPA-UNAM PAPIIT IN107311-3 project, CONACYT 182191 and CIC-UMSNH 2012-2013 and 2014-2015 Projects.

REFERENCES

[1] Pawelec, B., Navarro, R. M., Campos-Martin, J. M., Fierro, J. L. G. Towards near zero-sulfur liquid fuels: A perspective review. *Catal. Sci. Techn.* 2011, 1, 23-42.

[2] Topsøe, H., Clausen, B. S., Massoth, F. E. in Hydrotreating *Catalysts: Science and Technology* Anderson, J. R., Boudart, M., Eds., Springer, Berlin, Germany 1996, Vol. 11. 1-28.

[3] Breysse, M., Afanasiev, P., Geantet, Ch., Vrinat, M. Overview of support effects in hydrotreating catalysts. *Catal. Today* 2003, 86, 5-16.

[4] Iwamoto, R., Grimblot, J. Influence of phosphorous on the properties of alumina-based hydrotreating catalysts. *Adv. Catal.* 2000, 44, 417-503.

[5] Alibouri, M., Ghoreishi, S. M., Aghabozorg, H. R. Hydrodesulfurization of dibenzothiophene using CoMo/Al-HMS nanocatalyst synthesized by supercritical deposition. *J. Supercrit. Fluids.* 2009, 49, 239-248.

[6] Alibouri, M., Ghoreishi, S. M., Aghabozorg, H. R. Hydrodesulfurization activity of NiMo/Al-HMS nanocatalyst synthesized by supercritical impregnation. *Ind. Eng. Chem. Res.* 2009, 48, 4283-4292.

[7] Huirache-Acuña, R., Pawelec, B., Rivera-Muñoz, E., Nava, R., Espino, J., Fierro, J. L. G. Comparison of the morphology and HDS activity of ternary Co-Mo-W catalystssupported on P-modified SBA-15 and SBA-16 substrates. *Appl. Catal. B: Environ.* 2009, 92, 168-184.

[8] Cheng, C. F., Lin, Y. C., Cheng, H. H., Chen, Y. C. The effect and model of silica concentrations on physical properties and particle sizes of three-dimensional SBA-16 nanoporous materials. *Chem. Phys. Lett.* 2003, 382, 496-501.

[9] Stevens, W. J. J., Lebeau, K., Mertens, M., van Tendeloo, G., Cool, P., Vansant, E. F. Investigation of the morphology of the mesoporous SBA-16 and SBA-15 materials. *J. Phys. Chem. B* 2006, 110, 9183-9187.

[10] Gallo, J. M. R., Bisio, C., Marchese, L., Pastore, H. O. Surface acidity of novel mesostructured silicas with framework aluminum obtained by SBA-16 related synthesis. *Microp. Mesopor. Mater.* 2008, 111, 632-635.

[11] Pawelec, B., Damyanova, S., Mariscal, R., Fierro, J. L. G., Sobrados, I., Sanz, J., Petrov, L. HDS of dibenzothiophene over polyphosphates supported on mesoporous silica. *J. Catal.* 2004, 223, 86-97.

[12] Sayari, A. Catalysis by crystalline mesoporous molecular sieves. *Chem. Mater.* 1996, 8, 1840-1852.

[13] Park, K.-Ch., Yim, D.-J., Ihm, S.-K. Characteristics of Al-MCM-41 supported Pt catalysts: effect of Al distribution in Al-MCM-41 on its catalytic activity in naphthalene hydrogenation. *Catal. Today* 2002, 74, 281-290.

[14] Schmidt, R., Akporiayea, D., Stöcker, M., Ellestad, O. H. Synthesis of Al-containing MCM-41 materials: template interaction and removal. *Stud. Surf. Sci. Catal.* 1994, 8, 61-68.

[15] Ryoo, R., Kim, M. J. Generalised route to the preparation of mesoporous metallosilicates via post-synthetic metal implantation. *Chem. Commun.* 1997, 22, 2255-2226.

[16] Occelli, M. L., Biz, S., Auroux, A., Ray, G. J. Effects of the nature of the aluminum source on the acidic properties of some mesostructured materials. *Micropor. Mesopor. Mater.* 1998, 26, 193-213.

[17] Mokaya, R. The effect of particle size on aluminosilicate MCM-41 catalysts prepared via grafting routes. *J. Catal.* 1999, 186, 470-477.

[18] Trong On, D., Desplantier-Giscard, D., Danumah, C., Kaliaguine, S. Perspectives in catalytic applications of mesostructured materials. *Appl. Catal. A: Gen.* 2003, 253, 545-602.

[19] Sun, Y., Prins, R. Hydrodesulfurization of 4,6-dimethyldibenzothiophene over noble metals supported on mesoporous zeolites. *Angew. Chem. Int. Ed.* 2008, 47, 8478-8481.

[20] Venezia, A. M., Murania, R., la Parola, V., Pawelec, B., Fierro, J. L. G. Post-synthesis alumination of MCM-41: Effect of the acidity on the HDS activity of supported Pd catalysts. *Appl. Catal. A: Gen.* 2010, 383, 211-216.

[21] Zepeda, T. A., Pawelec, B., Fierro, J. L. G., Olivas, A., Fuentes, S., Halachev, T. Effect of Al and Ti content in HMS material on the catalytic activity of NiMo and CoMo

hydrotreating catalysts in the HDS of DBT. *Micropor. Mesopor. Mater.* 2008,111, 157-70.

[22] Chiranjeevi, T., Kumar, P., Rana, M. S., MuraliDhar, G., Prasada Rao, T. S. R. Physico-chemical characterization and catalysis on mesoporous Al-HMS supported molybdenum hydrotreating catalysts. *J. Mol. Catal. A: Chem.* 2002, 181, 109-117.

[23] Chiranjeevi, T., Kumar, P., Rana, M. S., Maity, S. K., Murali Dhar, G., Prasada Rao, T. S. R. Characterization and hydrodesulfurization catalysis on WS_2 supported on mesoporous Al-HMS material. *Micropor. Mesopor. Mater.* 2001, 44-45, 547-556.

[24] Halachev, T., de los Reyes, J. A., Araujo, C., Dimitrov, L., Córdoba, G., Dimitrov, L. Hydrodesulfurization of dibenzothiophene over Ni-Mo/(P)Ti-HMS catalysts. *Stud. Surf. Sci. Catal.* 1999, 12, 401-404.

[25] Alibouri, M., Ghoreishi, S. M., Aghabozorg, H. R. Effect of supercritical deposition synthesis on the dibenzothiophene hydrodesulfurization activity over CoMo/Al-HMS nanocatalyst. *Chem. Eng. Trans.* 2009, 17, 1005-1010.

[26] Klimova, T., Reyes, J., Gutiérez, O., Lizama, L. Novel bifunctional NiMo/Al-SBA-15 catalysts for deep hydrodesulfurization: Effect of support Si/Al ratio. *Appl. Catal. A: Gen.* 2008, 335, 159-171.

[27] Gómez-Cazalilla, M., Infantes-Molina, A., Moreno-Tost, R., Maireles-Torres, P. J., Mérida-Robles, J., Rodríguez-Castellón, E., Jiménez-López, A. Al-SBA-15 as a support of catalysts based on chromium sulfide for sulfur removal. *Catal. Today* 2009, 143, 137-144.

[28] Klimova, T., Lizama, L., Amezcua, J. C., Roquero, P., Torrés, E., Navarrete, J., Domínguez, J. M. New NiMo catalysts supported on Al-containing SBA-16 for 4,6-DMDBT hydrodesulfurization: Effect of the alumination method. *Catal. Today* 2004, 98, 141-150.

[29] Pauly, T. R., Liu, Y., Pinnavaia, T. J., Billinge, S. J. L., Rieker, T. P. Textural mesoporosity and the catalytic activity of mesoporous molecular sieves with wormhole framework structur. *J. Am. Chem. Soc.* 1999,121, 8835-8842.

[30] Rivera-Muñoz, E. M., Huirache-Acuña, R. Sol-gel-derived SBA-16 mesoporous materials, Inter. *J. Mol. Sci.* 2010, 11, 3069-3086.

[31] Amezcua, J. C., Lizama, L., Salcedo, C., Puente, I., Dominguez, J. M., Klimova, T. NiMo catalysts supported on titania-modified SBA-16 for 4,6-dimethyldibenzothiophene hydrodesulfurization. *Catal. Today* 2005, 107-108, 578-588.

[32] Zhao, D. Y., Huo, Q. S., Feng, J. L., Chmelka, B. F., Stucky, G. D. Nonionictriblock and star diblock copolymer and oligomeric surfactant syntheses of highly ordered, hydrothermally stable, mesoporous silica structures. *J. Am. Chem. Soc.* 1998, 120, 6024-6036.

[33] Zhao, D., Feng, J., Huo, Q., Melosh, N., Fredrickson, G. H., Chmelka, B. F., Stucky, G. D. Triblock copolymer synthesis of mesoporous silica with periodic 50 to 300 angstrom pores. *Science* 1998, *279*, 548-552.

[34] Stevens, W. J. J., Lebeau, K., Mertens, M., van Tendeloo, G., Cool, P., Vansant, E. F. Investigation of the morphology of the mesoporous SBA-16 and SBA-15 materials. *J. Phys. Chem. B* 2006, *110*, 9183-9187.

[35] Sakamoto, Y., Kaneda, M., Teresaki, O., Zhao, D. Y., Kim, J. M., Stucky, G., Shin, H. J., Ryoo, R. Direct imaging of the pores and cages of three-dimensional mesoporous materials. *Nature* 2000,408, 449-453.

[36] Beck, J. S., Vartul, J. C., Roth, W. J., Leonowicz, M. E., Kresge, C. T., Schmitt, K. D., Chu, C. T. W., Olsen, D. H., Sheppard, E. W., McCullen, S. B., Higgins, J. B., Schlenker, J. L. A new family of mesoporous molecular sieves prepared with liquid crystal templates. *J. Am. Chem. Soc.* 1992,*114*, 10834-10843.

[37] Kresge, C. T., Leonowicz, M. E., Roth, W. J., Vartuil, J. C., Beck, J. S. Ordered mesoporous molecular sieves synthesized by a liquid-crystal template mechanism. *Nature* 1992, *359*,710-712.

[38] Kim, T.-W., Ryoo, R., Kruk, M., Gierszal, K. P., Jaroniec, M., Kamiya, S., Teresaki, O. Tailoring the pore structure of SBA-16 silica molecular sieve through the use of copolymer blends and control of synthesis temperature and time. *J. Phys. Chem. B* 2004, *108*, 11480-11489.

[39] Wang, L., Fan, J., Tian, B., Yang, H., Yu, C., Tu, B., Zhao, D. Synthesis and characterization of small pore thick-walled SBA-16 templated by oligomeric surfactant with ultra-long hydrophilic chains. *Microporous Mesoporous Mater.* 2004, 67, 135-141.

[40] Chen, B.-C., Chao, M.-C., Lin, H.-P., Mou, C.-Y. Faceted single crystals of mesoporous SBA-16 from a ternary surfactant system: Surface roughening model. *Microporous Mesoporous Mater.* 2005,*81*, 241-249.

[41] Balangero Bottazzi, G. S., Martinez, M. L., Gomez Costa, M. B., Anunziata, O. A., Beltramone A. R. Inhibition of the hydrogenation of tetralin by nitrogen and sulfur compounds over Ir/SBA-16. *Appl. Catal. A: Gen.* 2011,404, 30-38.

[42] Huirache-Acuña, R., Pawelec, B., Loricera, C. V., Rivera-Muñoz, E. M., Nava, R., Torres, B., Fierro, J. L. G. Comparison of the morphology and HDS activity of ternary Ni(Co)-Mo-W catalysts supported on Al-HMS and Al-SBA-16 substrates. *Appl. Catal. B: Environ.* 2012, 125, 473- 485.

[43] Shah, A. T., Li, B., Abdalla, Z. E. A. Direct synthesis of Ti-containing SBA-16-type mesoporous materialby the evaporation-induced self-assembly method and its catalyticperformance for oxidative desulfurization. *J. Colloid. Interface Sci.* 2009, 336, 707-711.

[44] Guzmán, M. A., Huirache-Acuña, R., Loricera, C. V., Hernández, J. R., Díaz de León, J. N., de los Reyes, J. A., Pawelec, B. Removal of refractory S-containing compounds from liquid fuels over P-loaded NiMoW/SBA-16 sulfide catalysts. *Fuel* 2013,103, 321-333.

[45] Terrés, E., Montiel, M., le Borgne, S., Torres, E. Immobilization of chloroperoxidase on mesoporous materials for the oxidation of 4,6-dimethyldibenzothiophene, a recalcitrant organic sulfur compound present in petroleum fractions. *Biotechnol. Lett.* 2008, 30, 173-179.

[46] Pecoraro, T. A., Chianelli, R. R. Hydrodesulfurization catalysis by transition metal sulfides. *J. Catal.* 1981, 67, 430-445.

[47] Eijsbouts, S., de Beer, V. H. J., Prins, R. Periodic trends in the hydrodenitrogenation activity of carbon-supported transition metal sulfide catalysts. *J. Catal.* 1988, 109, 217-220.

[48] Vít, Z., Zdrazil, M. Simultaneous hydrodenitrogenation of pyridine and hydrodesulfurization of thiophene over carbon-supported platinum metal sulfides. *J. Catal.* 1989,119, 1-7.

[49] Cinibulk, J., Vít, Z. Hydrodenitrogenation of pyridine over alumina-supported iridium catalysts. *Appl. Catal. A: Gen.* 1999, 180, 15-23.

[50] Drahoradova, A. Hillerova, E. Hrzan, Z. Icanek, M. R. Vit, Z. *Chem. Prum.* 1993, 43, 93.

[51] Aguillón-Martínez, E. C., Melo-Banda, J. A., Guevara, L. A., Reyes. T. A., Ramos-Galván, C. E., Silva, R. R., Domínguez, J. M. Catalytic Properties of Ni-Mo Carbide and Nitride Phases supported on SBA-15 and SBA-16 in the Hydrodesulfurization of DBT. *Advanced Materials Research* 2010, 132, 111-125.

[52] Sheng-Fu Ma Na Ji, Pin Yi Wu, Li Hua Hu, Ping Ying Nie, Preparation, Characterization and Catalytic Performance for Hydrodesulfurization of WxC/SBA-16 Catalysts, *Acta Phys. Chim. August Sin.* 2007, 23(8):1189-1194.

In: Comprehensive Guide for Mesoporous Materials, Volume 4 ISBN: 978-1-63482-092-9
Editor: Mahmood Aliofkhazraei

Chapter 16

CATALYTIC APPLICATIONS OF MESOPOROUS MOLECULAR SIEVES IN SELECTIVE OXIDATION REACTIONS

Agata Wawrzynczak*[†], *Agnieszka Feliczak-Guzik* and *Izabela Nowak
Adam Mickiewicz University in Poznan, Faculty of Chemistry, Poznan, Poland

ABSTRACT

This overview covers recent achievements in the application of mesoporous molecular sieves as catalysts for oxidation reactions. Main structural and textural features are outlined in relation to the novel synthetic approaches to mesoporous molecular sieves and understanding of their properties. Different ways of introduction of active sites of mesoporous molecular sieve surfaces are discussed in view of their application as catalysts. A state-of-the-art review of transition metal containing mesoporous materials as promising redox, shape-selective catalysts is presented, focusing on the relationship of the preparation methods to the type of heteroatom species present and their special role in catalysis. The main part describes selected catalytic reactions catalyzed by mesoporous molecular sieves and it is based on very recent papers.

1. INTRODUCTION

Catalysis is the change in the rate of a chemical reaction achieved thanks to the presence of a substance called a catalyst. A catalyst may participate in multiple chemical transformations, however unlike the other reagents participating in the chemical reaction, it is not consumed in the reaction. Catalytic reactions have lower rate-limiting activation energy than the corresponding uncatalyzed reactions, resulting in higher reaction rate at the same temperature. Kinetically, catalytic reactions are typical chemical reactions; i.e., the reaction

[†] Corresponding author: Agata Wawrzyńczak, Adam Mickiewicz University in Poznań, Faculty of Chemistry, Umultowska 89b, PL61-614 Poznań, Poland. E-mail: agatha@amu.edu.pl.

rate depends on the frequency of contact of the reactants in the rate-determining step. Usually, the catalyst participates in this slowest step and the reaction rates are limited by the amount of the catalyst and its activity. Catalysis is principally divided in two branches: homogeneous catalysis, when the catalyst is in the same phase as the reaction mixture (typically, in liquid phase), and heterogeneous catalysis, when the catalyst is in a different phase (typically solid/liquid, solid/gas or solid/liquid/gas). Table 1 summarizes a comparison between the main features of homogeneous and heterogeneous catalysis. In principle, as seen from this table, in heterogeneous catalysis, the diffusion of reagents to the surface and diffusion of products from the surface can be the rate determining step. Analogous events, associated with substrate binding and product dissociation, apply to homogeneous catalysts. Most of heterogeneous catalysts are solids that act on substrates in a liquid or gaseous reaction mixture. Diverse mechanisms of reactions on surfaces are known, depending on how the adsorption takes place (Langmuir-Hinshelwood, Eley-Rideal, and Mars-van Krevelen) [1].

Table 1. Schematic comparison between homogeneous and heterogeneous catalysis

Feature		Homogeneous	Heterogeneous
Catalyst's	form	metal complex	solid, often metal oxide
	activity	high	variable
	selectivity	high	variable
	average time of life	variable	long
	sensitivity to poisons	low	high
	variations of steric/electronic features	possible	difficult
	recycling	difficult (and expensive)	easy
	separation from products	difficult	easy
Reaction conditions		mild	drastic
Problems of reagents/products diffusion		none	possible
Intelligibility of the mechanism		possible	difficult

2. Preparation of Single-site Heterogeneous Catalyst (SSHC) on Mesoporous Inorganic Supports

The host matrixes, where single-site centers can be uniformly dispersed, have inorganic, organic or composite nature. Commonly used inorganic matrices are typically oxidic solids (silicates, aluminosilicates, etc.), often with high porosity and very large specific surface area (in the order of 100 to 1000 $m^2 g^{-1}$), such as crystallographically ordered microporous molecular sieves (zeolites and zeotypes) or non-ordered mesoporous materials.

In the early 1990s, the invention of mesoporous materials (e.g. M41S) with tunable pore sizes of 2 to 30 nm has opened a way for the preparation of mesoporous silicates and metallosilicates. These mesoporous materials have: large pore sizes, large internal surface areas (>1000 $m^2 g^{-1}$), and a large number of surface hydroxyl groups (for metallosilicates). Some well-known mesoporous materials, that are ideal supports for SSHC, are e.g. MCM-41 & -48, HMS, MSU, KIT-6, SBA-1,-3,-15 or -16 (e.g. review paper [2]) and they may, in principle, have various potential applications in industrial catalytic reactions [3]. Indeed, the recent start up of production plants at large-scale level (up to 20 t year^{-1}) for the production of nanostructured mesoporous silicas, has made these materials right now suitable for a number of different applications in commodity and fine chemicals synthesis [4, 5].

Generally, single sites can have a structural role, when they are inserted into the framework of an ordered material via direct synthesis (Figure 1). They also can be added to an existing support by post-synthesis methods, such as deposition, tethering or linking to a parent matrix.

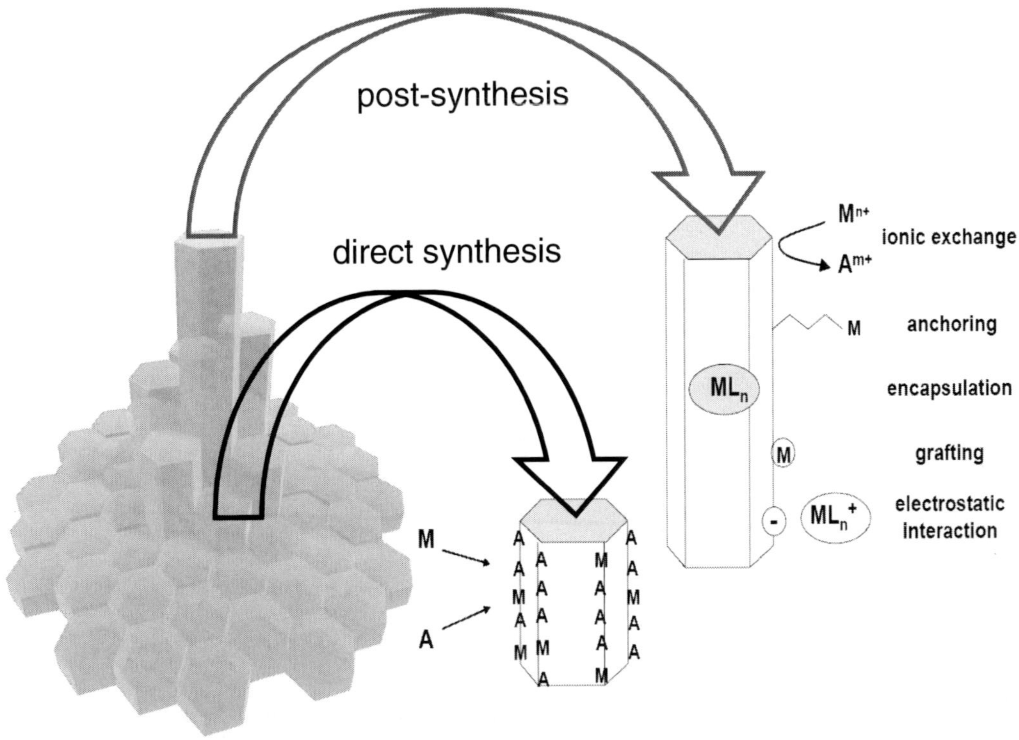

Figure 1. Functionalization of mesoporous inorganic oxide supports. Several strategies and pathways are possible to introduce novel functions into or onto molecular sieves. M: catalytically active metal (e.g., V, Ti, Mo, Cu, etc.); A: non-active atom (e.g., Si); L: ligand.

When inserted in direct synthesis, the catalytically active single sites are regularly dispersed in the matrix of the support and they are located at or adjacent to ions that have replaced the framework ions of the parent structure. The precursor of the active species is already present in the synthetic mixture of the final material, together with the other components, and the single sites are introduced. Thus, the chemical environment of the site is strongly dependent on the physicochemical characteristics of the matrix. As a main drawback, some centers can be buried within the bulk of the solid and they can be inaccessible and unavailable for catalysis. Therefore, it is difficult to calculate the TON and TOF numbers (TON - turnover number = molar ratio of converted substrate to catalyst. TOF - turnover frequency = TON per hour). Moreover, the presence of heteroatoms in the synthesis mixture may critically affect the synthesis pathway and then it is often necessary to set up a completely new protocol for the every new desired material. Such approach can be time-consuming, but the possibility of regeneration of these catalysts is generally better than that of the catalysts obtained via post-synthesis techniques (due to the calcination and severe washing treatment steps) [6].

When inserted by post-synthesis method (covalent deposition), it is possible to obtain single-site catalysts relatively easily, from well-known, predominantly commercially available, solids. The active center is generally added to the support as a precursor that can be deposited irreversibly (*anchored* or *grafted*) onto the surface as it is, by the formation of covalent bonds, or after a functionalization with a side chain (called a *tether*) [7]. In the *anchoring* technique, the covalent deposition takes place only at the opposite end of the tether. Thus, the active metal maintains the definite chemical surroundings as in the parent homogeneous precursor, with all the advantages of heterogeneous systems, namely an easy separation, recover and recyclability. Conversely, in the *grafting* technique, the active single site has a different chemical surrounding with respect to the parent precursor, since the coordination shell around the metal center is partially modified during the covalent deposition, and a new reactivity of the active species can be expected (Figure 2). Finally, the homogeneous precursors of the single-site active centers can be immobilized onto the surface of solid supports by non-covalent interactions, such as hydrogen bonds, weak van der Waals interactions or by mechanical confinement (*trapping* or *encapsulation*).

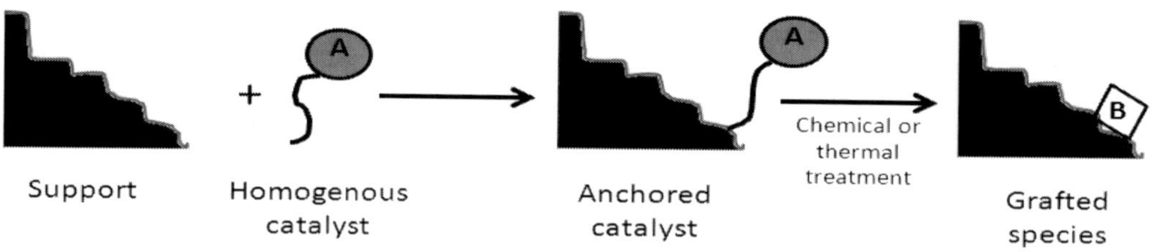

Figure 2. Post-synthesis covalent functionalization of supports: anchoring and grafting.

2.1. Preparation of Vanadium-containing Mesoporous Catalysts

The synthesis of [V]-MCM-41 was performed in 1994 for the first time and it is a nice example of the direct synthesis (Table 2). It has a hexagonal pore structure, very similar to that of MCM-41, with the Si/V ratio equal to 60 and is prepared in the absence of V_2O_5 domains. Cubic [V]-MCM-48 has also been prepared, as well as the hexagonal [V]-HMS, which has been prepared using primary amines as templates [15]. Additionally, a vanadoaluminosilicate with MCM-41 structure, [V, Al]-MCM-41, has been synthesized in 2002 [11].

In general, in the as-synthesized mesoporous VAPOs, V^{4+} ions occupy the positions of phosphorus atoms (Scheme 1, structure A). The calcination process leads to the total oxidation of V^{4+} to V^{5+}. This process is reversible by reduction under H_2 atmosphere. The structure of the vanadium ion in the calcined sample is reported to be V^{5+}, bearing the V=O bond (Scheme 1, structure C) to four –O–Al≡ bonds (Scheme 1, structure B) [14].

Table 2. Vanadium-containing molecular sieves prepared by one-pot syntheses with $VOSO_4$ as a V source

Material	Year of invention	Reference
[V]-MCM-41	1994	[7]
[V]-MTW	1995	[8]
[V]-HMS	1995	[9]
[V]-MCM-48	1996	[10]
[V, Al]-MCM-41	2002	[11]
[V, Al]-MWW - pillared material	2005	[12]
mesoporous VAPO	2005	[13]

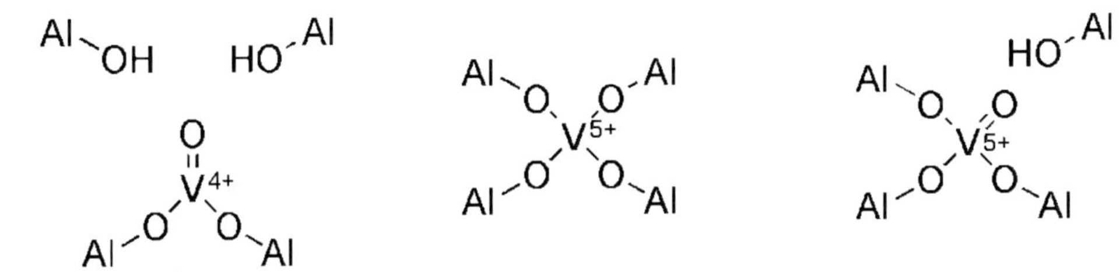

as synthesized (A) calcined (B) calcined (C)

Scheme 1. Possible structures of vanadium species in as-synthesized and calcined VAPO material.

However, in vanadium containing mesoporous molecular sieves a model of tetrahedral V^{5+} species was proposed (Scheme 2), besides extractable V_2O_5 domains (in the calcined sample). The proposed mechanism for oxidation/reduction of vanadium sites in the calcined samples is illustrated in the same scheme [15].

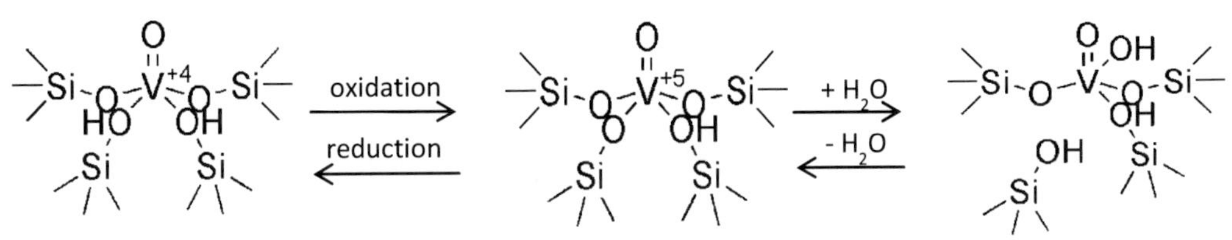

Scheme 2. The proposed model of oxidation/reduction of vanadium sites and coordination environment of V^{5+} species in the structure of silicate sieve.

Table 3. Vanadium containing molecular sieves prepared by methods other than one-pot syntheses

Material	Incorporation method	V source	Year of invention	Reference
V-MCM-48	wet impregnation	$V(O\text{-}C_3H_8)$	1996	[16]
	gas deposition	$VO(acac)_2$	1998	[17]
	ion-exchange	Na_3VO_4	2005	[18]
V-ALPO-5	wet impregnation	$VOSO_4$	2000	[19]
VO-MWW	ion-exchange	$VOSO_4$	2005	[20]

Besides one-pot syntheses, vanadium species may be introduced into molecular sieves by post-synthesis methods [15]. Some vanadium containing molecular sieves, prepared by such methods, including impregnation or chemical vapor deposition, are listed in Table 3.

Generally, a simple approach to post-synthesis vanadium incorporation is to proceed with ion-exchange of protons or sodium ions for vanadyl cations (VO^{2+}) [20]. The difficulty in this method is that the commonly used salt (vanadyl sulphate) undergoes hydrolyses in pH values higher than 2.8, leading to formation of V_2O_5 domains. On the other hand, low pH values lead to a competition between H^+ and VO^{2+} ions, which limit the amounts of incorporated vanadium. To achieve incorporation of higher amounts of metal the method of impregnation is another possibility of post-synthesis vanadium introduction, but this method leads to formation of V_2O_5 aggregates, which are often not interesting for catalysis. Still, another approach is based on methods that do not involve contact with an aqueous solution: *chemical vapor deposition* (CVD), *solid state ion-exchange* and *sublimation* [15]. In such methods the metallic species are vaporized at high temperatures and/or at low pressures and deposited on the material surface. The advantage is that they avoid the restricted diffusion of solvated oligomer species in aqueous exchange, which leads to poorly dispersed oxide domains at external surfaces. Three principal vanadium species in V-beta (α, β and γ), depending on the location of the atom in the sieve lattice (Scheme 3), were proposed.

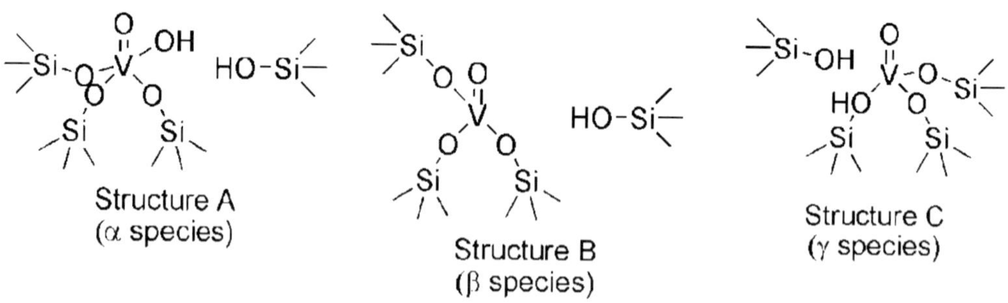

Scheme 3. Types of vanadium species correlated with the location of the atom in the sieve lattice.

2.2. Titanosilicates As Examples of Oxidation Materials

Up to now, many researchers have reported on the preparation, characterization and properties of selective catalytic oxidation of various Ti-containing mesoporous materials, such as [Ti]-MCM-41 [21], [Ti]-MCM-48 [22], [Ti]-SBA-15 [23], [Ti]-HMS [24], [Ti]-MSU [25] and MTS-9 [26]. So far, mainly two types of Ti(IV)-modified and ordered mesoporous silicate materials have been highlighted: in one of them the Ti(IV) centers are distributed throughout the pure silica framework by the addition of a suitable titanium precursor to the silica gel, while in the other one titanium species are grafted onto the surface of the pure silicate sieve by means of interaction between the surface hydroxyl groups and an organometallic precursor or inorganic metallic precursor. [Ti]-mesoporous molecular sieves prepared *via* a direct method are presented in Table 4.

Table 4. Titanium-containing molecular sieves prepared by one-pot syntheses

Material	Ti source	Si source	Method and surfactant type	Ti coordination	Reference
[Ti]-MCM-41	$Ti(OC_2H_5)_4$	SiO_2	direct; cationic	mainly located in the silica framework; no extra-framework TiO_2 was detected	[21, 27]
	$Ti(OC_2H_5)_4$	Na_2SiO_3	direct; cationic	mainly located in the silica framework, TiO_2 tetrahedron - not regular	[28]
	TPOT ($Ti(OC_3H_7)_4$)	TEOS	Si^+I^- and $Si^+X^-I^+$ assembly pathways (where Si^+ is a quaternary ammonium surfactant and I^- and I^+ are ionic silicon precursors)	predominantly tetrahedral coordination; some dehydrated five-and six-coordinated sites	[29]
	TBOT ($Ti(OC_4H_9)_4$)	several	direct; cationic	as above	[30]
	$Ti(OC_2H_5)_4$	methylated silicon source	direct; cationic	tetrahedral coordination	[31]
	$TiCl_3$	colloidal silica	direct; cationic	tetrahedral coordination; formation of anatase TiO_2 was avoided due to low hydrolysis rate of the inorganic $TiCl_3$ solution	[32]
	$TiCl_3$	TEOS	direct, cationic	H_2O_2 (30 wt.%) as a coordination agent - decrease the hydrolysis rate of inorganic titanium salts - formation of TiO_2 avoided	[33]
	Ti_2SO_4	Na_2SiO_3	direct, cationic	mainly tetrahedral	[34]
[Ti]-SBA-15	$TiOSO_4$	$Na_2Si_2O_5$	nonionic P123	mainly tetrahedral; part of Ti inaccessible	[35]
	different sources	TMOS	P123, addition of F	mainly tetrahedral	[36]
	titanium butoxide with acetylacetone	TEOS	P123, addition of H_2O_2	mainly tetrahedral	[37]
	titanocene dichloride	TEOS	P123	mainly tetrahedral	[38]

Table 4. (Continued)

Material	Ti source	Si source	Method and surfactant type	Ti coordination	Reference
[Ti]-SBA-1	different sources	-	cationic	mainly tetrahedral	[39, 40]
[Ti]-MSU-X, [Ti]-MSU-G, [Ti]-MSU-1 & [Ti]-MSU-2	titanium oxo-salt	TEOS	and silicon tetraethoxide	mainly tetrahedral	[41, 42]
[Ti]-MTS-9	nanosized titanosilicate precursors	TEOS	polymer surfactant, e.g. P123	mainly tetrahedral	[43]
[Ti]-TUD-1			triethanolamine (TEA),	mainly tetrahedral	[44, 45]
[Ti]-MCM-48			gemini surfactant	mainly tetrahedral	[46]
[Ti]-JLU			mixed surfactants: fluorocarbon-type and nonionic copolymer	mainly tetrahedral	[47]
[Ti]-HMS	Ti(iso-OC_2H_5)	TEOS	dodecylamine as surfactant	mainly tetrahedral	[48]
[Ti]-MWW	-	various	dry gel conversion method, special autoclave	mainly tetrahedral	[49]
[Ti]-MMM	-	various	unique process (addition of various organic templates at staggered times)		[50]

Table 5. Hydroxylation and oxidation reactions on metal-containing mesoporous molecular sieves

Catalytic reaction	Catalyst	Reference
olefins oxidation by H_2O_2 and TBHP	[Ti]-MCM-41	[59]
phenol hydroxylation by H_2O_2	[Ti]-MCM-41 [Ti]-HMS MTS-9	[60]
phenol hydroxylation by H_2O_2 2,3,6-trimethylphenol hydroxylation by H_2O_2	[Fe]-MCM-41 MFS-9	[61]
benzene hydroxylation	[V]-MCM-41 [V]-HMS	[10]
cyclohexane oxidation by H_2O_2 cyclodecane oxidation by H_2O_2	[V]-MCM-41 [V]-MCM-48	[62, 63]
propane oxidation by O_2	[V]-MCM-41 [V]-MCM-48 [K]-SBA-15	[64, 65]
methane partial oxidation by O_2	[V]-MCM-41 [Fe]-MCM-41	[66]
styrene oxidation by H_2O_2	[Cr]-MSU-1	[67]
tetralin oxidation by O_2	[Cr]-MSU-1 [Mn]-MSU-1 [Co]-MSU-1	[68]
gas phase cyclohexanol oxidation by air	[Mn]-MCM-41	[69]
gas phase toluene oxidation by air	[V]-HMS	[70]
cyclohexane photooxidation by O_2	[V]-HMS	[71]
photocatalytic partial oxidation of propane by O_2	[V]-MCM-41 [V, Ti]-MCM-41	[72]

It is worth emphasizing that the most active and selective sites in Ti-based epoxidation catalysts are those of isolated and four-coordinated Ti(IV) centers. However, aqueous synthesis conditions, for example, during the direct synthesis of Ti-mesoporous materials, would often promote agglomeration of metal oxide species [51].

Although the procedure of direct-synthesis is relatively simple, the efficiency of the synthesis is always low, i.e., only a small part of the heteroatoms can be added into the initial gels and introduced into the mesoporous products in aqueous solutions. To overcome these drawbacks, titanium-containing molecular sieves, obtained by a secondary synthesis, can be regarded as an efficient alternative. For instance, metallocene grafting did not affect the MCM-48 structural order, whilst transition metals introduced *via* hydrothermal synthesis route usually result in a partial loss of a long range order of the materials. The most recommended procedure is a grafting of titanium onto mesoporous materials in order to modify the inner mesopores with a titanium precursor. The advantages in using this grafting method are as follows: (1) greater accessibility of the grafted species on mesoporous silicates, and (2) tenability of titanium coordination, which can be adjusted by varying experimental conditions. The procedure is often carried out in organic solvents, like toluene and hexane, under an inert atmosphere [51]. Also "atom-planting" can be considered as a solid-gas reaction between highly siliceous sieves and metal chloride vapors [52]. These two methods, however, are substantially similar in principle. They all adopt the process of ion exchange with surface hydroxyl groups of calcined pure silica mesoporous sieves [53, 54]. A series of titanium compounds have been used as titanium precursors, such as tetrabutyl titanate, titanocene [55], and titanium isopropoxide [23].

3. Catalytic Activity

The understanding of the structure/reactivity/selectivity relationships in classical metal-based heterogeneous systems is complicated, mainly due to uncertainties in the active site structure(s) and the large variety of surface species, which are catalytically significant. In general, solid catalysts, working heterogeneously, possess a broad spectrum of active sites, each with their own energetics, activity and selectivity [1].

The performance of a catalyst is determined by its chemical composition, structure at the atomic scale and accessibility of active centers. In order to better understand a material's catalytic activity a broad range of physicochemical characterization techniques have been developed. Mostly, these characterization techniques aspire to unravel the elemental composition and structure of the catalytic material, preferentially at the atomic scale and under relevant conditions [1].

The high-resolution strategies for mapping chemocatalytic activities inside porous particles opened new perspectives for relating intraparticle mass transfer phenomena to the overall catalytic efficiency of a catalyst. The influence of such a transport phenomena on the catalytic process is sometimes dramatic, for instance when the diffusion of reagents inside the catalyst's pores towards the active sites becomes rate-limiting. In that case, most of the reagents will be converted before they reach the inner parts of the particles, and as a result only the outer shell of the catalyst will be used efficiently. Aside from the drop in catalytic efficiency, also the reaction selectivity can decrease, for instance when the slow transport of the formed product molecules inside the porous network causes undesired secondary reactions to occur. However, the identification and quantification of such intraparticle diffusion limitations is extremely challenging. A theoretical approach is given by the classical Thiele-Wheeler theory [56]. An important parameter here is the effectiveness factor, which is defined as the fraction of the catalyst that effectively contributes to the overall catalytic performance, and is determined by a combination of particle size, the intrinsic rate constant of the surface reaction and the intraparticle diffusion coefficient of the substrate molecules (D_{eff}). It was recently demonstrated that single, individual epoxidation events could be identified inside individual Ti-MCM-41 particles [57]. A red fluorescent BODIPY derivative was used as a reagent, which upon epoxidation is transformed into a yellow fluorescent product molecule. Accumulation of many such individual events (monitored in a wide-field fluorescence microscope) and their respective fitted positions finally yields the intraparticle concentration profile (usually presented as the histogram of the distances between the fitted position and the particle's edge), reflecting the reagent concentration profile inside the particles and can thus be directly related to the Thiele-Wheeler parameters. In steady-state conditions, the influx of reagents must equal this total rate in order not to have accumulation of reagents inside the particles. From this mass balance, a value for the effective intraparticle diffusion coefficient (D_{eff}) could be calculated, which was found to be in the order of 10^{-16} m^2 s^{-1} [58].

3.1. Olefins and Aromatics Oxidation and Hydroxylation

Modified mesoporous molecular sieves proved to be efficient catalysts in several hydroxylation and oxidation reactions. Examples selected from quite abundant literature are summarized in the Table 5 and briefly described in the text below.

[Ti]-MCM-41 material exhibits selective catalytic activity in the oxidation of olefins, however its intrinsic activity is lower than that of the Ti-beta and TS-1 materials, when using H_2O_2 as an oxidant. However, when using tertbutyl hydroperoxide as the oxidant, [Ti]-MCM-41 clearly has an advantage over other Ti-containing microporous zeolites with regard to the catalytic activity and selectivity involving large organic molecules [59].

Catalytic activities for the oxidation of aromatics by H_2O_2 over MTS-9, [Ti]-MCM-41, and TS-1 catalysts were designed by Xiao et al. [60]. In phenol hydroxylation, [Ti]-MCM-41 showed very low catalytic activity (2.5%), but MTS-9 exhibited very high catalytic activity, with a phenol conversion of 26% which is comparable with TS-1. In the same reaction, [Fe]-MCM-41 showed low activity, whereas MFS-9 (iron analogue of MTS-9) exhibited relatively high reactivity comparable with that of FeS-1 (iron analogue of TS-1).

In the 2,3,6-trimethylphenol hydroxylation, [Ti]-MCM-41 is inactive due to the relatively low oxidation ability of titanium species in the amorphous wall of Ti-MCM-41, also TS-1 is also inactive due to the inaccessibility of the small micropores of TS-1 to the large diameter of a bulky molecule like 2,3,6-trimethylphenol. However, MTS-9 was very active in this reaction with a conversion of 18.8%, indicating that MTS-9 is an effective catalyst for the oxidation of bulky molecules. Unfortunately, the molecular size of 2,3,6-trimethylphenol is larger than the pore size of FeS-1 catalyst, therefore it showed very low activity [61]. In contrast, MFS-9 exhibits very high catalytic activity, giving the 2,3,6- trimethylphenol conversion at 15.3%. As for [Fe]-MCM- 41, it exhibited much lower activity compared with MFS-9, although it also has large enough pore dimension and similar Fe content. Thus, the different activities of MFS-9 and Fe-MCM-41 arise from the different environments of Fe species in these samples. Due to the special preparation procedure, the Fe atoms introduced by preformed nanoclusters in MFS-9 locate at crystal-like rather than at amorphous sites, like in Fe-MCM-41. This may be responsible for the higher catalytic activity of MFS-9 material [61].

Vanadium-containing materials are often compared to their titanium-containing analogues. When used in benzene hydroxylation to phenol, for instance, [V]-MCM-41 was less active than [Ti]-MCM-41, but among the metal-HMS catalysts, the [V]-HMS material was the most active [10]. Unfortunately, in this reaction, selectivity to phenol reached 100% at low conversion values (~10%). It is also important to stress that usually the heteroatom source is an important factor for high activity. For example, Selvam et al. reported that vanadium incorporation varies according to its source: tetravalent salts seem to be more efficient [62]. For catalysis purposes, then, materials containing higher amounts of well dispersed vanadium ions, prepared from tetravalent sources, are more active. These materials, with vanadium amounts varying from 0.2 to 1.2%, were tested on cyclohexane oxidation. When tested in the same reaction, [V]-MCM-48 presented high conversions and a minimal vanadium leaching [63].

When compared to V_2O_5 supported on SiO_2 in propane selective oxidation, [V]-MCM-41 and [V]-MCM-48 containing up to 1% vanadium were much more selective [64]. Materials containing polymeric domains lead to formation of oxygenated products, such as acrolein, acetaldehyde and acetic acid. As a novel catalyst system for the selective oxidation of low alkanes, mesoporous SBA-15-supported potassium catalysts were also employed for the selective oxidation of propane to oxygenate by using molecular oxygen as oxidant. It was found, when compared with bare mesoporous SBA-15, that the selectivities to the oxygenates, including formaldehyde, acetaldehyde, acrolein and acetone, were remarkably enhanced over Kx/SBA-15(K:Si = x:100, mol) catalysts, and the main products were acrolein and acetone. At 500°C, the yield of the oxygenate can reach 4.64% over K3.0/SBA-15, which is the highest value over SBA-15-supported potassium catalyst [65].

A method developed by ion-exchange of the cetyltrimethylammonium template from MCM-41 with $VO(C_2O_4)$ and later calcination led to a material that showed higher efficiency in methane partial oxidation than [V]-MCM-41 prepared by one-pot synthesis [66].

The chromium-containing MSU-1 catalyst was highly active for the selective oxidation of styrene and the main reaction products over [Cr]-MSU-1 were benzaldehyde and phenylacetaldehyde. The catalytic performance of the catalyst remained stable within five repeated runs and no leaching was noticed for this chromium-based catalyst. With decreasing styrene/H_2O_2 molar ratio, the styrene conversion increased from 15.6% to 29.5% at the cost of H_2O_2 utilization. The selectivity to benzaldehyde increased along with the growing amount of by-products (such as styrene glycol and benzoil acid), indicating that excess amount of H_2O_2 favors the oxidation cleavage of C=C bonds to transfer into benzaldehyde and the formation of styrene glycol and benzoil acid from styrene oxide, respectively [67].

Also selective oxidation of tetralin by molecular oxygen over mesoporous chromium-containing MCM-41 molecular sieve catalyst, using supercritical carbon dioxide ($scCO_2$) solvent has been investigated. [Cr]-MCM-41 catalyst gave high selectivity (96.2%) and good yield (63.4%) of 1-tetralone. The presence of $scCO_2$ medium improved 1-tetralone selectivity and suppressed the leaching of chromium ions from the [Cr]-MCM-41. The activity over recycled catalyst remained nearly the same. [Cr]-MCM-41 material showed superior activity in comparison with other mesoporous catalysts, viz. [Mn]-MCM-41, [Co]-MCM-41, microporous CrAPO-5, CoMFI, and macroporous Cr/SiO_2 catalyst. In addition, this catalytic system was also applied to the oxidation of other benzylic compounds such as indane, fluorene, acenaphthene and diphenylmethane [68].

The Mn incorporation in MCM-41 creates acidity and redox properties. Appearance of both properties simultaneously favors the oxidative dehydrogenation reaction. It is believed that the reaction proceeds via two different mechanisms. One is a peroxy radical mechanism, in which the nascent oxygen formed from air at higher temperature forms peroxy radicals with manganese oxide. The oxygen is inserted in-between hydrogen and hydroxycarbon, and water is eliminated as a byproduct. The second is a carbocation mechanism (Scheme 4). The hydrogen from silanol group of Mn–MCM-41 material reacts with the hydroxyl group and water is eliminated as a byproduct. Another hydrogen from the alpha carbon goes to the catalyst and retains the proton [69].

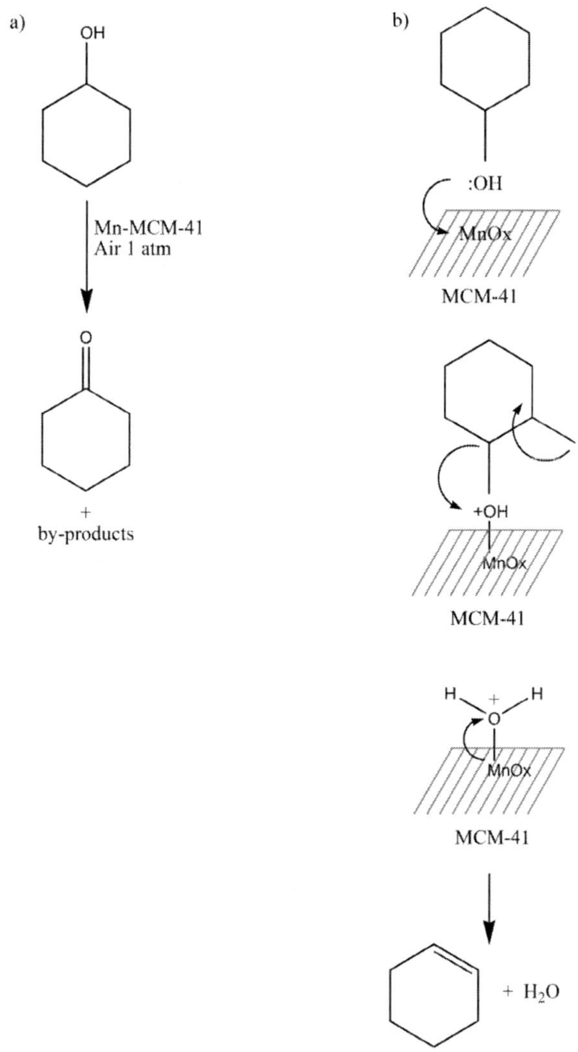

Scheme 4. Cyclohexanol oxidation (a) and dehydration (b) over Mn-MCM-41 catalyst.

The gas-phase cyclohexanol and air reaction over Mn-MCM-41 molecular sieve were also studied. The conversion is almost constant even up to 5h, and follows a parabolic behavior with respect to temperature. It reaches a maximum at 350°C and the selectivity towards cyclohexene is increased with temperature [69].

When prepared by one-pot synthesis, [V]-HMS offered higher selectivity to benzaldehyde in toluene oxidation when compared to V-HMS prepared by molecular designed dispersion [70]. The latter increases selectivity to carbon oxides.

Vanadium-containing mesoporous silica molecular sieves are also useful in photocatalysis. For instance, [V]-HMS materials with tetrahedrally coordinated V-oxide species (V^VO_4) have been successfully applied as catalysts in cyclohexane photooxidation by O_2 [71], while [V]-MCM-41 catalyst presented high conversion values in propane photooxidation. Moreover, the selectivity of the latter catalyst could be further improved when titanium ions were co-incorporated [72].

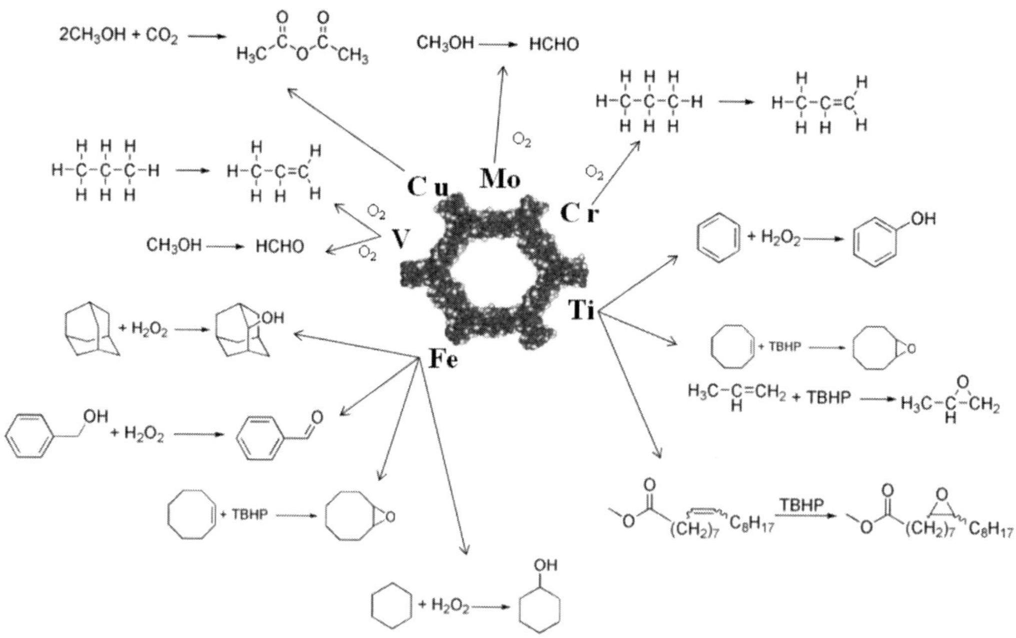

Figure 3. Single-site transition-metal ions grafted onto the surface of mesoporous silica.

By varying the catalytically-active centre, it is possible to find applications for SSHC in a large choice of selective oxidations, following the guidelines of clean and sustainable chemistry. Figure 3 presents a range of single-site transition-metal ions grafted onto the surface of mesoporous silica [2].

3.2. Epoxidation

Olefin epoxidation is of importance for a series of industrial processes, epoxides are for instance precursors in the synthesis of epoxy resins, plasticizers, perfumes, surfactants, etc.

SBA-3 mesoporous molecular sieves doped with transition metal ions (Fe, V, Nb and Ta) have been applied for selective oxidation of propene towards propylene oxide in the presence of N_2O as an oxidant. The kind and amount of applied modifiers significantly affected the catalytic activity. V/SBA-3 was found to be the most active among the catalyst under study. In spite of relatively high selectivity towards propylene oxide (reaching up to 23%), the main oxidation product was still propionaldehyde. Surprisingly, CO_x was not formed over V, Nb and Ta modified SBA-3 catalysts. A PO selectivity of about 20% could be achieved at a propylene conversion of 17% over mixed Fe/V/SBA-3 catalytic system. Additional modification of V containing samples (V/SBA-3) with iron complexes resulted in a further increase in the catalysts activity for epoxidation reaction. The addition of iron ions into Me/SBA-3 (Me = V, Nb and Ta) enhanced both propene conversion and selectivity towards propylene oxide [73].

In recent years, much attention has been focused on the solvent-free catalytic oxidations with O_2 or H_2O_2. Solvent-free oxidation of cycloolefins into ketones and/or epoxides over chromium containing mesoporous molecular sieve ([Cr]-MCM-41) catalyst at 1 atm O_2 has been reported. It was found that this catalyst is effective for oxidation of cycloolefins. A variety of benzylic compounds, such as cycloalkyl (tetralin, indane, diphenylmethane and fluorene) and alkylbenzenes (e.g., ethylbenzene, propylbenzene and n-butylbenzene) were studied. These compounds were selectively oxidized, giving the corresponding benzylic ketones with excellent selectivity (82-98%) and good yield (23-60%). Under these solvent-free reaction conditions, without the use of hydroperoxide oxidant, catalyst can be recycled [74].

Corma [75], Thomas [76], Pinnavaia [29] and others were quick to appreciate the advantages that large-pore silicas would confer in facilitating the preparation of single-site, metalcentered heterogeneous catalysts, aiming at obtaining solids with a good oxidation ability (as good as TS-1), but with high accessibility (as high as mesoporous silicas). A significant advance was made when Ti^{IV} active centers were grafted onto the inner walls of MCM-41 mesoporous silica using an organometallic precursor, in particular titanocene dichloride $[Ti(Cp)_2Cl_2]$ ($Cp = C_5H_5$), by Maschmeyer et al. [53]. This sort of Ti^{IV}-containing mesoporous silica proved to be active and selective in the epoxidation of cycloalkenes with high selectivity (always higher than 90%) and good activity (grafted catalysts showed a specific activity *ca.* 10 times higher than the in-matrix Ti-MCM-41 materials). Later, by optimizing the experimental conditions, titanocene-grafted silicas showed interesting results in the epoxidation of a wide variety of unsaturated substrates: alcoholic terpenes (e.g., carveol epoxide was obtained with 82% conversion and 73% selectivity in 24h) or unsaturated fatty acid methyl esters (e.g., castor oil methyl ester is epoxidized in 24h with 97% conversion and >98% selectivity) with *tert*-butylhydroperoxide (TBHP) [2] as well as in the oxidation of substituted phenols (99% yield in trimethylbenzoquinone from trimethylphenol after 1h) or in the epoxidation of cyclohexene (44% epoxide yield with >98% selectivity after 3h) in the presence of hydrogen peroxide [77, 78]. Similarly, Mayoral and co-workers have reported the synthesis of Ti^{IV}-silica catalysts by grafting $Ti(OiPr)_4$. This catalyst showed promising results in the epoxidation of non-functionalized alkenes, dienes and allylic alcohols using *tert*-butylhydroperoxide as oxidant [79]. It is also an efficient catalyst for epoxidation with dilute aqueous hydrogen peroxide under mild conditions (23% epoxide yield after 1h) [80, 81] and with slow dropwise addition of oxidant [82]. In the latter case, the contribution of direct epoxidation to the productive H_2O_2 consumption increases considerably with respect to one shot addition (from 21 to 56%), improving the selectivity.

[V]-MCM-41 samples with similar surface area and vanadium content but different pore diameters have been prepared. The effect of vanadium loading, surfactant chain length and type as well as pH has been considered for cyclohexene epoxidation. The pore wall radius of curvature effect in the liquid phase reaction was investigated. The initial reaction rates were observed to strongly correlate with the pore diameters: the larger the pore size, the higher the initial reaction rate due to higher accessibility of vanadium components in the pore walls. The overall turnover numbers for a long period were constant with respect to different pore sizes, which is probably due to a strong pore diffusion limitation in the liquid phase reaction. The physical structure of V-MCM-41 samples has been well maintained due to their high stability. Negligible amount of vanadium was lost because of leaching [83].

Especially active in the epoxidation reactions appeared to be [Ti]-MCM-41. There are no signs of Ti–O–Ti linkages, nor of any titanyl (TNO) groups of the kind proposed earlier, either of a three- or 5-coordinated kind. Detailed comparisons with several other local Ti^{IV}-centered oxygen environments leave no doubt that the active sites in the reported highly active catalyst (the so-called Ti↑MCM-41) are 4-coordinated [52, 84, 85]. There are no 6-

coordinated Ti^{IV}-centered sites present in the as-prepared catalyst. Thomas and Sankar also recorded operation of the Ti→MCM-41 catalyst in the epoxidation of cyclohexene to its epoxide with TBHP as a convenient test reaction [84, 86]. Significantly, the coordination shell of the original, 4-coordinated Ti^{IV} active sites expands to six during catalysis. This is compatible with the radical-free picture of epoxidation. Such materials have been recently shown to be ideally suited for the one-pot conversion of citronellal into isopulegol epoxide (Figure 4).

Oxidation of limonene yields a variety of products (Figure 5), including carveol and carvone (due to oxidation at allylic C–H bond) and 1,2-epoxide, 8,9-epoxide and 1,2 and 8,9-diepoxides (due to oxidation of the olefinic bond). Mn(Salen)Cl supported on propylthiol-functionalized SBA-15 yielded the 1,2-limonene epoxide with 100% chemo- and regioselectivity. Higher electron density at the site of Mn ions, and consequently lower redox potential of the Mn-complexes on immobilization, are the probable causes for their efficient and selective catalytic activity. The selectivity for 1,2-epoxide over different immobilized Mn-complexes in toluene medium increased in the following order: SBA-15-pr-NH_2-Mn(Salen)Cl (80.2%) < SBA-15-pr-SO_3H-Mn(Salen)Cl (82.0%) < SBA-15-pr-SHMn(Salen) Cl (100%). Interestingly, this variation in epoxide selectivity follows the variation in electron density (hyperfine coupling constant - A; EPR studies) and oxidizability (cyclic voltammetric studies) of the Mn ions. However, the extent of this decrease at the end of fourth recycle is lower for SBA-15-pr-SH-Mn(Salen)Cl (18%) than for SBA-15-pr-SO_3H-Mn(Salen)Cl (25%) and SBA-15-pr-NH_2-Mn(Salen) Cl (28%). Hence, on the basis of the observed conversions, it can be concluded that the extent of Mn leaching into the solution decreased in different catalysts in the following order: SBA-15-pr-NH_2-Mn(Salen)Cl > SBA-15-pr-SO_3HMn(Salen)Cl > SBA-15-pr-SH-Mn(Salen)Cl [87].

Selective catalytic oxidation of geraniol to citral with molecular oxygen in supercritical carbon dioxide (scCO_2) has been investigated. The catalyst used was a chromium containing mesoporous molecular sieve viz. CrMCM-41. Comparative studies were performed with CoMCM-41, PtMCM-41 and PdMCM-41 catalysts. Among the various catalysts studied, CrMCM-41 showed a high conversion of geraniol and an excellent selectivity to citral. In contrast, CoMCM-41, PtMCM-41 and PdMCM-41 catalysts exhibited low conversion of geraniol. However, all three compared catalysts showed similar selectivity to citral as CrMCM-41. The effect of CO_2 pressure and reaction temperature on geraniol oxidation was studied with CrMCM-41. Supercritical CO_2 medium was found to enhance the conversion of geraniol and/or yield of cytral (i.e., yield of citral from 35% without CO_2 to 52% at CO_2 pressure of 14 MPa). The catalyst can be recycled with negligible loss of conversion [88].

Also chromium(III) binaphthyl Schiff base complexes were immobilized onto the terminal NH_2 group of surface bound 3-aminopropyl tethering chains [89]. The asymmetric epoxidation of a series of unfunctionalized alkenes such as p-chlorostyrene, cis-methylstyrene, using PhIO as an oxidant was carried out in toluene. This catalyst exhibited significantly higher enantioselectivity than the same free complex. Its reusability was also examined to show that this catalyst could be reused at least two times without a decrease in activity and enantioselectivity.

Hutching et al. have reported the synthesis of a heterogeneous enantioselective epoxidation catalyst by combining manganese(III) salen with the mesoporous Al-MCM-41. This catalyst is effective as an enantioselective epoxidation catalyst for cis-stilbene. The trans epoxide is formed with an ee of 70%, which is similar to that observed for the equivalent homogeneous reactions. The turnover rates based on the amount of Mn-salen for the heterogeneous catalyst were found to be higher (e.g. 26) than those for homogeneously catalyzed reaction (e.g. 7). However, the Mn-MCM-41-salen catalyst suffered from a loss of enantioselectivity of 40% ee when reused for the first time [90].

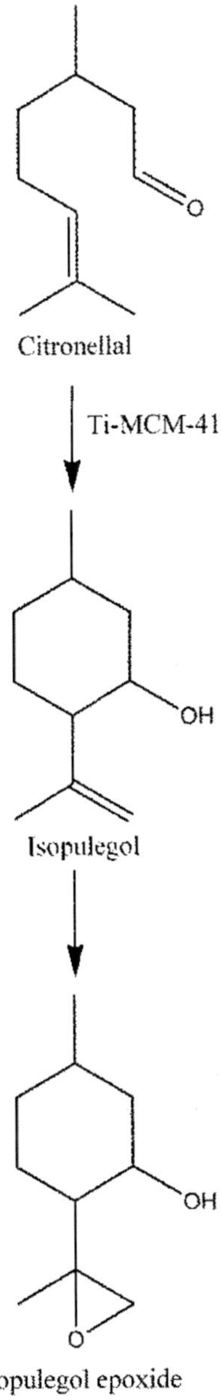

Figure 4. Conversion of citronellal into isopulegol epoxide over Ti-MCM-41.

In contrast to Ti^{IV}, octahedrally coordinated Mo^{VI} sites, known as catalytically active centers for olefin epoxidation cannot be easily incorporated into the tetrahedral positions of porous silicas (zeolites/mesoporous materials) [91, 92]. Therefore, the heterogenization of molybdenum-based organometallic or coordination compounds on the surface of mesoporous materials becomes a promising alternative.

The MCM-41 grafted complex exhibited high activity in the epoxidation of cyclooctene at 25°C (TOF: 306 mol $mol_{Mo}^{-1}h^{-1}$) and its kinetic profile was similar to that exhibited by the homogeneous complex. Alkenes such as cyclooctene, trans-2-octene, norbornene showed 80% of conversion with 100% of selectivity towards the epoxide after 6h. The more demanding R-(+)-limonene showed moderate conversions (ca. 78%), while α-pinene and styrene showed 20% of conversion and their epoxide selectivity remained at 30% after 6h. Stability tests confirmed that the reaction is mainly heterogeneous in nature. Besides the stability and recyclability, these systems were superior in epoxidation reactions over the $MoO_2Cl_2(thf)_2$ catalysts grafted on MCM-41 [93].

Tricarbonylcyclopentadienylmolybdenum complexes containing siloxane groups either attached at the cyclopentadiene ligand or directly at the molybdenum centre, grafted on MCM-41/48 have been also reported as active and robust catalysts (TOF: 2900-10200, selectivity: 73-100%) for cyclooctene epoxidation with TBHP [94].

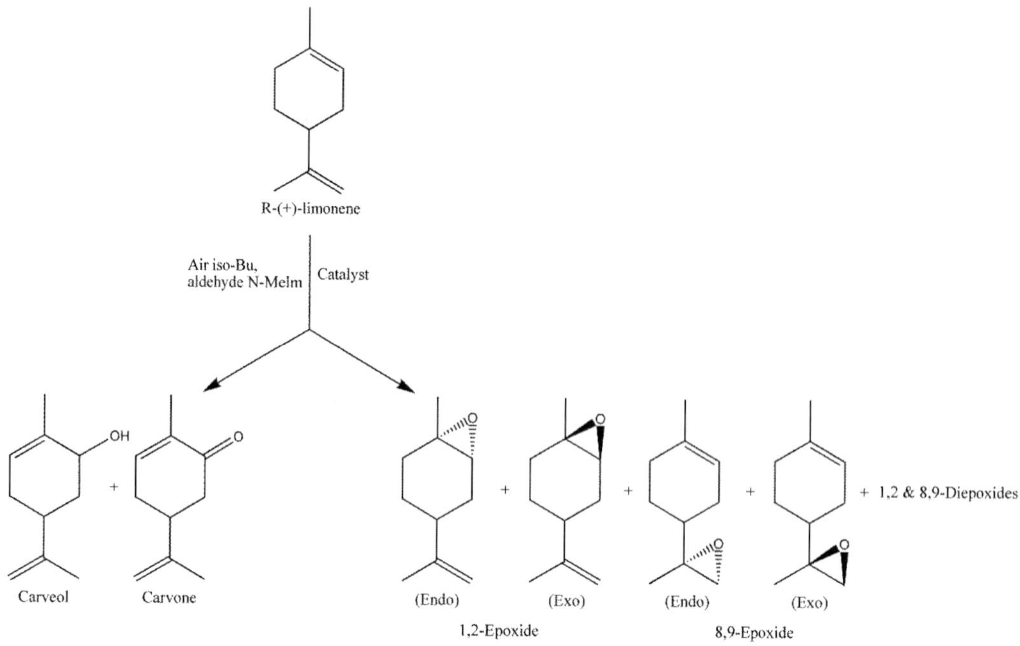

Figure 5. Limonene and its oxidation products.

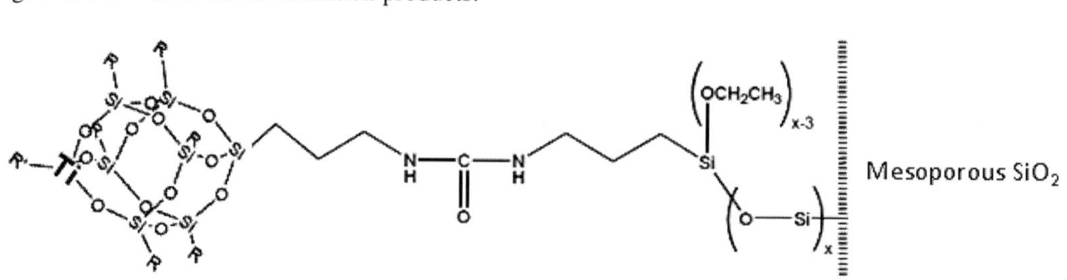

Figure 6. Anchoring of a Ti-containing polysilsesquioxane moiety onto the surface of SBA-15 mesoporous material; R - isobutyl and R' - isopropyl group.

It has been also shown that nanosized WO_3 particles supported on MCM-48 may act as a highly efficient and selective heterogeneous catalyst for the oxidation of olefins, sulfides, and cyclic ketones with hydrogen peroxide or peracetic acid as oxidants. Excellent catalytic properties of these materials can be related to the preparation method allowing the formation of amorphous WO_3 entities, which are well dispersed and tightly bound to the silica matrix. Koo et al. obtained nanosized WO_3 particles supported on mesoporous MCM-48 which are highly efficient and selective heterogeneous catalysts for olefin epoxidation, the oxidation of sulfides and the Baeyer-Villiger oxidation of cyclic ketones with hydrogen peroxide or peracetic acids. Their catalytic activities depend on the nature of the supports and the particle size of WO_3 [95]. Gao et al. have found that mesoporous WO_3/MCF materials show good activity and recyclability in the epoxidation of cycloocta-1,5-diene with hydrogen peroxide, which is mainly attributed to the presence of isolated tungsten species anchored on the support through W–O–Si covalent bonds [96].

Another complementary approach has been pioneered by Tilley and Bell and co-workers [97, 99, 100, 101]. A molecular precursor was exploited to prepare a series of active catalysts on mesoporous supports (Figure 3) containing metal centers such as Ti, Cr, Fe or VO. In this case, the desired atomic environment aimed at in the final catalyst (e.g., $Ti(OSi)_4$ or $Ti(OSi)_3$) was already present in the thermolytic precursor. Thus, for instance, starting from *tris*(*tert*-butoxy)siloxytitanium complex (iPrO)Ti-$[OSi(OtBu)_3]_3$, the local environment achieved at the end in the single-site catalyst was $Ti(OSi)_3$. Typical supports used by Tilley were the high-area mesoporous silicas MCM-41 and SBA- 15, the latter being, for this purpose, distinctly more thermally stable than the former [97]. Tilley et al. have also demonstrated that peroxo species, formed upon the reaction of TaSBA15 or Bu_{cap}TaSBA15 with H_2O_2, stoichiometrically oxidize cyclohexene to yield the epoxide as the only product [98].

Thanks to the anchoring strategy, large and bulky moieties can be covalently attached onto the surface of mesoporous molecular sieves, providing the mesopores large enough to accommodate the guest species. As an example, a Ti-containing polysilsesquioxane (Ti-POSS), a complex widely used as a soluble model for Ti single sites in a silica matrix, was recently anchored via covalent bonding to the surface of mesoporous SBA-15 silica (Figure 6). The final Ti-POSS-TSIPI-SBA-15 material revealed a good dispersion of Ti sites and an interesting catalytic activity in the liquid-phase epoxidation of limonene to limonene oxide (with selectivity as high as 88% to 1,2-limonene oxide after 24h) [102].

Analogously, Mn(salen) complexes, when immobilized inside siliceous nanopores, also exhibit the effect of enhancement of chirality, owing to the confinement effect in the enantioselective epoxidation of unfunctionalized olefins [103].

Moreover, the catalytic activity of (Ti, Al)-Beta/MCM-41 composite in oxidation reactions and titanium leaching from the framework has been tested by the epoxidation of cyclooctene using tert-butyl hydroperoxide as the oxidant. Titanium leaching from the composite was not observed [104].

Recently, a number of manganese-functionalized mesoporous materials have been prepared. For example, MCM-41 supported catalysts, prepared by the template ion-exchange method, show high activity for the epoxidation of stilbene [105, 106]. Gas-phase grafted Mn species provide higher activity in stilbene oxidation than impregnated Mn-MCM-41 when tert-butyl hydroperoxide was used as an oxidant. In contrast, Mn impregnated MCM-41 showed higher activity than gas-phase grafted Mn-MCM-41 in diphenylmethane oxidation using air as an oxidant. The differences were attributed to the nature of the Mn centers [107].

Table 6. Epoxidation reactions on metal-containing mesoporous molecular sieves

Catalytic reaction	Catalyst	Reference
selective oxidation of propene by N_2O	[V]-SBA-3 [Fe, V]-SBA-3 [Fe, Nb]-SBA-3 [Fe, Ta]-SBA-3	[73]
solvent-free oxidation of benzylic compounds by O_2	[Cr]-MCM-41	[74]
epoxidation of hex-l-ene by H_2O_2 and norbornene by TBHP	[Ti]-MCM-41	[75]
epoxidation of cyclohexene by H_2O_2	[Ti]-MCM-41 [Ti]-MCM-48	[53, 76, 78]
epoxidation of cyclohexene by H_2O_2	Ti↑MCM-41	[52, 86]
epoxidation of cyclohexene by H_2O_2 and TBHP	[V]-MCM-41	[83]
conversion of citronellal into isopulegol epoxide by TBHP	[Ti]-MCM-41	[86]
epoxidation of limonene by air	SBA-15-pr-NH_2-Mn(Salen)Cl SBA-15-pr-SO_3H-Mn(Salen)Cl SBA-15-pr-SHMn(Salen)Cl	[87]
selective oxidation of geraniol by O_2	[Cr]-MCM-41 [Co]-MCM-41 [Pt]-MCM-41 [Pd]-MCM-41	[88]
asymmetric epoxidation of alkenes by PhI = O	chromium binaphthyl Schiff base complex immobilized on MCM-41	[89]
epoxidation of (*Z*)-stilbene by PhI = O	Mn-exchanged Al-MCM-41 modified by chiral salen ligand	[90]
epoxidation of cis-cyclooctene by TBHP	MCM-41-MoO_2X_2 [X = Cl, Br] MCM-48-MoO_2X_2 [X = Cl, Br] η^5-CpMo$(CO)_3$R complexes grafted on MCM-41 or MCM-48	[93, 94]
epoxidation of olefins by H_2O_2	WO_3/MCM-48	[95]
epoxidation of cycloocta-1,5-diene by H_2O_2	WO_3/MCF	[96]
epoxidation of cyclohexene by H_2O_2, CHP or TBHP	Ti$[OSi(OtBu)_3]_4$, (iPrO)Ti$[OSi(OtBu)_3]_3$, $(tBuO)_3$TiOSi(OtBu) supported on MCM-41 or SBA-15 TaSBA15 Bu_{cap}TaSBA15	[97, 98]
epoxidation of limonene by TBHP	Ti-POSS-TSIPI-SBA-15	[102]
enantioselective epoxidation of unfunctionalized olefins by NaClO	chiral Mn(salen) complexes immobilized inside nanopores of MCM-41 and SBA-15	[103]

Catalytic reaction	Catalyst	Reference
epoxidation of cyclooctene by TBHP	(Ti,Al)-Beta/MCM-41 composite	[104]
epoxidation of stilbene by TBHP, PhI=O, H_2O_2 or O_2	Mn-MCM-41 prepared by the template ion-exchange (TIE) method Mn-MCM-41 prepared by the molecular organic chemical vapor deposition (MOCVD)	[105, 106, 107]

Olefin epoxidation is a substantial reaction, since epoxides find many applications in several industrial processes. Catalytic systems for epoxidation on mesoporous molecular sieves, briefly discussed in this chapter, have been also summarized in Table 6.

CONCLUSION AND OUTLOOK

Design and preparation of mesoporous materials with a large surface area and adjustable acidity or oxidation properties, for substitution of hazardous liquid acids, are of great importance in Green Chemistry. Acidic or oxidation strength, type and concentration as well as catalyst surface area and pore size are important factors for design and preparation of mesoporous solid acids. Ordered mesoporous materials exhibit good ability for catalytic conversion of bulky molecules. Mesoporous zeolites combine the advantages of both mesoporous materials (large surface area, mesopores, and fast diffusion) and zeolite crystals (strong acidity, high hydrothermal stability). Mesoporous acidic resins offer catalysts with both high content of acidic functionalized groups and large surface area. These novel mesoporous solid acids are potentially important for industrial applications in the near future.

REFERENCES

[1] Knözinger, H., Kochloefl, K. *Heterogeneous Catalysis and Solid Catalysts.* Ullmann's Encyclopedia of Industrial Chemistry, Wiley-VCH: Weinheim, 2002.

[2] Dal Santo, V., Liguori, F., Pirovano, C., Guidotti, M. *Molecules* 2010, *15*, 3829.

[3] Vartuli, J. C., Degnan, T. F. Jr., *Stud. Surf. Sci. Catal.* 2007, 168, 837.

[4] Wen, D. J., Min, S. C., Ling, Q., Ming, S. J., Zhong, X. X. *J. Porous Mater.* 2008, 15, 189.

[5] Ondrey, G. *Chem. Eng.* 2008, 13.

[6] Campbell, I. M. *The Constitution of the Catalytic Surfaces.* Catalysis at Surfaces, Chapman & Hall: London, 1988.

[7] Reddy, K. M., Moudrakovski, I., Sayari, A. *J. Chem. Soc. Chem. Commun.* 1994, 9, 1059.

[8] Bhaumik, A., Dongare, M. K., Kumar, R. *Microporous Mater.* 1995, 5, 173.

[9] Reddy, J. S., Sayari, A. *J. Chem. Soc. Chem. Commun.* 1995, 21, 2231.

[10] Zhang, W. Z., Wang, J. L., Tanev, P. T., Pinnavaia, T. J. *Chem. Commun.* 1996, 8, 979.

[11] Lee, D. S., Liu, T. K. *J. Sol-Gel Sci. Technol.* 2002, 23, 15.

[12] Albuquerque, A., Pastore, H. O., Marchese, L. *Stud. Surf. Sci. Catal.* 2005, 158, 901.
[13] Venkatathri, N., Srivastava, R. *Catal. Commun.* 2005, 6, 177.
[14] Prakash, A. M., Kevan, L. *J. Phys. Chem. B* 1999, 103, 2214.
[15] Teixeira-Neto, A. A., Marchese, A., Pastore, H. O. *Quim. Nova* 2009, 32 (2), 463.
[16] Morey, M., Davidson, A., Eckert, H., Stucky, G. *Chem. Mater.* 1996, 8, 486.
[17] Van der Voort, P., Morey, M., Stucky, G. D., Mathieu, M., Vansant, E. F. *J. Phys. Chem. B* 1998, 102, 585.
[18] Gomez, S., Garces, L. J., Villegas, J., Ghosh, R., Giraldo, O., Suib, S. L. *J. Catal.* 2005, 233, 60; *J. Catal.* 2005, 233, 478.
[19] Okamoto, M., Luo, L., Labinger, J. A., Davis, M. E. *J. Catal.* 2000, 192, 128.
[20] Albuquerque, A., Pastore, H. O., Marchese, L. *J. Catal.* 2006, 241, 367.
[21] Corma, A., Navarro, M. T., Perez-Pariente J. *Material de estructura tipo zeolita de poros ultragrandes con un red constituida por oxidos de silicio y titanio; su sintesis y utilizacion para la oxidacion selectiva de productos organicos.* ES 2 092 428, August 1, 1997.
[22] Morey, M., Davidson, A., Stucky, G. *Micropor. Mater.* 1996, 6 (2), 99.
[23] Morey, M. S., O'Brien, S., Schwarz, S., Stucky, G. D. *Chem. Mater.* 2000, 12, 898.
[24] Ahn, W. S., Lee, D. H., Kim, T. J., Kim, J. H., Seo, G., Ryoo, R., *Appl. Catal., A* 1999, 181, 39.
[25] Bagshaw, S. A. *Chem. Commun.* 1999, 271.
[26] Han, Y., Xiao, F. S., Wu, S., Sun, Y., Meng, X., Li, D., Lin, S. *J. Phys. Chem. B* 2001, 105, 7963.
[27] Blasco, T., Corma, A., Navarro, M. T., Perez-Pariente J. *J. Catal.* 1995, 156, 65.
[28] Alba, M. D., Luan, Z. H., Klinowski, J. *J. Phys. Chem.* 1996, 100, 2178.
[29] Zhang, W. Z., Froeba, M., Wang, J. L., Tanev, P. T., Wong, J., Pinnavaia, T. J. *J. Am. Chem. Soc.* 1996, 118, 9164.
[30] Koyano, K. A., Tatsumi, T. *Microporous Mater.* 1997, 10, 259.
[31] Corma, A., J. L., Navarro, M. T., Rey, F. *Chem. Commun.* 1998, 17, 1899.
[32] Yu, J. Q., Feng, Z. C., Xu, L., Li, M. J., Xin, Q., Liu, Z. M., Li, C. *Chem. Mater.* 2001, 13, 994.
[33] Luo, Y., Lu, G. Z., Guo, Y. L., Wang, Y. S. *Catal. Commun.* 2002, 3, 129.
[34] Kong, Y., Guo, X. F., Zhang, F. M., Jiang, S. Y., Wang, J., Lu, Y. N., Yan, Q. J. *Mater. Lett.* 2005, 59, 3099.
[35] Trtikhan, N. N., Romannikov, V. N., Shmakov, A. N., Vanina, M. P., Paukshtis, E. A., Bukhtiyarov, I., Kriveritsov, V. V., Danilov, I. Y., Kholdeeva, A. *Microporous Mesoporous Mater.* 2003, 59, 73.
[36] Zhang, W. H., Lu, J. Q., Han, B., Li, M. J., Xiu, J. H., Ying, P. L., Li, C. *Chem. Mater.* 2002, 14, 3413.
[37] Luan, Z. H., Kevan, L. *Microporous Mesoporous Mater.* 2001, 111 (44-45), 337.
[38] Melero, J. A., Arsuaga, J. M., Frutos, P. D., Iglesias, J., Sainz, J., Blazquez, S. *Microporous Mesoporous Mater.* 2005, 86, 364.
[39] Ji, D., Ren, T., Yan, L., Suo, J. S. *Mater. Lett.* 2003, 57, 4474.
[40] Ji, D., Zhao, R., Lv, G., Qian, G., Yan, L., Suo, J. *Appl. Catal., A* 2005, 281, 39.
[41] Bagshaw, S. A., Kemmit, T., Milestone, N. B. *Microporous Mesoporous Mater.* 1998, 22, 419.
[42] Wu, P., Sugiyama, H., Tatsumi, T. *Stud. Surf. Sci. Catal.* 2003, 146, 613.

[43] Meng, X. J., Li, D. F., Yang, X. Y., Yu, Y., Wu, S., Han, Y., Yang, Q., Jiang, D. Z., Xiao, F. S. *J. Phys. Chem. B* 2003, 107, 8972.
[44] Shan, Z. P., Maschmeyer, T., Jansen, J. C. *Inorganic oxides with mesoporosity or combined meso-and microporosity and process for the preparation thereof.* U.S. Patent 6,358,486, March 19, 2002.
[45] Shan, Z. P., Jansen, J. C., Marchese, L., Maschmeyer, T. M*icroporous Mesoporous Mater.* 2001, 48, 181.
[46] Monnier, A. *Science* 1993, 261, 1299.
[47] Yang, X. Y., Han, Y., Lin, K. K., Tian, G., Feng, Y. F., Meng, X. J., Di, Y., Du, Y. C., Zhang, Y. L., Xiao, F. S. *Chem. Commun.* 2004, 22, 2612.
[48] Gontier, S., Tuel, A. *Zeolites* 1995, 15, 601.
[49] Wu, P., Miyaji, T., Liu, Y., He, M. Y., Tatsumi, T. *Catal. Today.* 2005, 99, 233.
[50] Poladi, R. H. P. R., Landry, C. C. *Microporous Mesoporous Mater.* 2002, 52, 11.
[51] Hagen, A., Wei, D., Haller, G.L., Bonneviot, L., Beland, F., Danumah, C., Kaliaguine, S. *Stud. Surf. Sci. Catal.* 1998, 117, 191.
[52] Oldroyd, R. D., Thomas, J. M., Maschmeyer, T., MacFaul, P. A., Snelgrove, D. W., Ingold, K. U., Wayner, D. D. *Angew. Chem., Int. Ed. Engl.* 1996, 35, 2787.
[53] Maschmeyer, T., Rey, F., Sankar, G., Thomas, J. M. *Nature* 1995, 378, 159.
[54] Zheng, S., Gao, L., Zhang, Q. H., Guo, J. K. *J. Mater. Chem.* 2000, 10, 723.
[55] Kang, K. K., Ahn, W. S. *J. Mol. Catal. A: Chem.* 2000, 159, 403.
[56] De Cremer, G., Bartholomeeusen, E., Pescarmona, P. P., Lin, K., de Vos, D. E., Hofkens, J., Roeffaers, M. B. J., Sels, B. F. *Catal. Today* 2010, 157, 236.
[57] De Cremer, G., Roeffaers, M. B. J., Bartholomeeusen, E., Lin, K., Dedecker, P., Pescarmona, P. P., Jacobs, P. A., de Vos, D. E., Hofkens, J., Sels, B. F. *Angew. Chem., Int. Ed.* 2010, 49, 908.
[58] De Cremer, G., Sels, B. F., de Vos, D. E., Hofkens, J., Roeffaers, M. B. J. *Chem. Soc. Rev.* 2010, 39, 4703.
[59] Li, J., Zhou, Ch., Xie, H., Ge, Z., Yuan, L., Li, X. *J. Nat. Gas Chem.* 2006, 15, 164.
[60] Xiao, F. S., Han, Y., Yu, Y., Meng, X. J., Yang, M., Wu, S. *J. Am. Chem. Soc.* 2002, 124 (6), 888.
[61] Han, Y., Meng, X. J., Guan, H. B., Yu, Y., Zhao, L., Xu, X. Z., Yang, X. Y., Wu, S., Li, N., Xiao, F. S. *Microporous Mesoporous Mater.* 2003, 57, 191.
[62] Selvam, P., Dapurkar, S. E. *J. Catal.* 2005, 229, 64.
[63] Selvam, P., Dapurkar, S. E. *Appl. Catal., A* 2004, 276, 257.
[64] Pena, M. L., Dejoz, A., Fornes, V., Rey, E., Vazquez, M. I., Nieto, J. M. L. *Appl. Catal., A* 2001, 209, 155.
[65] Wang, H., Zhao, Z., Zhang, Z., Duan, A., Xu, Ch. *J. Porous Mater.* 2008, 15, 221.
[66] Zhang, Q. H., Yang, W., Wang, X. X., Wang, Y., Shishido, T., Takehira, K. *Microporous Mesoporous Mater.* 2005, 77, 223.
[67] Liu, H., Wang, Z., Hub, H., Liang, Y., Wang, M. *J. Solid State Chem.* 2009, 182, 1726.
[68] Dapurkar, S. E., Kawanami, H., Yokoyama, T., Ikushima, Y. *New J. Chem.* 2009, 33, 538.
[69] Santhanaraj, D., Suresh, C., Vijayan, P., Venkatathri, N., Shanthi K. *Reac. Kinet. Mech. Cat.* 2010, 99, 439.
[70] Williams, T., Beltramini, J., Lu, G. Q. *Microporous Mesoporous Mater.* 2006, 88, 91.

[71] Shiraishi, Y., Morishita, M., Teshima, Y., Hirai, T. *J. Phys. Chem. B* 2006, 110 (13), 6587.

[72] Hu, Y., Wada, N., Tsujimaru, K., Anpo, M. *Catal. Today* 2007, 120, 139.

[73] Held, A., Florczak, P. *Catal. Today* 2009, 142, 329.

[74] Dapurkar, S. E., Kawanami, H., Yokoyama, T., Ikushima, Y. *Catal. Commun.* 2009, 10, 1025.

[75] Corma, A., Navarro, M. T., Parente, J. P. *J. Chem. Soc. Chem. Commun.* 1994, 147.

[76] Sankar, G., Rey, F., Thomas, J. M., Greaves, G. N., Corma, A., Dobson, B. R., Dent, A. J. *J. Chem. Soc. Chem. Commun.* 1994, 2279.

[77] Kholdeeva, O. A., Ivanchikova, I. D., Guidotti, M., Pirovano, C., Ravasio, N., Barmatova, M. V., Chesalov, Y. A. *Adv. Synth. Catal.* 2009, 351, 1877.

[78] Guidotti, M., Pirovano, C., Ravasio, N., Lázaro, B., Fraile, J. M., Mayoral, J. A., Coq, B., Galarneau, A. *Green Chem.* 2009, 11, 1421.

[79] Cativiela, C., Fraile, J. M., Garcia, J. I., Mayoral, J. A. *J. Mol. Catal. A* 1996, 112, 259.

[80] Fraile, J. M., Garcia, J. I., Mayoral, J. A., Vispe, E. *J. Catal.* 2000, 189, 40.

[81] Fraile, J. M., Garcia, J. I., Mayoral, J. A., Vispe, E. *J. Catal.* 2001, 204, 146.

[82] Fraile, J. M., Garcia, J. I., Mayoral, J. A., Vispe, E. *Appl. Catal., A* 2003, 245, 363.

[83] Tang, Q., Wang, Ch., Hu, S., Sun, H., Chen, Y., Haller, G. L., Yang, Y. *Catal. Lett.* 2007, 117, 25.

[84] Thomas J. M., Sankar, G. *J. Synchrotron Radiat.* 2001, 8, 55.

[85] Thomas, J. M. *Faraday Discuss.* 1996, 105, 1.

[86] Thomas, J. M., Raja, R. *Chem. Commun.* 2001, 675.

[87] Saikia, L., Srinivas, D., Ratnasamy P. *Microporous Mesoporous Mater.* 2007, 104, 225.

[88] Dapurkara, S. E., Kawanamia, H., Chatterjeea, M., Rodeb, C. V., Yokoyamaa, T., Ikushimaa, Y. *Appl. Catal., A* 2011, 394, 209.

[89] Zhou, X. G., Yu, X. Q., Huang, J. S., Li, S. G., Li, L. S., Che, C. M. *Chem. Commun.* 1999, 1789.

[90] Piaggio, P., McMorn, P., Langham, C., Bethell, D., Bulman-Page, P.C., Hancock, F. E., Hutching, G. J. *New J. Chem.* 1998, 1167.

[91] Sheldon, R. A., Arends, I. W. C. E., Lempers, H. E. B. *Catal. Today* 1998, 41, 387.

[92] Sheldon, R. A., Wallau, M., Arends, I. W. C. E., Schuchardt, U., *Acc. Chem. Res.* 1998, 31, 485.

[93] Ferreira, P., Gonçalves, I. S., Kühn, F. E., Lopes, A. D., Martins, M. A., Pillinger, M., Pina, A., Rocha, J., Romão, C. C., Santos, A. M., Santos, T. M., Valente, A. A. *Eur. J. Inorg. Chem.* 2000, 2263.

[94] Sakthivel, A., Zhao, J., Hanzlik, M., Chiang, A. S. T., Herrmann, W. A., Kühn, F. E., *Adv. Synth. Catal.* 2005, 347, 473.

[95] Koo, D. H., Kim, M., Chang, S. *Org. Lett.* 2005, 7, 5015.

[96] Gao, R., Yang, X., Dai, W. L., Le, Y., Li, H., Fan, K. *J. Catal.* 2008, 256, 259.

[97] Jarupatrakorn, J., Tilley, J. D. *J. Am. Chem. Soc.* 2002, 124, 8380.

[98] Ruddy, D. A., Tilley, T. D. *J. Am. Chem. Soc.* 2008, 130, 11088.

[99] Fujdala, K. L., Tilley, T. D. *J. Catal.* 2003, 216, 265.

[100] Drake, I. J., Fujdala, K. L., Bell, A. T., Tilley, T. D. *J. Catal.* 2005, 230, 14.

[101] Bell, A. T. *Science* 2003, 299, 1688.

[102] Carniato, F., Bisio, C., Boccaleri, E., Guidotti M., Gavrilova, E., Marchese, L. *Chem. Eur. J.* 2008, 14, 8098.

[103] Zhang, H., Zhang, Y., Li, C. *J. Catal.* 2006, 238, 369.

[104] Mrak, M., Novak Tušar, N., Zabukovec Logar, N., Mali, G., Kljajić, A., Arčon, I., Launay, F., Gedeon, A., Kaučič, V. *Microporous Mesoporous Mater.* 2006, 95, 76.

[105] Yonemitsu, M., Tanaka Y., Iwamoto, M.J. *J. Catal.* 1998, 178, 207.

[106] Zhang, Q., Wang, Y., Itsuki, S., Shishido T., Takehira K., *J. Mol. Catal. A* 2002, 188, 189.

[107] Caps V., Tsang, S. C. *Catal. Today* 2000, 61, 19.

INDEX

#

α

A

B

F

G

H

I

Q

R

S

U

V

W